Practical Clinical Enzymology:
Techniques and Interpretations
and
Biochemical Profiling

STANFORD SERIES ON METHODS AND TECHNIQUES
IN THE CLINICAL LABORATORY

Edited by Paul L. Wolf, M.D.,
Director of Clinical Laboratory,
Clinical Pathology and School of Medical Technology,
Stanford University Medical Center,
Stanford, California

METHODS AND TECHNIQUES IN CLINICAL CHEMISTRY
  by Paul L. Wolf, Dorothy Williams
  Tashiko Tsudaka, and Leticia Acosta

PRACTICAL CLINICAL ENZYMOLOGY: TECHNIQUES
AND INTERPRETATIONS
  by Paul L. Wolf and Dorothy Williams
AND BIOCHEMICAL PROFILING
  by Elisabeth Von der Muehll
  and Paul L. Wolf

CLINICAL HEMATOLOGY PROCEDURES: INTERPRETATIONS
AND TECHNIQUES
  by Paul L. Wolf, Elisabeth Von der Muehll,
  Irma Mills, Patricia Ferguson, and Mary Thompson

METHODS, TECHNIQUES, AND INTERPRETATIONS OF
BLOOD TRANSFUSION
  by Paul L. Wolf, Betsy Hafleigh, and
  Gabriel Korn

Practical Clinical Enzymology:

Techniques and Interpretations

Paul L. Wolf, M.D.
Director of Clinical Laboratory,
Clinical Pathology and
School of Medical Technology
Associate Professor of Pathology

Dorothy Williams, M.T. (ASCP)
Chief Technologist
Clinical Chemistry Laboratory

and

Biochemical Profiling

Elisabeth Von der Muehll, M.T. (ASCP)
Education Coordinator
School of Medical Technology

Paul L. Wolf, M.D.
Director of Clinical Laboratory,
Clinical Pathology and
School of Medical Technology
Associate Professor of Pathology

Stanford University Medical Center
Stanford, California

a Wiley-Interscience Publication

JOHN WILEY & SONS

New York   **London**   **Sydney**   **Toronto**

Copyright © 1973, by John Wiley & Sons, Inc.

All rights reserved. Published simultaneously in Canada.

No part of this book may be reproduced by any means nor transmitted, nor translated into a machine language without the written permission of the publisher.

LIBRARY OF CONGRESS CATALOGING IN PUBLICATION DATA:

Wolf, Paul L.
Practical clinical enzymology.

Practical clinical enzymology

(Stanford series on methods and techniques in the clinical laboratory)

1. Clinical enzymology.   I. Williams, Dorothy
II. Von der Muehll, Elisabeth   III. Title
IV. Series   DNLM: 1. Enzyme tests.   QY 490 W855p 1973

RB48.W64   616.07'56   73-596
ISBN 0-471-95905-7

Printed in the United States of America

10 9 8 7 6 5 4 3 2 1

Preface

This book is intended as a review and reference text book for Clinical Pathologists and Medical Technologists who are concerned with daily problems in diagnostic Clinical Enzymology and Biochemical Profiling. We have cited a majority of the past and current technical information relevant to the determination of common Clinical Laboratory Enzymology procedures and their diagnostic interpretation. A great proliferation of knowledge has occurred in the last fifteen years in this field. The emphasis of this book is on the practical aspects of Clinical Enzymology and Biochemical Profiling with a tendency to refer to only a minimal amount of esoteric clinically irrelevant information. We have attempted to cover each Clinical Enzymatic determination performed in the Stanford Clinical Laboratory completely from a technical and clinical viewpoint. The book is organized such that, disease states can be quickly recognized by their chapter headings to enable the reader to quickly obtain the needed information. All of the important diagnostic enzymatic tests that are performed here as well as relevant clinical data is included in this book.

We thus hope that this book will provide the necessary current knowledge required by Clinical Pathologists, Medical Technologists, and Medical Technology Students for optimal care of their patients.

The Authors gratefully acknowledge the cooperation of Anne Burgwald, M.T. (ASCP), Dee Evans, M.T. (ASCP), Janice Gebhardt, M.T. (ASCP), Margaret Rivard, M.T. (ASCP), Nancy Conant, M.T. (ASCP), Leticia Acosta, M.S., Tashiko Tsudaka, M.T. (ASCP), and Suzanne Gibian, M.T. (ASCP), and others of our technical staff for their valuable contribution in compiling information and helping to revise these techniques.

We also gratefully acknowledge the excellent secretarial work of Miss Elisabeth Von der Muehll, M.T. (ASCP), Supervisor, School of Medical Technology, who typed and edited the entire manuscript.

THE AUTHORS

Dedication

To all of our devoted teachers and especially to the residents, medical students, and medical technology students from whom we receive continual inspiration and knowledge.

Contents

CHAPTER 1	Common Methods of Enzyme Analysis	1
	Acid Phosphatase Determination (Bessey-Lowry)	8
	Acid Phosphatase Determination (BMC Method)	12
	Prostatic Acid Phosphatase (Tartrate Inhibition)	15
	Aldolase (Bruns Ultra-Violet Method)	17
	Alkaline Phosphatase (Bessey-Lowry: Continuous Spectrophotometric Method)	20
	Amylase (Iodometric Method by Caraway)	23
	Ceruloplasmin	26
	Cholinesterase (BMC Method)	29
	Cholinesterase (Acholest Test Paper)	32
	Creatine Phosphokinase	34
	Gamma-Glutamyl-Transpeptidase	37
	Glucose Oxidase (GOD - Perid Method)	40
	Glucose-6-Phosphate Dehydrogenase (Ultra-Violet Method on Red Cell Hemolysate)	44
	Glutamate Dehydrogenase	47
	Alpha-Hydroxybutyrate Dehydrogenase	50
	Isocitrate Dehydrogenase (Ultra-Violet Methodology)	53
	Lactate Dehydrogenase (Wacker Method)	56
	LDH Isoenzymes by Electrophoresis	59
	Leucine Aminopeptidase (Goldbarg and Rutenburg: Modified by Sigma Chemical Company)	68
	Leucine Aminopeptidase (Kinetic Method after Method of Nagel)	71
	Lipase Determination (Turbidimetric)	73

Contents (Continued):

	Malate Dehydrogenase	76
	Pyruvate Kinase	79
	Sorbitol Dehydrogenase	83
	Glutamate-Oxaloacetate Transaminase	86
	Glutamate-Pyruvate Transaminase	89
	Triglycerides (BMC Method - Ultra-Violet Determination)	91
	Uric Acid (Uricase Method)	97
CHAPTER 2	Alkaline and Acid Phosphatase, 5-Nucleotidase, Leucine Aminopeptidase, and Gamma-Glutamyl-Transpeptidase	100
CHAPTER 3	Amylase and Lipase	129
CHAPTER 4	Glutamic Oxalacetic Transaminase and Glutamic Pyruvic Transaminase	139
CHAPTER 5	Lactic Dehydrogenase	154
CHAPTER 6	Creatine Phosphokinase and Aldolase	170
CHAPTER 7	Sorbitol, Glutamic, Isocitric, Glucose-6-Phosphate Dehydrogenases and Glucose-6-Phosphatase	181
CHAPTER 8	Enzyme Abnormalities in Neoplastic Disease	190
CHAPTER 9	Enzyme Patterns in Obstetric and Gynecologic Practice	206
CHAPTER 10	Enzyme Abnormalities in Liver Disease	215
CHAPTER 11	Enzyme Abnormalities in Pancreatic and Salivary Gland Disease	247
CHAPTER 12	Enzyme Abnormalities in Reticuloendothelial Diseases	263
CHAPTER 13	Enzyme Abnormalities in Lung and Heart Disease	286

Contents (Continued):

CHAPTER 14	Enzyme Abnormalities in Central Nervous System Disease	302
CHAPTER 15	Enzyme Abnormalities in Surgery and Following Therapeutic Procedures	313
CHAPTER 16	Enzyme Abnormalities in Genitourinary Tract Disease	323
CHAPTER 17	Enzyme Abnormalities in Bone Disease	342
CHAPTER 18	Enzyme Abnormalities in Skeletal Muscle Disease	354
CHAPTER 19	Enzyme Patterns and Abnormalities in the Pediatric Age Group	362
CHAPTER 20	Enzyme Abnormalities in Body Fluids	377
CHAPTER 21	Enzyme Histochemistry and Its Application to the Clinical Laboratory (Elisabeth Von der Muehll, M.T. (ASCP)	386
CHAPTER 22	Multiphasic Testing (Elisabeth Von der Muehll, M.T. (ASCP) and Paul L. Wolf, M.D.)	396
	Metabolic Acidosis - Anion Gap	401
	Diabetic Ketoacidosis	402
	Respiratory Alkalosis	403
	Respiratory Acidosis	404
	Metabolic Alkalosis	405
	Conn's Syndrome	406
	Cushing's Syndrome	407
	Excessive Diuretic Utilization	408
	Malabsorption - Chronic Diarrhea	409
	Utilization of Licorice	410
	Renal Tubular Acidosis	411
	Hypokalemic Periodic Paralysis	412
	Metabolic Acidosis - Hypercalcemia Parathyroid Adenoma or Hyperplasia	413
	Metabolic Alkalosis - Hypercalcemia Metastatic Cancer to Bone	414
	Addison's Disease	415
	Renal Insufficiency	416

Contents (Continued):

Prerenal Insufficiency	417
Hemolysis	418
Excessive Utilization of 5 Percent Glucose in Water	419
Inappropriate ADH	420
Hyperlipemia	421
Nephrotic Syndrome	422
Heart Failure	423
Dehydration	424
Cerebral Damage	425
Gastrointestinal Hemorrhage	426
Excessive Intravenous Glucose and Insulin	427
Excessive Insulin	428
Pregnancy - Third Trimester	429
Cirrhosis	430
Non-Fasting Specimen	431
Post-Mortem Vitreous Humor - Juvenile Diabetes	432
Post-Mortem Vitreous Humor - Acute Glomerulonephritis	433
Hyperparathyroidism	437
Cancer Metastatic to Bone	438
Multiple Myeloma	439
Hodgkin's Disease Involving Bone	440
Osteomalacia including Rickets	442
Milk Alkali Syndrome	443
Sarcoidosis	444
Hypervitaminosis D	446
Cushing's Syndrome	447
Addison's Disease	448
Thyrotoxicosis	449
Excessive Thiazide Diuretic Utilization	450
Intravenous Albumin from Placenta	451
Hyperalimentation	452
Dehydration	454
Evaporation of Specimen	455
Non-Fasting Lipemic Serum	456
Increased Calcium due to Increased Serum Protein	457
Frequent Renal Dialysis	459
Old Serum Artifact	461
Contamination by Detergent	462
Hypoparathyroidism	463

Contents (Continued):

Long Term Dialysis	464
Intravenous Glucose Administration of Several Days Duration	466
Malabsorption - Sprue	467
Acute Pancreatitis	469
Pseudohypoparathyroid	471
Pseudo-Pseudohypoparathyroid	473
Excessive Cortisone	474
Physiological Bone Growth	475
Acromegaly	476
Recent Bone Fracture	477
Nephrotic Syndrome	478
Uncontrolled Diabetes Mellitus	480
Renal Tubular Acidosis	481
Excessive Utilization of Antacid	482
Cirrhosis	483
Extensive Tissue Necrosis and Acute Hemolysis	485
Diabetes Mellitus with Ketoacidosis	487
Post-Mortem Vitreous Humor - Diabetes Ketoacidosis	488
Influence of Uremia on Glucose	489
Congestive Heart Failure	491
Recent Cerebral Damage	492
Drug Induced Liver Disease	493
Oral Contraceptive Usage	494
Thiazide Diuretic Usage	496
Excessive Insulin Utilization	497
Acute Leukemia	500
Strict Diet with Development of Ketoacidosis	502
Eclampsia	503
Hepatorenal Syndrome	504
Prerenal Azotemia	506
Acute Bacterial Infection	508
Vitreous Humor - Chronic Renal Insufficiency	510
Pregnancy - Third Trimester	511
Acute Gout	512
Myxedema	513
Psoriasis	514
Malignant Neoplasm	515
Allopurinol	516
Wilson's Disease	517

Contents (Continued):

Nephrotic Syndrome	519
Obstructive Jaundice	521
Hypothyroidism	523
Influence of Bilirubin on Cholesterol	524
Hyperthyroidism	525
Severe Anemia	526
Cirrhosis - Low Cholesterol	528
Polyclonal or Monoclonal Gammapathy	530
BSP Dye Artifact on Serum	532
Exudate	533
Transudate	535
Large Burns or Generalized Bullous Dermatitis	536
Hypogammaglobulinemia	538
Gilbert's Syndrome	539
Dubin-Johnson Syndrome	540
Recent Lung Infarct	541
Acute Viral Hepatitis	542
Infectious Mononucleosis with Acute Hepatitis	544
Influence of Bilirubin on Serum Albumin	546
Cancer Metastatic to Liver	547
Regan Enzyme - Undifferentiated Bronchogenic Cancer	548
Paget's Disease of Bone	549
Ulcerative Colitis	550
Hypophosphatasia	552
Utilization of Oxalate	553
Acute Renal Infarction	554
Acute Myocardial Infarction	555
Recent Cardiac Surgery	557
Postoperative State or Trauma	558
Pernicious Anemia	559
Gerlach's Ratio LDH/GOT	561
Acute Myositis or Primary Myopathy	562
Neurogenic Atrophy or Myasthenia Gravis of Skeletal Muscle	563
Long Term Renal Dialysis	564

CHAPTER 1

COMMON METHODS OF ENZYME ANALYSIS

Determination of the activities of serum enzymes in many clinical conditions has assumed an important position in recent years. The value of studying serum enzymes as a diagnostic tool was first demonstrated in the early 1900's when serum lipase was determined to aid in the assessment of disease of the pancreas.

Except for the digestive enzymes, the phosphatases and lipase were the first serum enzymes to find widespread diagnostic applications. That intracellular enzymes may be released into the circulation from damaged tissues has been confirmed by numerous investigators. The activities of many enzymes are now more or less routinely determined for diagnostic purposes, not only in myocardial infarction, but also in diseases of the liver, skeletal muscle, bone, and other tissues. Relatively few serum enzymes are specific indicators for pathology in a single tissue; gamma glutamyl transpeptidase is possibly the most specific of the enzyme tests commonly used. The distribution of enzymes varies from tissue to tissue, and careful selection of two or more tests will usually enable the source of the serum enzymes to be accurately defined. Simultaneous determination of several serum enzymes indicates that those of moderate or low specific activity may, in certain circumstances, be of diagnostic value. The value of one enzyme test, lactic dehydrogenase (LDH), has been greatly enhanced by discovery of the existence of the enzyme in multiple isoenzyme forms.

Duration of increased enzyme activity is frequently of diagnostic importance. After a myocardial infarction, the activity of creatine phosphokinase (CPK) in serum is markedly increased during the first 12 to 48 hours, while glutamic oxalacetic transaminase (GOT) reaches a peak after about 24 to 36 hours and returns to normal within 3 to 5 days.[2,12] Enhancement of LDH activity is less pronounced but more prolonged, but the alpha-hydroxybutyrate dehydrogenase (α-HBDH) activity, characteristic of heart muscle, is enhanced and usually does not return to normal in less than 10 days.[7]

Of the many serum enzymes showing increased activities in liver disease, the transaminases and alkaline phosphatase appear to be the most useful diagnostically. The massive increase in the activities of GOT and glutamic pyruvic transaminase (GPT) in serum that coincides with the onset of jaundice is almost pathognomonic of viral hepatitis; values up to 50 times the normal upper limit are common. GPT is usually greater than GOT in viral hepatitis. Moderate elevations occur in other liver diseases.[4]

Common Methods of Enzyme Analysis (Continued):

The activity of serum alkaline phosphatase is increased to a greater extent in obstructive hepatobiliary disease than in hepatitis. For many years this was believed to be the result of failure of the liver to excrete the enzyme in the bile; this was the basis of the retention theory.[5] A possible explanation for the greater serum alkaline phosphatase activity during biliary obstruction is foreshadowed by the findings that in the liver the enzyme is associated with the cell membrane and that in experimental bile duct ligation there is increased enzyme synthesis.[10,13] Increased pressure in the bile canaliculi exerts some effect on the cell membrane, inducing synthesis of the enzyme which causes some of it to enter the circulation.

Enzymes are proteins present in the globulin portion of serum. They are not confined solely to the serum but are present in various portions of cells distributed ubiquitously in many tissues. Differences exist in the amounts of enzymes in various portions of the cell, and different cells vary as to the amounts of specific enzymes. Enzymes are removed from the blood partly by excretion by different excretory organs and partly by metabolic degradation. Enzymes with highly specific functions have the greatest diagnostic usefulness, whereas those that function in intermediary metabolism are diagnostically less important.

The recognition of a diseased organ through serum enzymology would be easy if each organ had its own specific enzyme. This is not the case; most enzymes are shared, though in differing concentrations, among various types of cells. This differential concentration may indicate the cell of origin. When a cell is damaged, the intracellular enzymes leave the cell and enter the serum. If the cell contains more glutamic oxalacetic transaminase (GOT) than glutamic pyruvic transaminase (GPT), the serum concentration of the former will be relatively greater with cell death.

Approximately equal amounts of LDH are present per gram of heart, liver, and skeletal muscle. Thus, as a means of revealing a particular disease site, determination of total LDH is virtually useless. When one subjects LDH to isoenzyme differentiation, a more precise location of the organ pathology may be obtained (Table 1-1).

Common Methods of Enzyme Analysis (Continued):

TABLE 1-1
ENZYME CONTENT OF VARIOUS TISSUES
(%/100 Gm. Tissue)

Enzyme*	Heart	Liver	Muscle
GOT	80	100	40
GPT	20	100	20
LDH	90	95	100
CPK	10	2	100
GLDH	5	100	1
SDH	5	100	2
HBDH	100	10	0

*GOT (glutamic oxalacetic transaminase), GPT (glutamic pyruvic transaminase), LDH (lactic dehydrogenase), CPK (creatine phosphokinase), GLDH (glutamate dehydrogenase), SDH (sorbitol dehydrogenase), HBDH (hydroxybutyrate dehydrogenase).

Liver necrosis may be diagnosed by the simply performed, but uncommonly requested, enzyme determinations of sorbitol dehydrogenase (SDH) and glutamate dehydrogenase (GLDH).[1] Elevation of these enzymes is more specific for necrosis than that of the more commonly determined SGPT.

Both the transaminases GOT and GPT are found within the cell cytoplasm; in minor cellular damage the serum levels of both will be elevated. However, GLDH is solely a mitochondrial enzyme so that its increase indicates cell necrosis.[3]

In reversibly inflammatory processes characterized by increased membrane permeability, cell sap enzymes are more likely to be released into the circulation than are mitochondrial enzymes. In necrotic conditions, however, destruction of large numbers of cells will be followed by the appearance of mitochondrial enzymes GLDH or SDH in the serum.

GOT occurs in markedly different forms in the cell sap and the mitochondria; GPT is almost wholly confined to the soluble fraction of the liver. This led DeRitis to suggest the use of the GOT:GPT ratio as a means of distinguishing predominantly inflammatory lesions from necrotic processes. Values less than one indicated inflammation; while ratios greater than one suggested necrosis because GOT was released from mitochondria. Unfortunately our methods are not always sufficiently precise to permit the use of such ratios, and they have not found extensive application.[6]

Common Methods of Enzyme Analysis (Continued):

In inflammatory states the loss of soluble enzymes through an impaired membrane might act as a stimulus to increased synthesis. Certainly the amounts of transaminase released into the circulation during a moderate bout of hepatitis account for a substantial proportion of the total transaminase content of the liver. The possibility of increased enzyme synthesis in slightly damaged cells in an attractive hypothesis, but until recently there was little evidence to support it. The demonstration of increased serum alkaline phosphatase production in hepatobiliary diseases indicates that this suggestion is valid and it may be relevant to the transaminases and other enzymes.

The interpretation of serum enzyme patterns can be difficult if a disease leads to enzyme release from several organs. Typically, myocardial infarction causes GOT elevation. After a period of days, a picture of hepatic congestion may occur and is usually accompanied by an elevation of GPT and an increase in SDH levels. Serial determinations of CPK can be helpful in such instances in diagnosing myocardial damage because no CPK is found in the liver.

Different routes for enzyme excretion exist. Owing to their extremely high molecular weights (100,000 to 1,000,000), kidney excretion of enzymes is not the usual. Alkaline phosphatase and leucine aminopeptidase are partially excreted into the bile and thus into the gastrointestinal tract.

Rates of enzyme clearance also differ in hepatitis and after myocardial infarction; for example, the different clearance rates are widely used diagnostically. CPK is removed much faster than GPT. Variation in the rates of clearance has an important bearing on the use of enzyme tests in diagnosis.

Among the possible mechanisms in which serum enzymes may be removed are intravascular inactivation either through inhibition by small molecules, for which there is some evidence, or by antigen-antibody pathways.[11] Another possibility is that serum enzymes might be transported into the small intestine, where they would be digested like any other proteins.

Another method of degradation is suggested by some recent work on the relative rates of synthesis and destruction of the isoenzyme LDH_5 in rat heart and liver.[8,9] LDH_5 is characteristic of the liver, which was found not only to synthesize it faster than the heart, but also to destroy it more slowly. It is clear, therefore, that tissues

Common Methods of Enzyme Analysis (Continued):

have built-in mechanisms for degrading enzyme protein and it seems likely that highly vascular organs might well play a significant role in removing enzymes from the plasma.

The half-life of enzymes is important. If there is a delay between the onset of symptoms and the laboratory assay, an actual elevation of the enzyme activity may not be apparent. For example, CPK returns to normal levels within 3 to 4 days after acute myocardial infarction. It is important to assay enzymes soon after the serum is obtained. A rapid deterioration of CPK activity in the refrigerator may occur up to 50 per cent per day. A proteolytic enzyme in the serum may be responsible. In contrast, alkaline phosphatase activity rises with storage in the refrigerator with elevations of 30 to 50 per cent occurring in 24 hours. The half-life of GOT is 50 hours, GPT 75 hours, LDH 50 hours, and GLDH 60 hours in the serum.

An enzyme unit is that activity which transforms a micromole of substrate per minute under optimal conditions and at a defined temperature. Enzymes are assessed in the serum in terms of activity; they act on specific substrates yielding different products. The activity of enzymes is evaluated by the amount of enzyme-specific substrate utilized in the chemical reaction or by the amount of a substance produced, under specific conditions of amount of serum, period of time, and temperature. The determinations may be made on single or serial specimens. The degree of elevation of serum enzymes is a reflection of release from damaged cells or production by abnormal cells with seepage into the bloodstream. The degree of elevation is also subject to different inhibitory or accelerative factors present at the level of the cells or in the serum.

A convenient method utilizes the ultraviolet light Warburg technique with the hydrogen-transfer enzyme reaction mediated by the dehydrogenases in which coenzymes nicotinamide adenine dinucleotide (NAD) and its phosphate (NADP) participate. The coenzymes absorb ultraviolet light in the reduced but not in the oxidized state. The enzyme reaction may be followed with an ultraviolet spectrophotometer. The changes of extinction resulting from alteration of the oxidation-reduction state of the coenzyme are serially measured at regular time intervals. The velocity of this coenzyme reaction is directly related to the enzyme activity.

The measurement of enzyme activity is usually performed in the clinical laboratory by determining the turnover rate of the appropriate substrate. Factors which must be closely monitored are: (1) presence

Common Methods of Enzyme Analysis (Continued):

of inhibitors, (2) necessity for addition of coenzymes, (3) substrate concentration, (4) pH, and (5) temperature.

REFERENCES:

1. Asada, M., Galambos, J. T.: "Sorbitol Dehydrogenase and Hepatocellular Injury: An Experimental and Clinical Study". Gastroenterology 45:578, 1963.

2. Boyde, T. R. C.: "Serum Levels of the Mitochondrial Isoenzyme of Aspartate Aminotransferase in Myocardial Infarction and Muscular Dystrophy". Enzymol. Biol. Clin. 9:385, 1968.

3. Carlson, A. S., Siegelman, A. M., Robertson, T.: "Glutamic Dehydrogenase. II. Activity in Human Serums contrasted with that of Lactic Transaminase". Amer. J. Clin. Path. 38:260, 1962.

4. Clermont, R. J., Chalmers, T. C.: "The Transaminase Tests in Liver Disease". Medicine 46:197, 1967.

5. Clubb, J. S., Neale, F. C., Posen, S.: "The Behavior of Infused Human Placental Alkaline Phosphatase in Human Subjects". J. Lab. Clin. Med. 66:493, 1965.

6. DeRitis, F., Coltori, M., Giusti, G.: "An Enzyme Test for the Diagnosis of Viral Hepatitis: The Transaminase Serum Activities". Clin. Chim. Acta. 2:70, 1957.

7. Elliott, B. A., Jepson, E. M., Wilkinson, J. H.: "Serum Alpha-Hydroxy-butyrate: A New Test with Improved Specificity for Myocardial Lesions". Clin. Sci. 23:305, 1962.

8. Fritz, P. J., Vesell, E. S., White, E. L., Pruitt, K. M.: "The Roles of Synthesis and Degradation in Determining Tissue Concentrations of Lactate Dehydrogenase-5". Proc. Nat. Acad. Sci. 62:558, 1969.

9. Gay, R. J., McComb, R. B., Bowers, G. N., Jr.: "Optimum Reaction Conditions for Human Lactate Dehydrogenase Isoenzyme as they affect Total Lactate Dehydrogenase Activity". Clin. Chem. 14:740, 1968.

10. Kaplan, M. M., Righetti, A.: "Induction of Liver Alkaline Phosphatase by Bile Duct Ligation". Biochim. Biophys. Acta. 184:667, 1969.

Common Methods of Enzyme Analysis - References (Continued):

11. Posen, S.: "Turnover of Circulating Enzymes". Clin. Chem. 16:71, 1970.

12. Rosalki, S. B.: "Serum Alpha-hydroxybutyrate Dehydrogenase: A New Test for Myocardial Infarction". Brit. Heart J. 25:795, Vol. 2, 1963.

13. Wilkinson, J. H.: "Clinical Significance of Enzyme Activity Measurements". Clin. Chem. 16:882, 1970.

ACID PHOSPHATASE DETERMINATION
(Bessey-Lowry)

PRINCIPLE: Acid phosphatases are found in liver, muscle, spleen, and prostatic tissue as well as in erythrocytes. They may be isolated by hydrolysis at pH 4.8 - 4.9. The enzyme hydrolyzes p-nitrophenyl phosphate to produce p-nitrophenol and phosphate.

p-Nitrophenyl phosphate + H_2O $\xrightarrow{(p'tase)}$ p-nitrophenol + phosphate

The citric acid buffered substrate is incubated with serum for 30 minutes. Alkali is added, raising the pH and stopping enzyme action, and diluting up the end product to a convenient concentration. Because p-nitrophenol is an indicator which is yellow in alkaline solutions, blanks are prepared to correct for serum color and turbidity.

The absorbance of p-nitrophenol is measured at 405 nm. The maximum absorbance occurs at 400 nm., but at this wavelength the interference from unhydrolyzed reagent is considerably greater than at 405 nm.

Enzyme activity is expressed in Bessey-Lowry units. One Bessey-Lowry unit is equal to that amount of phosphatase which will liberate 1 millimole of p-nitrophenol per hour per liter of serum at $37°C$.

SPECIMEN: 0.4 ml. of unhemolyzed, non-fasting serum. Do not use hemolyzed serum. Serum acid phosphatase is very unstable at room temperature, demonstrating considerable loss of activity within 1 hour at room temperature. Serum or plasma separated rapidly and frozen until analysis will not lose appreciable activity. Specimens should be kept covered at all times due to pH change.

REAGENTS AND EQUIPMENT:

1. Water bath at $37°C$. constant temperature.

2. <u>Stock p-Nitrophenol Standard Solution</u>, 10 mmoles/Liter
   Sigma Reagent #104-1 (or):
   In a 1000 ml. volumetric flask dissolve 1.3911 gm. p-nitrophenol in distilled water and bring to volume. Stable for one year at $4°C$.

3. <u>Working p-Nitrophenol Solution</u>

Acid Phosphatase Determination (Continued):

Into a 100 ml. volumetric flask pipette 0.5 ml. stock p-nitrophenol standard and bring to volume with 0.02 N. NaOH. Mix thoroughly. Stable for one day.

4. Acid Buffer Reagent
   Sigma Reagent #104-4 (or):
   In a 1000 ml. volumetric flask, dissolve 18.907 gm. citric acid in 180 ml. 0.1 N. NaOH and 100 ml. 0.1 N. HCl. Bring to volume with distilled water. Add a few drops of chloroform. With the aid of a pH meter, check the pH of the buffer; this should be 4.8. Stable when stored at refrigerator temperatures.

5. p-nitrophenylphosphate, 0.04 gm./ml.
   In a 25 ml. volumetric flask, dissolve 0.1 gm. p-nitrophenyl phosphate in distilled water and bring to volume. The dry reagent is also available in pre-weighed capsules (Sigma) containing 0.1 gm. p-nitrophenyl phosphate; these are stable for 1 year in the freezer.

6. NaOH, 0.2 N.

7. NaOH, 0.02 N.

8. NaOH, 0.1 N.

9. Working Buffered Substrate
   Just prior to use mix equal parts of the acid buffer and the p-nitrophenyl phosphate substrate sufficient for a day's determinations.

10. Calibration Curve
    Pipette the solutions indicated in the following chart into six clean tubes, in duplicate. Mix the contents of each tube thoroughly.

	1	2	3	4	5	6
Ml. Working Standard	1.00	2.00	4.00	6.00	8.00	10.00
Ml. 0.02 N. NaOH	10.10	9.10	7.10	5.10	3.10	1.10
Equivalent Bessey-Lowry Units	0.28	0.56	1.12	1.67	2.23	2.80

Acid Phosphatase Determination (Continued):

Immediately read the absorbance of each tube and its duplicate, and determine the average at 405 nm. with 0.02 N. NaOH as a reference solution. Plot A. against equivalent units on graph paper.

PROCEDURE:

1. Label three clean test tubes as follows:
Reagent Blank, Control, Specimen.
Pipette 1.0 ml. Working Buffered Substrate into each.

2. Place in 37°C. water bath. Allow 5 minutes for temperature equilibration.

3. Pipette 0.2 ml. distilled water to tube labelled Reagent Blank. Mix by gentle lateral shaking. Replace in water bath. Start stop watch.

4. At 30 second intervals, add 0.2 ml. sera to control tube and specimen tubes.

5. After exactly 30 minutes incubation, add 4.0 ml. 0.1 N. NaOH. Cap with parafilm, and mix well by inversion.

6. Record the A. of the tests using the reagent blank as reference at 405 nm. on Gilford 300 N flow cell.

7. Since the yellow color of the serum also absorbs at 405 nm., prepare another set of tubes with 0.2 ml. serum plus 5.0 ml. 0.1 N. NaOH. Mix well by inversion. Read these serum blanks against 0.1 N. NaOH as reference. Record absorbance.

8. To calculate, subtract the Absorbance of the serum blank from the Absorbance of the incubated test. Obtain B-L units from the calibration curve.

9. Sera with activity greater than that of the calibration chart must be repeated on dilution.

Acid Phosphatase Determination (Continued):

NORMAL VALUES: 0.13 - 0.64 Bessey-Lowry units.

REFERENCES:

1. Lowry, et. al.: J. Biol. Chem., 20:207, 1954.

2. O'Brien & Ibbott: LABORATORY MANUAL OF PEDIATRIC MICRO CHEMISTRY, Ed. 3, pg. 245, 1967.

3. Sigma Chemical Company: Technical Bulletin #104, "Determination of Alkaline and Acid Phosphatases".

4. STANDARD METHODS OF CLINICAL CHEMISTRY, Vol. 5, pg. 2 and 11.

5. Henry, R. B.: CLINICAL CHEMISTRY: PRINCIPLES AND TECHNIQUES, Hoeber, pg. 482, 1964.

ACID PHOSPHATASE

PRINCIPLE: Acid phosphatase activity in serum and other biological materials is estimated by determining the rate of hydrolysis of various phosphate esters under specified conditions of temperature and hydrogen ion concentration. The substrate used in this method is p-nitrophenylphosphate. The formation of the colored product, p-nitrophenol is a measure of total acid phosphatase activity.

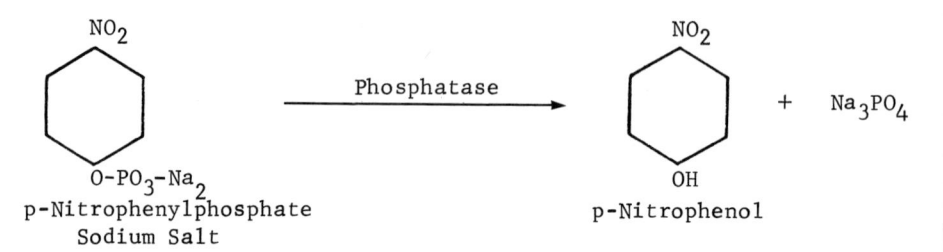

$$\text{p-Nitrophenylphosphate (Sodium Salt)} \xrightarrow{\text{Phosphatase}} \text{p-Nitrophenol} + Na_3PO_4$$

In healthy men and women, only a small amount of acid phosphatase is liberated by the platelets in the clotting process. It is also present in the liver, bone, spleen, kidney, and erythrocytes. Most of it is liberated by the prostate gland.

To differentiate prostatic acid phosphatase from other sources, the enzyme inhibitor used in this methodology is L (+) Tartrate.

A total serum acid phosphatase determination is performed on a given specimen, as well as a determination which includes tartrate. The difference between the two activities is due to prostatic acid phosphatase.

This determination is important for the diagnosis of metastatic prostatic carcinoma, or for following progress of hormone therapy in this disease.

SPECIMEN: 0.6 ml. of fresh, unhemolyzed serum or plasma is required Specimen should be processed as soon as possible since acid phosphatase rapidly loses activity at room temperature.

REAGENTS AND EQUIPMENT: (BioChemica Test Combination Colorimetric Method, Cat. No. 15988 TSAA)

1. <u>Buffer</u> (Bottle No. 1)
   0.05 M. Citrate buffer, pH 4.8

Acid Phosphatase (Continued):

2. Substrate (Bottle No. 2)
   0.05 M. Citrate buffer, pH 4.8; 0.0055 M. p-Nitrophenylphosphate, sodium salt

3. Inhibitor (Bottle No. 3)
   0.2 M. Sodium Tartrate

4. Sodium Hydroxide
   0.4 gm. Sodium Hydroxide, A. R., in 500 ml. of deionized water

5. Spectrophotometer, Gilford 300N, 405 nm.

COMMENTS:

1. Solutions No. 1 and No. 3 are stable for one year at +4°C.

2. Solution No. 2 is stable for one week at +4°C.

3. Sodium hydroxide is stable indefinitely at room temperature.

4. Dilute specimens with 5 percent albumin solution when optical densities are greater than 0.760.

PROCEDURE:

1. For each test, label three tubes, Sample 1 (Total Acid Phosphatase), Sample 2 (Non-Prostatic Phosphatase), and Blank (Serum Blank). Prepare a Blank for each specimen.

2. To Sample 1, add 1.0 ml. of substrate and 0.2 ml. of serum.

3. To Sample 2, add 1.0 ml. of substrate, 0.10 ml. of tartrate inhibitor, and 0.2 ml. of serum.

4. Blank tube gets only 1.0 ml. of substrate

5. Mix all tubes and incubate in 37°C. water bath for 30 minutes.

6. Stop the reaction by adding 10 ml. of NaOH to all tubes.

7. To each corresponding Blank tube, add 0.2 ml. of serum.

8. Set the machine to zero with the Blank, and read optical densities of the samples.

Acid Phosphatase (Continued):

9. CALCULATIONS:

    A. For Total Acid Phosphatase:

    $$\text{I.U.} = \frac{\Delta A \text{ Sample (1)}}{\epsilon \times d} \times 10^6 \times \frac{TV}{SV} \times \frac{1}{\text{Time}}$$

    ϵ = Molar Extinction Coefficient of p-Nitrophenol at 405 nm.
    = 18.8×10^3 Liter/Mole x cm.
    d = Diameter of the cuvette in cm.
    = 1.0 cm.
    TV = Total Volume
    = 11.3 ml.
    SV = Sample Volume
    = 0.2 ml.
    T = Time
    = 30 minutes

    10^6 converts Moles/Liter into micromoles/liter
    1 I.U. = 1 mU/ml.

    $$\text{I.U.} = \frac{\Delta A \text{ Sample (1)}}{18.8 \times 10^3} \times 10^6 \times \frac{11.3}{0.2} \times \frac{1}{30}$$

    I.U. = ΔA Sample (1) x 101

    I.U. = ΔA Sample (1) x 101 = mU/ml. of serum.

    B. Prostatic Acid Phosphatase:

    (ΔA Sample (1) — ΔA Sample (2)) x 101 = mU/ml. of serum

NORMAL VALUES: Normal values compiled by BMC at 37°C.
                 Total Acid Phosphatase: Up to 11 mU/ml. of seru
                 Prostatic Acid Phosphatase: Up to 4 mU/ml. of serum

    1.0 mU. corresponds to 0.06 mMole Units according to
            Bessey-Lowry.

REFERENCES:

1. Lowry, et. al.: J. Biol. Chem. 20:207, 1954.

2. Meites, S.: STANDARD METHODS OF CLINICAL CHEMISTRY, Vol. 5, Academic Press, New York, pg. 2 and 11, 1965.

3. Batsakis and Bierre: INTERPRETATIVE ENZYMOLOGY, C. Thomas, 1967.

PROSTATIC ACID PHOSPHATASE
(Tartrate Inhibition)

PRINCIPLE: In healthy men and women, the serum manifests only slight acid phosphatase activity, much of which is due to enzyme liberated by platelets in the clotting process. The liver, bone, spleen, kidney, platelets, and erythrocytes all exhibit acid phosphatase activity, but to a lesser extent than the prostate. It is occasionally desirable to distinguish between the amount of serum acid phosphatase contributed by the prostate gland and that which is of non-prostatic origin. Since a substrate which is specific for prostatic acid phosphatase has not been successfully achieved, the approach has been to utilize prostatic enzyme inhibitors.

The inhibitor employed in this methodology is L (+) tartrate. A total serum acid phosphatase determination is performed on a given specimen, as well as a determination which includes tartrate. The difference between the two activities is presumably due to acid phosphatase of prostatic origin.

SPECIMEN: 0.2 ml. of unhemolyzed serum. Serum acid phsophatase is unstable at room temperature; refrigerate quickly or freeze, if test is not to be run that day.

REAGENTS: All reagents used are the same as those for the total "Acid Phosphatase Determination" except for those indicated beneath.

1. Tartrate Acid Buffer, 0.04 M., pH 4.8
   Tartrate is in 0.09 M. Citrate and contains chloroform as a preservative. Store at 0 - 5°C. Available as Sigma #104-12.

2. Working Tartrate Substrate
   Prepare a 1:1 dilution of acid phosphatase substrate plus tartrate acid buffer.

PROCEDURE:

1. Allow two aliquots of serum from the same specimen. On one aliquot determine total acid phosphatase. On the second aliquot determine the activity with tartrate inhibition, as indicated beneath. The two determinations may be run simultaneously.

2. Prepare test tubes for reagent blank, control, and specimens. Pipette 1.0 ml. of working tartrate substrate into each tube and place in a water bath (37°C.) for 5 minutes.

Prostatic Acid Phosphatase (Continued):

3. Pipette 0.2 ml. of water into the test tube for the reagent blank and time with stopwatch. At exactly 30 second intervals deliver 0.2 ml. of control sera and specimens into their respective tubes. Mix each tube after addition, and incubate all for exactly 30 minutes at 37°C.

4. After exactly 30 minutes, remove the test tubes from the water bath and add 4.0 ml. of 0.1 N. NaOH. Mix well by inversion against parafilm.

5. Determine the absorbance of each specimen against the reagent blank at 405 nm. in a Gilford 300N flow cell.

6. CALCULATION:

    Subtract the prostatic test A from the A of the uncorrected total test. Determine units from the calibration chart.

NORMAL VALUES:

Male:	0.01 - 0.15 Sigma Units
Borderline:	0.15 - 0.20 Sigma Units

REFERENCES:

1. Fishman and Lerner: Journal of Biological Chemistry, 200:89, 1953.

2. Ozar and Issac: Journal of Urology, 74:150, 1955.

3. Andersch and Szezypinski: Am. J. Clinical Path., 17:571, 1947.

4. Batsakis and Brierre: INTERPRETIVE ENZYMOLOGY, C. Thomas, 1967.

5. Murphy, G., et. al.: Cancer, 23:1309, 1969.

SPECTROPHOTOMETRIC ALDOLASE ASSAY
(Bruns U. V. Method)

PRINCIPLE: The measurement of serum aldolase (Fructose-1, 6-diphosphate-O-glyceraldehyde-3-phosphate lyase) activity is performed rapidly and conveniently by spectrophotometric procedure. This is accomplished by coupling the aldolase reaction with that of a dehydrogenase acting upon one of the triosephosphates formed after splitting FDP. The latter reaction is accomplished by changes in NADH concentration which are measured spectrophotometrically at 340 nm.

1. $FDP \xrightleftharpoons{ALD} GAP + DAP$

2. $GAP \xrightleftharpoons{TIM} DAP$

3. $DAP + NADH + H^+ \xrightarrow{GDH-GI} \alpha\text{-Glycerophosphate} + NAD^+$

The disappearance of NADH is proportional to aldolase. Two moles of NADH are oxidized per mole of FDP hydrolyzed. See comments on calculation.

SPECIMEN: Specimen should be free of hemolysis - stable at $4°C$. up to about 5 days. Need 0.2 ml. serum.

COMMENTS ON PROCEDURE:

The following abbreviations are taken from Bergmeyer, Methods of Enzymatic Analysis, Academic Press, New York, 1965: ALD, Aldolase; DAP, dihydroxyacetonephosphate, FDP, Fructose-1, 6-diphosphate; GAP, glyceraldehydephosphate; GAPDH, glyceraldehydephosphate dehydrogenase; GDH glycerol-1-phosphate dehydrogenase; a-GP, glycerol-1-phosphate; PGA, 3-phosphoglyceric acid; TIM, triosephosphate isomerase.

REAGENTS AND EQUIPMENT:
   Biochemica Test Combination Kit Tc-D (Cat. No. 15974 TAAD).

1. Buffer - 0.056 M. Collidine Buffer; pH 7.4; 0.003 M. Monoiodacitrate; 0.003 M. FDP.

2. 0.020 M. NADH

Spectrophotometric Aldolase Assay (Continued):

3. **2.0 mg. GDH-TIM/ml.**

4. Dilute reagent No. 1 in 100 ml. distilled water. Dissolve reagent No. 2 in 2.0 ml. distilled water - add to reagent No. 1. Add reagent No. 3 to above. Rinse bottle. This makes the Working Substrate (enough for 30 tests). Good for 4 weeks refrigerated.

5. **DBG or Gilford 300 N**
Equipped with thermostated cuvette at 37°C. with recorder.

PROCEDURE:

1. Add 0.2 ml. (200 lambda) of serum to 2.8 ml. of prepared Working Substrate. Place in 37°C. water bath for 6 minutes.

2. Take first reading (A_1) at exactly 6 minutes.

3. Take second reading (A_2) at 16 minutes (10 minute time difference) at 340 nm.

   Method linear with A/10 minutes up to 0.500. If greater, dilute serum 1:10 with saline. Multiply units by 10.

4. CALCULATIONS:

$$\frac{\Delta A}{\epsilon x d} \times 10^6 \times \frac{TV}{SV} \times \frac{1}{Time} = I.U./Liter \text{ or } mU/ml.$$

ϵ of NADH at 340 nm. = 6.22×10^3 Liter/Mole x cm.

$$\frac{\Delta A}{6.22 \times 10^3 \times 1} \times 10^6 \times \frac{TV}{SV} \times \frac{1}{Time} \times \frac{1}{2} \quad \text{Since 2 Moles NADH are oxidized per Mole FDP hydrolyzed.}$$

$$\Delta A \times \frac{1}{6.22 \times 10^3 \times 1} \times 10^6 \times \frac{3.0}{0.2} \times \frac{1}{10} \times \frac{1}{2}$$

$\Delta A \times F = mU/ml.$

$F = 121$

NORMAL VALUES: 4 - 14 mU/ml.

Spectrophotometric Aldolase Assay (Continued):

REFERENCES:

1. Pinto, V. D., Kaplan, A., and Van Dual, P.: <u>Journal of Clin. Chem</u>. Vol. 15, May, 1969.

2. Bruns, F.: <u>Biochemische Zeitschrift</u>, Bd. 325, S. 156 - 162, 1954.

3. Bergmeyer: METHODS OF ENZYMATIC ANALYSIS, Academic Press, 1965.

ALKALINE PHOSPHATASE
(Bessey-Lowry: Continuous Spectrophotometric)

PRINCIPLE: The enzyme alkaline phosphatase hydrolyzes the substrate p-nitrophenyl phosphate to yield phosphoric acid and p-nitrophenol. With an excess of substrate and defined conditions as to pH, temperature, and buffer molarity, the rate of reaction is constant and is proportional to the concentration of the enzyme. The rate of reaction can be determined by measuring the change in absorbance at 405 nm. which is the absorbance maximum for the reaction product p-nitrophenol.

$$\underset{\substack{\text{p-Nitrophenyl} \\ \text{Phosphate}}}{\text{C}_6\text{H}_4(\text{O-PO}_3\text{H}_2)(\text{NO}_2)} + \text{H}_2\text{O} \xrightarrow{\text{Alkaline p'tase}} \underset{\text{p-Nitrophenol}}{\text{C}_6\text{H}_4(\text{OH})(\text{NO}_2)} + \text{H}_3\text{PO}_4$$

Elevated levels of serum alkaline phosphatase are usually due either to increased osteoblastic activity or to disease of the liver or bile ducts. The enzyme is normally elevated in pregnancy.

SPECIMEN: 25 microliters of serum. The anticoagulants EDTA and citrate may not be used, as they inhibit enzyme activity; heparin and oxalate - fluoride are not inhibitory for time periods up to 4 hours. Hemolysis does not interfere with the assay. Serum activity appears stable up to 16 months in frozen state. However, studies upon the stability at room or refrigerator temperatures have shown a 5 - 30% increase in enzyme activity within 24 hours, followed by decrease in activity within a few days.

REAGENTS AND EQUIPMENT:

1. 2-Amino-2-Methyl-1-Propanol Buffer, 0.625 M. pH 10.25
   This buffer is supplied as part of the SMA 12/60 Reagent System.

2. p-Nitrophenyl Phosphate
   The dry reagent is supplied as part of the SMA 12/60 Reagent System.

3. $MgCl_2$, 1.0 M.
   Solution is supplied as part of the SMA 12/60 Reagent System.

Alkaline Phosphatase, Bessey-Lowry Method (Continued):

4. Working Substrate
    AMP buffer, 0.625 M., pH 10.25; p-Nitrophenyl phosphate, 5 mM.; $MgCl_2$ 2 mM. Pipette 1.0 ml. of $MgCl_2$ into 499 ml. of AMP buffer and mix. Transfer 1.0 gm. of p-Nitrophenyl phosphate into the above solution, and mix until the reagent is in solution. Prepare fresh daily. The 12/60 Working Substrate may be used.

5. Ultramicro Dilutor:
    The sample syringe is set to take up 25 microliters of sample and the flush syringe is set to deliver 1.5 ml. of Working Substrate, (store dilutor filled with distilled water when not in use).

6. Spectrophotometer

    A. Gilford 300 N, 405 nm.
    This instrument is used when analyzing one sample at a time. The change in absorbance per minute is read from the Data Lister printout. The linearity of the reaction is visually monitored on the strip chart recorder. The instrument is used with the thermocuvette set at 30°C.

    B. Gilford 222, 405 nm.
    This instrument is used when analyzing four samples at a time. Temperature control is by means of a 30°C. circulating water bath. Rate of reaction is determined from the strip chart recorder set for 0.200 A. full scale and run at one inch per minute.

COMMENTS ON PROCEDURE:

1. Substrate exhaustion is rate-limiting; the reaction will be linear until this point.

2. Specimens with high enzyme activity may be monitored with the ratio switch of the Gilford 222 set at 0.5 or lower. The change in absorbance per minute must be multiplied by the appropriate factor.

3. One must be sure that the reaction mixture and cuvette chamber is at temperature before making a measurement.

Alkaline Phosphatase, Bessey-Lowry Method (Continued):

PROCEDURE:

1. With ultramicro dilutor, take up 25 microliters of serum, and flush with 1.5 ml. of Working Substrate. Mix well.

2. When using the Gilford 222, preincubate reaction mixture in 30°C. water bath to assure temperature equilibration. Preincubation is not necessary with the Gilford thermocuvette because the temperature equilibration and lag time are less than one minute.

3. Introduce reaction mixture into spectrophotometer, and record at least 2 minutes of linear reaction at 405 nm.

4. CALCULATION:

$$\frac{\Delta A}{\epsilon x d} \times 10^6 \times \frac{TV}{SV} \times \frac{1}{Time} = IU/Liter \text{ or } mU/ml.$$

ϵ p-Nitrophenol at 405 nm. = 18.6 x 10^3 Liter/Mole x cm.

$$\frac{\Delta A}{18.6 \times 10^3 \times 1} \times 10^6 \times \frac{TV}{SV} \times \frac{1}{Time}$$

$$\Delta A \times \frac{1}{18.6 \times 10^3 \times 1} \times 10^6 \times \frac{1.525}{0.025} \times \frac{1}{1}$$

$\Delta A \times F = mU/ml.$

$F = 3279 \text{ or } 3280$

NORMAL VALUES: 30 - 85 mU/ml.

REFERENCES:

1. Bessey, O., Lowry, O., Brock, M.: *J. Biol. Chem.*, 164:321,

2. Bowers and McComb: *Clin. Chem.*, 12:70-89, 1966.

3. Morgenstern, et. al.: *Clin. Chem.*, 11:876, 1965.

4. Henry, R. B.: CLINICAL CHEMISTRY: PRINCIPLES AND TECHNIQUES, Hoeber, pg. 491, 1965.

SERUM AND URINE AMYLASE
(Iodometric Method by Caraway)

PRINCIPLE: Alpha-amylase is an enzyme secreted by the pancreas and salivary glands. It hydrolyzes starch to the disaccharide maltose. Starch, but not maltose, forms a blue colloidal complex with iodine in solution, and the intensity of this color is directly proportional to the concentration of the starch. The blue color produced by the starch substrate when combined with iodine, is measured after incubation with serum and compared to a blank. The decrease in color is proportional to the amylase activity.

One amylase unit is the amount of enzyme that will hydrolyze 10 mg. of starch in 30 minutes to a stage at which no color is given by iodine.

SPECIMEN: When serum determination is made, pipettable 0.1 ml. of serum should be obtained. A fasting sample is not essential. Urines can be analyzed in the same manner as serum. For adequate interpretation, either a 1 hour or a 24 hour collection of urine should be obtained; random specimens have little diagnostic value. Other body fluids such as bile, pancreatic secretions, duodenal drainage, and so on, may also be analyzed in the same manner, inasmuch as the substrate is well buffered.

REAGENTS:

1. <u>Stable Buffered Starch Substrate</u> pH 7.0
   Dissolve 13.3 gm. of anhydrous disodium phosphate and 4.3 gm. of benzoic acid in about 250 ml. of water. Bring to boil. Mix separately 0.200 gm. of Merck's solution starch (Lintner) in 5.0 ml. of cold water and add it to the boiling mixture, rinsing beaker with additional cold water. Continue boiling for 1 minute. Cool to room temperature and adjust pH to 7.0. Dilute to 500 ml. with water. Stable at room temperature and should remain water clear.

2. <u>Stock Solution of Iodine</u>, 0.1 N.
   Dissolve 3.567 gm. potassium iodate (KIO_3) and 45 gm. of potassium iodate in approximately 800 ml. of water. Add slowly and with mixing, 9.0 ml. of concentrated hydrochloric acid (12 M.) and dilute to 1000 ml. with water.

3. <u>Working Iodine Solution</u>, 0.01 N.
   Dissolve 59 gm. of potassium fluoride ($KF \cdot 2H_2O$) in approximately 350 ml. water in a 500 ml. volumetric flask. Add 50 ml. of

Serum and Urine Amylase (Continued):

the stock solution of iodine and dilute to the mark. Solution is stable for 1 - 2 months, when stored in a brown bottle in the refrigerator.

PROCEDURE:

1. Run normal and abnormal controls with each set of determinations <u>Do not use a blow-out pipette</u>, as there is great amylase content in saliva.

2. Pipette 5.0 ml. (volumetric pipette) of starch substrate into each of 50 ml. graduated tubes marked "Test" and "Blank".

3. Place all the tubes in a water bath at 37°C. for <u>5 minutes</u> to warm the contents.

4. Pipette exactly 0.10 ml. of serum into the bottom of the tube labelled "Test", and allow the reaction to proceed for exactly $7\frac{1}{2}$ minutes. No serum is added to the "Blank" tube.

5. After $7\frac{1}{2}$ minutes, remove the tubes from the water bath; add deionized water up to the 40 ml. mark. Immediately add 5.0 ml. (volumetric pipette) of working solution of iodine to each tube, then dilute to 50 ml. with deionized water. Mix well by inversion and shaking.

6. Measure the % Transmission of the "Test" and "Blank" without delay against water at 660 nm. in a spectrophotometer using 12 mm. cuvettes.

7. CALCULATIONS:

Convert % T. readings to optical density.

$$\frac{\text{O.D. of Blank} - \text{O.D. of Test}}{\text{O.D. of Blank}} \times 800 = \text{Amylase units/100 ml.}$$

"800" indicates that complete hydrolysis of the starch would correspond to a serum amylase activity of 800 units per 100 ml.

8. If the activity of the amylase in serum exceeds 400 units, the test is repeated using a 5-fold dilution of the serum with 0.9% NaCl. Final results are corrected by multiplying by 5.

Serum and Urine Amylase (Continued):

NORMAL VALUES: Serum: 40 - 160 units/100 ml.
                Urine: 43 - 245 units/hour based on 6 to 24 hour collection.

REFERENCE: Caraway, Wendell, T.: <u>Amer. Jour. Clin. Path</u>., 32:97, 1959.

SERUM CERULOPLASMIN DETERMINATION

PRINCIPLE: Ceruloplasmin concentration is determined from the rate of oxidation of p-phenylenediamine at 37°C. and at pH 6.0. The rate of appearance of the purple oxidation product (Wurster's red), which has an absorption peak at 520 - 530 nm. is measured spectrophotometrically or photometrically.

SPECIMEN: 0.1 ml. serum or plasma.

REAGENTS AND EQUIPMENT:

1. **Beckman Model Du Spectrophotometer**
   Equipped with Thermospacers as the compartment temperature must be kept constant at 37°C.

2. **Acetate Buffer, 0.1 M. pH 6.0**
   Add 10 ml. 0.1 M. acetic acid (0.57 ml. glacial acid plus water to 100 ml.) to 200 ml. 0.1 M. sodium acetate (1.36 gm. $CH_3COONa \cdot 3H_2O$ per 100 ml.). The pH must be 5.95 - 6.00.

3. **Sodium Azide, 0.1% in 0.1 M. Acetate Buffer**
   The pH must be 5.95 - 6.00. Store in refrigerator.

4. **p-Phenylenediamine · 2HCl, 0.25% in 0.1 M. Acetate Buffer**
   Recrystallize commerical salt as follows: Dissolve in water, add Darco charcoal, warm in water bath at 60°C. with occasional mixing, and filter. Add acetone to filtrate until turbidity appears, refrigerate for several hours, filter off the p-phenylenediamine · 2HCl (PPD), and dry the crystals in the dark in a vacuum desiccator over anhydrous $CaCl_2$. Store in brown bottle. To prepare the 0.25% reagent, dissolve 12.5 mg. in 3.0 ml. acetate buffer and, using narrow range pH paper, adjust the pH to 6.0 by adding 1.0 N. NaOH dropwise from a 0.2 ml. serologic pipette (approximately 0.1 ml. required). Add acetate buffer to a final volume of 5.0 ml. This reagent can be used up to about 2 hours after preparation if kept in the dark.

PROCEDURE:

1. Set up the following in cuvets with 1.0 cm. light path:
   BLANK: 1.0 ml. azide reagent + 1.0 ml. buffer + 1.0 ml. PPD reagent.
   TEST: 2.0 ml. buffer + 1.0 ml. PPD reagent.

2. Place cuvets in compartment and allow 5 minutes for temperature equilibration.

Serum Ceruloplasmin Determination (Continued):

3. Add 0.1 ml. serum (heparinized or oxalated plasma is satisfactory) to each tube from a TC pipette, effecting mixing in the process.

4. Read absorbance of the Test against the Blank at 530 nm. at exactly 10 minutes and again at 40 minutes after addition of serum.

CALCULATION:

Ceruloplasmin Units = (A at 40 min. - A at 10 min.) x 1000

COMMENTS ON PROCEDURE:

1. Artificial Standard
   Of a number of dyes studied, pontacyl violet 6R (Du Pont) possesses an absorption curve closest to that of PPD oxidized by ceruloplasmin. The curves are not identical, however, and the concentration of dye given as a 400 unit standard was established for a Klett No. 54 filter and may not be valid for other photometers. It is not valid for a spectrophotometer at 530 nm.

2. Effect of Light
   Catalytic oxidation of PPD is increased by exposure to light. The test, therefore, must be carried out in the dark.

3. Variation in pH
   In the technic presented, the optimal pH for serum ceruloplasmin activity is at 6.0 and is fairly sharp. At pH 6.0 and above, the rates, after the lag phase, are linear to 60 minutes. At pH 5.8 and lower, however, a decrease in rate occurs during the 30 to 40 minute period. The cause for this is unknown.

4. Variation in Temperature
   In the method presented, Arrhenius plots of log of activity against reciprocals of absolute temperature give straight lines between 22 and 45°C. with a slope of —3700. This gives an activation energy, u, of 17,000 cal./mole and a Q_{10} (temperature coefficient) between 30 and 40°C. of 2.45.

5. Lag Phase
   Evidence indicates that PPD oxidized by ceruloplasmin is again reduced by the ascorbic acid present in the serum until all the ascorbic acid is used up. A lag period greater than 10 minutes

Serum Ceruloplasmin Determination (Continued):

has never been observed by our laboratory in the method presented, i.e., a linear rate is established by 10 minutes.

6. Hemolysis
   It has been reported that minimal hemolysis does not interfere. Hemoglobin has been added in our laboratory to sera to a concentration as high as 200 mg./100 ml. and no interference was observed.

7. Stability of Samples
   Stability at room temperature is somewhat variable, some sera showing no degradation in 2 days, others decreasing up to 15% in 24 hours. Samples are stable at least 2 weeks at 4°C. or in the frozen state. Ultraviolet light of 253.7 nm. wavelength inactivates the oxidase activity of ceruloplasmin by causing a splitting off of the copper.

ACCURACY AND PRECISION

The accuracy of any method for determination of the oxidase activity of ceruloplasmin is restricted by the absence of any reference ceruloplasmin standard. Oxidase activity can be plotted against ceruloplasmin concentration determined by immunochemical analysis but this is not feasible for most laboratories. The situation is further complicated by the fact that there are at least two, and possibly four or five ceruloplasmins in serum. In any event, the enzymatic activity observed in a method is governed not only by the enzyme concentration but also by the concentrations of various ions. Ferrous ion, and to a lesser extent ferric and other cations at low concentration, enhance the oxidase activity of purified ceruloplasmin. There is also evidence that albumin inhibits.

The precision of the test (95% limits) is about $\pm$ 5%.

NORMAL VALUES: The 95% adult limits are 280 to 570 Units

REFERENCE: Henry, J. B.: CLINICAL CHEMISTRY: PRINCIPLES AND TECHNIQUES, Hoeber, pg. 500 - 503, 1965.

CHOLINESTERASE

PRINCIPLE: In most of the quantitative methods, the enzymatic hydrolysis of an organic ester of choline under controlled conditions of sample size, substrate concentration, time, temperature, pH, ionic strength, and various salt concentrations results in the liberation of a choline salt and an organic acid. In this method, acetylthiocholine is used as a substrate; giving off hydrolysis products of thiocholine and acetic acid. The thiocholine is further reacted with dithiobisnitrobenzoic acid which gives off a colored product with maximum absorbance at 405 nm.

$$\text{Acetylthiocholine} \xrightarrow{\text{Cholinesterase}} \text{Thiocholine} + \text{Acetic Acid}$$

$$\text{Thiocholine (Sulfhydryl Group)} + \text{Dithiobisnitrobenzoic Acid} \longrightarrow \text{Color}$$

The product formed is proportional to the concentration of the enzyme. Cholinesterase is an enzyme found in the tissues of all animals and formed in the liver. When this enzyme is low, acetylcholine accumulates, resulting in the continuous stimulation of the parasympathetic system with undesirable symptoms. Cholinesterase activity is another diagnostic test for liver disease. It is very useful in detecting poisoning by organic phosphate insecticides or some drugs as in anesthesia.

SPECIMEN: 0.02 ml. of fresh, unhemolyzed serum is required. Plasma should not be used. Samples are stable for several days at 4°C. and for weeks in the freezer.

REAGENTS AND EQUIPMENT: (Biochemica Test Combination Colorimetric Method, Cat. No. 15984 TCAB).

1. Buffer (Bottle No. 1)
   0.05 M. Phosphate Buffer, pH 7.2, and 0.00025 M. Dithiobisnitrobenzoic acid.

2. Substrate
   0.156 M. Acetylthiocholine Iodide.

3. Spectrophotometer
   A. Gilford 300N, 405 nm.
      This instrument is used when analyzing one sample at a time. The change in absorbance per minute is read from the Data

Cholinesterase (Continued):

> Lister printout. The linearity of the reaction is visually monitored on the strip chart recorder. The instrument is used with the thermocuvette set at 25°C.

> B. Gilford 222, 405 nm.
> This instrument is used when analyzing four samples at a time. Temperature control is maintained by means of a 25°C. circulating waterbath. Rate of reaction is determined from the strip chart recorder set for 0.200 A. full scale and run at one inch per minute.

COMMENTS:

1. Reconstituted solutions are stable for six weeks at 4°C.

2. All specimens with optical density differences greater than 0.200/30 seconds should be diluted with physiological saline.

3. The amount of non-enzymatic hydrolysis of acetylthiocholine which may occur during the test should be determined for each new pack. This can be done by running it with deionized water in place of serum. The resulting optical density difference/30 seconds should be deducted from the optical density difference of each test result.

PROCEDURE:

1. Using pyrex cuvettes (10 mm. light path, 10 mm. wide), pipette 3.00 ml. of buffer solution, 0.02 ml. of serum and 0.10 ml. of acetylthiocholine iodide.

2. Dilute three more specimens in the same manner.

3. Place four specimens in the cuvette chamber of the Gilford 222.

4. Set the baseline of specimen No. 1 with the slit, and No.'s 2, 3, and 4 with the OFF-SET knobs. Switch to "AUTO" and scan all four channels.

5. Scan immediately, and record 2 minutes of linear reaction time at 405 nm. Determine $\Delta A/30$ seconds from the recorder chart.

6. CALCULATIONS:

Cholinesterase (Continued):

$$\text{I.U.} = \frac{\Delta A/30 \text{ Sec.}}{\epsilon x d} \times 10^6 \times \frac{TV}{SV} \times \frac{1}{\text{Time}}$$

- ϵ = Molar Extinction Coefficient of the product at 405 nm.
- = 13.3×10^3 Liter/Mole x cm.
- d = Diameter of the cuvette in cm.
- = 1.0 cm.
- TV = Total Volume
- = 3.12 ml.
- SV = Sample Volume
- = 0.02 ml.
- T = Time
- = 0.5 minutes

10^6 converts Moles/Liter into micromoles/liter
1.0 I.U./Liter = 1 mU/ml.

$$\text{I.U.} = \frac{\Delta A/30 \text{ Sec.}}{13.3 \times 10^3} \times 10^6 \times \frac{3.12}{0.02} \times \frac{1}{0.5}$$

= $\Delta A/30$ Sec. x 23400

$\Delta A/30$ Sec. x 23400 = mU/ml. of serum

Report results in I.U. at 25°C.

NORMAL VALUES: Normal Values compiled by BMC at 25°C.:
1900 - 3800 mU/ml. of serum.
2800 - 5200 mU/ml. whole blood.

REFERENCES:

1. Bergmeyer, H. U.: METHODS OF ENZYMATIC ANALYSIS, 2nd Ed., Academic Press, New York, pg. 771, 1965.

2. STANDARD METHODS OF CLINICAL CHEMISTRY, Vol. 3, pg. 93 - 98, 1961.

3. STANDARD METHODS OF CLINICAL CHEMISTRY, Vol. 4, pg. 47 - 56, 1963.

SERUM OR PLASMA CHOLINESTERASE
(Acholest Test Paper)

PRINCIPLE: Cholinesterase, an enzyme found in the tissues of all animals, formed in the liver, hydrolyzes acetylcholine to choline and acetic acid.

$$\text{Acetylcholine} \longrightarrow \text{Choline + Acetic Acid}$$

Acetylcholine stimulates the nerve impulses in the parasympathetic nervous system. When cholinesterase is low, acetylcholine accumulates, resulting in continuous stimulation of the parasympathetic system with undesirable symptoms.

The ACHOLEST TEST PAPER is a rapid, simplified screening test for the determination of plasma or serum cholinesterase. The kit consists of a test strip impregnated with a special substrate and a control strip (comparative color strip). The time required to reach the endpoint with the control strip is a measure of cholinesterase activity, described as "increased", "normal", "suspicious", or "decreased".

Cholinesterase activity is another diagnostic test for liver disease. It is very useful in detecting poisoning by organic phosphate insecticides or some drugs as in anesthesia.

SPECIMEN: 0.1 ml. of non-hemolyzed plasma or serum is required. Samples are stable for several days at 4°C. and for weeks in the freezer.

REAGENTS AND EQUIPMENT:

1. <u>ACHOLEST, cholinesterase Test Paper</u>
   Bottle I.

2. <u>Comparative Color Strip</u>
   Bottle II.

3. <u>Clean Slides</u>

4. <u>0.05 ml. Wiretrol Disposable Pipettes</u>

5. <u>Stop-Watch</u>

6. <u>Tweezers</u>

Serum or Plasma Cholinesterase (Continued):

The whole kit (for 30 determinations) is available and supplied by E. Fougera and Co., New York)

PROCEDURE:

1. Using Wiretrol disposable pipettes, pipette 0.05 ml. of plasma or serum on each of two clean slides. Label one slide "Test" and the other one "Control".
2. Cut the "Test" Strip and the "Control" Strip in half.
3. Using tweezers, place the Test Strip and Control Strip on the respective slides.
4. Cover with clean slides to ensure complete saturation of the paper.
5. Set the stop-watch. The Test Paper turns green, gradually developing into a yellowish color, which is the color of the Control Paper.
6. Cholinesterase activity is measured by the time required for the Test Paper to reach the color of the Control Paper.
7. From the moment of contact of the serum or plasma and Test Paper, to the point where the comparative tone of color has been reached, the following time values have been established:

Minutes	Activity of Plasma or Serum Cholinesterase
Below 5	"Increased"
5 - 20	"Normal"
20 - 30	"Suspicious"
30 and Longer	"Decreased"

NORMAL VALUES: 5 - 20 Minutes

REFERENCES:

1. STANDARD METHODS OF CLINICAL CHEMISTRY, 3:93 - 98, Academic Press, 1961.

2. STANDARD METHODS OF CLINICAL CHEMISTRY, 4:47 - 56, Academic Press, 1963.

CREATINE PHOSPHOKINASE

PRINCIPLE: Determination of the enzyme creatine phosphokinase (CPK, adenosine triphosphate, creatine phosphotransferase) in serum is of value in the investigation of skeletal muscle disease, possibly in the detection of carriers of muscular dystrophy, and in the diagnosis of suspected myocardial infarction. The CPK value is also elevated with muscle trauma, such as multiple intramuscular injections, and with severe exercise. CPK is found in highest activity in skeletal muscle, heart muscle and brain tissue, with minimal quantities found in lung, kidney, liver and red cells. For this reason, the CPK is helpful in interpreting elevated SGOT values in the presence of hepatic problems. The CPK activity also rises much more rapidly than SGOT following myocardial infarction.

This enzyme catalyzes the reversible formation of adenosine triphosphate and creatine from adenosine diphosphate and creatine phosphate CPK activity in serum is determined by the procedure in which adenosine triphosphate, liberated by the action of the enzyme, is linked to the reduction of nicotinamide-adenine dinucleotide phosphate and the formation of reduced nicotinamide-adenine dinucleotide phosphate followed spectrophotometrically at 340 nm.

1. ADP + CP $\longrightarrow$ ATP + Creatine

2. ATP + Glucose $\xrightarrow{\text{Hexokinase}}$ Glucose-6-Phosphate + ADP

3. Glucose-6-Phosphate + NADP $\longrightarrow$ 6-Phosphogluconate + NADPH

Increased reaction rate is achieved by including optimal amounts of Mg^{++} and glutathione as activators. AMP is included to inhibit the activity of any myokinase which may be present.

SPECIMEN: 50 microliters of unhemolyzed serum. The enzyme is extremely heat labile, and should be stored at refrigerator temperatures until one is prepared to make the determination.

REAGENTS AND EQUIPMENT:

The Reagents for the test are the Biochemica Test Combination supplied by Boehringer Mannheim Corporation: UV System CPK Single Test (Cat. #15790 - for 20 determinations). Volumes have been modified for use in this laboratory.

1. Pipette 2.5 ml. of Solution 2 into each Bottle No. 1. Prepare enough working substrate for one day run.

Creatine Phosphokinase (Continued):

Concentrations in the test volume:
a). 0.1 M. Triethanolamine buffer, pH 7.0, 20 mM. glucose
b). 10 mM. Mg-acetate, 1.0 mM. ADP, 10 mM. AMP
c). 0.6 mM. NADP, 35 mM. creatine phosphate, 50 ug. HK
d). 25 ug. G-6-PDH, 9.0 mM. glutathione

2. Spectrophotometer

   A. **Gilford 222**, 340 nm.
      This instrument is used when analyzing four samples at a time. Temperature control is by means of a 30°C. circulating water bath. Rate of reaction is determined from the strip chart recorder set for 0.200 A full scale and run at one inch per minute.

   B. **Gilford 300 N**, 340 nm.
      This instrument is used when analyzing one sample at a time. The change in absorbance per minute is read from the Data Lister print-out. The instrument is used with the thermocuvette set at 30°C.

COMMENTS ON PROCEDURE:

1. Monitor the working substrate for several minutes to check for substrate deterioration and/or instrument drift.

2. If the cuvette is being removed from a water bath to a cuvette well for determining absorbance values, one should work quickly to prevent temperature change in the reaction mixture.

3. Sera with low activity will show a longer lag time than sera with high activity.

4. With an initial A reading over 0.800, or ΔA/minute greater than 0.060 (300 mu./ml.), test must be repeated using a lesser volume of sera. Multiply result by dilution factor.

PROCEDURE:

1. Pipette 1.5 ml. of working CPK substrate into a 1.0 cm. square cuvette and equilibrate to 30°C. for 5 minutes.

2. Using a 50 microliter pipette (calibrated "to deliver"), deliver serum into the cuvette with substrate and mix well by inversion. Return to water bath. Start stop watch.

Creatine Phosphokinase (Continued):

3. After a lag time of approximately 5 minutes (longer for low activity), place cuvette into a 30°C. thermostated cuvette well and monitor the reaction until at least 2 consecutive minutes of linearity are recorded. Spectrophotometer set at 340 nm.

4. CALCULATION:

   Calculation is based on ΔA of 1 minute. Light path is 1.0 cm. in diameter.

   ϵ of NADPH at 340 nm. = 6.22 x 10^3 Liter/Mole x cm.

   $$\frac{\Delta A}{\epsilon x d} \times 10^6 \times \frac{TV}{SV} \times \frac{1}{Time} = \text{I.U./Liter or mU./ml.}$$

   $$\Delta A \times \frac{1}{6.22 \times 10^3 \times 1} \times 10^6 \times \frac{1.55}{0.05} \times \frac{1}{1}$$

   ΔA x F = mU./ml.

   F = 5000

   NORMAL VALUES: Males   5 - 90 I.U./Liter or mU./ml.
                          Females 5 - 70 I.U./Liter or mU./ml.

NOTE:

A laboratory must determine its own normal range of values due to variation type of methodology used and reagent concentration and purity.

REFERENCES:

1. Rosalki, S. B.: *Journal of Laboratory and Clinical Medicine*, Vol. 4, April, 1967.

2. Bergmeyer: METHODS OF ENZYMATIC ANALYSIS, Academic Press, 1965.

SERUM GAMMA-GLUTAMYL-TRANSPEPTIDASE

PRINCIPLE: Serum γ-glutamyl-transpeptidase hydrolyzes the substrate γ-glutamyl-p-nitroanilide to yield p-nitroaniline and γ-glutamyl compounds.

```
   CO-NH-⟨    ⟩-NO₂                   NH₂
   CH₂                                 |
   CH₂           γ-GT        ⟨    ⟩        +   γ-Glutamyl
 H C NH₂         ———→                              Compounds
   COOH                          NO₂
L-γ- Glutamyl-p-Nitroanilide   p-Nitroaniline
```

Since the cleavage product p-nitroaniline has a maximum absorption at 405 nm. (substrate is transparent at this wavelength), its rate of formation can be utilized to determine the activity of the enzyme.

γ-glutamyl transpeptidase activity in human serum has value in differential diagnosis of liver diseases. This enzyme is particularly elevated in obstructive jaundice, hepatic carcinoma and chronic alcoholism.

SPECIMEN: 0.1 ml. of serum is required. Serum should be separated from cells as soon as possible. Only minimum hemolysis allowed. A slight inhibition of the enzyme occurs when oxalate, citrate, and fluoride are present in the usual concentrations. Heparin and EDTA do not inhibit. Serum is stable for over a week at 4°C.

REAGENTS AND EQUIPMENT:

BioChemica Test Combination C-System Kit, 15794 TMBG

1. Substrate (Bottle No. 1)
   4 mM γ-glutamyl-p-nitranilide; 40 mM glycilglycine.

2. Buffer (Bottle No. 2)
   185 mM Tris buffer, pH 8.25.

3. Working Substrate
   Warm the buffer (Solution No. 2) for a few minutes at 50 - 60°C. Add 3.0 ml. of the warmed buffer to each bottle (Bottle No. 1) of substrate. Combine all substrate into a 125 ml. Erlenmeyer flask and connect to the microdilutor.

Serum Gamma-Glutamyl-Transpeptidase (Continued):

4. Ultramicro Dilutor
   The sample syringe is set to take up 100 microliters of sample, and the flush syringe is set to deliver 1.5 ml. of substrate.

5. Spectrophotometer

   A. Gilford 300 N, 405 nm.
      This instrument is used when analyzing one sample at a time. The change in absorbance per minute is read from the Data Lister Print-out. The linearity of the reaction is visually monitored on the strip chart recorder. The instrument is used with the thermocuvette set at 30°C.

   B. Gilford 222, 405 nm.
      This instrument is used when analyzing four samples at a time. Temperature control is maintained by means of a 30°C. circulating water bath. Rate of reaction is determined from the strip chart recorder set for 0.200 A full scale and run at one inch per minute.

COMMENTS ON PROCEDURE:

1. Working substrate is stable for 24 hours at 20 - 25°C.

2. Dilute all specimens when the change in rate is greater than 0.150 ΔA/minute.

3. Substrate exhaustion is rate-limiting; the reaction will be linear for 10 minutes.

4. Reaction mixture and cuvette chamber must be at temperature before making a measurement.

PROCEDURE:

1. With the ultramicro dilutor, take up 100 microliters of serum and flush with 1.5 ml. of working substrate. For the Gilford 222, use Pyrex cuvettes 10 mm in width and a 10 mm. light path. Mix well.

2. Dilute up three more specimens in the same manner.

3. Place 4 specimens in the cuvette chamber of the Gilford 222.

Serum Gamma-Glutamyl-Transpeptidase (Continued):

4. Set the baseline of Specimen No. 1. with the slit, and No.'s 2,3 and 4 with the Off-Set knobs. Switch to "Auto" and scan all 4 channels.

5. Record 2 - 3 minutes of linear reaction time at 405 nm. Determine the mean ΔA/minute from recorder chart.

6. CALCULATIONS:

$$\frac{\Delta A}{\epsilon x d} \times 10^6 \times \frac{TV}{SV} \times \frac{1}{Time} = I.U./Liter \text{ or } mU/ml.$$

ϵ of p-nitroaniline at 405 nm. = 9.9×10^3 Liters/Mole x cm.

d = Diameter of light path = 1.0 cm.
TV = Total Volume
SV = Sample Volume
Time = 1 Minute

10^6 converts Moles/Liter into Micromoles/ml.

$$\frac{\Delta A}{1} = \frac{1}{9.9 \times 10^3 \times 1} \times 10^6 \times \frac{1.6}{0.1} \times \frac{1}{1} = mU/ml.$$

ΔA x F = mU/ml.
F = 1616

NORMAL VALUES:

Normal values should be established by each laboratory running the test. A normal typical value for 25°C. would be 0 - 11 I.U.

REFERENCES:

1. Szasz, G.: Clin. Chem., 15:124, 1969.

2. Zein, M., Discombe, G.: The Lancet, pg. 748 - 750, Oct., 1970.

GLUCOSE OXIDASE
(GOD - Perid Method)

PRINCIPLE: Glucose is oxidized by the specific enzyme glucose oxidase to gluconolactone which, in aqueous solution, is converted to gluconic acid.

$$\text{Glucose} + O_2 + H_2O \xrightarrow{\text{GOD*}} \text{Gluconic acid} + H_2O_2$$

In the presence of peroxidase, the hydrogen peroxide produced in the above reaction oxidizes with the formation of a green color.

$$H_2O_2 + \underset{\text{(reduced-colorless)}}{\text{Hydrogen donor dye}} \xrightarrow[\text{POD***}]{\text{ABTS**}} \underset{\text{(oxidized-green)}}{\text{Hydrogen donor dye}} + 2H_2O$$

The intensity of the dye is proportional to the glucose concentration.

SPECIMEN: 10 microliters of serum. The specimen must be spun down or chilled to $4°C$. within 15 minutes after collection.

REAGENTS AND EQUIPMENT:

Reagents available as Biochemica Test combination for Blood Sugar GOD - Perid Method, 15756 TEAP, 3 x 175 tests, Boehringer Mannheim Corporation.

1. <u>Stock Glucose Standard</u>, 10 mg./ml.
   Place 1000 mg. of dessicated glucose (dextrose) into a 100 ml. volumetric flask. Dissolve and bring to volume with saturated benzoic acid.

   Glucose oxidase is specific for beta-D glucose; solutions of glucose freshly prepared from the dry chemical should be allowed to stand for at least two hours to insure that mutarotation has reached a state of equilibrium. No such precautions are necessary when preparing dilutions from a Stock Standard solution.

2. <u>Working Glucose Standards</u>

   A. 50 mg%. glucose: Quantitatively transfer 5.0 ml. of stock glucose standard into a 100 ml. volumetric flask, and dilute with saturated benzoic acid.

Glucose Oxidase (Continued):

- B. 100 mg%. glucose: Quantitatively transfer 10 ml. of stock glucose standard into a 100 ml. volumetric flask, and dilute with saturated benzoic acid.

- C. 300 mg%. glucose: Quantitatively transfer 30 ml. of stock glucose standard into a 100 ml. volumetric flask, and dilute with saturated benzoic acid.

3. Working Substrate
One bottle containing 100 mM. phosphate buffer, pH 7.0; 20 micrograms POD/ml., 180 micrograms GOD/ml., 1.0 mg. ABTS/ml. is dissolved with deionized water and brought to 1000 ml. in a volumetric flask. Keep in a dark bottle. Stable for 6 weeks at 4°C.

4. Ultramicro Dilutor
Sample syringe set to take up 10 microliters of specimen, and flush syringe set to deliver 5.0 ml. of substrate. The reagent is stored in the dilutor with a dark bottle. The dilutor is stored in a light-tight cupboard.

5. Gilford 300 N
Equipped with either regular or microaspiration cuvette.

COMMENTS ON PROCEDURE:

1. It is essential that all glassware and reagents used in the procedure be chemically clean. Chemical contamination may interfere with the enzyme system.

2. The Working Substrate must be kept light-tight for reproducible results within a series; the reagent darkens progressively upon exposure to light.

3. Stability of the Working Substrate has been proven to be good up to 10 days at room temperature.

4. The determination is carried out at room temperature; it is not critical that the temperature be exact as long as the temperature is *stable* during the analytical run. Temperatures of greater than 38°C. may inactivate the enzyme. The analysis may also be performed at other temperatures, but the reaction time is affected.

Glucose Oxidase (Continued):

5. The green color which is measured attains its maximum absorption after 20 minutes at room temperature.

6. Adult serum separated within 30 minutes of drawing of the blood sample has a glucose level identical with that of heparinized plasma analyzed as soon as possible. Once the serum has been removed from the cells, the glucose concentration remains constant for several days at 25 - 30°C. without the addition of a preservative. However, when serum is left in contact with the cells, disappearance of glucose commences within a few minutes. The specimen must be spun down or chilled to 4°C. within 15 minutes after collection. The rate of glycolysis is higher in newborn infants than in adults. So it is important that the cells are separated immediately. (See Table I)

TABLE I

RECOVERY OF GLUCOSE FROM BLOOD SPECIMEN
NOT SEPARATED FROM CELLS

Time Specimen Left Standing	1 Hour	2 Hours	3 Hours	4 Hours
Chilled	100%	100%	100%	100%
Room Temperature	90%	84%	84%	63%

7. Hemoglobin and Bilirubin do not seem to interfere, although elevated uric acid values can lower the results. (See Table II)

TABLE II

RELATIONSHIP OF CONCENTRATION OF URIC ACID IN SERA
WITH AMOUNT OF GLUCOSE RECOVERED

Concentration Uric Acid	34 Mg%.	19 Mg%.	11 Mg%.	7.6 Mg%.	4.0 Mg%.
% Glucose Recovered	51%	77%	83%	90%	Reference 100%

PROCEDURE:

1. Dispense 5.0 ml. of substrate into a new 16 x 100 test tube; label as "Reagent Blank".

2. At ½ minute intervals, take up 10 microliters of Standards, Controls, and Specimen; dispense into new tubes with 5.0 ml. of substrate. After each dilution, rapidly vortex the tube.

Glucose Oxidase (Continued):

3. Allow to stand 20 minutes, and read samples at ½ minute intervals.

4. Determine the absorbance of the Reagent Blank in the Gilford 300 N at 650 nm. Record absorbance of Reagent Blank. Then zero the machine with the Reagent Blan, and read samples.

5. Calculate glucose concentration of the Controls and samples from the linear graph paper.

6. Plot quality controls.

7. Be sure to leave enough substrate at room temperature for the next analysis.

NORMAL VALUES:

1. Newborn
   6 - 59 mg%. (BMC) increasing to 13 - 75 mg%. by the sixth day.
   20 - 80 mg%. (O'Brien & Ibbott)
   30 - 100 mg%. (Cornblath & Schwartz)

2. Normal Fasting Levels
   50 - 95 mg% (Boehringer-Mannheim Corp.)

REFERENCES: (Method)

1. Werner, W., Rey, H., Rey, G., and Wiebinger, H.: Z. Anal. Chem., 252:224, 1970.

2. Schmidt, F. H.: Internist, 4:554, 1963.

ABBREVIATIONS USED:

   *GOD (Glucose Oxidase)
   **ABTS (2,2' Azinodiethylbenzthiazoline-Sulfonic Acid)
   ***POD (Peroxidase)

GLUCOSE-6-PHOSPHATE DEHYDROGENASE
(U. V. Method on Red Cell Hemolysate)

PRINCIPLE: The enzyme G-6-PDH catalyzes the reaction which takes place when glucose-6-phosphate is converted to 6-phosphogluconic acid. In carbohydrate metabolism, this enzyme introduces the reaction which starts the pentose phosphate (hexose monophosphate) pathway. For this reason, an active and adequate concentration of G-6-PDH must be present in the red cell under certain conditions of stress. Low levels of G-6-PDH in the red cell have been associated for some time with hemolytic episodes in individuals following exposure to agents such as primaquine (and some other aromatic, heterocyclic structured medications) and fava beans.

G-6-PDH catalyzes the hydrogen transport reaction indicated beneath The rate of formation of $NADPH_2$, which absorbs strongly at 340 nm. and 366 nM., is utilized as a measure of enzyme activity.

$$\text{Glucose-6-Phosphate} + NADP \xrightleftharpoons{\text{G-6-PDH}} \text{6-Phosphogluconate} + NADPH_2$$

SPECIMEN: Collect whole blood in sodium citrate or sodium heparin anticoagulant. This may be stored at refrigerator temperatures for up to four days. (See specimen preparation under "Comments on Procedure.")

REAGENTS AND EQUIPMENT:

Available as Biochemica Test Combination TC-W 15993 (20 Determinations).
1. Triethanolamine Buffer, 0.05 M., pH 7.6
    Dissolve the contents of Bottle #1 in 100 ml. of redistilled water. Also contains 0.005 M. EDTA. Stable for one year at room temperature.

2. NADP, 0.01 M.
    Dissolve the contents of Bottle #2 in 2.0 ml. of redistilled water. Stable for four weeks at approximately 4°C.

3. Glucose-6-Phosphate, 0.031 M.
    Dissolve the contents of bottle #3 in 1.5 ml. of redistilled water. Stable for four weeks at approximately 4°C.

4. Digitonin, approximately 0.02%
    Use the solution in bottle #4 undiluted. Stable for one year at room temperature.

Glucose-6-Phosphate Dehydrogenase (Continued):

COMMENTS ON PROCEDURE:

1. Specimen Preparation
    a. Wash 0.2 ml. of whole blood three times with 2.0 ml. physiological saline. Centrifuge after each washing for ten minutes at approximately 3000 rpm.
    b. Do a red count on 20 lambda of the packed cell button.
    c. Suspend the washed button in 0.5 ml. digitonin and allow to stand at 4°C. for 15 minutes. Recentrifuge for clear hemolysate.

2. Derivation of Factor (at 340 nm.):

$$\frac{0.001}{6.22} \times 10^6 \times \frac{1}{1} \times \frac{3.25}{0.10} \times \frac{6}{1} \times \frac{5}{1} = 15661$$

3. Refer to original methodology for G-6-PDH determination on serum.

PROCEDURE:

1. Into a 1.0 cm. square glass cuvette, pipette the following solutions: 3.0 ml. of triethanolamine buffer, 0.1 ml. of NADP, and 0.1 ml. of hemolysate.

2. Mix the contents of the cuvette well by inversion and incubate in the water bath at 25°C. for approximately five minutes.

3. Pipette 0.05 ml. of glucose-6-phosphate into the cuvette, mix, and determine the absorbance at 340 nm. or 366 nm. Immediately start stopwatch and record absorbance change in the temperature controlled cuvette at 1, 2, and 3 minutes. (Or scan for three minutes on a recorder.)

4. Determine the mean absorbance change per minute (Δ E/min.) against an air reference. Absorbance differences greater than 0.060/min. at 340 nm. or 0.030/min. at 366 nm. require a dilution of 1:10 with physiological saline.

5. CALCULATION:

$\Delta E_{340 \text{ nm.}}$/min. x 15661 = milli-units (mU)/#RBC in 1.0 ml. blood.

$\Delta E_{366 \text{ nm.}}$/min. x 29600 = milli-units (mU)/#RBC in 1.0 ml. blood.

Glucose-6-Phosphate Dehydrogenase (Continued):

NORMAL VALUES: 120 - 240 mU./10^9 erythrocytes.

REFERENCES:

1. Kornberg, A., Colowick and Kaplan: METHODS IN ENZYMOLOGY, Vol. 1, Academic Press, pg. 322, 1955.

2. Bergmeyer: METHODS OF ENZYMATIC ANALYSIS, Academic Press, 1965.

3. Bishop: J. Lab. and Clin. Med., 68:149, 1966.

4. Batsakis and Brierre: INTERPRETIVE ENZYMOLOGY, C. Thomas, 1967.

GLUTAMATE DEHYDROGENASE (GLDH)

PRINCIPLE: Glutamate dehydrogenase is exclusively a mitochondrial enzyme, principally found in the cells of the liver, heart, and kidney. Trace amounts are found in sera of healthy, human beings. Glutamate dehydrogenase (GLDH) catalyzes the reaction:

$$\alpha\text{-Oxoglutarate} + \text{DPNH} + \text{NH}_4^+ \xrightleftharpoons{\text{GLDH}} \text{Glutamate} + \text{DPN}^+ + \text{H}_2\text{O}$$

The decrease in absorbance at 340 nm. due to the oxidation of DPNH to DPN^+ is a measure of the activity of the enzyme.

SPECIMEN: 1.0 ml. of fresh serum is required. Hemolysis does not interfere in the test because erythrocytes contain no GLDH. Serum can be stored in the refrigerator for at least 48 hours without loss of activity.

REAGENTS AND EQUIPMENT: (BioChemica Test Combination UV Method, 159995 TGAD)

1. <u>Buffer</u> (Bottle No. 1)
   0.05 M. Triethanolamine Buffer, pH 8.0; 0.004 M. Ethylene-diamine-tetraacetate.

2. <u>Ammonium Acetate</u> (Bottle No. 2)
   3.2 M. Ammonium Acetate

3. <u>NADH</u> (Bottle No. 3)
   0.012 M. reduced Diphosphopyridine Nucleotide

4. <u>Substrate</u> (Bottle No. 4)
   0.41 M. α-oxoglutarate

5. <u>Spectrophotometer</u> - Gilford 300N, 340 nm.
   This instrument is used to analyze one sample at a time. The change in absorbance per minute is read from the Data Lister printout. The linearity of the reaction is visually monitored on the strip chart recorder. The thermocuvette is set at 25°C.

COMMENTS:

1. Solutions are quite stable. Bottles No.'s 1 and 2 are stable for a year at room temperature.

2. Solutions No.'s 3 and 4 are stable for 4 weeks at refrigerator temperature.

Glutamate Dehydrogenase (GLDH) (Continued):

3. Dilute all specimens with physiological saline when optical density differences are greater than 0.600/10 minutes.

PROCEDURE:

1. Using disposable culture tubes (12 x 75 mm.), set up a Blank and Test for each specimen.

2. To both tubes, pipette 1.0 ml. of buffer, 0.5 ml. of serum, 0.05 ml. of ammonium acetate and 0.025 ml. of NADH.

3. Mix and incubate the tubes in a 25°C. water bath for approximately 20 minutes.

4. Set the machine to zero with deionized water. Aspirate the Blank and read the optical density (E_1). Take another reading after 10 minutes (E_2). In most cases, $E_1 = E_2$.

5. To the Test tube, add 0.025 ml. of α-oxoglutarate. Mix, aspirate into the cuvette and read optical density (E_3). Take another reading after 10 minutes (E_4).

6. CALCULATIONS:

    To obtain $\Delta A/10$ minutes, subtract optical density differences $E_1 - E_2$ from optical density difference $E_3 - E_4$.

    $$\text{I.U.} = \frac{\Delta A/10 \text{ Min.}}{\epsilon \times d} \times 10^6 \times \frac{TV}{SV} \times \frac{1}{\text{Time}}$$

    ϵ = Molar Extinction Coefficient of NADH at 340 nm.
       = 6.22×10^3 Liter/Mole x cm.
    d = Diameter of the cuvette in cm.
       = 1.0 cm.
    TV = Total Volume
       = 1.6 ml.
    SV = Sample Volume
       = 0.5 ml.
    T = Time
       = 10 minutes

    10^6 converts Moles/Liter into micromoles/liter
          1 I.U./Liter = 1 mU/ml.

Glutamate Dehydrogenase (GLDH) (Continued):

$$I.U. = \frac{\Delta A/10 \text{ Min.}}{6.22 \times 10^3} \times 10^6 \times \frac{1.6}{0.5} \times \frac{1}{10}$$

I.U. = $\Delta A/10$ minutes x 51.44

$\Delta A/10$ minutes x 51.44 = mU/ml. of serum.

Report results in I.U. at 25°C.

NORMAL VALUES: Normal values compiled by BMC at 25°C.
Up to 0.9 mU/ml. of serum.

REFERENCES:

1. Carlson, A. S., et. al.: <u>Amer. J. Clin. Path.</u>, 38:260, 1962.

2. Frieden: <u>J. Biol. Chemistry</u>, 234:809, 1959.

ALPHA-HYDROXYBUTYRATE DEHYDROGENASE
(α-HBD)

PRINCIPLE: Advantage is taken of the clinical differentiation of the serum lactic dehydrogenase (LDH) isoenzymes. It has been shown that the faster moving isoenzymes reduce α-ketobutyrate while the slower moving ones show negligible activity. Damage to heart muscle, altered erythropoiesis and extreme damage to other tissue could lead to increased HBD activity.

The HBD measurement in relation to total LDH becomes of great value in the measurement of the "heart" fractions of LDH isoenzymes.

This test has the advantage of being performed quite rapidly and simply.

The following reaction takes place:

$$\alpha\text{-oxobutyrate} + NADH_2 \xrightleftharpoons{HBD} \alpha\text{-hydroxybutyrate} + NAD$$

SPECIMEN: A pipetable 100 microliters (0.10 ml.) of serum, free from hemolysis, is required. This specimen is stable for several days at room temperature or 4°C.

REAGENTS AND EQUIPMENT:

Biochemica Kit TC-HD, Cat. No. 15953 THAD for 60 determinations.

Bottle No. 1 0.05 M. Phosphate Buffer, pH 7.4
Bottle No. 2 0.004 M. NADH
Bottle No. 3 0.10 M. alpha-oxobutyrate

1. Dissolve contents of Bottle No. 1 in 200 ml. deionized water.

2. Dissolve contents of Bottle No. 2 in 2.5 ml. deionized water.

   Mix these reagents together in a brown glass bottle. This makes Solution 1. (Stable four weeks at 4°C.)

3. Dissolve the contents of Bottle No. 3 in 6.5 ml. of deionized water. This makes Solution 2. (Stable three months at 4°C.)

   Do _not_ mix the two solutions.

Alpha-hydroxybutyrate Dehydrogenase (Continued):

4. **DBG or Gilford 300 N Spectrophotometer**
   An instrument with temperature controlled cuvette well and recorder should be used.

5. **Eppendorf Pipette**
   Use a 100 microliter Eppendorf pipette for dispensing serum sample.

PROCEDURE:

Place into a 10 mm. square cuvette:

1. 3.1 ml. of Solution 1.

2. 0.10 ml. (100 microliters) of serum.

   Mix and place into a 30°C. water bath for approximately 15 minutes.

3. Add 0.10 ml. (100 microliters) of Solution 2. Mix. Place cuvette in cuvette well and record A. change for two minutes of linearity.

4. CALCULATIONS:

   Based on $\Delta A/1$ minute and a light path of 1.0 cm.

   (ϵ NADH at 340 nm. = 6.22 Liter/Mole x cm.)

   $$\frac{\Delta A}{\epsilon x d} \times 10^6 \times \frac{TV}{SV} \times \frac{1}{Time} = I.U./L. \text{ or } mU/ml.$$

   $$\Delta A \times \frac{1}{6.22 \times 10^3 \times 1} \times 10^6 \times \frac{3.3}{0.1} \times \frac{1}{1} = mU/ml.$$

   $\Delta A \times F = mU/ml.$

   $F = 5305$

NORMAL VALUES: Up to 168 mU/ml. at 30°C.

Alpha-hydroxybutyrate Dehydrogenase (Continued):

REFERENCES:

1. Rosalki, S. B. and Wilkinson, J. H.: Nature, 188:1110, 1960.

2. Rosalki, S. B.: British Heart Journal, 25:795, 1963.

3. Preston, J. A., Batsakis, J. G., and Briere, R. O.: Amer. J. Clin. Path., 41:237, 1964.

ISOCITRATE DEHYDROGENASE
(U. V. Methodology)

PRINCIPLE: Mammalian tissue contains two isocitrate dehydrogenase enzymes; the one which is of clinical interest is that enzyme linked to NADP. The enzyme shows molecular heterogeneity, with four distinct isoenzymes being separated following electrophoresis. The activity of I.C.D. in man is found in greatest concentration in the liver, then heart, tumors and muscle. Abnormal I.C.D. activity is best considered as a relatively sensitive indicator of hepatocellar damage. Elevations of serum I.C.D. may also be found with myocardial infarction, but not consistently so, due to the rapid denaturation or "clearing" of the heart isoenzyme fraction in the body. Red cells and platelets also display significant I.C.D. activity.

The spectrophotometric determination is based upon the reaction indicated beneath. The reduction of NADP to $NADPH_2$ is followed spectrophotometrically at 340 nm. or 366 nm. $NADPH_2$ is strongly absorbant at the preceding wavelengths and its increase is proportional to I.C.D. concentration.

$$\text{Isocitrate} + \text{NADP} \xrightleftharpoons[Mn^{++}]{I.C.D.} \text{Oxalosuccinate} + NADPH_2$$

Mn^{++} is an activator and is added to the reaction mixture along with the substrate.

SPECIMEN: 0.5 ml. of **unhemolyzed** serum. Do not use plasma specimens; anticoagulants interfere with enzyme activity.

REAGENTS AND EQUIPMENT:

Reagents for U. V. Method are available as Biochemica Test Combination TC-ID #15933 TIAA. One kit is sufficient for 25 determinations.

1. <u>Triethanolamine Buffer</u>, 0.1 M., pH 7.5 and <u>DL-Isocitrate</u>, 0.0046 M.
   Dissolve contents of bottle #1 in 75 ml. redistilled water. Stable for 3 months at approximately 4°C. Bottle also contains 0.052 M. NaCl.

Isocitrate Dehydrogenase (Continued):

2. NADP, 0.0091 M. and MnSO$_4$, 0.12 M.
   Dissolve contents of Bottle No. 2 in 3.0 ml. of redistilled water. Stable for four weeks at approximately 4°C.

3. Water Bath, at 30°C.

4. Spectrophotometer, with temperature controlled cuvette well (30°C.)

PROCEDURE:

1. Into a glass cuvette of 1.0 cm. light path, pipette 2.5 ml. of buffer and isocitrate reagent.

2. Deliver 0.5 ml. of serum specimen into the cuvette containing reagent. Mix the contents and place the cuvette in a 30°C. water bath for 5 minutes.

3. Deliver 0.1 ml. of NADP-MnSO$_4$ reagent into the cuvette and mix well by inversion.

4. Place cuvette in thermostated cuvette well and start recorder.

5. Monitor the change in absorbance for at least two consecutive minutes of linearity.

6. CALCULATIONS:

    Calculations based on mean A change of 1 minute.

    NADPH$_2$ at 340 nm. = 6.22 x 10^3 Liters/mole x cm.

$$\frac{\Delta A}{\epsilon x d} \times 10^6 \times \frac{TV}{SV} \times \frac{1}{Time} = I.U./L. \text{ or } mU/ml.$$

$$\frac{\Delta A}{1} \times \frac{1}{6.22 \times 10^3 \times 1} \times 10^6 \times \frac{3.1}{0.5} \times \frac{1}{1} = mU/ml.$$

$\Delta A \times F = mU/ml.$
$F = 997$

NOTE: Sera with absorbance changes per minute greater than 0.100 at 340 nm. should be diluted 1:10 with physiological saline and repeated on 0.5 ml. of this dilution. Multiply final answer x 10.

Isocitrate Dehydrogenase (Continued):

NORMAL RANGE: Up to 11 mU/ml.

REFERENCES:

1. Batsakis and Briere: INTERPRETIVE ENZYMOLOGY, C. Thomas, 1967.

2. Ochoa, in Colowick & Kaplan: METHODS IN ENZYMOLOGY, Vol. 1, Academic Press, pg. 699.

3. Wolfson: Proc. Soc. Exp. Biol. & Med., 92:231, 1957.

4. Wolfson, Ann: N. Y. Acad. Sci., 75:260, 1958.

5. Henry, J. B.: WORKSHOP ON CLINICAL ENZYMOLOGY; PRE-WORKSHOP MANUAL, ASCP, 1964.

6. Clin. Chem., 6:208, 1960 (Colorimetric Determination)

7. J. Lab. & Clin. Med., 62:148, 1963 (Colorimetric Determination)

8. Sterkel, R. & Wolfson: J. Lab. Clin. Med., 52:176, 1958.

9. Cohen: Ann. of Int. Med., 55:604, 1961.

LACTATE DEHYDROGENASE
(Wacker Method)

PRINCIPLE: LDH catalyzes the following reaction:

$$\text{Lactate} + NAD^+ \rightleftharpoons \text{Pyruvate} + NADH + H^+$$

The reduction of NAD^+ proceeds at the same rate as the oxidation of of lactate and in equimolar amounts. The rate at which NADH is formed can be determined by the increase in absorbance at 340 nm. This is the forward reaction according to Wacker utilizing lactate as the substrate at an alkaline pH.

SPECIMEN: 0.050 ml. (50 microliters) of hemolysis-free serum (or body fluid) is required. Serum should be separated from cells soon after clotting takes place. Specimens for LDH may be stored at room temperature or 4°C. for several days.

REAGENTS AND EQUIPMENT:

1. Working Substrate
   Biochemica Test Combination LDH-L 10 Test reagents are used.

   Contents of 10 Test System LDH-L vial:
   a). 150 mg. NAD/vial (7.10 mM after reconstitution)
   b). Stabilized 0.05 M. pyrophosphate buffer, pH 8.6; 0.045 M. L-Lactate.

2. Spectrophotometer

   A. Gilford 222, 340 nm.
      This instrument is used when analyzing four samples at a time. Temperature control is by means of a 30°C. circulating water bath. Rate of reaction is determined from the strip chart recorder set for 0.200 A. full scale and run at one inch per minute.

   B. Gilford 300 N, 340 nm.
      This instrument is used when analyzing one sample at a time The change in absorbance per minute is read from the Data Lister print-out. The instrument is used with the thermo-cuvette set at 30°C.

3. Bailey Microdilutor
   The sample syringe is set to pick up 0.050 ml. (50 microliters) and the reagent syringe is set to dispense 1.5 ml. of substrate

LDH (Continued):

COMMENTS ON PROCEDURE:

1. Substrate exhaustion is rate limiting. The reaction will be linear for only a few minutes. DO NOT read after 5 minutes.

2. The reaction mixture and cuvette chamber must be at temperature before taking a reading.

3. During the run, the substrate and cuvettes are kept at 30°C. in the water bath.

4. Dilute all specimens when ΔA is greater than 0.080/ minute.

PROCEDURE:

1. With a microdilutor, take up 50 microliters of specimen, and flush with 1.5 ml. of pre-incubated substrate. Mix well and return to the water bath.

2. Quickly dilute up 3 more specimens in the same manner.

3. When 4 specimens have been diluted, immediately dry cuvettes and place in Gilford 222 (normally takes about 45 seconds).

4. Quickly set the baseline of Specimen No. 1 with the slit, and No.'s 2, 3, and 4 with their Off-Set knobs. Switch to "AUTO" and scan all 4 channels.

5. Record 2 to 3 minutes of linear reaction time at 340 nm. Determine ΔA/minute from recorder chart.

6. CALCULATIONS:

$$I.U. = \frac{\Delta A}{\epsilon \times d} \times 10^6 \times \frac{TV}{SV} \times \frac{1}{Time}$$

ϵ = Molar Extinction Coefficient

ϵ of NADH at 340 nm. = 6.22×10^3 Liter/Mole x cm.

d = Diameter of light path (1.0 cm.)

TV = Total Volume: 1.55 ml.

SV = Sample Volume: 0.05 ml.

LDH (Continued):

T = Time in minutes.

10^6 converts Moles/Liter (or mMoles/ml.) into Micromoles/L. (or Millimicromoles/ml.)

1 I.U./Liter = 1 mU/ml.

THEREFORE:

$$\frac{\Delta A}{6.22 \times 10^3 \times 1} \times 10^6 \times \frac{1.55}{0.05} \times \frac{1}{1} = mU/ml.$$

When all conditions of the assay remain constant, a factor may be derived:

$$\frac{\Delta A}{1} \times \frac{1}{6.22 \times 10^3 \times 1} \times 10^6 \times \frac{1.55}{0.50} \times \frac{1}{1} = mU/ml.$$

$\Delta A \times F = mU/ml.$ at 30°C.

F = 4984

WE report results as mU/ml. at 37°C.

I. U. at 37°C. = mU/ml. at 30°C. x Temperature Conversion Factor
Temperature Conversion Factor = 2 (For this Laboratory)

F = (9968 Actual) We round off to 10,000.

NORMAL VALUES: 85 - 200 mU/ml.   (For this Laboratory)

REFERENCES:

1. Wacker, W. E. C., Ulmer, D. D., and Valu, B. L.: New Eng. J. Med., 255:449, 1956.

2. Amador, C. L., Dorfman, D., and Wacker, W. E. C.: Clin. Chem., 9:391, 1963.

3. Gay, R. J., McComb, R. B., and Bowers, C. N., Jr.: Clin. Chem., 14:740, 1968.

LDH ISOENZYMES BY ELECTROPHORESIS

PRINCIPLE: The isoenzymes of lactate dehydrogenase are separated by electrophoresis on agarose, thin-gel plates. Following electrophoresis, a reagent film consisting of sodium lactate and the coenzyme, nicotinamide adenine dinucleotide (NAD$^+$), is spread over the gel surface. According to the following reaction, NAD is reduced to NADH, which is fluorescent.

$$\underset{\text{Lactate}}{\begin{matrix} COO^- \\ CHOH \\ CH_3 \end{matrix}} + NAD^+ \xrightleftharpoons{\text{LDH}} \underset{\text{Pyruvate}}{\begin{matrix} COO^- \\ C=O \\ CH_3 \end{matrix}} + NADH + H^+$$

The intensity of fluorescence of each of the five fractions is directly proportional to the concentration. The relative fluorescence of the fractions is determined by scanning the gel strips in a specially adapted door for the Turner Fluorometer.

SPECIMEN: 1.0 microliter of unhemolyzed serum. Should be stored at room temperature if unable to perform analysis on day of specimen collection. Analysis should be made within three days of specimen collection. Do not freeze samples.

EQUIPMENT AND SUPPLIES:

Unless otherwise specified, the following equipment is obtained through Analytical Chemists, Inc., Palo Alto, California. The ACI part numbers are indicated in parentheses.

1. <u>Cassette Electrophoresis Cell and Power Supply</u> (#1-3300)
   1 set required.

2. <u>Quantitative Microliter Sample Dispenser</u> (Elevitch), (#1-4100) a modified 10 microliter Hamilton syringe. 1 each required. Disposable Sample Tips (#1-4110) (100 tips/vial)

3. <u>Electrophoresis Buffer</u>, Barbital, pH 8.6 (#1-5100) 0.05 M. with 0.035% EDTA. 2 sets required.

4. <u>Film Cutter</u> (#1-4300)
   1 each required.

LDH Isoenzymes (Continued):

5. Incubator/Oven (#1-3500)

6. Agarose UNIVERSAL Electrophoresis Film^R (#1-1000-96)
   12 Films (96 Determinations) per package.

7. Fluorometric Lactate Dehydrogenase Isoenzyme Substrate (#1-1500)
   One package is enough for 96 determinations.

8. Sta-Moist Paper (#1-3550)
   100 Sheets/package, for use in incubator.

9. Turner Fluorometer (#111 - G. K. Turner Associates, 2524 Pulgas
   Ave. Palo Alto, California)

   Prepare the fluorometer as follows:
   Filter Selection:
       Primary: 365 nm.
       Secondary: 410 nm.

10. Turner Strip Scanning Door, Automatic (#110-525)
    Designed to fit the above fluorometer.

11. Varian Chart Recorder, Model G-22, 10 mv. - Varian Associates
                                                      Palo Alto, California.

12. Chart Paper for above Recorder #5A (#00-940507-01)
    Varian Aerograph 2700 Mitchell Dr., Walnut Creek, California.

13. Tablet of scratch paper, 8½ x 11 inches, lined.

14. Labelling Tape (3 MM white pressure sensitive tape 1 inch width

PROCEDURE:

1. Determine Total LDH value for all specimens. Samples with a
   Total LDH of 200 I.U. yield optimal tracings. Dilute samples
   with values greater than 200 I.U. with saline or isoenzyme
   buffer so that the dilution falls in the optimal range. The
   dilution need not be exact; drop-counting with a Pasteur pipett
   is satisfactory. If the value is much lower than 100 I.U., a
   double application can be made.

LDH Isoenzymes (Continued):

2. Each film has eight (8) numbered strip positions. In this laboratory, we conventionally run all samples in duplicate (reporting averaged results) and reserve positions No. 1 and No. 8 for Control Serum. Assign strip positions to each specimen and record these together with pertinent sample identification information in a log or workbook.

3. The plastic bag enclosing each box of films contains a small amount of EDTA solution which acts as a bacteriostatic and to maintain moisture. Save the tape on the plastic bag and use it to reseal the package. Remove the box of films carefully so as not to spill the EDTA. Handle the individual films carefully, avoiding pressure on the soft (film) side. If necessary, rinse the film with a small amount of distilled water and drain and wipe gently. Reseal the box of unused films into the plastic bag.

4. Arrange the work area for application: have on hand
    a). A clean pad of lined tablet paper
    b). Appropriately diluted samples
    c). Hamilton or Elevitch syringe with a clean disposable tip for each sample.

5. Peel the film from its rigid plastic backing, which may be discarded. Place the film, agarose side up, on the lined paper. The film may be secured in place by taping to the paper at the corners. The numbered edge of the film with the application troughs is the cathode (negative) edge. (See Diagram Below)

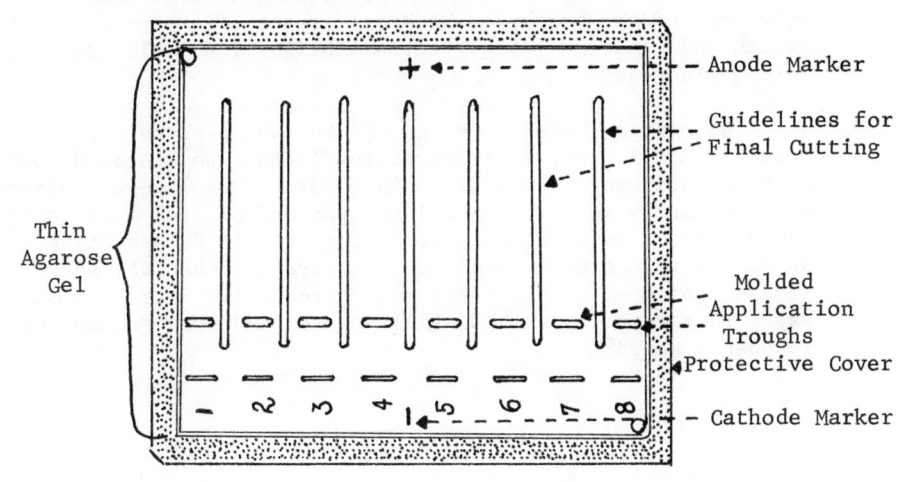

LDH Isoenzymes (Continued):

6. Apply 1.0 microliter of appropriately diluted specimen to each of the assigned troughs. Dispense the aliquot of sample in several consecutive portions, allowing each portion to soak into the gel. This will minimize the application artifact. Fill each position; if there are extra spaces, fill these with repeat samples or additional controls.

7. While waiting for the application to be completely absorbed into the gel, prepare the electrophoresis box. Connect the cassette to the power supply with the banana plugs. Measure 190 ml. of buffer with a graduated cylinder and add to the cassette, tilting the cassette to equilibrate the liquid level between the chambers. Cover with a pair of discarded film backings to protect from evaporation and contamination.

8. When the application troughs look "dry", pick up the film by the protective backing and flex into a "U" or trough shape with the protective cover on the outside and the agarose gel on the inside or concave aspect. Anode and Cathode markers will be opposite each other. Insert the flexed film into the trough holder in the electrophoresis box lid, being certain that the + and - markers on the film correspond to the + and - markings on the lid. (The lid markings are engraved on the outside of the lid. We have found it useful to clearly mark the inside of the lid with a bright wax marker crayon). Make sure that the film is snuggly inserted and that its edges are caught and held in place by the lip or "gutter" of the holder.

9. Invert the lid over the box and position gently to avoid splashing of the buffer. Placement of the lid trips the "ON" switch and the signal light turns a bright red. Electrophorese for 35 minutes.

10. Ten (10) minutes before the end of the run, prepare the substrate. Add 2.0 ml. of refrigerated lactate to frozen dispenser bottle containing 10 mg. NAD. Make sure to replace the dispenser cap firmly so that it will not pop off or leak when using the dispenser as a drop-bottle. Store the light-sensitive substrate in the dark until ready to use. (Substrate saved frozen from previous few days may be reused). Soak STA-MOIST paper to saturation with distilled water and pre-warm in the incubator at 38°C.

LDH Isoenzymes (Continued):

11. At the end of the 35 minute run, remove the lid vertically; again avoid splashing of the buffer. Place on a piece of absorbant toweling to drain. At the end of the run, the anode pH is about 8.2 and the cathode pH about 9.0 in contrast to the initial uniform pH of 8.6. Before removing film from lid, blot excess liquid from edges with a piece of clean tissue, but do not scar the agarose gel. Remove from lid maintaining the "U"-shape and release gradually to avoid spraying of condensation from plastic protector. Blot away any excess liquid, but do not touch the agarose gel.

12. Place the film, agarose side up, on a clean lined paper pad with the cathode side away from the operator. Tape the anode (near) edge to the paper, aligning the film edge with the lines on the paper which serve as a guide during application of the substrate.

13. Turn on the fluorometer and recorder to warm up at this time.

14. Using the dropper bottle in which it has been prepared, distribute 16 - 20 drops of the NAD-Lactate Substrate (about 0.6 ml.) near to and parallel to the anode edge of the agarose. Lay a plastic 5.0 ml. serological pipette along the drops, parallel to the edge and resting lightly on the agarose, so that the drops coalesce into a line along and ahead of the pipette. Check visually to see that the meniscus formed is continuous from side to side. Holding the pipette ends lightly between the fingertips, raise the pipette slightly off the agarose, but not enough to break the meniscus or separate the substrate from the agarose. Very slowly advance the pipette towards the cathode, pushing the substrate forward without touching the gel with the pipette. Watch the lateral edges to check that the meniscus is showing through the pipette. Keep the pipette parallel to the lines on the paper and advance right off the cathode end onto the paper. The process should be slow enough so that only a small amount of substrate is pushed off, and the bulk of it is absorbed by the end of the application.

15. As soon as the substrate is absorbed, place the film, agarose side up, on the wet filter paper on the incubator shelf. The shelf is removable for convenience. "Float" the film onto saturated filter paper laying down first at one edge to avoid trapping air bubbles (insulators). Uniform heating is essential. Plug in shelf for 15 minutes. During this time turn on the oven to pre-heat.

LDH Isoenzymes (Continued):

16. When incubation is complete, remove film and dry the plastic side before transferring it to the removable oven shelf. Lay the film on the shelf so that the anode-cathode axis is at right angles to a line extended from the banana plugs on the shelf. In this position the film can be held in place by elastic bands placed over the film edges and the shelf (the plastic tends to flex). Dry for 12 minutes. Remove to a dark, dust-free area until ready to scan.

17. View the film under an ultra-violet light source to locate the extremes of the fluorescing patterns; mark the edges on the protective backing; trim along mark leaving a clearance of about 2.0 cm. from the pattern ends. (Careful trimming and taping will permit four pattern strips to be loaded onto the scanning drum simultaneously). Take a strip of 1-inch white paper tape the same length as the protective backing (4½ inch) and place it so that ¼ inch of its width adheres to the film (the unused application troughs make a handy guide for positioning). Place film and tape on the cutter and split the tape in half lengthwise. Position the trimmed-off strip of tape on the opposite side of the film, again allowing ¼ inch to overlap and press firmly into place. When the film is positioned gel-side up and with the application slots towards the operator, the left-most side of the film is position No. 1. It is advisable to number the positions on the labelling tape before cutting apart.

18. Line up the untaped edge of position No. 1 on the cutter plate and trim away just the protective backing without cutting into the agarose. Roll the cutter blade firmly and allow the tape to hold the film firmly in place. Release the film and tape carefully so that the agarose is not pulled away from the backing. Realign the film so that the interstrip division line is on the edge of the cutter plate and trim off the first pattern strip. Leaving the rest of the film in position on the cutter, take strip position No. 1 and apply it to the rotating drum of the fluorometer.

19. Hold the strip so that the application trough is on the right and position it on the drum so that the right-hand tape margin lines up with the marker "1A", on the vertical face of the drum. The drum width is exactly wide enough to seat the film, so it is important that the sides of the strips be cut parallel. Neither film nor tape must be allowed to project over the metal

LDH Isoenzymes (Continued):

rim of the drum channel; the clearance as the drum rotates is extremely narrow and any projections will catch and tear or crumple the film.

20. Cut the next strip, No. 2 and again keeping the application-trough end to the right, position the strip on the drum to the left of Strip No. 1, allowing the tape to overlap somewhat. In this manner four trimmed strips will fit onto the drum which normally will hold only three full-length strips.

21. The Turner Fluorometer should be allowed to warm-up for half an hour; after turning Power to "ON" and activating the "START" toggle switch, check visually for the blue light of the mercury vapor lamp. The Varian recorder is warmed up with the voltage switch (left hand toggle switch) at "STANDBY". The recorder pen is activated by the "RECORD" switch on the fluorometer. Chart speed is controlled by the right-hand toggle switch on the recorder and is left at "LOW". Chart motion is stopped and started by the left-hand toggle switch of the recorder and is used in the "HIGH" position. The scanning drum may be rotated manually in either direction by means of the knurled knob underneath the drum housing. When scanning automatically the drum moves counterclockwise until it hits the automatic stop. While scanning, the red indicator light in front of the drum is ON, when stopped, the light goes OUT. To reactivate scanning, the drum is rotated manually counterclockwise (approximately 3/4 turn of the knurled knob). In the starting position, the notation "OA" will be visible in the window on the top of the drum cover. Rotation is stopped or started by turning the "SAMPLE" toggle switch on the fluorometer to "ON".

22. After loading the strips onto the drum, replace the drum cover. Rotate the drum until the letters "OA" appear in the window. Set all switches on the fluorometer and recorder to "running" positions. The light on the drum should be "ON". Rotate the drum manually, clockwise, until the light goes out and then back again, counterclockwise until the light just comes ON again. This places the drum in the starting position. One complete rotation takes approximately ten minutes.

23. Monitor the recorder during the scan for pattern and sensitivity. The pattern should stay between 0 - 10 on the recorder scale. Generally, if the specimens have been prediluted, no adjustment of the sensitivity will be needed. Label the tracings as they appear.

LDH Isoenzymes (Continued):

24. If a pattern goes off scale, the sensitivity must be readjusted and the tracing repeated. Sensitivity is controlled by the knurled knob on the top left of the drum housing. If the setting of this knob is changed, the baseline adjustment (Blank-Adjust knob on top of fluorometer) must also be reset.

EVALUATION OF TRACINGS:

1. Draw the baseline. Application artefacts and various chemicals in medications can cause irregularities.

2. Draw a perpendicular line from each peak to the baseline; measure the height in mm. and record. The five peaks are equally based isosceles triangles, so that their comparative areas are proportional to their heights.

3. CALCULATIONS:

    Calculate the per cent of each fraction by the following formula:

    $$\frac{\text{Height of fraction in mm.}}{\text{Total of all fractions in mm.}} \times 100 = \text{Fraction \%}$$

    Average the duplicates and then multiply each percentage by the total LDH. This gives International Units per fraction. Report both the International Units and the percentage of Total.

4. INTERPRETATION:

    Fractions are numbered from one-to-five, beginning at the anode. Fraction one is generally considered LDH of cardiac origin. Fractions four and five are primarily of liver origin.

NORMAL VALUES:

	Fraction 1	Fraction 2	Fraction 3	Fraction 4	Fraction 5
%:	17 - 27%	28 - 38%	19 - 27%	5 - 16%	5 - 16%
LDH:	Up to 54 I.U.	76 I.U.	54 I.U.	32 I.U.	32 I.U.

REFERENCES:

1. Methodology available through Analytical Chemist, Inc.

LDH Isoenzymes (Continued):

2. Elevitch, F.: PROGRESS IN CLINICAL PATHOLOGY, Ed. by
   Stefanini, Grune and Stratton, 1966.

LEUCINE AMINOPEPTIDASE (LAP)
(Goldbarg and Rutenburg: Modified by Sigma Chemical Company)

PRINCIPLE: LAP is an enzyme which catalyzes the following reaction:

1-Leucyl-β-Naphthylamide + H_2O $\xrightarrow{LAP}$ Leucine + β-Naphthylamine

β-Naphthylamine + $NaNO_2$ + Dye Base $\longrightarrow$ Blue Dye

The absorbance of the dye produced is proportional to the amount of enzyme present.

SPECIMEN: 0.1 ml. serum added to 4.9 ml. H_2O for a 1:50 dilution. The serum may be stored up to 7 days at 4°C. without significant loss of activity. Serum bilirubin up to 20 mg%. and serum hemoglobin up to 0.2% do not affect LAP activity.

REAGENTS:

(Sigma Chemical Company)

1. **LAP substrate**, Stock #251-1
   1-leucyl-β-Napthylamide HCl. Store at 4°C.

2. **Sodium Nitrite**, Stock #251-4 (2.0 mg. Tablets)
   Each tablet is dissolved in 2.0 ml. H_2O to make a 0.1% solution. Prepare fresh daily.

3. **Ammonium Sulfamate**, Stock #251-3
   0.5% Solution. Store at 4°C.

4. **N-(1-Napthyl)-Ethylenediamine-2-HCl**, Stock #251-5
   Pre-weighed bottle. 0.05% Alcoholic Solution. To the pre-weighed bottle add 110 ml. Ethyl Alcohol. Store at 4°C. Discard when solution becomes cloudy.

5. **TCA**, 25%

6. **TCA**, 13.3%
   For calibration curve only. Add 9.0 ml. H_2O to 10 ml. of 25% TCA.

7. **LAP Calibration Standard**, Stock #251-10
   A Standard Solution of β-Napthylamine for calibration.

LAP, Goldbarg and Rutenburg Method (Continued):

PROCEDURE - CALIBRATION CURVE:

1. Preparation of Calibration Curve

Tube No.	Ml. of Stock No. 251-10	Ml. of 13.3% TCA	G-R Units LAP/ml. of 1:50 serum
1	0	1.0	0
2	0.1	0.9	108
3	0.2	0.8	216
4	0.4	0.6	432
5	0.6	0.4	648
6	0.8	0.2	864

2. Add 1.0 ml. of 0.1% sodium nitrite to each tube, mix, and allow to stand for exactly 3 minutes at room temperature.

3. Add 1.0 ml. of ammonium sulfamate to each tube, mix, and allow to stand for exactly 2 minutes at room temperature.

4. Add 2.0 ml. of alcoholic solution Stock No. 251-5   1-Naphthyl - Ethylenediamine to each tube, mix, and allow to stand for 30 minutes at room temperature.

5. Transfer to 1.0 cm. square cuvettes and read A on the Gilford 222, using water as the reference solution at 580 nm. Plot the calibration curve on graph paper.

COMMENTS ON PROCEDURE:

After a 60 minute incubation of serum and substrate at 37°C., hydrolysis is terminated by protein precipitation with TCA. Diazotization of β-naphthylamine is accomplished by the addition of $NaNO_2$. Excess $NaNO_2$ is decomposed by the addition of ammonium sulfamate. The addition of N-(1-Naphthyl) ethylenediamine dihydrochloride produces a blue azo-dye which is proportional to the amount of β-naphthylamine liberated. This is measured spectrophotometrically at 580 nm.

PROCEDURE:

1. Dilute serum 1:50 (0.1 ml. + 4.9 ml. deionized water)
2. Prepare three tubes as follows:

	Reagent Blank	Serum Blank	Test
Deionized Water	1.0 ml.	1.0 ml.	—
Substrate 251-1	1.0 ml.	—	1.0 ml.
Diluted Serum	—	1.0 ml.	1.0 ml.

LAP, Goldbarg and Rutenburg Method (Continued):

3. Incubate at 37°C. for exactly 1 hour. Stop reaction by adding 1.0 ml. 25% TCA. Invert with parafilm. Let stand at room temperature for 10 minutes and then centrifuge.

4. Transfer 1.0 ml. supernatant into another series of test tubes. Add the following reagents at exact time increments:

   A. 1.0 ml. NaNO$_2$. Mix by lateral tapping. Wait 3 minutes.

   B. Add 1.0 ml. ammonium sulfamate. Mix. Wait 2 minutes.

   C. Add 2.0 ml. N-(1-Naphthyl)-Ethylenediamine Dihydrochloride. Mix by inversion with parafilm. Let color develop for 30 minutes.

5. Read A against deionized water in the Gilford 222 at 580 nm. using 1.0 cm. square cuvettes.

6. CALCULATIONS:

   Subtract the A of the Reagent Blank plus the A of the Serum Blank from the A of the Test. Determine the corrected LAP G-R units from the Calibration Curve.

NORMAL VALUES: 70 - 200 G-R units
      Borderline: 200 - 250 G-R units
      Elevated: Over 250 G-R units

REFERENCES:

   1. Goldbarg, Julius A. and Rutenburg, Alexander: Cancer, 11:283, March-April, 1958.

   2. Sigma Technical Bulletins No. 250 and 251.

   3. Bergmeyer: METHODS OF ENZYMATIC ANALYSIS, Academic Press, 1965.

LEUCINE AMINO PEPTIDASE (LAP)
(Kinetic Method After Method of Nagel)

PRINCIPLE: The ability of LAP to split the substrate leucine-p-nitroanilide leucine and p-nitroaniline is utilized. With the LAP activity being proportional to the intensity of the yellow color of p-nitroanilien.

The absorbance of p-nitroaniline is very high at 405 nm. (9.9×10^3 moles/Liter x cm.), while the substrate has minimum absorbance at this wavelength, making possible the direct kinetic measurement of LAP.

SPECIMEN: 0.1 ml. of serum. May be stored at $4^\circ C$. up to 5 days without significant loss of activity.

REAGENTS: (Biochemica Test Combination Kit, TC-LAP, Cat. No. 15952)

1. Stock Buffer (Bottle No. 1)
   0.05 M. phosphate buffer, pH 7.2.

2. L-leucyl-p-nitroanalide Substrate (Bottle No. 2)
   0.25 M. leucine-p-nitroanalide

3. Working Buffer
   Dissolve contents of Bottle No. 1 in 100 ml. of redistilled water. Stable for 1 year at $4^\circ C$.

4. Working Substrate
   Dissolve contents of Bottle No. 2 in 3.0 ml. of methanol. Stable for 6 months at $4^\circ C$. in a brown bottle.

5. Buffered Substrate
   Add 30 parts of Solution in Step 3 to 1 part of Step 4, should be discarded after 1 day's use. Very sensitive to light.

   Solution in Step 4 can easily be made by dissolving 125.6 mg. of anhydrous leucine-p-nitroanalide (mol. wt. 251.3) in 20 ml. of methanol. Enough for approximately 200 tests.

PROCEDURE:

1. Dispense 3.0 ml. buffered substrate into 10 mm. square cuvettes. Allow to equilibrate to $30^\circ C$. in a water bath for 5 minutes.

LAP, Kinetic Method (Continued):

2. Add 100 microliters of serum. Mix by inversion. Place in a 30°C. thermostated cuvette well. Instrument should be set at 405 nm.

3. Record the reaction until at least two consecutive minutes are linear, using either the Beckman DBG with recorder or the Gilford 300 N with digital print-out.

    Method linear up to 100 mU/ml. if greater, dilute 1:10 with saline. Multiply results x 10.

4. CALCULATION:

    Calculations are based on mean ΔA/minute.

    $$\frac{\Delta A}{\epsilon x d} \times 10^6 \times \frac{TV}{SV} \times \frac{1}{Time} = I.U./Liter \text{ or } mU/ml.$$

    ϵ of p-nitroaniline at 405 nm. = 9.9×10^3 Liters/Mole x cm.

    $$\Delta A = \frac{1}{9.9 \times 10^3 \times 1} \times 10^6 \times \frac{3.1}{0.1} \times \frac{1}{1}$$

    $\Delta A \times F = mU/ml.$
    $F = 3131$

NORMAL VALUES: 15 - 33 mU/ml. at 30°C.

REFERENCES:

1. Nagel, W., Willig, F. and Schmidt, F. H.: <u>Klin. Wachs.</u>, 42:447, 1964.

2. Szasz, G.: <u>Am. Jour. Clin. Path.</u>, 47:607 - 613, May, 1967.

SERUM LIPASE DETERMINATION
(Turbidimetric)

PRINCIPLE: The serum lipases are enzymes which hydrolyze long chain fatty acid esters of glycerin. The methodology measures enzyme concentration via its activity upon a nearly pure triglyceride emulsion. The amount of turbidity emulsion is made with purified olive oil and utilizing a natural emulsifying agent in the bile salt, sodium deoxycholate. Olive oil is reported to be the most specific substrate for "pancreatic" lipase activity. The pH 9.1 of Tris buffer, as well as the concentration of the bile salt present, is optimal for the activity of "pancreatic" lipase, but not for normal serum lipase. The original method indicates a reaction temperature of 38°C., however, 37°C. may be used with temperature correction.

SPECIMEN: 0.2 ml. of unhemolyzed serum. Hemolysis lowers the results. Plasma from calcium-binding anticoagulants should not be used; EDTA is inhibitory to lipase of pancreatic origin; the other anticoagulants are uncertain as to effect.

REAGENTS:

1. 0.05 M. Tris Buffer, pH 9.1

2. Buffer Diluent
   This contains 0.35% sodium deoxycholate in the above buffer; re-pH, after dissolving the bile salt, to 9.1 at 38°C. if being made from dry reagents. Buffer diluent also available as Harleco Item No. 64196-B. Good for one month or until pH varies by 0.3 units of stated value.

3. Purified Olive Oil, 1% Solution (Alcoholic)
   Available as Harleco Item No. 64196-A

4. 0.04% Triglyceride Emulsion
   Measure 25 ml. of buffer diluent into a wide-mouth flask. While mixing the buffer on a magnetic mixer, add 1.0 ml. of the 1% Olive oil dropwise. The emulsion is usable as long as the initial absorbance has not decreased by more than 10% of its original value. Use for one day only.

COMMENTS ON PROCEDURE:

1. The original methodology recommends making the turbidity determination at 400 nm. However, the specific absorbance peak must be determined for the instrument being used. (Beckman instruments use 450 nm.)

Serum Lipase Determination (Continued):

2. In a triglyceride emulsion undergoing the action of lipase, the rate of removal of triglyceride will decrease as the reaction proceeds. As fatty acids are individually removed from triglyceride, the enzyme may also act on di-and monoglyceride in the presence of the inhibitory effect of fatty acids. For this reason, the reaction is a non-linear one, and further dilutions must be made on sera of high activity.

3. The 1:5 serum dilution, reaction temperature, pH, and deoxycholate concentration should significantly diminish normal serum lipase activity, making the method more sensitive to that lipase present in pancreatitis.

4. The original methodology utilizes "lipase units" derived from the absorbance change (A) multiplied by 1000. The Harleco method utilizes a factor (34 or 35 depending upon reaction temperature) which attempts to correlate the turbidimetric procedure values with those of the Cherry-Crandall Method.

5. Stability of substrate (1% alcoholic olive oil) is a problem.

PROCEDURE:

1. Dilute all serum specimens (1:5) by delivering 0.2 ml. of serum into 0.8 ml. of the buffer diluent.

2. Pipette 4.0 ml. of triglyceride substrate (0.04%) into sufficient tubes for reagent blank, sera and controls. Bring all tubes to temperature for five minutes in a 37°C. water bath (or a 38°C. water bath if available).

3. Remove the reagent blank from the water bath, mix <u>gently</u> by inversion and determine the absorbance against <u>buffer</u> diluent at 450 nm. DO NOT SHAKE WHEN MIXING. Replace tube in water bath.

4. Remove a tube containing substrate from the water bath, deliver 0.2 ml. of buffered serum into it and mix gently by inversion. DO NOT SHAKE THE TUBE. Determine the absorbance <u>immediately</u> against buffer diluent and return tube to water bath. Note time

5. At timed intervals, remove the other tubes from the water bath and repeat Step No. 4.

6. At exactly 20 minutes incubation time, mix and determine the absorbance value on the first serum specimen against buffer

Serum Lipase Determination (Continued):

   diluent. Determine the absorbance values on all subsequent specimens in timed sequence.

7. Mix and determine the absorbance value for the reagent blank against buffer diluent.

8. If the absorbance of any of the tubes has dropped by more than one-third the initial absorbance, the determination must be rerun on a dilution. Prepare dilution by delivering 0.2 ml. of the original serum-buffer diluent into 0.8 ml. of buffer diluent. The final lipase units must be multiplied by 5.

9. CALCULATION:

   A. $$\frac{(\text{Absorbance of initial reading}) - (\text{Absorbance of second reading})}{\text{Absorbance of initial reading}}$$

   B. The above factor is multiplied by 34 (at 38°C.) or 35 (at 37°C.) to get Units of Lipase activity.

NORMAL VALUES: 0.1 - 1.5 Units

Most normal sera will give no reaction with this test. A few normal sera will give up to 1 Unit of lipase activity.

NOTE: Values in the lower range correspond to Cherry-Crandall Units: Values greater than 10 Units appear elevated in relation to the Cherry-Crandall Units.

REFERENCE: Vogel, W. C., and Zieve, L.: Clin. Chem., 9:168, 1963.

MALATE DEHYDROGENASE

PRINCIPLE: The enzyme malic dehydrogenase (MDH) catalyzes the reaction:

$$\text{Aspartate} + \text{Ketoglutarate} \underset{}{\overset{\text{GOT}}{\rightleftharpoons}} \text{Oxalacetate} + \text{Glutamate}$$

$$\text{Oxalacetate} + \text{NADH} + \text{H}^+ \underset{}{\overset{\text{MDH}}{\rightleftharpoons}} \text{Malate} + \text{NAD}^+$$

The MDH activity is measured by determining the decrease in absorbance per minute at 340 nm. which results when NADH is oxidized to NAD^+.

Malic dehydrogenase occurs in animal and plant tissues and in microorganisms. Its greatest concentration occurs in the liver, followed by heart, skeletal muscle and brain. The determination of this enzyme is of great importance, since its elevation is indicative of diseases of the liver.

SPECIMEN: 0.1 ml. of fresh, unhemolyzed serum is required. Test should be done as soon as possible, since 17 percent of its activity is lost in 24 hours at room temperature, 11 percent at refrigerator temperature, and 2 percent lost in the frozen state.

REAGENTS AND EQUIPMENT: (BioChemica Test Combination UV Method, Cat. No. 15981 TMAA)

1. Buffer (Bottle No. 1)
   0.1 M. Phosphate buffer, pH 7.4; 0.042 M. Aspartate.

2. Substrate (Bottle No. 2)
   0.065 M. Alpha-oxoglutarate

3. NADH (Bottle No. 3)
   0.012 M. reduced diphosphopyridine nucleotide

4. GOT (Bottle No. 4)
   0.1 mg. GOT/ml.

5. Spectrophotometer
   A. Gilford 300N, 340 nm.
      This instrument is used when analyzing one sample at a time. The change in absorbance per minute is read from the Data Lister printout. The linearity of the reaction is visually monitored on the strip chart recorder. The instrument is used with the thermocuvette set at 25°C.

Malate Dehydrogenase (Continued):

    B. Gilford 222, 340 nm.
       This instrument is used when analyzing four samples at a time. Temperature control is maintained by means of a 25°C. circulating water bath. Rate of reaction is determined from the strip chart recorder set for 0.200 Å. full scale and run at one inch per minute.

COMMENTS:

1. Solution is stable for one year at room temperature. Solution No. 4 is also stable for one year but at refrigerator temperature.
2. Solutions No.'s 2 and 3 are stable for 4 weeks at refrigerator temperature.
3. Dilute all specimens with physiological saline when optical density differences are greater than 0.060/minute.

PROCEDURE:

1. Using pyrex cuvettes (10 mm.), pipette 3.0 ml. of buffer, 0.05 ml. of α-oxoglutarate, 0.05 ml. of NADH, and 0.05 ml. of GOT.
2. Prepare three more tubes in the same manner.
3. Mix and incubate the tubes in 25°C. water bath for at least 5 minutes.
4. Add 0.10 ml. of test serum to all tubes.
5. Place 4 specimens in the cuvette chamber of the Gilford 222.
6. Set the baseline of specimen No. 1 with the slit, and No.'s 2, 3, and 4 with the OFF-SET knobs. Switch to "AUTO", and scan all four channels.
7. Record 3 minutes of linear reaction time at 340 nm. Determine ΔA/minute from recorder chart.
8. CALCULATIONS:

$$\text{I.U.} = \frac{\Delta A/\text{Minute}}{\epsilon \times d} \times 10^6 \times \frac{TV}{SV} \times \frac{1}{\text{Time}}$$

Malate Dehydrogenase (Continued):

ϵ = Molar Extinction Coefficient of NADH at 340 nm.
  = 6.22×10^3 Liter/Mole x cm.
d = Diameter of the cuvette in cm.
  = 1.0 cm.
TV = Total Volume
  = 3.25 ml.
SV = Sample Volume
  = 0.10 ml.
T = Time
  = 1 minute

10^6 converts Moles/Liter into micromoles/liter
 1 I.U./Liter = 1 mU/ml.

I.U. = $\dfrac{\Delta A/\text{Minute}}{6.22 \times 10^3}$ x 10^6 x $\dfrac{3.25}{0.10}$ x $\dfrac{1}{1}$

I.U. = ΔA/minute x 5225

I.U. = ΔA/minute x 5225 = mU/ml. of serum.

Report results in I.U. at 25°C.

NORMAL VALUES: Normal values compiled by BMC at 25°C.
 48 - 96 mU/ml. of serum.

REFERENCES:

1. Bing, R. J., Castellanos, A., Siegel, A.: **J. Amer. Med. Assoc.** 164:647-650, 1957.

2. Bergmeyer, H. U.: METHODS OF ENZYMATIC ANALYSIS 2nd. Ed., Academic Press, New York, pg. 757, 1965.

PYRUVATE KINASE

PRINCIPLE: Under the controlled conditions of enzymatic reactions (sample size, substrate concentration, time, temperature, pH, ionic strength, and various salt concentrations), pyruvate kinase catalyzes the conversion of phosphoenolpyruvate to pyruvate, which is further converted to lactate, as shown in the following reaction:

$$ADP + PEP \xrightleftharpoons{PK} ATP + Pyruvate$$

Adenosine   Phosphoenol-   Adenosine
Diphosphate   Pyruvate   Triphosphate

$$Pyruvate + DPNH + H^+ \xrightleftharpoons{LDH} Lactate + DPN^+$$

The decrease in absorbance of DPNH due to its oxidation to DPN^+ is a direct measurement of the concentration of pyruvic kinase.

One of the metabolic abnormalities of the red cell associated with hemolytic disease is due to pyruvic kinase deficiency. The determination of this enzyme is of utmost value in cases of hemolytic anemia.

SPECIMEN: 1.0 ml. of fresh serum, free from hemolysis and 0.2 ml. of blood for hemolysate.

REAGENTS AND EQUIPMENT: (Biochemica Test Combination UV Method, Cat. No. 15985 TPAB)

1. Buffer (Bottle No. 1)
   0.16 M. Triethanolamine buffer, pH 7.5; 0.12 M. KCl; 0.021 M. $MgSO_4$; 0.0013 M. Ethylene Diamine Tetraacetate

2. Stock NADH (Bottle No. 2)
   0.006 M. NADH; 0.0325 M. Phosphoenolpyruvate

3. LDH (Bottle No. 3)
   0.5 mg. LDH/ml.

4. ADP (Bottle No. 4)
   0.1 M. Adenosine Diphosphate

5. Saline
   0.9 gm. Sodium Chloride, A. R. in 100 ml. of deionized water

6. Spectrophotometer, Gilford 300N, 340 nm.
   This instrument is used to analyze one sample at a time. The change in absorbance per minute is read from the Data Lister

Pyruvate Kinase (Continued):

printout. The linearity of the reaction is visually monitored on the strip chart recorder. The thermocuvette is set at 25°C.

COMMENTS:

1. Solutions No.'s 1 and 3 are stable for a year at room temperature.
2. Solutions No.'s 2 and 4 are stable for three weeks at refrigerator temperatures.
3. Make dilutions with optical density differences above 0.600/10 minutes.

PROCEDURE:

A. Hemolysate
   1. Determine the number of erythrocytes/ml. of blood.
   2. Wash 0.20 ml. of blood 3 times with 2.0 ml. of physiological saline each.
   3. After each wash, centrifuge for 10 minutes at 3000 r.p.m.
   4. Suspend washed and centrifuged erythrocytes in 2.0 ml. of deionized water.
   5. Allow to stand for 15 minutes in the refrigerator.
   6. Centrifuge. Use supernatant for determination of enzyme activity.

B. Enzyme Activity
   1. Using disposable culture tubes (12 x 75 mm.), set up 2 tubes, Blank and Test, for serum and also 2 tubes, Blank and Test, for hemolysate.
   2. To the serum tubes (Blank and Test), add 2.50 ml. of buffer, 0.10 ml. of NADH-PEP, 0.50 ml. of serum, and 0.05 ml. of LDH.
   3. To the hemolysate tubes (Blank and Test), add 2.0 ml. of buffer, 0.90 ml. of deionized water, 0.10 ml. of NADH-PEP, 0.10 ml. of hemolysate and 0.05 ml. of LDH.

Pyruvate Kinase (Continued):

4. Mix and incubate all tubes in a 25°C. water bath for 5 minutes.

5. Set the machine to zero with deionized water. Aspirate the serum Blank and measure the optical density (S) E_1. Aspirate the hemolysate Blank and measure the optical density (H) E_1. Take another reading after 10 minutes (S) E_2 and (H) E_2.

6. To the serum Test and Hemolysate Test, add 0.10 ml. of adenosine diphosphate.

7. Mix and obtain optical density of both serum (S) E_3 and hemolysate (H) E_3. Take another reading after 10 minutes (S) E_4 and (H) E_4.

8. To obtain $\Delta A/10$ minutes, subtract optical density difference $(E_1 - E_2)$, from optical density difference $(E_3 - E_4)$ for both serum and hemolysate.

9. CALCULATIONS:

   A. Serum

   $$I.U. = \frac{\Delta A/10 \text{ Min.}}{\epsilon \times d} \times 10^6 \times \frac{TV}{SV} \times \frac{1}{Time}$$

   ϵ = Molar Extinction Coefficient of NADH at 340 nm.
     = 6.22×10^3 Liter/Mole x cm.
   d = Diameter of the cuvette in cm.
     = 1.0 cm.
   TV = Total Volume
     = 3.25 ml.
   SV = Sample Volume
     = 0.5 ml.
   T = Time
     = 10 minutes

   10^6 converts Moles/Liter into micromoles/liter
     1 I.U./Liter = 1 mU/ml.

   $$I.U. = \frac{\Delta A/10 \text{ min.}}{6.22 \times 10^3} \times 10^6 \times \frac{3.25}{0.50} \times \frac{1}{10}$$

   I.U. = $\Delta A/10$ minutes x 104.5

       $\Delta A/10$ minutes x 104.5 = mU/ml. of serum.

Pyruvate Kinase (Continued):

    B. <u>Hemolysate</u>

        Calculated the same way as for the serum.

        $\Delta A/10$ minutes x 5486 = mU/number of erythrocytes in 1.0 ml. of blood.

        Report results in I.U. at 25°C.

NORMAL VALUES: Normal Values compiled by BMC at 25°C.
        <u>Serum</u>:   Up to 26 mU/ml. of serum

        <u>Erythrocytes</u>: 60 - 220 mU/10^9 erythrocytes

REFERENCES:

1. Tanaka, K. R., Valentine, W. N., and Miwa, S.: <u>Blood</u> 19:267, 1962.

2. Bergmeyer, H. U.: METHODS OF ENZYMATIC ANALYSIS, 2nd Ed., Academic Press, New York, pg. 573, 1965.

SORBITOL DEHYDROGENASE

PRINCIPLE: The enzyme, sorbitol dehydrogenase (SDH) catalyzes the following reaction:

$$\begin{array}{c} HO-C-H_2 \\ C=O \\ HO-C-H \\ H-C-OH \\ H-C-OH \\ CH_2OH \end{array} + DPNH + H^+ \underset{}{\overset{SDH}{\rightleftharpoons}} \begin{array}{c} HO-C-H_2 \\ H-C-OH \\ HO-C-H \\ H-C-OH \\ H-C-OH \\ CH_2OH \end{array} + DPN^+$$

D-Fructose   D-Sorbitol

The SDH activity is measured by determining the ΔO.D. 340 per minute which results when DPNH is oxidized to DPN.

Sorbitol dehydrogenase has been found in the liver and the genital organs of many mammals. The determination of SDH activity is of clinical importance, because healthy individuals have very little activity, but it is apparent in liver cell damage (infectious, toxic, or hypoxic in origin). Therefore, it is a fairly specific indicator of liver cell damage.

SPECIMEN: 1.0 ml. of fresh, unhemolyzed serum is required.

Test should be done immediately, as serum SDH is very labile. Activity drops approximately 1.0 percent per hour at room temperature and 0.5 percent per hour at 0°C. SDH activity is inhibited by ethylenediamine-tetraacetate and mercury ions; heparin does not substantially affect the activity.

REAGENTS AND EQUIPMENT: (BioChemica Test Combination UV Method, 15960 TSAB).

1. <u>Buffer</u> (Bottle No. 1)
   0.2 M. Triethanolamine buffer, pH 7.4

2. <u>Stock NADH (DPNH)</u> (Bottle No. 2)
   15 mM reduced diphosphopyridine nucleotide

3. <u>Substrate</u> (Bottle No. 3)
   1.65 M. Fructose

83

Sorbitol Dehydrogenase (Continued):

4. Spectrophotometer
   A. Gilford 300N, 340 nm.
      This instrument is used when analyzing one sample at a time. The change in absorbance per minute is read from the Data Lister printout. The linearity of the reaction is visually monitored on the strip chart recorder. The instrument is used with the thermocuvette set at 30°C.

   B. Gilford 222, 340 nm.
      This instrument is used when analyzing four samples at a time. Temperature control is maintained by means of a 30°C. circulating waterbath. Rate of reaction is determined from the strip chart recorder set for 0.200 A. full scale and run at one inch per minute.

COMMENTS:

1. Solutions are very stable. Bottle No. 1 is stable for a year at room temperature, while bottles No. 2 and No. 3 are stable for one month at 4°C.

2. Dilute all specimens with physiological saline when optical density differences are greater than 0.060 per minute.

PROCEDURE:

1. Using pyrex cuvettes (10 mm. light path, 10 mm. wide), pipette 2.0 ml. of buffer, 1.0 ml. of serum and 0.05 ml. of NADH.

2. Dilute up three more specimens in the same manner.

3. Mix and incubate the tubes in a 30°C. water bath for 45 minutes.

4. Add 0.20 ml. of fructose solution to all tubes.

5. Place 4 specimens in the cuvette chamber of the Gilford 222.

6. Set the baseline of specimen No. 1 with the slit, and No.'s 2, 3, 4 with the off-set knobs. Switch to "Auto", and scan all 4 channels.

7. Record 5 to 8 minutes of linear reaction time at 340 nm. Determine ΔA/min. from recorder chart.

Sorbitol Dehydrogenase (Continued):

8. CALCULATION:

$$I.U. = \frac{\Delta A}{\epsilon \times d} \times 10^6 \times \frac{TV}{SV} \times \frac{1}{Time}$$

ϵ = Molar Extinction Coefficient of NADH
  = 6.22×10^3 Liter/Mole x cm.
d = Diameter of the cuvette in cm.
  = 1 cm.
TV = Total Volume
  = 3.25 ml.
SV = Sample Volume
  = 1.0 ml.
Time = 1 minute

10^6 converts Moles/Liter into micromoles/liter
  1.0 I. U./Liter = 1 mU/ml.

$$I.U. = \frac{\Delta A}{6.22 \times 10^3} \times 10^6 \times \frac{3.25}{1} \times \frac{1}{1}$$

I. U. = ΔA/Min. x 5225

Report results in I. U. at 30°C.

NORMAL VALUES: The SDH activity in the serum of healthy people is less than 1.0 unit/ml.

REFERENCES:

1. Blakley, R. L.: <u>Biochemistry J.</u> 49:259, 1951.

2. Secchi, G. C. and Ghidoni, A.: <u>Enzymol. Biol. Clin.</u> 2:99-107, 1962/63.

3. Bergmeyer, H. U.: METHODS OF ENZYMATIC ANALYSIS 2nd. Ed., Academic Press, New York, pg. 761, 1965.

GLUTAMATE-OXALOACETATE TRANSAMINASE

PRINCIPLE: Glutamate-oxaloacetate transaminase (GOT) catalyzes the reaction:

1). L-Aspartate + alpha-Oxoglutarate $\rightleftarrows$ L-Glutamate + Oxaloacetat

2). Oxaloacetate + NADH + H$^+$ $\overset{MDH}{\rightleftarrows}$ Malate + NAD$^+$

The activity of the transaminase is measured by the conversion of L-aspartate and alpha-oxoglutarate into glutamate and oxaloacetate. In an indicator reaction with malic dehydrogenase, oxaloacetate and NADH are converted into malate and NAD. Enzyme activity is measured by the rate of decrease in absorbance of NADH.

SPECIMEN: 0.5 ml. of serum. Serum should be separated from cells as soon as possible. Only minimum hemolysis allowed. Specimen stable at 4°C. up to 72 hours.

REAGENTS AND EQUIPMENT:

(Biochemica Test Combination Kit - TC Al - 15971 TGAE)

1. <u>Stock Buffer + L-Aspartate</u> (Bottle No. 1)
   0.1 M. Phosphate buffer, pH 7.4, and 0.04 M. L-Aspartate.

2. <u>Stock NADH</u> (Bottle No. 2)
   0.012 M. NADH.

3. <u>Malate Dehydrogenase/Lactate Dehydrogenase</u> (Bottle No. 3)
   0.25 mg. each of MDH and LDH.

4. <u>Alpha-Oxoglutarate</u> (Bottle No. 4)
   0.25 M. alpha-oxoglutarate.

5. <u>Working Buffer</u> (Solution 1)
   Dissolve contents of Bottle No. 1 in 100 ml. redistilled water in clean glass container. Dissolve Bottle No. 2 in 1.5 ml. redistilled water. Use Bottle No. 3 as is (liquid). Combine the three bottles to make "Working Buffer". Be sure to rinse contents of bottles.

6. <u>Working Substrate</u>, alpha-Oxoglutarate (Solution 2)
   Dissolve contents of Bottle No. 4 in 3 ml. redistilled water (directly into bottle). Do not add to Working Buffer.

Glutamate-Oxaloacetate Transaminase (Continued):

>NOTE: Solutions 1 and 2, while kept separate and at 4°C., are stable for 4 weeks. Do not pipette from either of these. Use Cornwall syringe for dispensing Solution 1 and Eppendorf pipette with clean disposable tip for solution 2. These reagents are sufficient for 25 determinations.

7. Beckman DBG
   Instrument with thermostated cuvette well at 37°C.; Beckman 10 inch recorder with expanded scale.

PROCEDURE:

1. To 3.1 ml. of buffered substrate (Solution 1), add 0.5 ml. of serum with 500 lambda Eppendorf pipette.

2. Mix. Allow to come to temperature for 5 minutes in a 37°C. water bath.

3. Add 0.1 ml. of alpha-oxoglutarate (Solution 2). Mix by gentle inversion with parafilm.

4. Place in DBG, and using recorder, scan for 1 minute of linearity.

5. CALCULATION:

   ϵ NADH at 340 nm. = 6.22 x 10^3 Liter/Mole x cm.

   $$\frac{\Delta A}{\epsilon x d} \times 10^6 \times \frac{TV}{SV} \times \frac{1}{Time} = IU/Liter \text{ or } mU/ml.$$

   $$\frac{\Delta A}{1} \times \frac{1}{6.22 \times 10^3 \times 1} \times 10^6 \times \frac{3.7}{0.5} \times \frac{1}{1} = mU/ml.$$

   $\Delta A \times F = mU/ml.$
   $F = 1190.$

NORMAL VALUES: 5 - 40 mU/ml.

REFERENCES:

1. Karmen, A., et al: J. Clin. Invest., 34:1261, 1955.

2. Bergmeyer, H. U.: METHODS OF ENZYMATIC ANALYSIS, Academic Press, 1968.

Glutamate-Oxaloacetate Transaminase (Continued):

3. Henry, R. J., Chiamari, N., Golub, O. J., and Berkman: Amer. J. Clin. Path., 34:381, 1960.

GLUTAMATE-PYRUVATE TRANSAMINASE

PRINCIPLE: GPT converts L-alanine and alpha-oxoglutarate into glutamate and pyruvate. In an indicator system with LDH, pyruvate and NADH are converted into lactate and NAD. GPT is measured by the decrease in absorbance of NADH at 340 nm. (ϵ NADH at 340 nm. is 6.22×10^3 liter/mole x cm.)

SPECIMEN: 0.5 ml. serum. Serum should be separated from cells as soon as possible. Only minimum hemolysis allowed. Specimen stable at 4°C. up to 72 hours.

REAGENTS AND EQUIPMENT:

(Biochemica Test Combination Kit, TC-HI 15978 TGAH)

1. Stock Buffer + DL-Alanine (Bottle No. 1)
   Buffer, pH 7.4, 0.08 M. DL-alanine, 0.1 M. Phosphate

2. Stock NADH (Bottle No. 2)
   0.012 M. NADH

3. LDH (Bottle No. 3)
   0.25 mg. LDH

4. Alpha-Oxoglutarate (Bottle No. 4)
   0.25 M.

5. Working Buffer (Solution 1)
   Dissolve bottle No. 1 in 100 ml. redistilled water (use clean glass container). Dissolve contents of Bottle No. 2 in 1.5 ml. water; add to buffer (rinse bottle); add contents of bottle No. 3 to buffer (rinse bottle). Mix by swirling.

6. Working Substrate, alpha-oxoglutarate (Solution 2)
   Add 3.0 ml. of redistilled water to bottle No. 4. Add directly to bottle; mix by inversion. Do NOT add to the working buffer.
   NOTE: Solutions 1 and 2, while kept separate and at 4°C., are stable for 4 weeks. Do not pipette from either of these.

   Use Cornwall syringe for dispensing Solution 1. Use Eppendorf pipette with disposable tip for Solution 2.

   These reagents sufficient for 25 determinations.

Glutamate-Pyruvate Transaminase (Continued):

7. <u>Beckman DBG</u>
   Instrument with thermostated cuvette well at 37°C. Beckman 10 inch recorder with expanded scale.

PROCEDURE:

1. Into 10 mm. square civette add 3.1 ml. of Working Buffer (Solution 1) + 0.5 ml. of serum with a 500 microliter Eppendorf pipette).

2. Mix. Place in 37°C. water bath for 5 minutes.

3. Add 0.1 ml. of alpha-oxoglutarate (Solution 2). Mix by inverting with parafilm.

4. Place in DBG. Record reaction for 1 minute of linearity using recorder.

5. CALCULATIONS:

$$\frac{\Delta A}{\epsilon x d} \times 10^6 \times \frac{TV}{SV} \times \frac{1}{Time} = IU/Liter \text{ or } mU/ml.$$

NADH at 340 nm. = 6.22×10^3 Liter/Mole x cm.

$$\frac{\Delta A}{6.22 \times 10^3 \times 1} \times 10^6 \times \frac{TV}{SV} \times \frac{1}{1}$$

$$\frac{\Delta A}{1} \times \frac{1}{6.22 \times 10^3 \times 1} \times 10^6 \times \frac{3.7}{0.5} \times \frac{1}{1}$$

$\Delta A \times F = mU/ml.$
$F = 1190$

NORMAL VALUES: 5 - 35 mU/ml.

REFERENCES:

1. Wroblewski, F. and LaDue, J.S.: <u>Ann. Intern. Med.</u> 45:80, 1956.

2. Wroblewski, F. and LaDue, J.S.: <u>Proc. Soc. Exp. Biol. Med.</u>, 91:569, 1966.

TRIGLYCERIDES - BOEHRINGER MANNHEIM DETERMINATION
OF GLYCEROL AND TRIGLYCERIDES IN SERUM
(U.V. Method with NADH)

PRINCIPLE: Serum lipids consist mainly of triglycerides (neutral fats), cholesterol and phospholipids. In this method serum is saponified with alcoholic potassium hydroxide to hydrolyze triglyceride to fatty acids and glycerol. (Phospholipids are hydrolyzed to alpha and beta-glycerophosphates, stable in alkaline solution. Since the enzyme glycerokinase (GK) is specific for glycerol and does not act upon alpha and beta-glycerophosphates there is no interference by phospholipid.)

The saponification splits triglyceride into fatty acids and free glycerol. Glycerol is phosphorylated with adenosine-5'-triphosphate (ATP) in a reaction catalyzed by the enzyme glycerokinase (GK) to result in glycero-1-phosphate and adenosine-5'-diphosphate (ADP):

$$\text{Glycerol} + \text{ATP} \rightleftharpoons \text{Glycero-1-phosphate} + \text{ADP}$$

In an auxiliary reaction with the enzyme pyruvate kinase (PK), the ADP and phsophoenolpyruvate (PEP) are transferred into ATP and pyruvate.

$$\text{ADP} + \text{PEP} \rightleftharpoons \text{ATP} + \text{Pyruvate}$$

In the indicator reaction the enzyme LDH catalyzes the reaction of pyruvate and NADH to lactate and NAD.

$$\text{Pyruvate} + \text{NADH} + H^+ \rightleftharpoons \text{Lactate} + NAD^+$$

The amount of NADH consumed during the reaction is equivalent to the amount of glycerol present in the sample. NADH can be measured by its absorbancy at 340 nm. The method described is based on studies of M. Eggstein and F. H. Kreutz [3,4].

SPECIMEN: Use fresh serum free from hemolysis. The determination of triglyceride should always be preceded by a determination of free glycerol. Strongly lipemic sera should be diluted (1 + 9 with physiological saline) prior to the glycerol determination. 200 microliters of serum is used for the saponification. 0.5 ml. of serum is used for determining free glycerol.

Triglycerides (Continued):

REAGENTS: Biochemica Test Kit TNAA (Cat. No. 15989)

Solution 1. Buffer (0.1 M. Triethanolamine, pH 7.6; 0.004 M. $MgSO_4$); dissolve the contents of bottle No. 1 with 150 ml. redistilled water. The solution is stable for approximately 2 months at room temperature.

Solution 2. NADH/ATP/PAP (0.006 M. NADH, 0.33 M. ATP, 0.011 M. PEP); reconstitute 1 bottle of solution No. 2 with 2.0 ml. of redistilled water. The solution is stable for approximately 2 weeks at +4°C. (Stability of this solution may be prolonged by aliquoting and freezing.)

Solution 3. LDH/PK (2.0 mg. LDH/ml.; 1.0 mg. PK/ml.); Use suspension of bottle No. 3 undiluted. Stable for approximately 1 year at +4°C.

Solution 4. GK (2.0 mg. GK/ml.); Use suspension of bottle No. 4 undiluted. Stable for approximately 1 year at +4°C.

Solution 5. Alcoholic potassium hydroxide (0.5 N.); Dissolve 3.3 gm. potassium hydroxide pellets, A. R. free from glycerol, in approximately 10 ml. redistilled water. After cooling dilute to 100 ml. with denatured ethyl alcohol.
NOTE: Methanol may not be used.

Solution 6. Magnesium sulfate (0.15 M.); Dissolve 3.7 gm. $MgSO_4 \cdot 7H_2O$ A. R. in 100 ml. of distilled water.

COMMENTS ON PROCEDURE:

A. If solution No. 2 is improperly stored, the ATP may deteriorate. This would result in a considerable reduction of NADH prior to the glycerol reaction. The glycerol reaction will then be limited to the amount of remaining NADH and lead to false low values. After completion of the reaction, a drop of diluted glycerol added to the cuvette should cause further decrease in optical density.

B. Some sera will result in a continuous linear decrease of optical density after the reaction is completed. In these

Triglycerides (Continued):

cases, after the 10 minutes, take additional readings at minute intervals for 2 or 3 minutes. Multiply the O.D./minute of this "creep" by 10 (the reaction time) and subtract this figure from the ΔO.D. ($E_1 - E_2$) to obtain the "corrected" ΔO.D.

C. For absolute accuracy a reagent blank should be determined once for each Lot Number. The determination is carried out by using water instead of serum. The optical density difference resulting after the addition of Solution No. 4 is to be deducted from the optical density difference obtained with serum.

D. The three critical areas in pipetting are the adding of serum and alcoholic KOH in the saponification step and the addition of serum or supernatant in the glycerol determination. This laboratory uses Eppendorf pipettes and syringes with Cheney Adaptors.

PROCEDURE:

Set up two sets of centrifuge tubes for each serum, reagent blank, and standards. First set is for the determination of total glycerol which requires saponification. Second set is for the determination of "free" glycerol which needs no saponification.

A. Saponification

1. Pipette into centrifuge tube:

	Blank	Test
Distilled water	0.2 ml.	-
Serum	-	0.2 ml.
Alcoholic KOH	0.5 ml.	0.5 ml.

Mix, close centrifuge tube with clean stopper or parafilm and allow to stand in water bath for 30 minutes at 70°C., then allow to cool to room temperature in cold water. Add 1.0 ml. $MgSO_4$, mix and spin.

2. Into the second set of centrifuge tubes pipette:

	Blank	Test
$MgSO_4$ Solution	1.0 ml.	1.0 ml.
Serum	-	0.2 ml.
Distilled water	0.2 ml.	-

Triglycerides (Continued):

3. Mix well, and then add cold

	Blank	Test
Alcoholic KOH	0.5 ml.	0.5 ml.

   to each tube, one at a time, stopping to mix each tube at once. Stopper and centrifuge. Use 0.5 ml. of the clear supernatant fluid respectively for glycerol assay.

B. <u>Determination of Glycerol</u>

1. Pipette into cuvette: Glass cuvette: 1.0 cm. light path.

	Routine Assay
Solution 1.	2.50 ml.
Solution 2.	0.10 ml.
Serum or supernatant fluid	0.50 ml.
Solution 3.	0.02 ml.

2. Cover with parafilm and mix -- allow to stand for 10 minutes at room temperature; measure optical density E_1. Wavelength 340 nm. Blank machine with air. Read reagent blank and record. Set machine to 1.0 O.D. with R. B. then read tests.

   Solution 4.   0.02 ml.

3. Mix and wait until reaction stops (approximately 10 minutes) or take 3 - 5 readings at 2 minute intervals and extrapolate E_2 to the time of the addition of suspension 4; refer to "Biochemica - Test Combinations, Principles and Practice".

$$E_1 - E_2 = \Delta E$$

4. With optical density differences exceeding 0.800 (340 nm.), mix 0.1 ml. serum or filtrate with 0.9 ml. physiological saline and repeat assay with 0.5 ml. of this dilution. Result must be multiplied by 10.

5. CALCULATION

   The ratio of assay volumes between routine and semi-micro assays is not exactly 2.5:1. The difference, however, is less than 1.0% and can be neglected. For both procedures the same factors can be applied.

Triglycerides (Continued):

Free Glycerol: If a 0.5 ml. serum sample is used directly.

$$\frac{\Delta O.D. \times Va \times S_{M.W.}}{x\ E. \times Vs} = \text{micrograms/ml.}$$

Va = Total assay volume - 3.14 ml.
Vs = Sample volume - 0.50 ml.
E = Light path - 1.0 cm.
    Absorbancy coefficient - 340 nm. = 6.22
M.W. = Molecular Weight glycerol = 92
To convert micrograms/ml. to mg%., a factor of 1/10
    (100/1000) must be added.

Resolving the above formula, the factor for free glycerol then is:

340 nm.: O.D. x 9.28 = mg%. free glycerol in serum.

Total Glycerol: Due to saponification and neutralization, the sample is diluted 1 to 8.5.

The factor for total glycerol is then:

$$\frac{\Delta O.D. \times 3.14 \times 92}{6.22 \times 1 \times 0.5 \times 1/8.5} = \text{micrograms/ml.}$$

340 nm.: O.D. x 78.6 = mg%. Total glycerol in serum.

Triglycerides:

Total Glycerol - mg%. (Serum with saponification)
Free Glycerol - mg%. (Serum without saponification)
"Glyceride glycerol" - mg%.

The mg%. of "Glyceride Glycerol" (glycerol derived from triglycerides) is now converted into mg%. of triglycerides. This is accomplished by equating the average molecular weight of triglycerides (885) to the molecular weight of glycerol (92).

$$\frac{885}{92} = \frac{X\ (mg\%.\ \text{triglyceride})}{1.0\ mg\%.\ \text{glycerol}}$$

X = 9.62

Triglycerides (Continued):

Mg%. "Glyceride - glycerol" x 9.62 = mg%. Triglycerides (neutral fat).

Example:

0.5 ml. serum each was determined prior and after saponification
Wavelength: 366 nm. lightpath 1.0 cm.

	Free Glycerol	Total Glycerol After saponification
Prior to addition of GK: E_1	0.748	0.695
10 minutes after addition of GK: E_2	0.665	0.560
11 minutes after addition of GK:	0.663	0.557
12 minutes after addition of GK:	0.661	0.554
13 minutes after addition of GK:	0.659	0.551
Δ O.D./minute of "creep"	0.002	0.003
$E_1 - E_2$	0.083	0.135
10 minute "creep (10 x Δ O.D./minute creep)	0.020	0.030
"Corrected" Δ O.D. for calculation	0.063	0.105

NORMAL VALUES IN SERUM:

Free Glycerol: 0.5 - 1.7 mg%.
Glyceride - Glycerol: 7.7 - 17.9 mg%.
Neutral Fat: 40 - 150 mg%.

REFERENCES:

1. Eggstein, M.: Klin. Waschr., 44:276, 1966.

2. Eggstein, M. and F. H. Dreutz: Klin. Waschr., 44:262, 1966.

3. Schmidt, F. H., et al.: Klin. Chem. and Klin. Biochem., 6:156, 1968.

4. Pinter, J.K., et al.: Arch. Biochem. & Biophys., 121:404, 1967.

URIC ACID DETERMINATION
(Uricase Method)

PRINCIPLE: The enzyme uricase as a substrate is specific for uric acid, and catalyzes the following reaction:

Uric Acid + O_2 + $H_2O \longrightarrow$ allantoin + H_2O_2 + CO_2

Uric acid has an absorbance peak at 293 nm.; the methodology takes advantage of this fact, measuring the decrease in absorbance due to the oxidation of uric acid in the sample by the uricase added. The absorbance decrease is proportional to the amount of uric acid present in the sample, the reaction being allowed to go to completion. At the dilutions at which one is working, the method is specific for uric acid and is not interfered with by substances in the sample.

SPECIMEN: 50 microliters of serum. Plasma may not be used for this procedure. 50 microliters of centrifuged urine which has been diluted 1:10 (1.0 ml. + 9.0 ml.) with physiological saline. If original urine is cloudy, warm to 60°C. for 10 minutes, mix, then centrifuge. Fresh serum and urine samples are preferred if possible.

REAGENTS:

Boehringer Mannheim "Uric Acid Test Combination", Catalogue #15986 THAA (U.V. Method); for approximately 30 determinations. The test combination includes the following:
1. **0.2 M. Borate Buffer**, pH 9.5 (Bottle #1)
   Dissolve the contents of bottle #1 in 100 ml. of redistilled water. Stable for several months at room temperature; protect from evaporation.

2. **Uricase**, 2.0 mg./ml. (Bottle #2)
   The contents of this bottle are used undiluted. Store at refrigerator temperature; stable for several months.

3. **Uric Acid Stock Standard**, 1.0 mg./ml.
   Into a 100 ml. volumetric flask, place 100 mg. uric acid, and 60 mg. lithium carbonate (Li_2CO_3). Add approximately 50 ml. re-distilled water and dissolve contents at 60°C. Allow flask to cool and bring to volume with re-distilled water. Stable up to one year at refrigerator temperature.

4. **Uric Acid Working Standard**, 0.1 mg./ml.
   Place 10 ml. of stock standard in a 100 ml. volumetric flask and bring to volume with re-distilled water.

Uric Acid Determination (Continued):

COMMENTS ON PROCEDURE:

1. The ten minute timing need not be exact to the second, but should not be extended beyond thirty seconds.

2. The reference against air should be checked before each reading.

3. The reaction appears to be linear at 15 mg%. uric acid and beyond.

4. Factor Explanation:
   With a 1.0 cm. light path, the oxidation of 1 microgram of uric acid per ml. corresponds to an absorbance decrease of 0.075 at 293 nm.

   $$\frac{\Delta A}{0.075} \times \frac{TV}{SV} = \text{ug. Uric Acid/ml.}$$

   $$\Delta A \times \frac{1}{.075} \times \frac{3.07}{0.05} = \text{ug./ml.}$$

   Multiplication by 0.1 = Concentration in mg./100 ml.

   $$\Delta A \times F = \text{Mg./100 ml.}$$

   $$F = 81.8 \text{ for serum}$$

   $$F = 818 \text{ for 1:10 dilution of urine}$$

5. If one wishes to check the accuracy of the enzymatic system, a series of standards may be pipetted into the buffer and run as specimens.

Working Standard	5 Lambda	25 Lambda	50 Lambda
Buffer	3.0 ml.	3.0 ml.	3.0 ml.
Concentration	1.0 mg%.	5.0 mg%.	10.0 mg%.

PROCEDURE:

A. Determining the absorbance of the Uricase. (A_e)
   1. The following procedure is carried out <u>once</u> with each new "test combination kit" in order to determine the absorbance contributed by that lot of enzyme.

Uric Acid Determination (Continued):

    2. Place 3.0 ml. of borate buffer, Bottle No. 1, into two quartz cuvettes. Reserve one cuvette as a reference cell.

    3. Pipette 20 microliters of uricase, Bottle No. 2, into the other cuvette and mix thoroughly. Determine the absorbance of this solution against the buffer alone at 293 nm.

B. <u>Analysis of Samples</u>.
1. Pipette 3.0 ml. of borate buffer into a quartz cuvette; prepare one cuvette for each sample to be analyzed.

    2. Deliver 50 microliters of serum or diluted urine into a cuvette with a T.C. pipette and mix well by inversion. Determine the initial absorbance of the uric acid against air at 293 nm. (A_1)

    3. After taking the initial reading, deliver 20 microliters of uricase into the cuvette, mix again, and allow to stand at room temperature (20° - 25°C.) for ten minutes.

    4. Take the second absorbance reading (A_2) against air at 293 nm.

    5. CALCULATION:

        ($A_1 + A_e$) - A_2 = ΔA for the sample.

        for serum: ΔA x 81.8 = mg%. uric acid
        for urine: ΔA x 818   = mg%. uric acid

NORMAL VALUES:

    Serum: 1. Men   2.6 - 6.8 mg%.
               2. Women 2.0 - 6.3 mg%.

    Urine: 0.25 - 0.75 gm./24 hours.

REFERENCES:

    1. Praetorius, Elith, in Bergemeyer: METHODS OF ENZYMATIC ANALYSIS, Academic Press, pg. 500 - 501, 1965.

    2. Henry, R. B.: CLINICAL CHEMISTRY: PRINCIPLES AND TECHNIQUES, Hoeber, pg. 280, 283 - 287, 1964.

CHAPTER 2

ALKALINE AND ACID PHOSPHATASE, 5′-NUCLEOTIDASE,
LEUCINE AMINOPEPTIDASE, AND GAMMA GLUTAMYL TRANSPEPTIDASE

Alkaline Phosphatase

The phosphatases are generally located in the cytoplasm of the cell, with alkaline phosphatase being a microsomal enzyme and acid phosphatase a lysosomal one. They are hydrolytic and act on phosphoric esters with liberation of inorganic phosphate from various substrates. The phosphatases are ubiquitous in distribution. The location of alkaline phosphatase in the growing individual is in the osteoblasts and the chondroblasts of the skeleton. Thus, most of the alkaline phosphatase is present in the bones in this age group. In the adult the gastrointestinal mucosa and liver contain the largest amount of enzyme; lung and spleen have a large amount of enzyme; other sites for enzyme production include vascular endothelium, renal tubules, thyroid epithelium, bile duct epithelium, liver parenchymal cells, pancreatic parenchymal cells, placenta, and the enzyme is also found in the cells of the myeloid series in the peripheral blood and the bone marrow.

The alkaline phosphatase activity of human serum has been separated into different isoenzymes. The term isoenzyme is utilized to specify different molecular forms of an enzyme. Electrophoresis may be utilized with different media to separate the isoenzymes. In the adult, alkaline phosphatase is primarily derived from the liver and gastrointestinal tract, in contrast to a bone origin in the growing child. (Table 2-1)

TABLE 2-1

SERUM ALKALINE PHOSPHATASE IN PEDIATRIC PATIENTS

1 to 3 months	73 to 256 I.U.
3 to 10 years	73 to 151 I.U.
11 to 16 years	73 to 256 I.U.

The normal values vary according to the method employed to test for the enzyme. With the Bodansky method, the normal serum adult activity is from 1.5 to 4.0 Bodansky units. For the King-Armstrong method, one must usually multiply the Bodansky range of normal by a factor of three, giving a normal range for the King-Armstrong method

Alkaline Phosphatase (Continued):

of 4.0 to 13.5. With the newer method of kinetic analysis in International Units, the normal values are from 20 to 85 I.U. This new system has been adopted to standardize testing and reporting of enzyme assays.

The level of alkaline phosphatase is higher in growing children as previously mentioned than adults. There is a marked rise during the neonatal period with an elevation to three times the adult level at the end of the first month of life. This level is maintained until approximately the age of 8 years, at which time it is approximately twice the adult level. It remains at this level until the onset of puberty at which time, with the spurt of bone growth, the alkaline phosphatase again may attain levels of three times the adult range.[40] Generally males tend to exhibit somewhat higher levels of alkaline phosphatase than do women. This may be related to differences in skeletal mass, but it also is related to greater physical activity of the male which serves as a stimulus for an increase in bone formation.[52] Serum alkaline phosphatase in women is higher when the endometrium is in a proliferative phase; this change in activity may be related to estrogenic stimulation. The cyclic increase of leukocyte alkaline phosphatase in females is related to estrogenic stimulation during the proliferative phase. Leukocyte alkaline phosphatase may serve as a test for intra and extra-uterine pregnancy because it also is elevated in these conditions, presumably on the basis of stimulation by estrogen.[10] Alkaline phosphatase is elevated in the serum in the pregnant female, especially during the second and third trimester. Serum alkaline phosphatase begins to rise at the fourth month of gestation. This elevation in the blood in the pregnant female is a reflection of placental alkaline phosphatase entering the maternal blood, since the human trophoblastic cells are rich in alkaline and acid phosphatase.[4,11] Adults who are cachectic and malnourished tend to have slight lowering of the serum alkaline phosphatase level. Alkaline phosphatase tends to increase following a carbohydrate or a fatty meal.[62]

It is essential that the sample of serum sent to the clinical laboratory for the determination of alkaline phosphatase be properly processed since great variation may occur in the test results if stringent adherence to certain precautions is not observed. The level of alkaline phosphatase activity may increase if the serum is kept at 20°C. for several hours. Thus, if it is not analyzed as soon as possible, it may show an increase in activity. This is especially true if the specimen is kept in a refrigerator for 12 to 24 hours. Some specimens have shown a 10 to 30 percent rise because of storage

Alkaline Phosphatase (Continued):

for 24 hours in a refrigerator. Another factor which may affect the alkaline phosphatase activity on storage is a loss or an increase in the carbon dioxide in the serum. Loss of carbon dioxide would increase the alkalinity of the specimen; gain of carbon dioxide would make the serum more acid and would inhibit the enzymatic activity.[26] It is important not to add excessive oxalate or fluoride to the specimen since both inhibit the activity of alkaline phosphatase.[2,52]

Causes for an elevated serum alkaline phosphatase are:

Physiologic:
1. Infant and children bone growth
2. Pregnancy

Pathologic:
1. Obstructive hepatobiliary tract disease, infiltrative lesions of liver
2. Osteoblastic lesions of bone
3. Primary or secondary hyperparathyroidism
4. Metastatic cancer, lymphoma, histiocytic lesion involving bone
5. Paget's disease
6. Healing bone fractures
7. Gastrointestinal lesions:
   Malabsorption
   Peptic ulcer
   Ulcerative colitis
8. Necrosis of lung, kidney
9. Neoplasms producing alkaline phosphatase isoenzyme of Regan
10. Intravenous use of plasma expander produced from human placenta
11. Hemolysis, slight
12. Diseases of the spleen, including infarction

The clinician usually considers the possibility of hepatic obstructive jaundice (Fig. 2-1) or osteomalacia of bone as the primary causes for elevation of serum alkaline phosphatase. Other causes are neoplastic diseases in which a unique isoenzyme termed the "Regan Isoenzyme" (Fig. 2-2) is produced by the neoplastic cells,[55] gastrointestinal disease such as peptic ulcer or ulcerative colitis, in which the enzyme is liberated from the diseased mucosa.[24,44] Intestinal alkaline phosphatase decreases in the serum during periods of fasting and increases after a fatty meal.[47] Intestinal alkaline phosphatase occurs mostly in the serum of individuals who are Group B and

Alkaline Phosphatase (Continued):

O secretors; 80 percent of these secretors contain serum intestinal alkaline phosphatase.[36] Prominent increases in serum alkaline phosphatase may occur in patients receiving intravenous serum protein.[4] This commercial protein is prepared from blood derived from human placenta which contains much alkaline phosphatase.

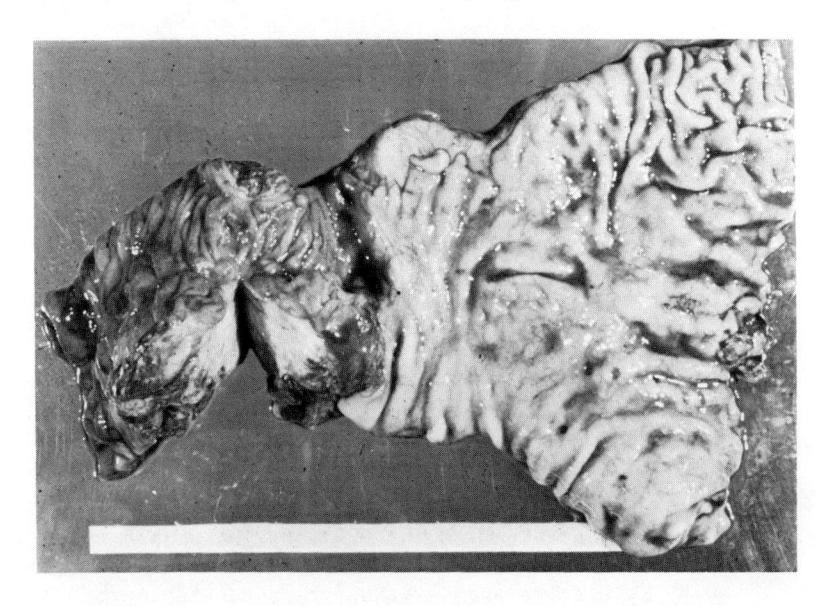

Fig. 2-1 Gross photograph of carcinoma of head of pancreas which caused obstructive jaundice with serum alkaline phosphatase of 450 I.U.

Alkaline Phosphatase (Continued):

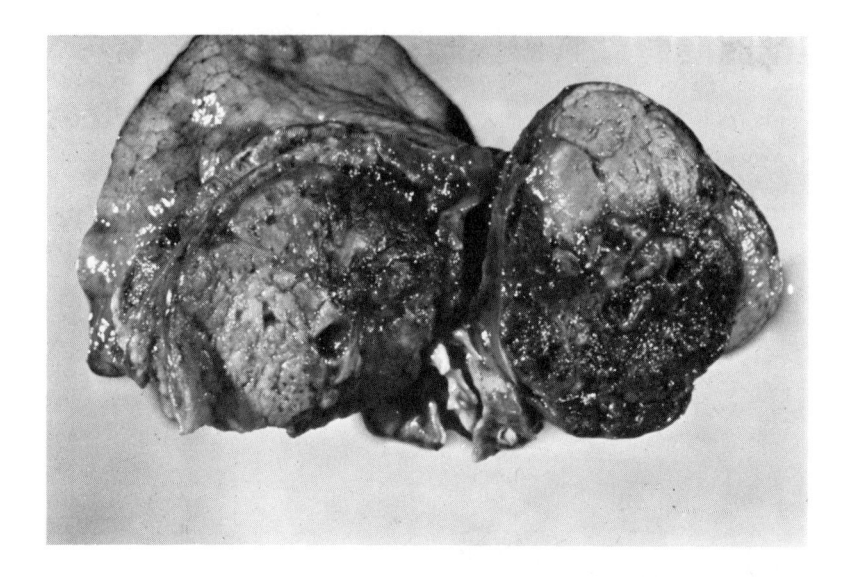

Fig. 2-2   Gross photograph of undifferentiated bronchogenic
           carcinoma which caused a serum alkaline phosphatase
           of 215 I.U. The enzyme was heat stable and inhibited
           by phenylalanine and was thus the Regan isoenzyme.

In considering elevations of alkaline phosphatase, one must always bear in mind the age of the patient because as previously noted, there are normal elevations of the enzyme in growing children and also that pregnant females in the second and third trimesters of pregnancy normally have an elevated serum alkaline phosphatase.

On electrophoresis, liver alkaline phosphatase migrates with the serum beta globulins and beta lipoproteins. Kidney alkaline phosphatase migrates in the haptoglobin zone. Bone alkaline phosphatase migrates in the beta globulin zone while intestinal alkaline phosphatase migrates in the beta globulin and beta lipoprotein zones. Patients with bone disease show alkaline phosphatase electrophoretic bands which migrate more slowly than the alkaline phosphatase derived from liver disease.[23]

Bone and kidney alkaline phosphatase are inhibited by urea; while placental and gastrointestinal are not. Liver alkaline

Alkaline Phosphatase (Continued):

phosphatase shows decreased activity when it is exposed to urea of less than 1.5 M concentration. The simplest method of differentiating the isoenzymes is to observe the stability of the enzyme after incubating it at 56°C. for 15 minutes. (Table 2-2)

TABLE 2-2

DIFFERENTIATION OF ALKALINE PHOSPHATASES

Source	Phenylalanine Inhibition	Heat Stability
Liver	Not Inhibited	Moderately Stable
Intestine	Inhibited	Moderately Stable
Placenta	Inhibited	Stable
Bone	Not Inhibited	Labile

The isoenzymes of alkaline phosphatase derived from liver and gastrointestinal sources are moderately heat stable, whereas the isoenzyme derived from the placenta and Regan tumor is markedly heat stable; that from a bone source is very heat labile.[29,45] Fitzgerald states that if the alkaline phosphatase level in the serum after incubation at 56°C. for 15 minutes is diminished by more than 60 percent, this is a demonstration of heat lability and suggests that the source of the enzyme was from bone.[18]

Another method for differentiating the isoenzymes of alkaline phosphatase is the utilization of phenylalanine to inhibit the action. Alkaline phosphatase from a placental source is inhibited by phenylalanine as is intestinal alkaline phosphatase. The Regan isoenzyme is produced by neoplastic cells such as bronchogenic carcinoma, ovarian tumors, or other malignant cells including lymphomas such as Hodgkin's disease. It is thought that the neoplastic cells produce this isoenzyme and secrete it into the blood stream which then causes the patient to present with an elevated alkaline phosphatase. The Regan isoenzyme is similar to the placental one in that it is heat stable and inhibited by phenylalanine. Isoenzymes derived from liver and bone sources are not inhibited by phenylalanine.[31]

In bone disease with an elevation of alkaline phosphatase, the pathophysiologic mechanism is injury to bone and a reparative process with proliferation of osteoblasts producing alkaline phosphatase which is secreted into the blood stream. An enzyme increase may signify the presence of osteomalacia with inadequate mineralization and an active attempt to repair the injury. Osteopetrosis causes an

Alkaline Phosphatase (Continued):

elevation of serum acid phosphatase but not of alkaline phosphatase.[17] The outstanding causes for elevated serum alkaline phosphatase with bone disease are hyperparathyroidism related to primary hyperplasia (Fig. 2-3), adenoma (Fig. 2-4), or carcinoma of the parathyroid glands, rickets, Paget's disease, metastatic carcinoma to bone from primary sources such as breast, lung, kidney, prostate or thyroid, and primary malignancy of bone. In addition, trauma and fracture of the skeletal system with osteoblastic repair will produce alkaline phosphatase elevation.[6,61] Surgical repair and manipulation of bone for example, insertion of intramedullary nails, will elevate alkaline phosphatase. Determination of the enzyme in the work-up of a patient with vitamin D deficiency leading to rickets is one of the important aids in the assessment of this disease and the level serves as an index of the severity of the disorder.[5] Clinicians have noted that a progressive and rapid decline in elevated serum alkaline phosphatase is indicative of effective therapy. The serum alkaline phosphatase is consistently elevated when rickets is active. Prominent elevation of the enzyme level above 200 I.U. has been reported. The level begins to decline within several days after vitamin D therapy is instituted.

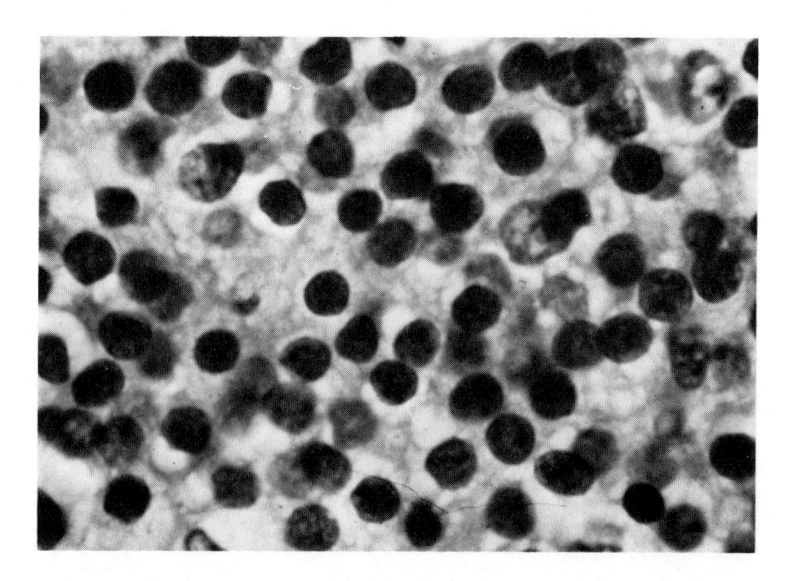

Fig. 2-3 Photomicrograph of primary hyperplasia of parathyroid gland causing a heat labile serum alkaline phosphatase of 135 I.U.

Alkaline Phosphatase (Continued):

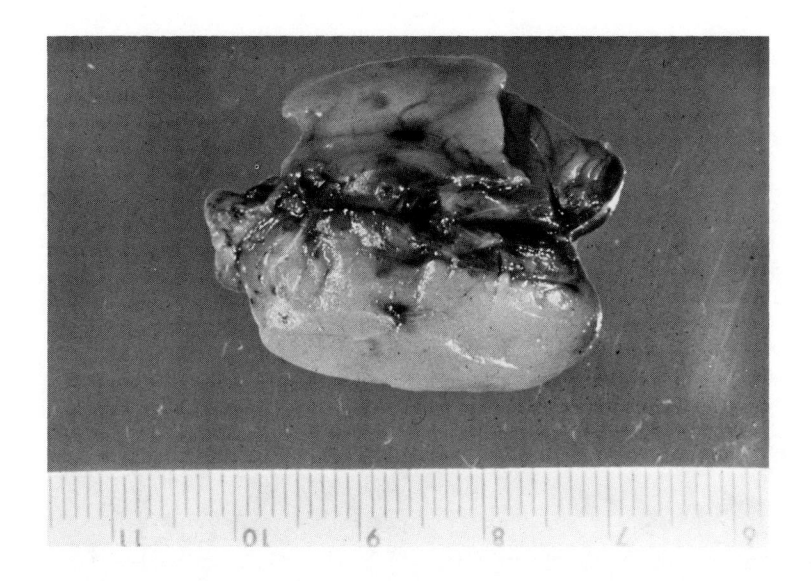

Fig. 2-4   Gross photograph of a parathyroid adenoma which
          caused an elevated serum alkaline phosphatase of
          115 I.U.

Extreme elevations of alkaline phosphatase are constantly seen in osteitis deformans (Paget's disease) (Fig. 2-5). Among the bone diseases, the highest values for this enzyme have been seen in Paget's disease.[38] Higher values are generally seen in younger individuals with Paget's disease; patients who are elderly may show only slight elevations. The degree of elevation can also be utilized in a prognostic way since polyostotic disease shows greater activity than the monostotic form. Slightly elevated levels are usually associated with quiescent stages of the disease.

Alkaline Phosphatase (Continued):

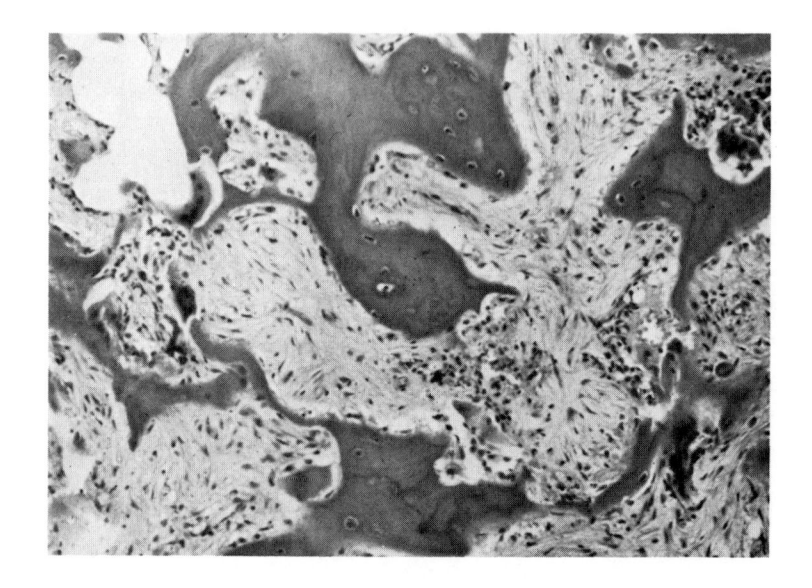

Fig. 2-5 Photomicrograph of section of bone demonstrating mosaic pattern of Paget's disease. The serum alkaline phosphatase in the patient was 240 I.U. and was heat labile.

Subtotal gastric resection and malabsorption leads to osteomalacia and an elevated serum alkaline phosphatase. The cause is lower intake of calcium and vitamin D.[65]

Various other bone disorders cause elevation of the serum alkaline phosphatase. Metastatic carcinoma involving the bones and primary malignant neoplasms may elevate the serum alkaline phosphatase. In addition, the histiocytic diseases and healing of bone fractures are associated with elevated alkaline phosphatase. In these condition a response to appropriate therapy is usually associated with an initia rise in the enzymes shortly after therapy is begun which reflects the successful reparative process. In contrast, osteoporosis is usually associated with normal serum phosphatase.

Alkaline Phosphatase (Continued):

Serum alkaline phosphatase may be elevated with various diseases of the spleen since the enzyme is present in large amounts in the spleen especially in vascular endothelium. Likewise, pulmonary infarction will result in liberation of endothelial enzyme, especially during the phase of organization. Infarction of the kidney will cause an elevation of serum alkaline phosphatase from necrotic renal tubules.

Causes for a decreased serum alkaline phosphatase levels are:
1. Hypophosphatasia
2. Magnesium deficiency
3. Cachexia
4. Pernicious anemia
5. Hypothyroidism
6. Anticoagulants
   Fluorides
   Oxalates
   EDTA
7. High serum phosphate
8. Glycine buffer in chemical testing

The major cause for a low serum alkaline phosphatase is hypophosphatasia.[57] In this inherited condition, there is a marked reduction or absence in the alkaline phosphatase in myeloid leukocytes, the serum, and the tissues including bone. Clinically the condition resembles rickets.[43] The disease varies in severity. Patients with a mild condition may survive into adult life, but they may have deformities of their bones with increased fragility resulting in frequent fractures. Hypercalcemia with calcinosis of the soft tissues may result. In contrast, the condition may be extremely severe at the time of birth, and death may result shortly after birth or in early childhood. The diagnostic features in the blood are low leukocyte alkaline phosphatase and marked depression of the alkaline phosphatase in the serum. The presence of low leukocyte and serum alkaline phosphatase quickly differentiates this condition from rickets. Another characteristic biochemical abnormality is the excretion of phosphoethanolamine in the urine. This substance is a natural substrate for alkaline phosphatase; because of the deficiency in the enzyme, the substrate accumulates and is excreted in the urine.

Another cause of a lower serum alkaline phosphatase is magnesium deficiency which is at times associated with 1). gastrointestinal disease such as chronic diarrhea or malabsorption, 2). uncontrolled diabetes mellitus in which parenteral fluids have been administered without the addition of magnesium ion, or 3). other conditions in

Alkaline Phosphatase (Continued):

which parenteral fluids have been utilized without replacement of magnesium leading to a prominent magnesium deficiency. Magnesium is one of the ions necessary for the function of alkaline phosphatase; a magnesium deficiency results in a lowered serum alkaline phosphatase. A lower serum level is also present in malnourished cachectic individuals and in cases of pernicious anemia. It has been postulated that an inadequate amount of vitamin B_{12} impairs osteoblastic activity and this results in the lower production of alkaline phosphatase.[60] Some investigators have noted a lower serum alkaline phosphatase in patients who are hypothyroid. This may result from stunted bone growth.[9] The alkaline phosphatase activity of the thyroid gland appears to be related to the functional activity of the gland. Greater alkaline phosphatase activity is present in hyperplastic glands and a lower amount in atrophic glands. In hypothyroidism, there is an underlying deficiency in osteoblastic activity resulting in a deficiency of the serum alkaline phosphatase level. A decrease in endometrial alkaline phosphatase is present in endometrial carcinoma. The serum and leukocyte alkaline phosphatase also falls.

Acid Phosphatase

Acid phosphatase is present in many tissues. The greatest amount is found in the prostate gland which contains approximately a hundred times more acid phosphatase than do other tissues.[7,32]

Acid phosphatase is found in greater amounts in serum than in plasma because the enzyme is liberated from platelets.

Acid phosphatase is unstable above pH 7.0 and also at warm temperatures. When frozen, it is stable for 3 months and is stable for 2 weeks at 0 to 4°C.

Cerebrospinal fluid does not normally contain acid phosphatase but may show activity in patients with osteolytic vertebral cancer, metastases or multiple myeloma.

Causes for an elevated serum acid phosphatase level are:
1. Physiological
   Bone growth
   Pregnancy
2. Metastatic carcinoma of the prostate
3. Bone lesions, osteolytic and osteoblastic
   Hyperparathyroidism
   Paget's disease
   Primary and secondary bone cancer

Acid Phosphatase (Continued):

4. Hepatic disease
   Metastatic cancer
   Viral hepatitis
   Cirrhosis
5. Myocardial infarction
6. Renal disease
7. Hematologic disorders
   Hemolytic anemia
   Multiple myeloma
   Thrombocytopenia
   Thromboembolism
8. Reticuloendothelial disease
   Gaucher's disease
   Niemann-Pick disease
   Eosinophilic granuloma
   Histiocytic lymphoma
   Hodgkin's disease

Causes for low serum acid phosphatase levels are:
1. Anticoagulants
   Oxalate
   Fluoride
   Heparin
2. Fever due to various etiologies

Determination of serum acid phosphatase is a useful procedure primarily in the assessment of metastatic carcinoma of the prostate. Approximately 20 percent of patients with this disease will at times fail to show elevation of serum acid phosphatase.[7,22] In order to stimulate production of the enzyme by these metastases the Sullivan Technique is rarely employed. In this procedure 25 mg. of testosterone per day is administered for 5 days.[8] Slight increase in the serum acid phosphatase after such treatment may be of diagnostic value. Ninety percent of patients with metastatic prostatic carcinoma had elevated acid phosphatase when there was metastases to distinct sites (Fig. 2-6). Tartrate-inhibited acid phosphatase has been considered specific for metastatic carcinoma of the prostate, but a recent report has challenged this. A new substrate, thymolphthalien monophosphate has marked specificity for the prostatic isoenzyme.

Acid Phosphatase (Continued):

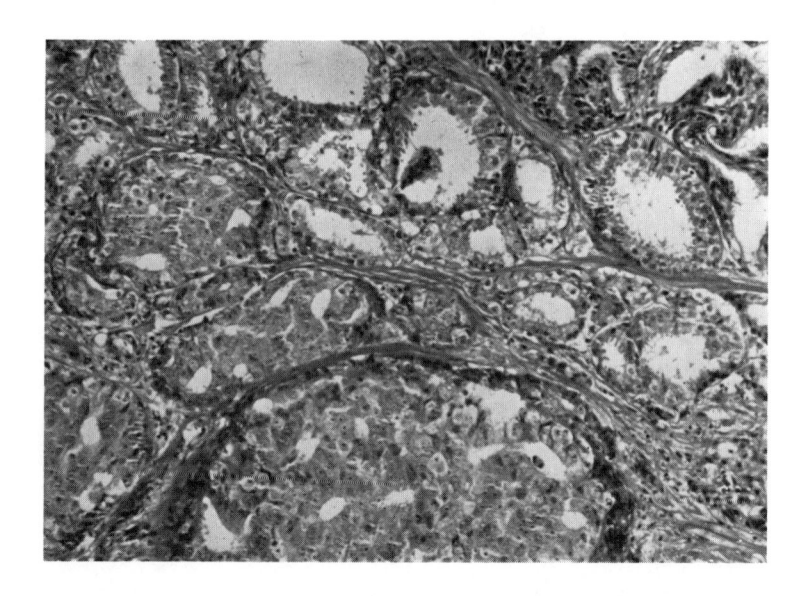

Fig. 2-6 Photomicrograph of carcinoma of the prostate. The lesion metastasized to bone and the patient presented with a tartrate inhibited acid phosphatase of 1.7 I.U. and a heat labile alkaline phosphatase of 220 I.U.

It is important to follow the effectiveness of therapy in patients who have carcinoma of the prostate by serial determinations of the serum acid phosphatase. When the enzyme level has been elevated prior to an orchiectomy, a marked decrease in the level should be seen approximately 4 to 6 weeks after the surgery. Estrogen therapy will also induce a decrease. If acid phosphatase levels remain elevated following castration, estrogen therapy, or both, the therapy has not been successful.[15,27] If therapy is successful and there is a relapse of the carcinoma, serum acid phosphatase may again rise. In hyperplasia of the prostate gland, simple manipulation of the gland may cause an elevation of the serum acid phosphatase. Elevation of the enzyme level following a rectal examination was more frequent in patients with cancer of the prostate than in patients without malignancies.[28] Rectal examination and palpation of the nondiseased prostate gland may produce a slight increase in the

Acid Phosphatase (Continued):

tartrate inhibited fraction of the acid phosphatase.[12,16,19] Rarely, infarction of the prostate has been associated with a rise in the acid phosphatase.[54] In addition, mechanical trauma to the prostate by either surgery or catheterization may produce transient elevation of the enzyme level.

The increase in serum acid phosphatase may be summarized in the following manner as it relates to carcinoma of the prostate.

1. It may corroborate the diagnosis of metastatic prostatic carcinoma.

2. It may provide initial evidence of metastases from a prostatic carcinoma. It may indicate the prostatic origin of a metastatic lesion when the primary site of the malignancy is not clinically detectable. Serum acid phosphatase should in most patients be elevated and it should regress after successful therapy is instituted.

3. It is a useful test to follow the course of treatment in a patient with metastatic carcinoma of the prostate.

Other diseases may cause a rise in the serum acid phosphatase,[34] for example, prominent destruction of blood platelets,[67,41] as is seen in acute thrombocytopenic purpura. The determination of acid phosphatase in patients with thrombocytopenia may be important.[41] Those patients exhibiting thrombocytopenia with high serum acid phosphatase have been found to have normal or elevated numbers of megakaryocytes in the bone marrow (Fig. 2-7). In contrast, those patients with thrombocytopenia with diminished numbers of megakaryocytes in the bone marrow will not develop an elevated serum acid phosphatase.

Acid Phosphatase (Continued):

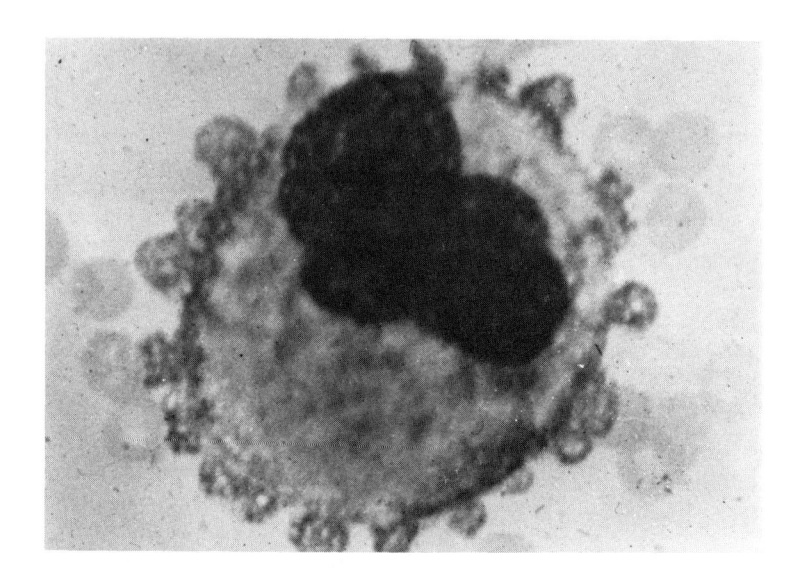

Fig. 2-7   Photomicrograph of abnormal megakaryocyte found in smear of bone marrow in idiopathic thrombocytopenic purpura. The platelet count was 5,000 per cu. mm. and the serum acid phosphatase was 1.9 I.U.

Another clinical situation in which acid phosphatase is elevated is Gaucher's disease, [59] in which there is a cerebrosidase deficiency with histiocytes containing kerasin. The acid phosphatase that is present in the serum is not inhibited by L-tartrate or formaldehyde. The elevated serum enzyme most likely arises from the Gaucher cells (Fig. 2-8). Gaucher-like cells are also found in chronic myelogenous leukemia and thalassemia.

Acid Phosphatase (Continued):

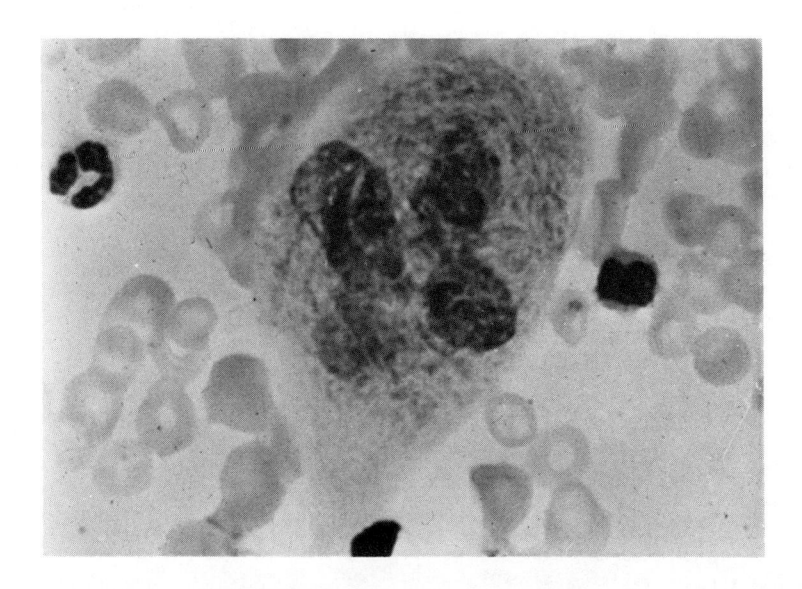

Fig. 2-8 Photomicrograph of smear of bone marrow from a patient with Gaucher's disease. The marrow was filled with Gaucher cells and the serum acid phosphatase was 2.3 I.U., which was not inhibited by L-tartrate.

Acid phosphatase is also elevated when there is necrosis of tissue following thromboembolism. The acid phosphatase in this situation may be derived from necrotic tissue such as necrotic myocardial cells in a myocardial infarct or from the breakdown of red blood cells and platelets in the area of infarction.

Patients with hemolytic anemia may have an elevation in the serum acid phosphatase. Red blood cells contain acid phosphatase which is liberated from the hemolyzed cells.

Various types of bone disease such as Paget's disease may be associated with an elevated serum acid phosphatase especially if the disease is polyostotic.[20] In addition, patients with extensive bone destruction such as multiple myeloma or bone cancer, may present

Acid Phosphatase (Continued):

with an elevated serum acid phosphatase. Normally the cerebrospinal fluid does not have acid phosphatase but it is elevated in the cerebrospinal fluid in multiple myeloma or metastatic cancer to the vertebrae.

Rarely acid phosphatase may be elevated in patients with liver or kidney disease. The acid phosphatase is liberated from necrotic renal tubules, and if there is oliguria or anuria, there will be inability to excrete the enzyme. In liver disease such as viral hepatitis, cirrhosis, or metastatic cancer, there are reports of elevated serum acid phosphatase; the acid phosphatase is most likely being liberated from necrotic liver cells.

Patients with trisomy have increased acid phosphatase in their abnormal polymorphonuclear leukocytes.

The physiological causes for elevated serum acid phosphatase are bone growth (the serum level is twice the adult level in the growing child) and pregnancy due to production and liberation of the enzyme from the placenta.

The chemical techniques used to differentiate the isoenzymes of acid phosphatase are as follows:[1,13,48]

1. L-tartrate inhibits 95 percent of prostatic acid phosphatase and 70 to 90 percent of liver, kidney, and spleen acid phosphatase.

2. Formaldehyde inhibits erythrocyte acid phosphatase but does not affect the prostatic, hepatic, renal, or splenic isoenzyme.

3. Ethanol inhibits the prostatic and erythrocyte isoenzymes but not the hepatic or splenic ones.

4. Urine acid phosphatase arises from prostate and kidney and is inhibited by L-tartrate.

5. A new substrate thymolphthalein monophosphate is hydrolyzed only by the prostatic isoenzyme and is specific for prostatic disease.[48]

The enzyme is relatively unstable in the presence of heat and alkaline pH. The effect of normal body temperature and pH are considered sufficient to maintain tissue and serum enzyme equilibrium

Acid Phosphatase (Continued):

in health. This accounts for inactivation of small amounts released by non-neoplastic abnormal prostate glands or by other diseased tissues containing acid phosphatase. An interesting association of fever and elevations of serum acid phosphatase has been observed. In patients with elevated serum acid phosphatase, the onset of fever will cause a fall in the level of the enzyme.[63] As the fever regresses, the serum acid phosphatase will rise again.

5'-Nucleotidase

5'-nucleotidase is a hepatobiliary tract enzyme and has specificity for nucleoside monophosphate substrates.[30] It should be determined when there is a question of hepatobiliary disease and when there is elevation of serum alkaline phosphatase. The 5'-nucleotidase activity has a normal value of up to 1.6 units, with the same normal level in children, adolescents, and adults. This enzyme is not elevated in bone disease. Thus, when there is an elevation of alkaline phosphatase, and it is necessary to differentiate between bone and liver disease,[14] determination of the 5'-nucleotidase is recommended.

Elevated levels are usually present in obstructive disease of the hepatobiliary system.[51] Young studied 125 jaundiced patients and found that 90 percent of the patients with jaundice on a hepatic basis had enzyme activities lower than 10 units whereas, 95 percent of the obstructive jaundice cases had activity greater than 10 units.[66] The serum level is slightly raised in advanced hepatocellular disease associated with cirrhosis. 5'-nucleotidase is usually elevated in neoplastic disease involving the liver. In a recent study, 5'-nucleotidase levels were elevated in obstruction of the hepatobiliary tree by intrahepatic or extrahepatic conditions, primary or secondary cancer of the liver, hepatitis, or cirrhosis.[3]

Phosphate and EDTA, which inhibit alkaline phosphatase have no effect on 5'-nucleotidase. Pregnancy and the early postpartum period do not alter or raise 5'-nucleotidase activity whereas both non-specific alkaline phosphatase and leucine aminopeptidase are increased at these times.

Alkaline phosphatase and 5'-nucleotidase have been found to be elevated in patients with lymphoma without evidence of liver involvement; actually hepatic involvement might be subclinical.

Leucine Aminopeptidase

Aminopeptidases hydrolyze amino acids containing alpha amino groups. The highest concentration of leucine aminopeptidase is usually found in the pancreas and liver. The normal adult ranges for LAP are: 75 - 230 units for men and 80 - 210 units for women.[56]

Increased leucine aminopeptidase levels are found in:
1. Carcinoma of the head of the pancreas
2. Diffuse carcinoma, lymphoma, and granulomata involving the liver
3. Choledocholithiasis
4. Acute pancreatitis causing obstruction of the common bile duct
5. Pregnancy

The measurement of serum leucine aminopeptidase activity was introduced in the late 1950's by Rutenberg et. al. who utilized the substrate 2-leucyl-β-naphylamide.[49] The enzyme is found in renal tubules, mucosa of small intestine, colon, stomach, bile duct epithelium and pancreatic acini. Originally the determination for leucine aminopeptidase was proposed as a laboratory test for the detection of pancreatic carcinoma. However, it was subsequently observed that only carcinoma of the head of the pancreas, and not the body or the tail, produced an elevation of this enzyme in the serum. It was also found that any condition causing biliary obstruction, either intrahepatic or extra-hepatic would cause an elevated serum leucine aminopeptidase level.[30] Thus, metastatic carcinoma to the liver would also elevate the level. Since the enzyme is not affected by bone disease, this determination is useful when there is an elevated serum alkaline phosphatase and one is not certain whether bone or liver disease is present.[3,25,46]

The chief value of determining LAP is its ability to discriminat between hepatobiliary tract and other diseases. LAP elevations are limited to hepatobiliary disease and pregnancy. Thus, LAP determination has advantages over tests for BSP, glutamic oxaloacetic transaminase, glutamic pyruvic transaminase, the flocculation tests and alkaline phosphatase.

The determination of leucine aminopeptidase has a special place in the jaundiced newborn for the differentiation of neonatal hepatitis (Fig. 2-9) and biliary atresia (Fig. 2-10). In this clinical situation, the enzyme determination ranks with the ^{131}I labeled Rose Bengal test as a diagnostic test to differentiate biliary atresia

Leucine Aminopeptidase (Continued):

from neonatal hepatitis presenting with jaundice. In patients with neonatal hepatitis, the values for leucine aminopeptidase are generally lower than 500 I.U.; those with biliary atresia have levels greater than 500 I.U.

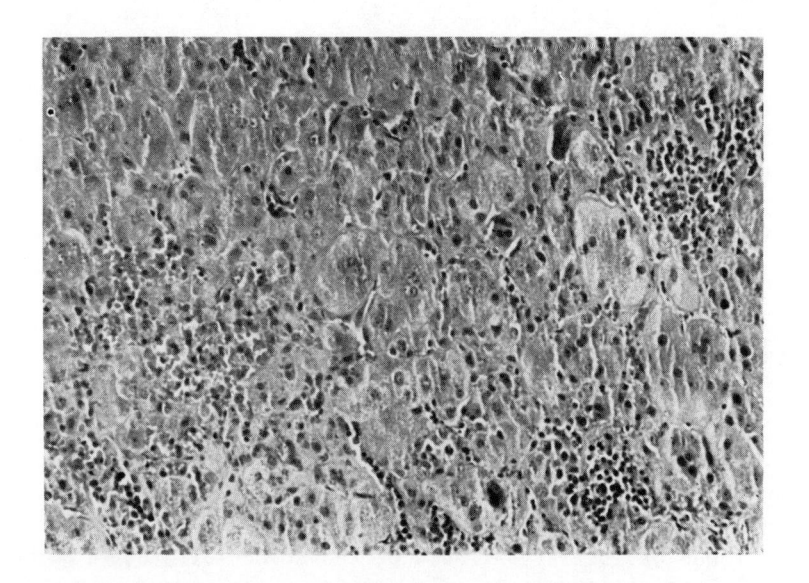

Fig. 2-9　Photomicrograph of section of liver of infant affected by viral giant cell hepatitis. The serum leucine aminopeptidase was elevated to 385 I.U.

The determination of the leucine aminopeptidase isoenzymes may also be helpful in diagnosis of biliary atresia.[39] In neonatal giant cell hepatitis, electrophoresis gives one zone of LAP activity corresponding to the post-albumin zone. In contrast in biliary atresia, electrophoresis gives two zones, an intensified zone as in giant cell hepatitis which extends to the alpha globulin zone, and a second band between the α_2 and globulin zones.

Leucine Aminopeptidase (Continued):

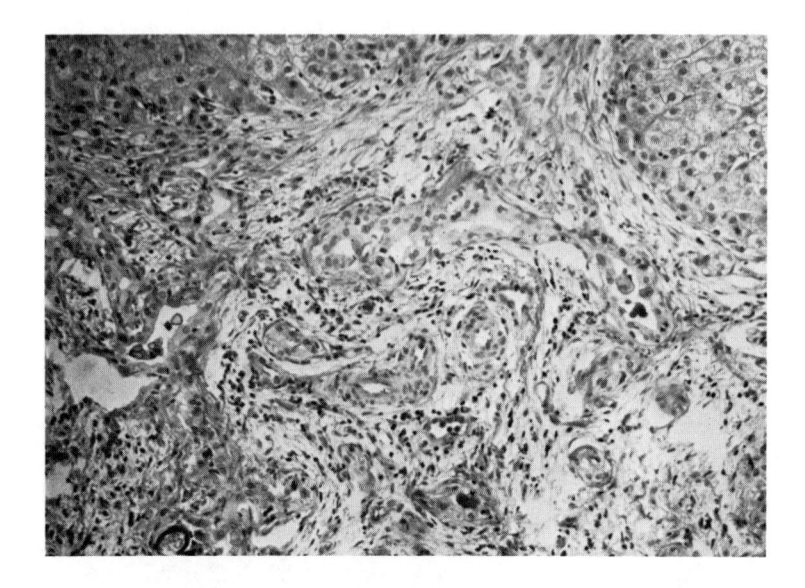

Fig. 2-10 Photomicrograph of section of liver from an infant who had biliary atresia of the common bile duct. Note the prominent bile stasis in the liver. The serum leucine aminopeptidase was 595 I.U.

Leucine aminopeptidase will at times be elevated before an elevation of alkaline phosphatase is observed in patients suffering from obstructive jaundice. It may be elevated without a simultaneous elevation of alkaline phosphatase in non-jaundiced individuals; in this situation, carcinoma of the head of the pancreas is strongly suggested. Patients with carcinoma of the body or tail of the pancreas usually do not demonstrate any serum LAP elevation.

The gastrointestinal tract contains alkaline phosphatase, 5' nucleotidase, and leucine aminopeptidase. Ulcerative colitis may be associated with ascending cholangitis, and patients with this combination would have an elevation of all three enzymes. With severe and diffuse gastrointestinal disease only alkaline phosphatase would be elevated.

Leucine Aminopeptidase (Continued):

Two aminopeptidases may be elevated in the serum of pregnant women, especially in the latter stages of pregnancy. These are leucine aminopeptidase and cystine aminopeptidase,[37] also known as oxytocinase. A number of recent investigations however, have challenged the findings that LAP activity increases during pregnancy. Page and Siegel for example, failed to find an increase.[42,53] Others have found that rise in activity is most likely due to enhanced and combined activity of both LAP and CAP. Thus, an apparent increase in serum LAP activity during pregnancy may be a summation of the activity of both CAP and LAP and can be differentiated only by the employment of specific substrates for these two enzymes. Two forms of CAP exist and have been differentiated by starch gel electrophoresis. They are designated cystine aminopeptidase$_1$ and cystine aminopeptidase$_2$. CAP$_1$ has its greatest activity early in pregnancy but declines as the pregnancy progresses; in contrast, CAP$_2$ has its greatest activity during the third trimester.

The increasing activity of CAP throughout pregnancy has pointed to a possible relation of this enzyme to the quiescence of the uterus during this period. Hilton has speculated that a decrease of CAP prevents the inactivation of oxytocin, which thereby initiates labor.[23] This idea is supported by his finding of a rapid decline of CAP at the onset of labor.[58] CAP is absent in fetal serum.

Page has utilized the determination of serum CAP as a pregnancy test; using this technique pregnancy can be diagnosed as early as 4 weeks following conception. Melander has reported an accuracy of 80 percent 16 days after conception and 100 percent at the end of the second month.[37]

In summary, leucine aminopeptidase is a valuable enzyme to determine obstructive hepatobiliary disease, which is caused by extra-hepatic causes such as common duct calculus, or carcinoma of the head of the pancreas. Intra-hepatic obstructive disease due to malignant neoplasms, malignant lymphoma, or granulomatous disease may also cause leucine aminopeptidase to be elevated. Leucine aminopeptidase is also elevated in the blood of pregnant females as is cystine aminopeptidase. The source of the enzyme in pregnant females is the placenta.

Gamma Glutamyl Transpeptidase

This enzyme is usually elevated in patients with hepatobiliary tract disease. Elevated levels are usually associated with metastatic

Gamma Glutamyl Transpeptidase (Continued):

cancer to the liver and common bile duct obstruction due to neoplasms. Icteric or non-icteric patients may have high values if there is a hepatobiliary intra or extra-hepatic obstruction. The increased levels are partly due to interference with excretion. Various other liver diseases may be associated with increased values. Recently high values have been found in post-myocardial infarction patients, and in patients with neurologic disease in which highly vascular or necrotic changes in vascular endothelium have been demonstrated.[21]

REFERENCES

1. Abul-Fadl, M. A. M., King, E. J.: "Properties of the Acid Phosphatases of Erythrocytes and the Human Prostate". Biochem. J. 45:51, 1949.

2. Bahr, M., Wilkinson, J. H.: "Urea as a Selective Inhibitor of Human Tissue Alkaline Phosphatases". Clin. Chim. Acta 17:367, 1967.

3. Bardawil, C., Chang, C.: "Serum Lactic Dehydrogenase, Leucine Aminopeptidase, and 5'Nucleotidase Activities: Observations in Patients with Carcinoma of the Pancreas and Hepatobiliary Disease". Canadian Med. Assn. J. 89:755, 1962.

4. Bark, C. J.: "Artificial Serum Alkaline Phosphatase from Placental Albumin". J. Amer. Med. Assoc. 207:953, 1969.

5. Bodansky, A., Jaffe, H. L.: "Phosphatase Studies V. Serum Phosphatase as a Criterion of the Severity and Rate of Healing of Rickets". Amer. J. Dis. Child. 48:1268, 1934.

6. Botterell, E. H., King, E. J.: "Phosphatase in Fractures". Lancet 1:1267, 1935.

7. Cantarow, A., Trumper, M.: CLINICAL BIOCHEMISTRY, Saunders Co., pg. 458, 1965.

8. Cantarow, A., Trumper, M.: CLINICAL BIOCHEMISTRY, Saunders Co., pg. 459, 1965.

9. Cassar, J., Joseph, S.: "Alkaline Phosphatase Levels in Thyroid Disease". Clin. Chim. Acta. 23:33, 1969.

References (Continued):

10. Climie A. R. W., et. al.: "Neutrophilic Alkaline Phosphatase Test: A Review with Emphasis on Findings in Pregnancy". Am. J. Clin. Path. 38:95, 1962.

11. Colombo, J. P., Richterich, R., Rossi, E.: "Serum-Kreatin-Phosphokinase: Bestimmung and diagnostische Bedeutung". Klin. Wschr. 40:37, 1962.

12. Daniel, O., Van Zyl, J. S.: "Rise of Serum Acid Phosphatase Level following Palpation of Prostate". Lancet 1:998, 1952.

13. Delory, G. E., Sweetser, T. H., White, T. A.: "The Use of Formalin and Alcohol in the Estimation of Prostatic Phosphatase". J. Urol. 66:724, 1951.

14. Dixon, T. F., Furdom, M.: "Serum 5'-Nucleotidase". J. Clin. Path. 7:341, 1954.

15. Doe, R. P., Seal, U. S.: "Acid Phosphatase in Urology". Surg. Clin. N. Amer. 45:1455, 1965.

16. Dybkaer, R., Jensen, G.: "Acid Phosphatase Levels following Massage of the Prostate". Scand. J. Clin. Lab. Invest. 10:349, 1958.

17. Ensign, D. C.: "Serum Phosphatase in Osteopetrosis". J. Lab. Clin. Med. 32:1541, 1947.

18. Fitzgerald, M., Fennelly, J. J., McGenney, K.: "The Value of Differential Alkaline Phosphatase Thermostability in Clinical Diagnosis". Amer. J. Clin. Path. 51:194, 1969.

19. Glenn, J. F., Spanel, D. L.: "Serum Acid Phosphatase and Effect of Prostatic Massage". J. Urol. 82:240, 1959.

20. Gutman, A. B., Gutman, E. B., Robinson, J. N.: "Determination of Serum Acid Phosphatase Activity in differentiating Skeletal Metastases secondary to Prostatic Carcinoma from Paget's Disease of Bone". Amer. J. Cancer 38:103, 1940.

21. Gambino, S. R., Lum, G.: "Serum Gamma-Glutamyl Transpeptidase Activity as an Indicator of Disease of Liver, Pancreas, or Bone". Clin. Chem. 18:358, 1972.

References (Continued):

22. Hill, J. H.: "Prostatic Serum Acid Phosphatase in Patients with Localized Prostatic Cancer". Amer. J. Clin. Path. 26:120, 1956.

23. Hilton, J. G., Johnson, R. F.: "Changes in Blood Oxytocinase during Postpartum Period". Amer. J. Obstet. & Gynec. 78:479, 1959.

24. Hodson, A. W., Latner, A. L., Raine, L.: "Isoenzymes of Alkaline Phosphatase". Clin. Chim. Acta. 7:255, 1962.

25. Hoffman, E., Nachlas, M. M., Gaby, S. D., Abrams, S. J., Seligman, A. M.: "Limitations in the Diagnostic Value of Serum Leucine Aminopeptidase". New Eng. J. Med. 263:541, 1960.

26. Horne, M., Cornish, C. J., Posen, S.: "Use of Urea Denaturation in the Identification of Human Alkaline Phosphatases" J. Lab. Clin. Med. 72:905, 1968.

27. Huggins, C., Hodges, C. V.: "Studies on Prostatic Cancer. Effect of Castration, of Estrogen, and of Androgen Injection on Serum Phosphatases in Metastatic Carcinoma of Prostate". Cancer Res. 1:293, 1941.

28. Kendall, A. R.: "Acid Phosphatase Elevation following Prostatic Examination in the Earlier Diagnosis of Prostatic Carcinoma". J. Urol. 86:442, 1961.

29. Kerkhoff, J. F.: "A Rapid Serum Screening Test for Increased Osteoblastic Activity". Clin. Chim. Acta. 22:231, 1968

30. Kowlessar, O. D., Haeffner, L. J., Riley, E. M.: "Localization of Serum Leucine Aminopeptidase, 5'-Nucleotidase and Non-specific Alkaline Phosphatase by Starch-gel Electrophoresis. Clinical and Biochemical Significance in Disease States". Ann. N. Y. Acad. Sci. 94:836, 1961.

31. Kreisher, J. H., Close, V. A., Fishman, W. H.: "Identification by means of L-phenylalanine Inhibition of Intestinal Alkaline Phosphatase Components separated by Starch-gel Electrophoresis of Serum". Clin. Chim. Acta. 11:122, 1965.

References (Continued):

32. Kutscher, W., Worner, A.: "Prostatic Phosphatase. Mitteilung Hoppe-seyler Zeschr". Physiol. Chem. 239:109, 1936.

33. Latner, A. L., Skillen, A. W.: ISOENZYMES IN BIOLOGY AND MEDICINE. Academic Press, 1968.

34. Lepow, H., Schoenfeld, M. R., Messeloff, C. R., Chu, F.: "Nonprostatic Causes of Acid Hyperphosphatasemia: Report of a Case due to Multiple Myeloma". J. Urol. 87:991, 1962.

35. McCance, R. A., Morrison, A. B., Dent, C. E.: "The Excretion of Phosphoethanolamine and Hypophosphatasia". Lancet 1:131, 1955.

36. Marcus, D. M.: "The ABO and Lewis Blood-group System: Immunochemistry, Genetics and Relation to Human Disease". New Eng. J. Med. 280:994, 1969.

37. Melander, S.: "Plasma Oxytocinase Activity. A Methodological and Clinical Study with Special Reference to Pregnancy". Acta. Endocr. (Kobenhavn) Suppl. 96:94, 1965.

38. Nagant de Deuxchaisnes, C., Krane, S. M.: "Paget's Disease of Bone: Clinical and Metabolic Observations". Medicine 43:233, 1964.

39. Natoli, G., Natoli, V., Lapi, A. S., Mancini, G., Renzulli, F.: "LAP Isoenzymes in Neonatal Hepatitis and Biliary Atresia". Lancet 2:209, 1969.

40. O'Brien, D., Ibbott, F. A., Rodgerson, D. O.: LABORATORY MANUAL OF PEDIATRIC MICROBIOCHEMICAL TECHNIQUES. 4th Ed. Harper & Row, 1968.

41. Oski, F. A., Norman, J. L., Diamond, L. K.: "Use of Plasma Acid Phosphatase Value in the Differentiation of Thrombocytopenic States". New Eng. J. Med. 268:1423, 1963.

42. Page, E. W., Titus, M. A., Mohun, G., Glending, M. B.: "The Origin and Distribution of Oxytocinase". Amer. J. Obstet. & Gynec. 82:1090, 1961.

References (Continued):

43. Pimstone, B., Eisenberg, E., Silverman, S.: "Hypophosphatasia: Genetic and Dental Studies". Ann. Intern. Med. 65:722, 1966.

44. Posen, S.: "Alkaline Phosphatase". Ann. Intern. Med. 67:183, 1967.

45. Posen, S. Neale, F. C., Clubb, J. S.: "Heat Inactivation in the Study of Human Alkaline Phosphatases". Ann. Intern. Med. 62:1234, 1965.

46. Pruzanski, W., Fischi, J.: "The Evaluation of Serum Leucine Aminopeptidase Estimation". Amer. J. Med. Sci. 248:581, 1964.

47. Pulverataft, C. N., Luffman, J. E., Robson, E. B., Harris, H., Langman, M. J. S.: "Isoenzymes of Alkaline Phosphatase in Patients operated upon for Peptic Ulcer". Lancet 1:237, 1967.

48. Roy, A. V., Brower, M. E., Hayden, J. E.: "Sodium Thymolphthalein Monophosphate: A New Acid Phosphatase Substrate with Greater Specificity for the Prostatic Enzyme in Serum". Clin. Chem. 17:1093, 1971.

49. Rutenberg, A. M., Goldbarg, J. A., Pineda, E. P.: "Leucine aminopeptidase Activity Observations in Patients with Cancer of the Pancreas and other Disease". New Eng. J. Med. 259:469, 1958.

50. Schapira, G., Dreyfus, J. C.: "The Serum Enzymes and Diseases of Striated Muscle". PROCEEDINGS FIRST EUROPEAN SYMPOS. MED. ENZYMOLOGY. Milan. Academic Press, pg. 119, 1961.

51. Schwartz, M. K., Bodansky, O.: "Serum 5 Nucleotidase in Patients with Cancer". Cancer 18:886, 1965.

52. Searcy, R. L.: DIAGNOSTIC BIOCHEMISTRY. McGraw-Hill, 1969

53. Siegel, I. A.: "Leucine Aminopeptidase in Pregnancy". Amer. J. Obstet. & Gynec. 14:488, 1959.

References (Continued):

54. Steward, C., Sweetser, T. H., Delory, G. E.: "A Case of Benign Prostatic Hypertrophy with Recent Infarcts and Associated High Serum Acid Phosphatase". J. Urol. 63:128, 1950.

55. Stolbach, L. L., Krant, M. J., Fishman, W. H.: "Ectopic Production of Alkaline Phosphatase Isoenzyme in Patients with Cancer". New Eng. J. Med. 281:757, 1969.

56. Szasz, C.: "Serum Leucine Aminopeptidase Activity in Acute Lesions of the Liver Parenchyma". Lancet 1:441, 1964.

57. Teree, T. M., Klein, L.: "Hypophosphatasia: Clinical and Metabolic Studies". J. Pediatrics 72:41, 1968.

58. Titus, M. A., Reynolds, D. A., Glending, M. B., Page, E. W.: "Plasma Aminopeptidase Activity (Oxytocinase) in Pregnancy and Labor". Amer. J. Obstet. & Gynec. 80:1124, 1960.

59. Tuchman, L. R., Goldstein, G., Clyman, M.: "Studies on the Nature of the Increased Serum Acid Phosphatase in Gaucher's Disease". Amer. J. Med. 27:959, 1959.

60. Van Dommelan, C. K. V., Klaassen, C. H. L.: "Cyanocobalamin Dependent Depression of the Serum Alkaline Phosphatase Level in Patients with Pernicious Anemia". New Eng. J. Med. 271:541, 1964.

61. Wilkins, W. E., Regen, E. M.: "Course of Phosphatase Activity in Healing of Fractured Bones". Proc. Soc. Exp. Biol. Med. 32:1373, 1935.

62. Wilkinson, J.: "Clinical Significance of Enzyme Activity Measurements". Clin. Chem. 16:11 & 882, 1970.

63. Woodard, H. W.: "A Note on the Inactivation by Heat of Acid Glycerophosphatase in Alkaline Solution". J. Urol. 65:688, 1951.

64. Woodard, H. W.: "Changes in Blood Chemistry Associated with Carcinoma Metastatic to Bone". Cancer 6:1219, 1953.

References (Continued):

65. Yong, J. M.: "Cause of Raised Serum Alkaline Phosphatase after Partial Gastrectomy and in other Malabsorption States". Lancet 1:1132, 1966.

66. Young, I. I.: "Serum 5'-Nucleotidase: Characterization and Evaluation in Disease States". Ann. N. Y. Acad. Sci. 75:357, 1958.

67. Zucker, M. B., Woodard, H. Q.: "Elevation of Serum Acid Glycerophosphatase Activity in Thrombocytosis". J. Lab. Clin. Med. 59:760, 1962.

CHAPTER 3

AMYLASE AND LIPASE

Amylase

The origin of serum amylase is enigmatic although most investigators consider that the major but not the sole source is the pancreas. Experimental evidence in dogs demonstrates that normal serum amylase levels are maintained even after total removal of the pancreas. The liver provides most of the amylase in normal human serum. The amount of amylase in normal serum is not markedly affected by pancreatectomy or excision of the salivary glands. When a patient with a pancreatic lesion develops an elevated serum amylase, the electrophoresis of the amylase demonstrates that it travels in the gamma globulin area of the serum proteins.

The serum amylase level is low in early infancy, is first demonstrable at 2 months of age, and attains the adult level at the age of 1 year.

Causes of an elevated serum amylase level are:
1. Acute pancreatitis
2. Perforated peptic ulcer penetrating the pancreas
3. Intestinal obstruction
4. Pancreatic duct obstruction by carcinoma or calculi
5. Opiates, codeine or methycholine spasm of Spincter of Oddi
6. Pancreatic stimulation by secretin or pancreozymin
7. Parotid gland disease, mumps, suppurative parotitis, and calculi in the parotid duct
8. Renal failure with oliguria or anuria, causing retention of amylase due to insufficient clearance by the kidney
9. Hepatic necrosis
10. Ectopic pregnancy
11. Macroamylasemia due to combination of amylase with an immunoglobulin
12. Ectopic production of amylase by extrapancreatic carcinoma
13. Hemolysis

The commonest cause for a high serum amylase is acute pancreatitis which may be secondary to various causes (Fig. 3-1).

Amylase and Lipase (Continued):

Fig. 3-1 Gross Photograph of necrotic pancreas from an alcoholic patient who died from acute pancreatitis. The serum amylase on the day of admission was 760 Somogyi Units.

Destruction of pancreatic acinar tissue, with or without obstruction of flow of pancreatic secretion, results in the escape of various pancreatic enzymes into the pancreas and into the peritoneal cavity, consequently there is absorption of these pancreatic enzymes into the blood with subsequent prominent elevation of serum amylase. Persistence of elevated circulating amylase usually depends on continued secretion of the enzyme. In acute pancreatitis the serum amylase invariably increases almost simultaneously with the onset of symptoms. It frequently rises above 500 Somogyi Units. The peak level is usually reached within 24 hours, after which there is usually a prominent drop in the enzyme level with a return to normal values within 2 to 4 days.[1,23] Absence of an increase in the amylase level within the first 24 hours after the onset of symptoms is suggestive evidence against the diagnosis of acute pancreatitis.

Amylase and Lipase (Continued):

An elevation in serum amylase is accompanied by an elevation in urine amylase. Elevated urine amylase with a normal serum amylase may occur in acute pancreatitis because of the rapid clearance of amylase by the kidney. Thus, the urine amylase may be higher and may last longer than serum amylase elevation in acute pancreatitis.[22] In this condition, renal clearance of amylase may increase three-fold. The amount of urinary amylase in a 1 hour specimen should be requested,[12] with the normal being up to 160 Somogyi Units per hour. Repeated serum and urinary enzyme measurements should be performed.

If one has missed an increase in serum amylase, it is recommended that urine amylase levels be determined. As previously mentioned, normally approximately 160 Somogyi Units of amylase are excreted into the urine on an hourly basis, with a total amount of up to 4000 Somogyi Units per 24 hours. Thus, many clinicians believe that the assay of urinary amylase in patients with suspected pancreatitis is more useful than that of serum amylase.[9,10]

The height of serum amylase activity may be inversely related to the severity of the pancreatic disease. When the amylase is over 1000 Units, a surgically correctable lesion is present. When the amylase is between 200 and 500 Units, a severely necrotizing or hemorrhagic pancreatitis is present. An abrupt fall is an ominous prognostic sign.

Another procedure to detect the presence of acute pancreatitis is abdominal paracentesis with serial determinations of amylase in the peritoneal fluid.[8,17,19,21] Elevation of amylase in the fluid is strongly suggestive of acute pancreatitis. Other conditions such as ruptured ectopic pregnancy and perforated small intestine will also elevate peritoneal fluid amylase.[13]

An increase in the blood amylase results from an acutely inflamed pancreas with the enzyme entering the circulation through the portal vein. The thoracic duct also contains an elevated amount of enzyme.[20] Serum amylase may also become elevated when there is spasm of the Sphincter of Oddi following the administration of opiates. Similar elevations occur due to obstruction of the common bile duct at the Sphincter of Oddi by gallstones, carcinoma of the bile duct or ampulla and carcinoma of the head of the pancreas.

When a peptic ulcer perforates into the pancreas causing acute pancreatitis, the blood amylase levels will show a prominent rise. Disease of the small or large bowel may also result in elevations of amylase on the basis of the liberation of the enzyme from the colonic

Amylase and Lipase (Continued):

or small intestinal mucosa. An elevation in serum amylase following obstruction and perforation of the small intestine is usually secondary from the resulting pancreatitis, but it may result from liberation from the intestinal tract mucosa.

Another acute abdominal emergency in which there is an elevated serum amylase is rupture of an ectopic pregnancy (Fig. 3-2). The cause for this elevation is liberation of amylase from the mucosa of the fallopian tube.

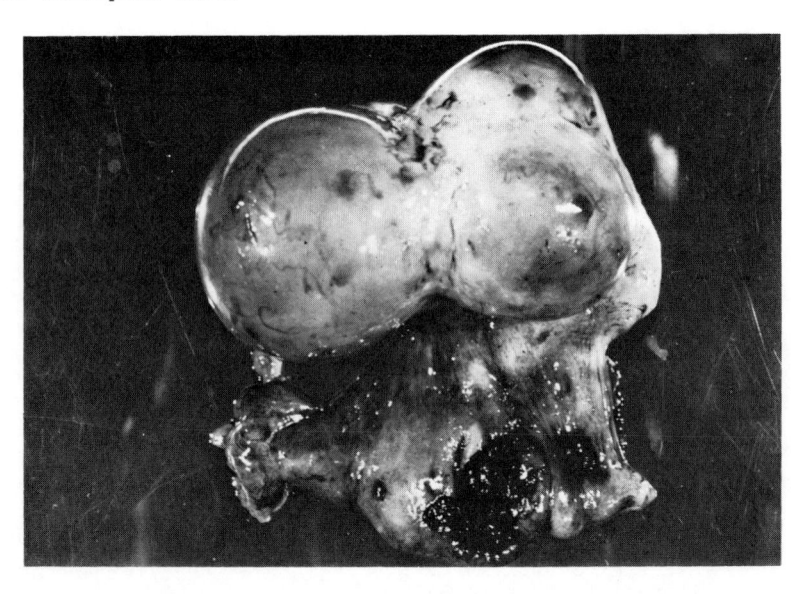

Fig. 3-2 Gross photograph of a ruptured ectopic pregnancy from the right oviduct. The serum amylase was 415 Somogyi Units. The elevated serum amylase was derived from the oviduct mucosa.

Amylase has also been shown to be present in the parenchymal cells of the liver. In massive hepatic necrosis elevated serum amylase levels are seen (Fig. 3-3). Some observers believe that this elevation is usually from the renal failure which accompanies the hepatic necrosis. Amylase is excreted primarily by the kidneys into the urine, and with renal insufficiency there is poor renal clearance of the enzyme. It is common to observe prominent elevation

Amylase and Lipase (Continued):

of serum amylase in patients who have renal insufficiency with oliguria or anuria.[16] Recently there has been conjecture concerning modest elevation in viral hepatitis without renal insufficiency. In the presence of massive necrosis of the liver one would expect elevation of amylase from necrotic hepatic cells.

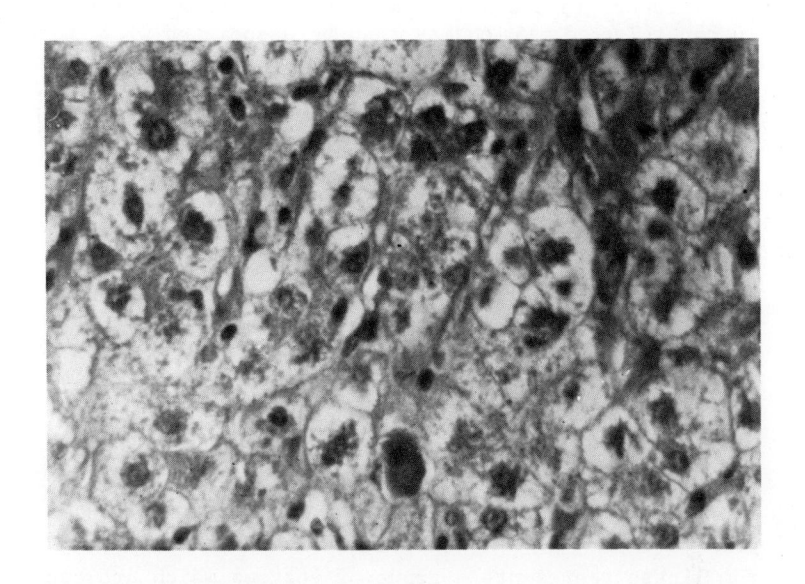

Fig. 3-3 Photomicrograph of section of liver from a patient who died from massive liver necrosis due to acute viral hepatitis. The serum amylase was 865 Somogyi Units.

Gamma globulin binds amylase to form a macroamylase. In macroamylasemia, an unusual condition, the amylase is bound to a IgA or IgG immunoglobulin.[5] This condition has been associated with malabsorption in which biopsy demonstrates lymphoma. The lymphoma causes malabsorption with the production of an IgA or IgG immunoglobulin which binds to serum amylase giving the macromolecule. The presence of macroamylase in large amounts causes an elevation in serum amylase simulating that of pancreatic disease. Unlike elevations of nonbound amylase, macroamylasemia will not produce an elevated urinary amylase. This is due to the non-clearance of the large molecule by the kidney.[7]

Amylase and Lipase (Continued):

The parotid glands are a rich source of amylase. Infections of the parotid glands such as mumps, parotid sialoadenitis, and calculi in the parotid duct will cause high serum amylase levels.[6]

Trauma to the abdomen may result in damage to the pancreas causing an increase in the serum amylase.[15] Trauma to the brain may result in hyperglycemia which may then be associated with minor elevations of the serum amylase. This response may be related to the hyperglycemia or to the stress induced by the cerebral trauma. Another rare cause for increased serum amylase is the production of the enzyme by malignancies that are extrapancreatic.

Elevations of serum amylase can be seen after alcohol ingestion, alcohol being a powerful stimulator of both parotid and pancreatic secretion.

Decreased serum levels of amylase are found in:
1. Citrate anticoagulant [11]
2. Oxalate anticoagulant [11]
3. Cachexia
4. Marked destruction of pancreas
5. Administration of glucose, insulin or cortisone

Lipase

Serum lipase represents a group of enzymes which are capable of hydrolyzing triglycerides. The clinical interest in them is in the assessment of acute pancreatitis. Lipases are present in various tissues, in the largest amounts in the pancreas, but also in the stomach, small intestine, and leukocytes such as monocytes.

NORMAL VALUES: 0.1 - 1.5 Units

Lipase is present in fewer tissues than is amylase, thus elevation of this enzyme are more specifically related to pancreatic disease.

Causes of an elevated serum lipase are:
1. Acute pancreatitis
2. Conditions involving the pancreatic duct
   Carcinoma of the pancreas
   Carcinoma of the Ampulla of Vater
   Calculus in the Ampulla of Vater
   Peptic ulcer penetrating into the pancreas
   Spasm of the Sphincter of Oddi by opiates

Amylase and Lipase (Continued):

       3. Fat embolism to the lungs
       4. Intestinal obstruction
       5. Renal insufficiency
       6. Multiple sclerosis
       7. Bacterial contamination of specimen

Elevations of serum lipase will persist up to a week longer than those of serum amylase. Thus, hyperlipasemia is invariably regarded as a sign of acute pancreatic disease.[4,14] Elevations occur in the serum by the same mechanism that causes a rise in the serum amylase when there is obstruction to flow of the pancreatic enzymes by various mechanisms. In acute pancreatitis, serum lipase activity may not correlate with the severity of the disease. The elevation in serum lipase lags behind serum amylase by 24 hours. Also, because lipase is excreted by the kidney, there will be a rise in the serum lipase level when there is renal failure with oliguria or anuria. Urine lipase determination should not be requested in contrast to urine amylase. The urine lipase excretion is variable and unrelated to serum lipase.[3,18] Fat embolism to the lung causes a release of lipase from injured pulmonary parenchyma (Fig. 3-4).[2]

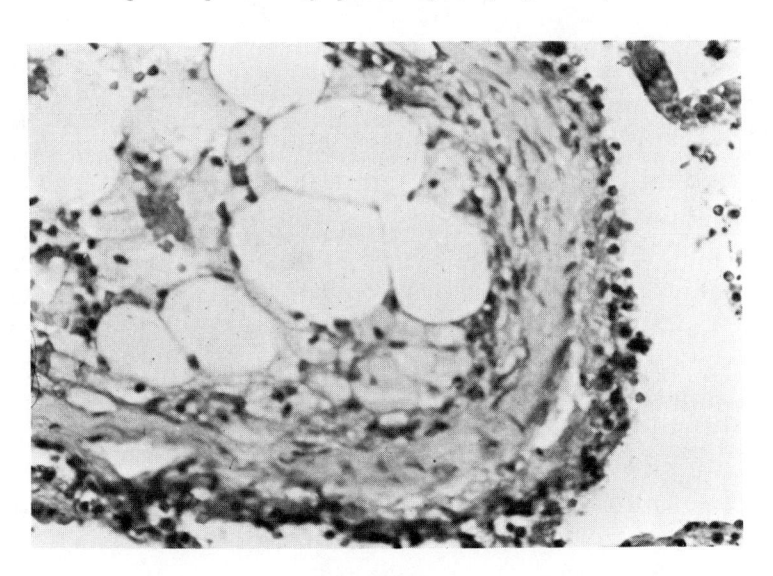

Fig. 3-4 Photomicrograph of fat embolus to lung in a patient who died following multiple bone fractures sustained in an automobile accident. The fat embolus caused release of lung lipase with elevation of serum lipase to 3.2 Units.

Amylase and Lipase (Continued):

Lipase activity is decreased when there is prominent destruction of the pancreatic gland with diminished production of the enzyme. This is found in patients with chronic pancreatitis and marked fibrosis of the gland. Severely malnourished patients have been shown to have decreased lipase activity, and hemolysis inhibits function of the enzyme.

REFERENCES

1. Adams, J. T., Libertino, J. A., Schwartz, S. I.: "Significance of an Elevated Serum Amylase" Surgery 63:877, 1968.

2. Adler, F., Peltier, L. F.: "The Laboratory Diagnosis of Fat Embolism" Clin. Orthopedics 21:226, 1961.

3. Ambrovage, A. M., Howard, J. M., Pairent, F. W.: "The Twenty-four Hour Excretion of Amylase and Lipase in the Urine: Correlation with the Functional State Operative Injury of the Pancreas" Ann. Surgery 167:539, 1968.

4. Berk, J. E.: "Serum Amylase and Lipase" Journal of Amer. Med. Assoc. 199:98, 1967.

5. Berk, J. E., Kizu, H., Wilding, P., Searcy, R. L.: "Macroamylasemia: A Newly Recognized Cause for Elevated Serum Amylase Activity" New Eng. Jour. Med. 277:941, 1967.

6. Berk, J. E., Searcy, R. L., Hayashi, S., Ujihira, I.: "Distribution of Serum Amylase in Man and Animals. Electrophorectic and Chromatographic Studies" Jour. of Amer. Med. Assoc. 192:389, 1965.

7. Berk, J. E., Searcy, R. L., Wilding, P., Kizu, H. and Svoboda, A. C.: "Macroamylase: A New Cause for Elevated Serum Amylase Activity" Jour. of Amer. Med. Assoc. 200:545, 1967.

8. Bolooki, H., Gliedman, M. L.: "Peritoneal Dialysis in the Treatment of Acute Pancreatitis" Surgery 64:466, 1968.

9. Budd, J. J., Jr., Walter, K. E., Harris, M. L., Knight, W. A., Jr.: "Urine Disease in the Evolution of Pancreatic Disease" Gastroenterology 36:333, 1959.

References - Amylase and Lipase (Continued):

10. Calkins, W. G.: "Study of Urinary Amylase Excretion in Normal Persons" J. Gastroenterology 46:407, 1966.

11. Christian, D. G.: "Drug Interference with Laboratory Blood Chemistry Determinations" Am. J. Clin. Path. 54:118, 1970.

12. Gambill, E. E., Mason, H. L.: "One Hour Value for Urinary Amylase in 96 Patients with Pancreatitis: Comparative Diagnostic Value of Tests of Urinary and Serum Amylase and Serum Lipase" J. Am. Med. Assoc. 186:24, 1963.

13. Green, C. L.: "Identification of Alpha-amylase as a Secretion of the Human Fallopian Tube, and 'Tube-like' Epithelium of Mullerian and Mesonephric Duct Origin" Am. J. Obstetrics and Gynecology 73:402, 1957.

14. Johnson, T. A. and Bockus, H. L.: "Diagnostic Significance of Determinations of Serum Lipase" Arch. Internal Medicine 66:62, 1940.

15. Jones, R. C. and Shires, G. T.: "The Management of Pancreatic Injuries" Arch. Surgery 90:502, 1965.

16. Levitt, M. D., Rapport, M., Cooperband, S. R.: "The Renal Clearance of Amylase in Renal Insufficiency, Acute Pancreatitis, and Macroamylasemia" Ann. Internal Medicine 71:919, 1969.

17. MacFate, R. P.: "Amylase in Biological Fluids" Sunderman, F. W. and Sunderman, F. W., Jr. (Eds.) MEASUREMENTS OF EXOCRINE AND ENDOCRINE FUNCTIONS OF THE PANCREAS Lippincott, pg. 14 - 31, 1961.

18. Pfeffer, R. B., Dishman, A., Cohen, T., Tesler, M., Aronson, A. R.: "Urinary Lipase Excretion in Pancreatic and Hepatobiliary Disease Surgery Gynecology and Obstetrics 124:1071, 1967.

19. Pfeffer, R. B., Mixter, G., Jr., Hinton, W. J.: "Acute Haemorrhagic Pancreatitis: A Safe Effective Technique for Diagnostic Paracentesis Surgery 43:550, 1958.

20. Popper, H. L., Necheles, H.: "Pathways of Enzymes into the Blood in Acute Damage of the Pancreas" Proc. Soc. Exp. Biol. Med. 43:220, 1940.

References - Amylase and Lipase (Continued):

21. Rott, H. D., Hauser, C. W., McKinley, C. R., LaFave, J. W., Mendiola, R. P.: "Diagnostic Peritoneal Lavage" Surgery 57:633, 1965.

22. Saxon, E. I., Hinkley, W. C., Vogel, W. C. and Zieve, L.: "Comparative Value of Serum and Urinary Amylase in the Diagnosis of Acute Pancreatitis" Arch. Internal Medicine 99:607, 1957.

23. Zieve, L.: "Clinical Value of Determinations of Various Pancreatic Enzymes in Serum" Gastroenterology 46:62, 1964.

CHAPTER 4

GLUTAMIC OXALACETIC TRANSAMINASE AND GLUTAMIC PYRUVIC TRANSAMINASE

The greatest amount of glutamic oxalacetic transaminase (GOT) is present in the liver followed by lesser amounts in heart muscle and skeletal muscle. A small amount is present in the kidney, pancreas, red blood cells, brain and skin. The activity in red blood cells is about ten times the normal serum level. Thus, in hemolytic anemia there may be an elevation of GOT.

Glutamic pyruvic transaminase (GPT) is found in greatest amounts in the liver with lesser amounts in various other tissues. The usual reason for a request for a GPT determination is the possibility of hepatocellular disease.

Normal GOT and GPT values in serum and spinal fluid are as follows (Table 4-1).

TABLE 4-1

NORMAL TRANSAMINASE VALUES IN SERUM AND SPINAL FLUID IN I.U.

	SERUM		SPINAL FLUID
	GOT	GPT	GOT
Infants	Up to 67	27 to 54	1 to 7
Older Children	Up to 40	Up to 35	15 to 27
Adults	Up to 40	Up to 35	15 to 27

Most of the transaminases are present within the cytoplasm of the cell with only small amounts in plasma and cerebrospinal fluid. Various observers have discerned that species differences are present. Enzyme activity is associated electrophoretically with a fast-moving anodic component and a slow-moving cathodic component.[6] Only one electrophoretic band of transaminase activity is detectable in human serum with mobility between the alpha$_2$ and beta globulins. A second anodic band of transaminase activity has been observed in patients with a myocardial infarct.[7]

An efficient barrier between the blood and cerebrospinal fluid prevents the transfer of the transaminase enzymes between the two.[25] Thus, there is very little correlation between spinal fluid and blood

GOT and GPT (Continued):

transaminase values.[19] Transaminases may be detected in the urine; activity of the urine reflects the amount of enzyme in the blood. For example in patients who have had a myocardial infarct, when there is an elevated serum GOT level, the urinary enzyme levels also are elevated. Some investigators have noted that different types of meals give significant rises in the blood GOT activity; for example, increased vitamin B_6 intake results in significant elevation of blood transaminase.[22] This is disputed by other observers who find no variations in GOT values through the day.

Prolonged and severe exercise tends to raise the serum GOT levels.[29] Healthy males tend to exhibit slightly higher transaminase levels than females. There is a slight decrease in levels during normal pregnancy.[33,39] When serum transaminase levels are elevated, this results from cellular destruction or a change in the cell membrane permeability with release of the transaminase of the cellular cytoplasm into the blood.

Causes of an elevated serum GOT level are:
1. Acute hepatitis
2. Acute myocardial infarction
3. Active cirrhosis of liver
4. Infectious mononucleosis with acute hepatitis
5. Toxic hepatic necrosis due to carbon tetrachloride or other hepatotoxins
6. Carcinoma metastatic to liver or leukemic infiltration of liver
7. Acute pancreatitis
8. Trauma to skeletal muscle: surgical, accidental, crush injury, or irradiation
9. Pseudohypertrophic muscular dystrophy
10. Dermatomyositis
11. Acute hemolytic anemia
12. Acute renal necrosis, specifically recent renal infarction
13. Severe burns with extensive epidermal injury
14. Cardiac catheterization and angiography with injury of heart muscle
15. Recent brain injury with brain necrosis
16. Hemorrhagic pulmonary infarction

Causes of an elevated serum GPT level are:
1. Acute viral hepatitis
2. Less marked elevation in necrosis of skeletal muscle, heart, kidney, or pancreas

GOT and GPT (Continued):

        3. Infectious mononucleosis with hepatitis
        4. Acute myocardial infarction

The commonest causes for elevated serum transaminase levels are heart and liver diseases. GOT activity in the blood rises within the first 18 hours following an acute myocardial infarction (Fig. 4-1).[1,9] Maximum levels are reached within 48 hours and return to normal within 3 to 5 days. The degree of elevation of GOT is related to the amount of myocardial muscle that has become necrotic. An elevation of serum GOT is particularly helpful to the clinician diagnostically when the electrocardiogram does not indicate the presence of myocardial infarct.[41] Patients who have angina pectoris do not have an increase in GOT. Extensions of a myocardial infarct will cause persistent and recurrent elevation in the GOT levels. Occasionally elevated serum transaminase levels may be seen in patients with severe tachycardia or arrhythmia with a cardiac rate greater than 160 beats per minute. Active myocarditis secondary to either viral, metabolic, rheumatic, or a bacterial cause is associated with elevated transaminase levels. Here again, the severity of the myocarditis influences the level of the GOT. Patients who undergo cardiac catheterization and angiography may have elevations in transaminase levels. Presumably the cardiac muscle is injured during the procedure, with release of GOT into the blood. Use of opiates or direct-current electrical counter-shock will elevate GOT.[14,23,27,31] Patients who develop congestive heart failure will have GOT and GPT elevations on the basis of hepatic congestion and central lobular necrosis of the liver.[7,38] The GOT elevation in the urine of patients with acute myocardial infarction usually correlates with elevated blood levels. In some patients, the urinary concentrations remain elevated for a longer time than in the serum, simulating the situation of urinary amylase in acute pancreatitis.[36]

GOT and GPT (Continued):

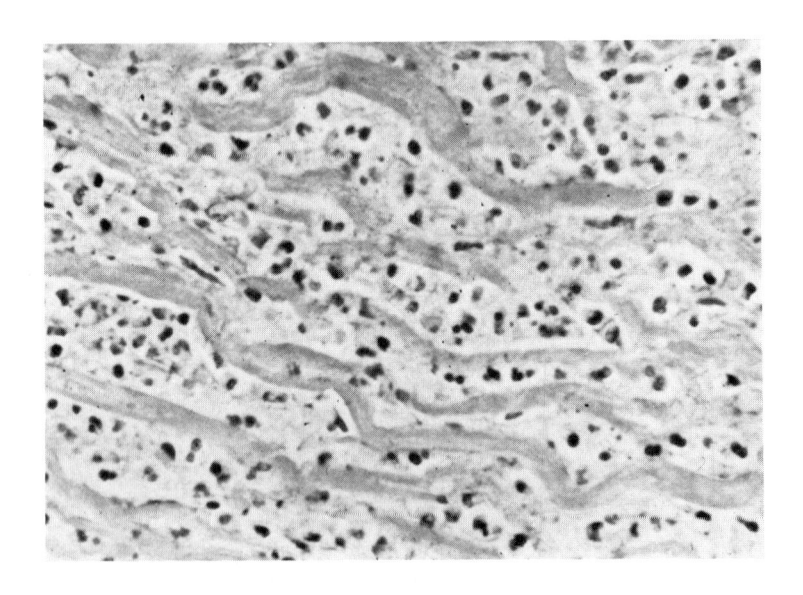

Fig. 4-1 Photomicrograph of section of heart from patient who died 30 hours following onset of substernal chest pain associated with acute myocardial infarction. Note the necrosis of myocardial cells and PMN cellular infiltrate. The SGOT was 380 I.U. shortly before death.

In acute myocardial infarction the level usually rises over 300 I.U., and at times there is a great increase reaching 1000 I.U., especially if there is associated heart failure with centrilobular liver necrosis. When infarction of the myocardium is suspected, daily measurements of GOT activity are more useful than a single determination which may not detect the transient peak elevation.[18,30]

Pericarditis usually does not elevate the GOT levels. If GOT is elevated in patients with acute pericarditis, there most likely is subepicardial necrosis of the heart or perhaps congestive heart failure secondary to the pericarditis.

GOT and GPT (Continued):

The value of determining GOT in heart disease may be summarized as follows:
1. Differentiation between an acute myocardial infarct and angina pectoris; in the latter no increase occurs.
2. Diagnosis of acute myocardial infarction when the electrocardiogram does not indicate the infarct, or when the electrocardiogram is difficult to interpret because of the presence of a previous myocardial infarct, a left bundle branch block, digitalis effect, or the Wolff-Parkinson-White syndrome.
3. Diagnosis of extension of a myocardial infarction.

The myocardium contains approximately 16 times more GOT than GPT. Thus a ratio of the GOT to GPT levels, the DeRitis quotient, may be helpful in elucidating the source and etiology of transaminase elevation.[10] In myocardial disease, the DeRitis quotient is greater than one whereas in hepatic disease this quotient is less than one.

The liver is the richest source of both GOT and GPT. Thus, any damage to liver parenchymal cells such as occurs in neoplastic, inflammatory, or degenerative lesions of the liver, will result in elevations of both these enzymes. The extent of the rise usually reflects the severity of the hepatic damage. The highest values have been obtained in acute hepatic necrosis which is associated with chemical poisoning such as carbon tetrachloride.[2,43] In these patients the rise might be as much as 25,000 I.U. In acute viral hepatitis the rise may be up to 5,000 I.U. (Fig. 4-2). Increase in serum GOT and GPT may be present in a hepatitis patient for as long as 1 to 4 weeks before the appearance of clinical jaundice. Thus, the transaminase levels may be used as biochemical indicators or detectors of the carrier state of the anicteric patient.[3] One or both elevated enzymes may be found in 25 percent of asymptomatic individuals in the first 5 months of the post-convalescent period.[5,28] Elevated GOT, when present unaccompanied by symptoms or other significant laboratory findings such as increased BSP, does not warrant restriction of activity in the convalescent phase.

GOT and GPT (Continued):

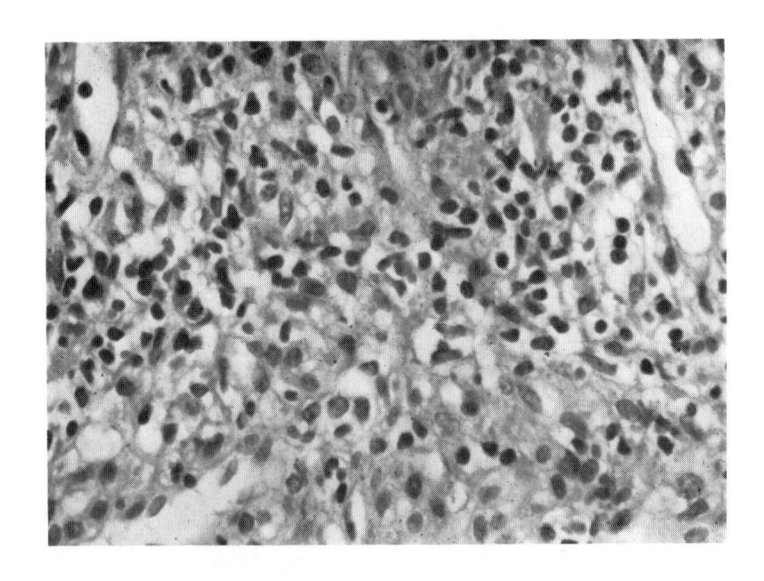

Fig. 4-2   Photomicrograph of section of liver from a patient who died from acute serum viral hepatitis induced by blood transfusion for saphenous vein myocardial revascularization. The SGOT was 2,650 I.U. shortly before death.

Cerebrospinal fluid GOT remains normal in viral hepatitis even though serum GOT is extremely high.[11,16] Injection of large amounts of GOT into the dog subarachnoid space does not cause an elevation in the serum.[26]

Transaminase levels in the range of 100 to 500 I.U. are usually found in cirrhosis due to alcoholism, in biliary cirrhosis, and in mild toxic hepatitis due to drugs. Acute alcoholic hepatitis may cause anorexia, abdominal pain, fever, jaundice, hepatomegaly, and GOT and GPT may be elevated.[40] The increase in GPT usually exceeds that of GOT. With extremely severe liver damage, the rise in GOT may not be sustained and may fall because of great damage to the cell which interferes with continued and persistent enzyme production.

GOT and GPT (Continued):

Patients with neoplasms metastatic to the liver may have increases in GOT activity due to the necrosis of liver cells in and surrounding the metastatic cancer.[34] Serum GOT levels are elevated only in sporadic instances in cases of obstructive liver disease. GOT is excreted in the bile, and increased biliary pressure may stimulate hepatic cells to release GOT. Biliary obstruction tends to elevate GOT only in post-cholecystectomy patients.[4,15,21,24,37] This may be in part due to biliary pressure being exerted against the liver cells in the absence of the gallbladder.

The transaminase elevation in both GPT and GOT seen in infectious mononucleosis or viral hepatitis is presumably due to the hepatic damage, elevations of GPT being more striking than GOT.[13] Thus, the DeRitis ratio is usually less than one. If the DeRitis ratio is greater than one, necrosis of the liver has occurred with leakage of GOT from mitochondria.

Maximal values for transaminase levels are reached within 7 to 10 days after liver disease begins; elevations may persist up to 6 weeks. Secondary rises suggest a relapse. Persistent elevations suggest a continuing active hepatitis.

Several other important causes for elevations of serum transaminase are recognized. One is acute infarction of the kidney with necrosis of renal tubular cells (Fig. 4-3).[12] Renal insufficiency without infarction will not cause an elevated serum transaminase.

GOT and GPT (Continued):

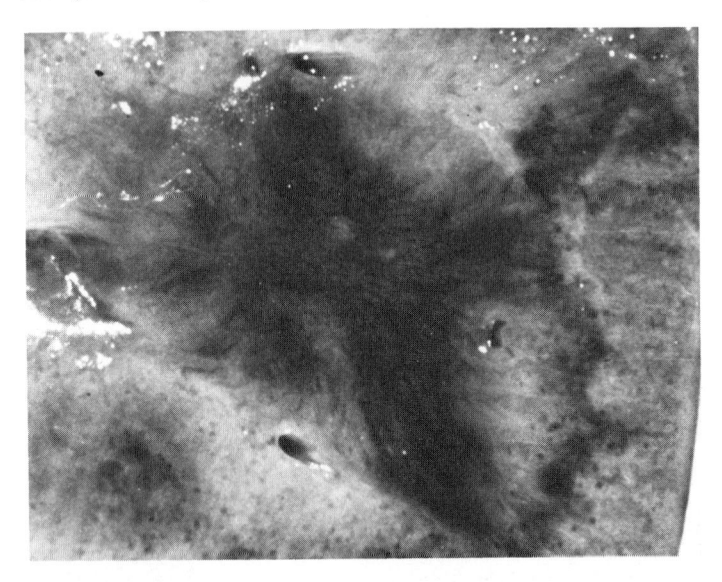

Fig. 4-3   Gross Photograph of recent infarcts of kidney which caused the SGOT to be elevated to 250 I.U.

Transaminase is present in pancreatic acinar cells. In patients with acute pancreatitis, there may be a moderate rise in the GOT levels in the blood.[35]

GOT is present in red cells and white cells in the peripheral blood. Patients with hemolytic anemia or leukemia will have a rise in GOT levels.[42] In leukemia the rise results from proliferation of the leukemic cells with production of GOT by them and also from liberation of GOT from leukemic cells which are undergoing destruction.

An early rise in GOT following a burn has been observed.[2,20] Burns which extend over 10 percent of the body surface may liberate GOT from damaged epidermal cells. Blister fluid from burns contains a large amount of GOT. GPT may also be elevated in patients who have extensive burns, suggesting possible liver damage secondary to the burn. A small amount of GPT may also be liberated from the damaged epidermal cells.

GOT and GPT (Continued):

GOT elevations are frequently observed following surgery. The GOT is probably liberated from the surgical wound especially from skeletal muscle.[44] Patients who have undergone cardiac surgery invariably show GOT elevations for a number of days following the surgical procedure.[32]

GOT is also present in the brain.[19] In patients with cerebral disease, the GOT in the cerebrospinal fluid and in the blood increases. There is some evidence that the transaminases do not pass the blood-brain barrier easily, so that elevation in the cerebrospinal fluid is greater than in the blood.

Skeletal muscle is a rich source of transaminases, specifically GOT. After severe muscle trauma, such as accidents or surgery, elevations of GOT are found. The levels bear a direct relation to those of creatine phosphokinase (CPK) and aldolase levels, indicating a skeletal muscle origin. In the muscular dystrophies, marked elevations of both GOT and CPK are found; 78 percent of approximately 500 patients with progressive muscular dystrophy had elevated transaminase levels.[42] However, it should be noted that the transaminases are less useful than CPK or aldolase in the diagnosis of muscular disease (Fig. 4-4).

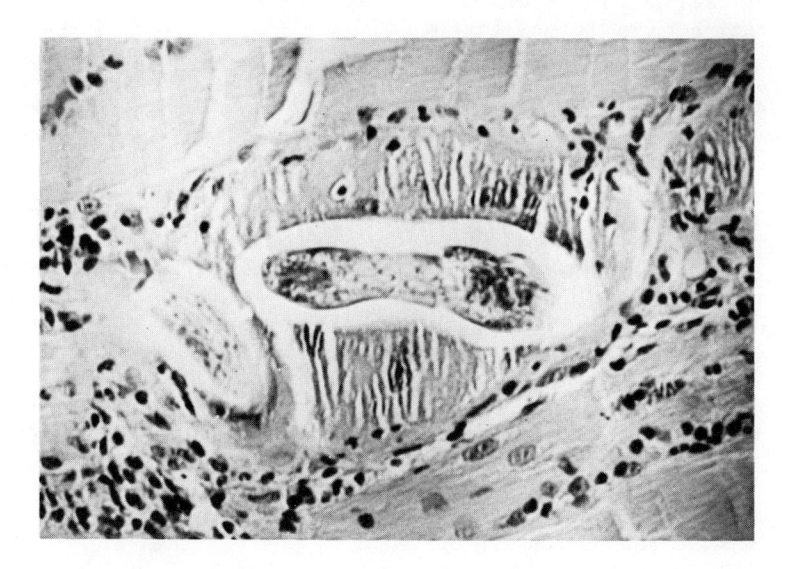

Fig. 4-4   Photomicrograph of section of skeletal muscle biopsy from patient with trichinosis. Note necrosis of muscle fibers, inflammatory cell infiltrate, and sarcolemmal cell proliferation which surround the trichina. The SGOT was 360 I.U.

GOT and GPT (Continued):

Drug Effects

A whole host of drugs can give spurious elevations of GOT.[8] These include:

Amantadine	Coumadin	Methotrexate
Ampicillin	Cycloserine	Methyldopa
Anabolic agents	Erythromycin	Nafcillin
Androgens	Ethionamide	Nalidixic Acid
Cephalothin	Gentamycin	Opiates
Chloroquine	N-Hydroxyacetamide	Oxacillin
Clofibromate	Indomethacin	Polycillin
Cloxacillin	Isoniazid	Phenothiazines
Chlorpromazine	Lincomycin	Progesterone
Colchicine		

Some of the GOT elevation is caused by hepatic parenchymal cell damage due to the drugs while some drugs may cause cholestasis. Some drugs, specifically erythromycin and the tranquilizer group such as phenothiazines, interfere with the method of GOT analysis. These two drug groups cause formation of a hydrazone without the presence of liver damage. The hydrazone method entails usually transformation of hydrazine to hydrazone in the presence of GOT. In the presence of an erythromycin or phenothiazine drug, a hydrazone forms from the combination of the drug and hydrazine, giving a false color reaction. This effect can be shown to be spurious if the GOT level is assayed by the NAD-NADH method. Discontinuance of the drug leads to normal levels within 24 hours.

GOT may be elevated in ketoacidosis associated with starvation or diabetic acidosis. The ketoacids combine with the colorimetric dye utilized in the SMA 12/60 instrument giving a spurious elevation.

The oral contraceptives may cause liver damage; if jaundice and elevated enzymes occur, they are seen in the first three drug cycles.

Causes of a decrease in GOT level are:[46]
1. Elevated pyruvate and lactate levels found in beriberi, diabetic ketoacidosis, and liver disease
2. Pregnancy
3. Prolonged renal dialysis
4. Low pyridoxine

GOT and GPT (Continued):

REFERENCES

1. Agress, C. M., Kim, J. H. C.: "Evaluation of Enzyme Tests in the Diagnosis of Heart Disease". Amer. J. Cardiol. 6:641, 1960.

2. Allegra, F. G.: "Serum Glutamic Pyruvic Transaminase in Cardiac and Hepatic Disease". Ital. Gen. Review. Derm. 101:126, 1960.

3. Alsever, J. B.: "The Blood Bank and Homologous Serum Jaundice. A Review of Medicolegal Considerations". New Eng. J. Med. 261:383, 1959.

4. Ballard, H., Bernstein, M., Farrar, J. T.: "Fatty Liver Presenting as Obstructive Jaundice". Amer. J. Med. 30:196, 1961.

5. Bang, N. V., Ruegsegger, P., Ley, A. B., LaDue, J. S.: "Detection of Hepatitis Carrier by Serum Glutamic Oxalacetic Transaminase Activity". J. Amer. Med. Assoc. 171:2303, 1959.

6. Block, W. D., Carmichael, R. L., Jackson, C. E.: "Quantitative Determination of Isoenzymes of Glutamic Oxalacetic Transaminase in Human Serum". Proc. Soc. Exp. Biol. Med. 115:941, 1964.

7. Boyde, T. R. C., Latner, A. L.: "Glutamic Oxalacetic Transaminase Isoenzymes". Biochem. J. 82:51, 1962.

8. Christian, D. G.: "Drug Interference with Clinical Interpretations". Amer. J. Clin. Path. 54:1, 1970.

9. Clermont, R. J., Chalmers, T. C.: "The Transaminase Tests in Liver Disease". Medicine 46:197, 1967.

10. DeRitis, F., Coltorti, M., Giusti, G.: "Diagnostic Value and Pathogenic Significance of Transaminase Activity changes in Viral Hepatitis". Minerva Med. 47:167, 1956.

11. Fleisher, G. A., Wakin, K. G.: "Transaminase in Canine Serum and Cerebrospinal Fluid after Carbon Tetrachloride Poisoning and Injection of Transaminase Concentrates". Mayo Clin. Proc. 31:640, 1956.

GOT and GPT References (Continued):

12. Frahm, C. J., Folse, R.: "Serum Oxalacetic Transaminase Levels following Renal Infarction. Report of a Case and Experimental Observations following Ligation of the Renal Arteries". J. Amer. Med. Assoc. 180:209, 1962.

13. Gelb, D., West, M., Zimmerman, H. J.: "Serum Enzymes in Diseases IX. Analysis of factors responsible for elevated Values in Infectious Mononucleosis". Amer. J. Med. 33:249, 1962.

14. Hunt, D., Bailie, M. J.: "Enzyme Changes following Direct Current Counter Shock". Amer. Heart J. 76:340, 1968.

15. Jones, W. A., Tisdale, W. A.: "Posthepatic Cirrhosis Clinically simulating Extrahepatic Biliary Obstruction (so-called 'Primary Biliary Cirrhosis')". New Eng. J. Med. 268:629, 19

16. Katzman, R., Fishman, R. A., Goldensohn, E. S.: "Glutamic Oxalacetic Transaminase Activity in Spinal Fluid". Neurology 7:853, 1957.

17. Killip, T., Payne, M. A.: "High Serum Transaminase Activity in Heart Disease Circulatory Failure, and Hepatic Necrosis" Circulation 21:646, 1960.

18. LaDue, J S., Wroblewski, F., Kormer, A: Serum Glutamic Oxalacetic Transaminase Activity in Human Acute Transmural Infarction". Science 120:497, 1954.

19. Lieberman, J, Daiber, O., Dulkin, S. I., Lobstein O. E., Kaplan, M. R.: "Glutamic Oxalacetic Transaminase in Serum and Cerebrospinal Fluid of Patients with Cerebrovascular Accidents". New Eng. J. Med. 257:1201, 1957.

20. Lieberman, J., Lasky, I. I., Dulkin, S. L., Lobstein O. E.: "The Serum Transaminases in Burned Patients". Ann. Intern. Med. 46:485, 1957.

21. Linde, S.: "On the Mechanism of the Elevation of Serum Glutamic Oxalacetic Transaminase in Obstructive Jaundice". Scan. J. Clin. Lab. Invest. 10:308 1959.

22. Marsh, M. E., Greenberg, L. D., Rinehart, J. F.: "Relation of Pyridoxine to Glutamic Oxalacetic Transaminase". J. Nutrition 56:115, 1955.

GOT and GPT References (Continued):

23. Mossberg, S. M.: "Myocardial Infarction-like Syndrome in Cholecystectomized Patients given Narcotics". Brit. Med. J. 1:948, 1964.

24. Mossberg, S. M., Ross, G.: "High Serum Transaminase Activity Associated with Extra-hepatic Biliary Disease". Gastroenterology 45:345, 1963.

25. Myerson, R. M., Hurwitz, J. K., Sall, T.: "Serum and Cerebrospinal Fluid Transaminase Concentrations in Various Neurologic Disorders". New Eng. J. Med. 257:273, 1957.

26. Munck, B. G., Kjerulf, K.: "Comparative Investigations on the Activity of Leucine Aminopeptidase, Glutamic Oxalacetic Transaminase, and Alkaline Phosphatase in Serum". Gut 2:225, 1961.

27. Orm, S., Davies, J. P. H., Weinbaum, I., Taggert, P., Kitchen, L. D.: "Conversion of Atrial Fibrillation in Sinus Rhythm by Direct Current Shock". Lancet 2:159, 1963.

28. Prince, A. M., Gershon, R. K.: "The Use of Serum Enzyme Determinations to Detect Anicteric Hepatitis". Transfusion 5:120, 1965.

29. Remmers, A. R., Kaljot, V.: "Serum Transaminase Levels: Effect of Strenuous and Prolonged Physical Exercise on Healthy Young Subjects". J. Amer. Med. Assoc. 185:968, 1963.

30. Resnick, W. H.: "Preinfarction Angina". Mod. Conc. Cardiovasc. Dis. 31:751, 1962.

31. Shuster, F., Napier, E. A., Jr., Henley, K. S.: "Serum Transaminase Activity following Morphine, Meperidine, and Codeine in Normals". Amer. J. Med. Sci. 246:714, 1963.

32. Snyder, D. D., Barnard, C. N., Varco, R. L., Lillehei, C. W.: "Serum Enzymes Following Cardiac Surgery". Surgery 44:1083, 1958.

33. Stone, M. L., Lending, M., Slobody, M. D., Mestern, J.: "Glutamic Oxalacetic Transaminase and Lactic Dehydrogenase in Pregnancy". Amer. J. Obstet. & Gynec. 80:104, 1960.

GOT and GPT References (Continued):

34. Tan, C. O., Cohen, J., West, M., Zimmerman, H. J.: "Serum Enzymes in Disease. XIV. Abnormality of Levels of Transaminases and Glycolytic and Oxidative Enzymes and of Liver Functions as related to the Extent of Metastatic Carcinoma of the Liver". Cancer 16:1373, 1963.

35. Ticktin, H. E., Trujillo, N. P.: "Serum Enzymes in Diagnosis". Disease-a-Month, 1966.

36. Wacker, W. E. C., Rosenthal, M., Snodgrass, P. J., Adamor, E.: A Triad for the Diagnosis of Pulmonary Embolism and Infarction". J. Amer. Med. Assoc. 178:8, 1961.

37. Weinstein, B. R., Korn, R. J., Zimmerman, H. J.: "Obstructive Jaundice as a Complication of Pancreatitis". Ann. Intern. Med. 58:245, 1963.

38. West, M., Gelb, D., Pilz, C. G., Zimmerman, H. J.: "Serum Enzymes in Disease. VII. Significance of Abnormal Serum Levels in Cardiac Failure". Amer. J. Med. Sci. 241:350, 1961.

39. West, M., Zimmerman, H. J.: "Lactic Dehydrogenase and Glutamic Oxalacetic Transaminase in Normal Pregnant Women and Newborn Children". Amer. J. Med. Sci. 235:443, 1958.

40. West, M., Zimmerman, H. J.: "Serum Enzymes in Hepatic Disease". Med. Clin. N. Amer. 43:371, 1959.

41. Woods, J. D., Lauri, W., Smith, W. G.: "The Reliability of the Electrocardiogram in Myocardial Infarction". Lancet 2:265, 1963.

42. Wroblewski, F.: "The Clinical Significance of Transaminase Activities of Serum". Amer. J. Med. 27:911, 1959.

43. Wroblewski, F., LaDue, J. S.: "Serum Glutamic Pyruvic Transaminase (SGPT) in Hepatic Disease. A Preliminary Report". Ann. Intern. Med. 45:801, 1956.

44. Wroblewski, F., LaDue, J. S.: Serum Glutamic Pyruvic Transaminase in Cardiac and Hepatic Disease". Proc. Soc. Exp. Biol. Med. 91:569, 1956.

GOT and GPT References (Continued):

45. Wolf, P. L., Langston, C., Potolsky, A. I., Williams, D. J.: "Ketosis Causing Spurious Elevation of SGOT Values". Clin. Chem. 17:341, 1971.

46. Wolf, P. L., Williams, D. J., Coplon, N., Coulson, A. S.: "Low Aspartate Transaminase Activity in Serum of Patients Undergoing Chronic Hemodialysis". Clin. Chem. 18:567, 1972.

CHAPTER 5

LACTIC DEHYDROGENASE

Lactic dehydrogenase (LDH) acts in the glycolytic cycle to catalyze the conversion of lactic acid to pyruvic acid. It is widely distributed throughout the body. The isoenzymes of lactic dehydrogenase are composed of two types of subunits or monomers. Each isoenzyme contains four subunits in one of five different combinations. The percentage distribution of the LDH isoenzymes in various tissues is related to the metabolic activity of the tissue. The isoenzyme that exhibits optimal activity at high levels of lactic acid predominates in those cells where this compound tends to be present. The isoenzyme that acts on pyruvic acid is usually present in greater amounts in tissues with a richer supply of oxygen. LDH_1 is the usual isoenzyme in aerobic tissues while LDH_5 predominates in anaerobic tissues.

The five LDH isoenzymes differ in their mobility in an electrophoretic field (Fig. 5-1). A sixth isoenzyme has been identified in testicular tissue.[13] When the isoenzymes are electrophoretically separated, LDH_1 travels between albumin and $alpha_1$ globulin; LDH_2 with $alpha_1$ globulin; LDH_3 with beta globulin; LDH_4 with the fast gamma globulin; and LDH_5 with the slow gamma globulin.[24]

Each isoenzyme of LDH is a tetramer containing varying combinations of two monomers, A and B [31] or H and M [48]. A random genetically controlled combination of the monomers appears to be the mechanism whereby isoenzymes are produced. Each isoenzyme has the same molecular weight of approximately 135,000. A regularity in the spacing of the five LDH isoenzymes occurs in the electrophoretic pattern.[48]

Lactic Dehydrogenase (Continued):

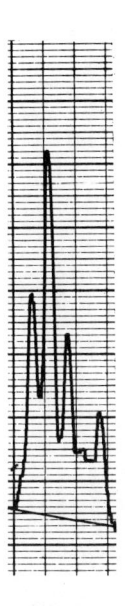

Fig. 5-1 Electrophoresis of serum isoenzymes of LDH showing a normal pattern.

The heat stability of LDH isoenzymes are of diagnostic importance. LDH_5 is heat labile at 65°C. for 30 minutes in contrast to heat stability of LDH_1. Stability during storage is also important. With storage LDH_5 disappears easily while LDH_1 concentration remains constant at 25°C., 4°C., and -20°C. for 1 month. At -20°C., there is rapid loss of LDH_4 and LDH_5 within days. After 8 to 10 days at 25°C., LDH_2 and LDH_3 decreases. The best way to store sera is at room temperature or at 4°C. avoiding excessive heat and bacterial contamination.[12,20,22]

Lactic dehydrogenase is an intracellular enzyme. Elevations occur in the LDH serum level when there is cellular death and leakage of enzyme from the cell. When neoplastic cells proliferate, serum LDH also will be elevated due to increased production. Strenuous exercise may increase serum LDH due to leakage from skeletal muscle. The enzyme may also be elevated in the postpartum period due to the muscle exertion during labor.

Lactic Dehydrogenase (Continued):

Levels of serum lactic dehydrogenase are not influenced by meals.[51] LDH is somewhat higher in infants and children than in adults. Hemolysis will increase the level because a large amount of enzyme is present within red blood cells. Gerlach's ratio is useful in hemolysis. The normal ratio of LDH to GOT is 5:1. In hemolysis the ratio becomes 10:1. The activity remains stable in stored serum at room temperature.[25]

Oxalate inhibits LDH,[9] thus it is advisable to use serum rather than plasma for the enzyme determination. Inhibitors of LDH are present in urine and these should be removed by dialysis before assaying in urine.

Causes for an increase in serum lactic dehydrogenase are:
1. Acute myocardial infarction
2. Acute leukemia and lymphoma
3. Pernicious anemia
4. Acute pulmonary infarction
5. Malignant neoplasms
6. Acute renal infarction
7. Hepatic necrosis
8. Sprue due to megaloblastic anemia
9. Skeletal muscle necrosis, inflammation or dystrophy
10. Shock with necrosis of various major organs
11. Necrosis of brain
12. Hemolytic anemia

In order to determine which tissue is diseased when total lactic dehydrogenase is elevated, it should be separated into its isoenzymes. This is best accomplished by electrophoresis, but other methods are also available. The heat stability-lability test can be used to differentiate LDH_1 from LDH_5. The serum is diluted with buffer at pH 7.4 and incubated for 30 minutes at 65°C. If LDH persists, the presence of LDH_1 is indicated since LDH_5 is labile.[36] In contrast, another method of differentiation is based on the fact that LDH_1 reacts with the substrate for alpha hydroxybutyrate dehydrogenase, whereas the other isoenzymes show progressively decreasing activity. Another less-used procedure for the chemical fractionation of LDH is based on the fact that LDH_1 reacts best at low substrate concentration and is not inhibited by urea.[27] In contrast, LDH_5 reacts best at high substrate concentrations and is inhibited by urea.

The major localization of the various isoenzymes is shown in Table 5-1.

Lactic Dehydrogenase (Continued):

TABLE 5-1

LDH_1	LDH_2	LDH_3	LDH_4	LDH_5
Heart	Red Blood Cells	Lung	Kidney	Liver
Red Blood Cells	R. E. System	Placenta	Medulla	Kidney
R. E. System	Kidney Cortex		Skeletal	Medulla
Kidney Cortex	Lung		Muscle	Skeletal
Muscular Dystrophy			Pancreas	Muscle
			Placenta	Pancreas

Acute myocardial infarction causes an LDH elevation within the first 12 hours after onset of the infarct. A peak is reached within 72 hours and there is persistence of elevation for 7-10 days. Alpha hydroxybutyrate dehydrogenase elevation may persist for 2 weeks.[18] Thus, if the elevation of GOT, LDH, or CPK is missed and a myocardial infarct is suspected, alpha hydroxybutyrate dehydrogenase should be determined. Generally coronary insufficiency without infarction does not cause LDH elevation. The increase in serum LDH is proportional to the size of the myocardial infarction. It may also increase to a higher level if congestive heart failure associated with the myocardial infarct occurs with leakage from the liver.[21]

Use of myocardial enzymes as indices for recognition of rejection of human heart transplants is controversial. Barnard felt that the enzymes were of no predictive value. Cooley indicated the first evidence of acute rejection is a significant rise in LDH_1 and LDH_2. The LDH_1 rise preceeds other clinical signs by 24 hours. Barnard believes this rise is related to the megaloblastic erythropoiesis accompanying azothioprine therapy.[7]

Stinson relates the rise of CPK, GOT and LDH to inflammation of skeletal muscle with injection of antilymphocytic globulin.[40]

In acute pulmonary infarction, elevation of LDH_2 and LDH_3 occurs (Fig. 5-2). Generally GOT and CPK levels are normal. In some series all patients with pulmonary infarcts had increased LDH and normal CPK but in others only half the patients with pulmonary infarcts had LDH elevation and normal or slightly increased CPK.[1, 11, 14, 34, 39]

Lactic Dehydrogenase (Continued):

Fig. 5-2   Photomicrograph of a hemorrhagic pulmonary infarct which was associated with total serum LDH of 350 I.U. Isoenzyme LDH_3 was prominently elevated.

In patients with liver disease, total LDH will increase with hepatocellular necrosis. Elevation is usually LDH_4 and LDH_5,[40,52] especially the latter. LDH_5 elevation is seen in congestive heart failure with necrosis of the hepatic centrilobular cells, in acute and chronic active hepatitis, and in carcinoma metastatic to the liver (Fig. 5-3). LDH_5 is also increased with necrosis of skeletal muscle.[3,15] The necrosis may result from trauma, dystrophic, or inflammatory lesions. Recently LDH_1 and LDH_2 isoenzymes have also been shown to be increased in skeletal muscle necrosis. This is probably because of the alteration in the isoenzyme resulting from the pathologic lesion affecting the skeletal muscle.[32,41]

Lactic Dehydrogenase (Continued):

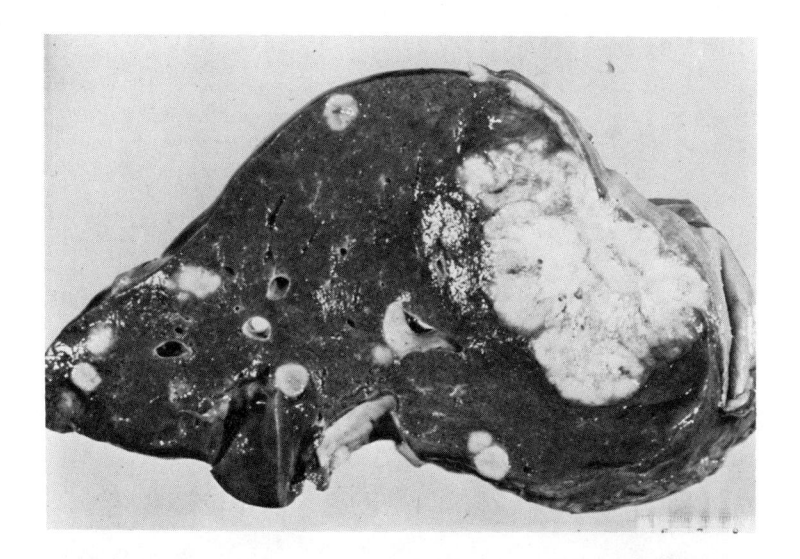

Fig. 5-3 Gross photograph of liver demonstrating metastatic carcinoma from sigmoid colon to liver. The serum LDH was 450 I.U. and indicated tumor infiltrate in the liver.

Patients with infarction of the renal cortex have an increased serum LDH usually LDH_1 and LDH_2. Increased LDH in the urine may result from infarction of the renal cortex or carcinoma of the kidney.[2,17] Other urinary tract conditions that cause an elevated urine LDH are acute cystitis, acute pyelonephritis, and acute glomerulonephritis.[35] When there is rejection of a kidney transplant, serum LDH may rise.

Malignant neoplasms result in elevated LDH_2, LDH_3, LDH_4 and LDH_5. The increased activity is usually due to proliferation and necrosis of neoplastic cells containing the enzymes. A relationship exists between the serum elevation and the rapidity of growth of the malignancy. Serum LDH_2, LDH_3, and LDH_4 rise in patients with carcinoma of the ovary. Patients with seminoma, dysgerminoma of the ovary, or malignant teratoma of the testes or ovary show LDH elevations. Patients with advanced carcinoma of the prostate may present with

Lactic Dehydrogenase (Continued):

LDH$_5$ elevation. The enzyme decreases when the carcinoma is successfully treated by orchiectomy or stilbesterol or both. Patients undergoing x-ray or chemotherapy for malignancy may have a decrease in serum LDH. Patients with leukemia, either acute or chronic (Fig. 5-4), may present with an elevated serum LDH.[6,38,45,46,53] Various lymphomas also may present with elevated serum LDH. The use of LDH as a screening test for leukemia or cancer is not entirely reliable; various series differ in the incidence of elevated LDH in cancer patients from 40 to 90 percent.

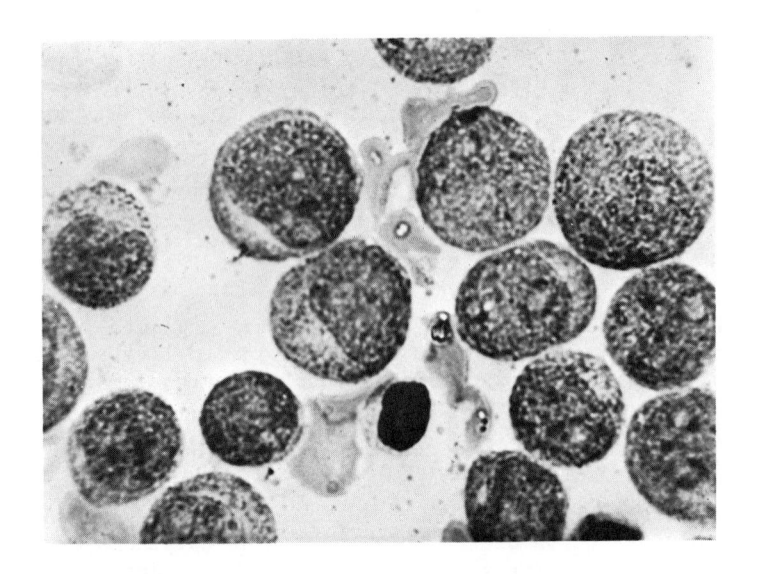

Fig. 5-4 Photomicrograph of bone marrow smear from a patient with acute myelogenous leukemia in which the total serum LDH was 540 I.U. LDH$_1$ and LDH$_2$ were both elevated.

Various anemias, especially megaloblastic ones such as pernicious anemia result in an elevated LDH (Fig. 5-5). The enzyme is produced and released from the megaloblasts.[26,30] Destruction of the megaloblasts may account for elevation of LDH by utilizing the tetrazolium formazan technique and staining smears of pernicious anemia bone marrows. It has been shown that LDH is present in large amounts in

Lactic Dehydrogenase (Continued):

the cytoplasm of the megaloblasts. Treatment with vitamin B_{12} causes a rapid decrease in serum LDH. Patients with hemolytic anemia may also have elevated serum LDH since there is an abundant amount of the enzyme in red blood cells.

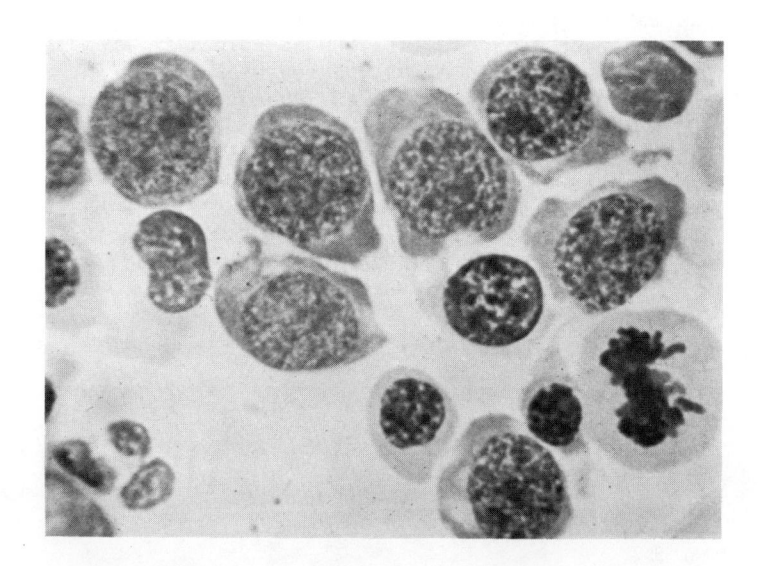

Fig. 5-5 Photomicrograph of bone marrow smear from a patient with megaloblastic pernicious anemia. The total LDH was elevated to 410 I.U. with a prominent elevation of LDH_1.

Only a minimal elevation of LDH occurs during pregnancy,[5,29] and this is found only during and shortly after labor. The elevation occurs for approximately 2 days after delivery, and may be related to the increased muscular activity during labor or to necrosis of the placenta. Isoenzymes LDH_3 and LDH_4 may be elevated. LDH activity may be elevated if there is an abruptio placenta.[8,28] This elevation results from necrosis of the placenta and also from the blood clot associated with the abruptio placenta. An elevation of LDH has been found in patients with choriocarcinoma and hydatidiform mole.[47] LDH activity of umbilical cord blood is greater than that of blood of the normal adult.[23] The activity in the umbilical cord blood is

Lactic Dehydrogenase (Continued):

higher in jaundiced babies. Exchange transfusion in erythroblastosis causes a decline in LDH associated with the decrease in jaundice.

LDH_2 and LDH_3 are elevated in the spinal fluid when there are destructive lesions of the central nervous system.[16,49] Cerebrospinal fluid LDH is also elevated in Tay-Sachs disease, with the level reaching its peak in the second year of the disease and then declining to normal.[4] In contrast, spinal fluid LDH is normal in Niemann-Pick disease.[3] Cerebrospinal fluid LDH becomes elevated in patients who have infarction of the cerebral cortex, tuberculous leptomeningitis, convulsive disorders, and hemorrhage into the cerebral cortex.

LDH is elevated in the gastric juice of patients with pernicious anemia and those with carcinoma of the stomach (Fig. 5-6), but not in patients with benign peptic ulceration.[33,37] Blood in the stomach will cause a high LDH in the gastric juice.

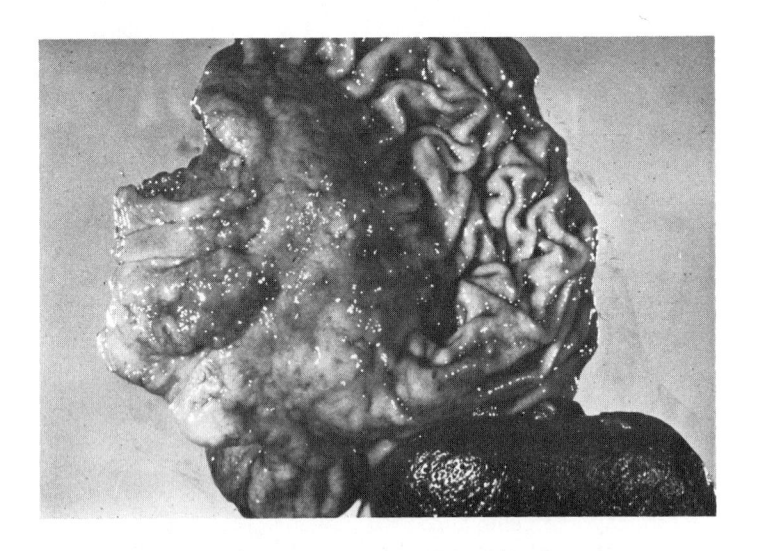

Fig. 5-6 Gross photograph of carcinoma of stomach in which the LDH of the gastric juice was elevated. Benign gastric ulcer does not cause an increase in gastric juice LDH.

Lactic Dehydrogenase (Continued):

Lactic dehydrogenase is present in increased amounts in synovial fluid in patients who have rheumatoid arthritis (Fig. 5-7), due to the increased number of cells in the fluid producing the enzyme. Conditions in which there is an increased cell count in the synovial fluid will result in an elevated LDH.[10,42]   In osteoarthritis, the LDH content of the synovial fluid is normal.

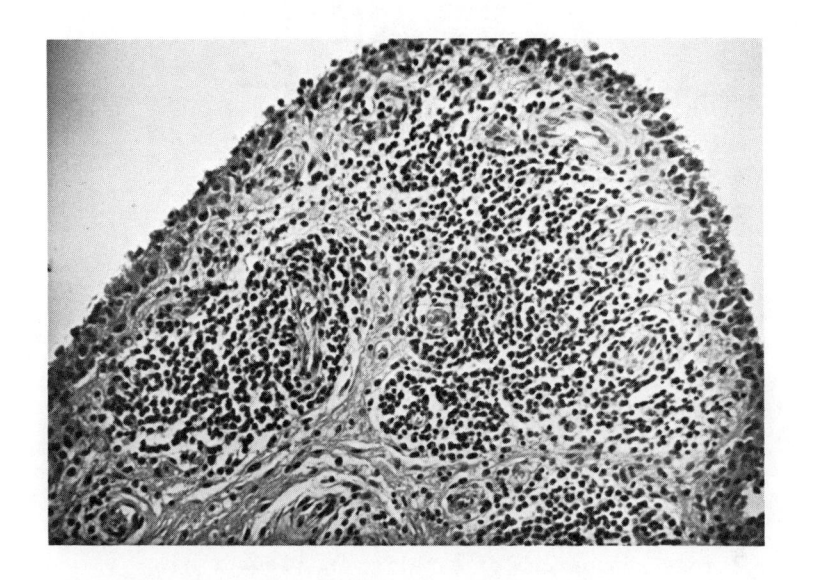

Fig. 5-7   Photomicrograph of synovial tissue involved with rheumatoid arthritis. The synovial fluid contained an elevated LDH in contrast to the synovial fluid from osteoarthritis which contains a normal LDH.

LDH is present in elevated amounts in effusions that are exudates.[19]   Determination of LDH in an effusion is one of the better methods of differentiating a transudate from an exudate (Fig. 5-8). The increased LDH level reflects the greater cell count of an exudate over that of a transudate. Effusions which are exudates are caused by a malignant neoplasm, primary or metastatic in the mesothelial surface or by the presence of an inflammatory, immunologic, or disease associated with tissue necrosis, characterized by increased numbers

Lactic Dehydrogenase (Continued):

of cells in the effusion with a resultant increased LDH level in the fluid. In a malignant effusion, the effusion-serum LDH ratio is greater than one. Transudates may have a slight rise in LDH secondary to an increase in serum LDH.

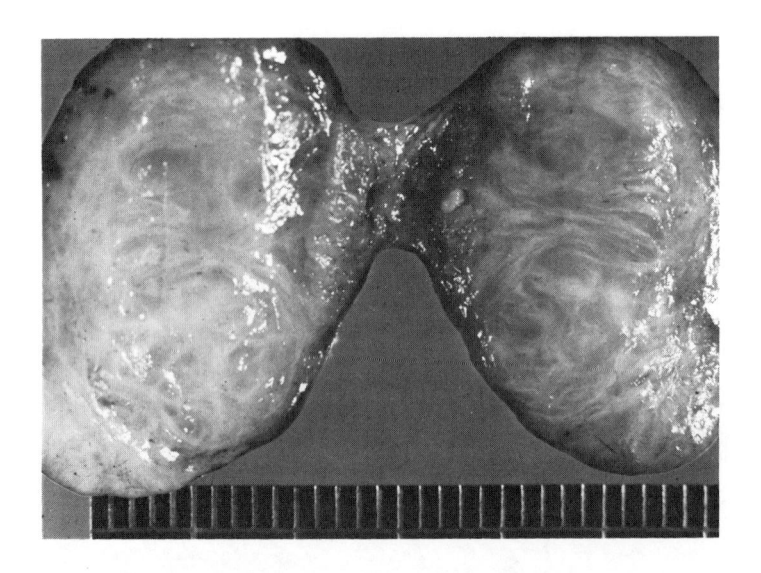

Fig. 5-8 Gross photograph of a malignant pleural mesothelioma. The LDH of the fluid was 850 I.U. emphasizing the prominent elevation of LDH in an exudative fluid.

Causes of a decreased serum LDH level are:
1. Clofibrate utilization
2. Oxalate anticoagulant
3. Muscular dystrophy causing LDH_5 to change to LDH_1

Lactic Dehydrogenase (Continued):

REFERENCES

1. Amador, E., Potchen, E. J.: "Serum Lactic Dehydrogenase Activity and Radioactive Lung Scanning in the Diagnosis of Pulmonary Embolism". Ann. Intern. Med. 65:1247, 1966. (Also Letter to the Editor. Ibid. 66:1292, 1967).

2. Amador, E., Zimmerman, T. S., Wacker, W. E. C.: "Urinary Alkaline Phosphatase Activity. I. Elevated Urinary LDH and Alkaline Phosphatase Activities for the Diagnosis of Renal Adenocarcinoma". J. Amer. Med. Assoc. 185:133, 1963.

3. Aronson, S. M.: "Enzyme Determinations in Neurologic and Neuromuscular Diseases of Infancy and Childhood". Pediat. Clin. N. Amer. 7:527, 1960.

4. Aronson, S. M., Saifer, A., Volk, B. W.: "Serial Enzyme Studies of Serum and Cerebrospinal Fluid in Amaurotic Family Idiocy". Amer. J. Dis. Child 97:684, 1959.

5. Atuk, N. O., Wax, S. H., Word, B. H., McGaughey, H. S., Corey, E. L., Wood, J. E.: "Observations of the Steady State of Lactic Dehydrogenase Activity across the Human Placental Membrane". Amer. J. Obstet. & Gynec. 82:271, 1961.

6. Bierman, H. R.: "Correlation of Serum Lactic Dehydrogenase Activity with Clinical Status of Patients with Cancer, Lymphomas, and the Leukemias". Cancer Res. 17:660, 1957.

7. Barnard, C. N.: "Human Heart Transplantation. The Diagnosis of Rejection". Amer. J. Cardiology 22:811, 1968.

8. Boutselis, J. G., Wensinger, J. A., Sollarsk, J.: "Serum Lactic Dehydrogenase in Abruptio Placentae". Amer. J. Obstet. & Gynec. 86:762, 1963.

9. Christian, D. G.: "Drug Interference with Laboratory Blood Chemistry Determinations". Amer. J. Clin. Path. 54:118, 1970.

10. Cohen, A. S.: "Lactic Dehydrogenase (LDH), and Transaminase (GOT) Activity of Synovial Fluid and Serum in Rheumatic Disease States, with a Note on Synovial Fluid LDH Isoenzymes". Arthritis Rheum. 7:490, 1964.

Lactic Dehydrogenase References (Continued):

11. Coodley, E. L.: "Enzyme Profiles in the Evaluation of Pulmonary Infarction". J. Amer. Med. Assoc. 207:1307, 1969.

12. Dawson, D. M., Goodfriend, T. L., Kaplan, N. O.: "Lactic Dehydrogenases: Functions of the Two Types". Science 143:929, 1964.

13. Eliasson, R., Haggmann, K., Wiklund, B.: "Lactate Dehydrogenase in Human Seminal Plasma". Scand. J. Lab. Invest. 20:353, 1967.

14. Elliott, B. A., Jepson, E. M., Wilkinson, J. H.: "Serum Alpha-hydroxybutyrate Dehydrogenase, a new Test with Improved Specificity for Myocardial Lesions". Clin. Sci. 23:305, 1962

15. Emery, A. E. H.: "Electrophoretic Pattern of Lactate Dehydrogenase in Carriers and Patients with Duchenne Muscular Dystrophy". Nature 201:1044, 1964.

16. Fleisher, G. A., Wakin, K. G., Goldstein, N. P.: "Glutamic Oxalacetic Transaminase and Lactic Dehydrogenase in Serum and Cerebrospinal Fluid of Patients with Neurologic Disorders". Mayo Clin. Proc. 32:188, 1957.

17. Gault, M. H., Steiner, G.: "Serum and Urinary Enzyme Activity after Renal Infarction". Canad. Med. Assoc. J. 93:1101, 1965.

18. Gudbjarnson, S., Cowan, C., Braasch, W., Bing, R. J.: "Changes in Enzyme Pattern of Infarcted Heart Muscle during Tissue Repair". Cardiologia 51:148, 1967.

19. Horrocks, J. E., King, J., Waind, A. P. B., Ward, J.: "Lactate Dehydrogenase Activity in the Diagnosis of Malignant Effusions". J. Clin. Path. 15:57, 1962.

20. Kaplan, N. O.: "Nature of Multiple Molecular Forms of Enzymes". Ann. N. Y. Acad. Sci. 151:382, 1968.

21. Killen, D. A.: "Serum Enzyme Elevations: A Diagnostic Test for Myocardial Infarction during the Early Postoperative Period". Arch. Surg. 96:200, 1968.

22. Kreutzer, H. H., Fenis, W. H. S.: "Lactic Dehydrogenase Isoenzymes in Blood Serum after Storage at Different Temperatures". Clin. Chim. Acta. 9:64, 1964.

Lactic Dehydrogenase References (Continued):

23. Lapan, B., Friedman, M. A.: "A Comparative Study of Fetal and Maternal Serum Enzyme Levels". J. Lab. Clin. Med. 54:417, 1959.

24. Lauderback, A. L., Shanbron, E.: "Lactic Dehydrogenase Isoenzyme Electrophoresis". J. Amer. Med. Assoc. 205:294, 1968.

25. Levitan, R., Golub, M., Setzel, L.: "LDH Activity in Saliva, Bile, Gastric and Duodenal Contents". Amer. J. Dig. Dis. 5:458, 1960.

26. Libnoch, J. A., Yakulis, V. J., Heller, P.: "Lactate Dehydrogenase in Magaloblastic Bone Marrow". Amer. J. Clin. Path. 45:302, 1966.

27. Lindy, S., Konttinen, A.: "Urea-stable Lactate Dehydrogenase as an Index of Cardiac Isoenzymes". Amer. J. Cardiol. 19:563, 1967.

28. Little, W. A.: "Serum Enzyme Alterations in Abruptio Placentae". Surg. Forum 10:716, 1959.

29. Little, W. A.: "Serum Lactic Dehydrogenase in Pregnancy". Amer. J. Obstet. & Gynec. 13:152, 1959.

30. McCarthy, C. F., Fraser, I. D., Read, A. E.: "Plasma Lactate Dehydrogenase in Megaloblastic Anemia". J. Clin. Path. 17:51, 1966.

31. Markert, C. L.: "Lactate Dehydrogenase Isoenzymes: Dissociation and Recombination of Subunits". Science 140:1329, 1963.

32. Perkoff, G. T., Hardy, P., Velez-Garcia, E.: "Reversible Acute Muscular Syndrome in Chronic Alcoholism". New Eng. J. Med. 274:1277, 1966.

33. Piper, D. W., Macoun, M. L., Broderick, F. L., Fenton, B. H., Builder, J. E.: "The Laboratory Diagnosis of Gastric Juice". Gastroenterology 45:614, 1963.

34. Polachek, A. A., Zoneraich, S., Zoneraich, O., Sass, M.: "Pulmonary Infarction and Serum Lactic Dehydrogenase". J. Amer. Med. Assoc. 204:811, 1968.

Lactic Dehydrogenase References (Continued):

35. Riggins, R. A., Kiser, W. S.: "A Study of Lactic Dehydrogenase in Urine and Serum of Patients with Urinary Tract Disease". J. Urol. 90:594, 1963.

36. Rosalki, S. B.: "Serum Hydroxybutyrate Dehydrogenase: A New Test for Myocardial Infarction". Brit. Heart J. 25:795, 1963.

37. Smyrniotis, F., Schenker, S., O'Donnel, J., Schiff, L.: "Lactic Dehydrogenase Activity in Gastric Juice for the Diagnosis of Gastric Cancer". Amer. J. Dig. Dis. 7:712, 1962.

38. Spector, I., McFarland, W., Trujillo, N. P., Ticktin, H. E.: "Bone Marrow Lactic Dehydrogenase in Hematologic and Neoplastic Disease". Enzym. Biol. Clin (Basel) 7:78, 1966.

39. Stevens, L. E., Burdette, W. J.: "Enzymatic Levels in Pulmonary Infarction". Arch. Surg. 88:705, 1964.

40. Stinson, E. B., Dong, E., Bieber, C. P., Schroeder, J. S., Shumway, N. E.: "Cardiac Transplantation in Man. I. Early Rejection". J. Amer. Med. Assoc. 207:2233, 1969.

41. Trujillo, N. P., Nutter, D., Evans, J. M.: "The Isoenzymes of Lactic Dehydrogenase. II. Pulmonary Embolus, Liver Disease, the Postoperative State, and other Medical Conditions". Arch. Intern. Med. 119:333, 1967.

42. Vesell, E. S., Osterland, K. C., Bearn, A. G., Kunkel, H. G.: "Isoenzymes of Lactic Dehydrogenase. Their Alterations in Arthritic Synovial Fluid and Sera". J. Clin. Invest. 41:2012, 1962.

43. Wacker, W. E. C., Rosenthal, M., Snodgrass, P. J., Amador, E. "A Triad for the Diagnosis of Pulmonary Embolism and Infarction". J. Amer. Med. Assoc. 178:8, 1961.

44. Wacker, W. E. C., Snodgrass, P. J.: "Serum LDH Activity in Pulmonary Embolism Diagnosis". J. Amer. Med. Assoc. 174:2142, 1960.

45. Walsh, P. N., Kissane, J. M.: "Nonmetastatic Hypernephroma with Reversible Hepatic Dysfunction". Arch. Intern. Med. 122:214, 1963.

Lactic Dehydrogenase References (Continued):

46. West, M., Heller, P., Zimmerman, H. J.: "Serum Enzymes in Disease. II. Lactic Dehydrogenase and Glutamic Oxalacetic Transaminase in Patients with Leukemia and Lymphoma". Amer. J. Med. Sci. 235:689, 1958.

47. West, M., Zimmerman, H. J.: "Lactic Dehydrogenase and Glutamic Oxalacetic Transaminase in Normal Pregnant Women and Newborn Children". Amer. J. Med. Sci. 235:443, 1958.

48. Wilson, A. C., Chan, R. D., Kaplan, N. O.: "Functions of the Two Forms of Lactic Dehydrogenase in the Breast Muscle of Birds". Nature 197:331, 1963.

49. Wroblewski, F., Decker, B., Wroblewski, R.: "The Clinical Implications of Spinal Fluid Lactic Dehydrogenase Activity". New Eng. J. Med. 258:635, 1958.

50. Wroblewski, F., Gregory, K. F.: "Lactic Dehydrogenase Isoenzymes and their Distribution in Normal Tissues and Plasma". Ann. N. Y. Acad. Sci. 94:212, 1961.

51. Wroblewski, F., LaDue, J. S.: "Serum Lactic Dehydrogenase". Proc. Soc. Exp. Biol. Med. 90:210, 1955.

52. Zimmerman, H. J.: "Serum Enzymes in the Diagnosis of Hepatic Disease". Gastroenterology 46:613, 1964.

53. Zondag, H. A., Klein, F.: "Clinical Applications of Lactate Dehydrogenase Isoenzymes: Alterations in Malignancy". Ann. N. Y. Acad. Sci. 151:578, 1968.

CHAPTER 6

CREATINE PHOSPHOKINASE AND ALDOLASE

Creatine Phosphokinase

The isoenzymes of creatine phosphokinase (CPK) vary in tissue distribution. The enzyme is found primarily in the brain, heart muscle, and skeletal muscle. It is absent in the erythrocyte, liver, and kidney. The greatest amount is present in skeletal muscle, followed next by heart muscle and brain.[13,36] A small amount is present in the adrenals, thyroid, and lung. The normal serum values are: Males up to 90 I.U. and females up to 70 I.U.

CPK is an unstable enzyme.[21] When cysteine is added to the serum, 90 percent of activity persists after 4 days of storage at 4°C. If no cysteine is added, there will be a 50 percent loss of activity after 4 to 6 hours at room temperature or 24 hours at 4°C.[42]

If the thiol-stimulated reaction systems for CPK are utilized, there is extension of the activity span for CPK beyond the usual time. Persistence of activity with such stimulated reactions may extend for 8 to 9 days.

Human tissues contain three naturally occurring CPK isoenzymes. The enzyme may be a dimer of two subunits with an intermediate unit which may be a hybrid of the two parent forms. The electrophoretically rapid-moving isoenzyme is the brain type, whereas the slow-moving one is from muscle. Skeletal muscle contains two isoenzymes, one slow-moving and the other intermediate. The brain isoenzyme travels in the area of albumin; skeletal muscle isoenzyme in the area of gamma globulin; and heart muscle isoenzyme in the area of beta and gamma globulin.

    Causes of an elevated serum CPK level are:
1. Acute myocardial infarction
2. Muscular dystrophy
3. Hypothyroidism with skeletal muscle myopathy[20,30]
4. Infarction of the cerebral cortex of the brain
5. Schizophrenia
6. Severe exercise
7. Injections of drugs into skeletal muscle
8. Postpartum period
9. Minimal elevation in pulmonary infarction
10. Diabetic ketoacidosis, especially in the recovery perio
11. Malignant hyperthermia
12. Heart transplant rejection [4,9,10,40]
13. Clofibrate Therapy

Creatine Phosphokinase and Aldolase (Continued):

An increase in serum CPK occurs within 12 hours after acute myocardial infarction; the peak being reached within 24 to 36 hours (Table 6-1). The level usually returns to normal within four days after onset of the infarction.[22] CPK determination can also be helpful in the evaluation of other cardiac or lung conditions. Serum GOT elevation may be present in congestive heart failure due to liver congestion; since CPK is not present in the liver, there will be no CPK elevation in congestive heart failure. CPK will be elevated in such cases only when myocardial damage has occurred.[43] To distinguish between acute myocardial infarction and pulmonary infarction, it is important to request both LDH and CPK. Both enzymes are increased in myocardial infarction but only LDH is prominently increased in pulmonary infarction. CPK may rarely be elevated in pulmonary infarction.[44]

TABLE 6-1

ACTIVITY OF MYOCARDIAL SERUM ENZYMES

	CPK	GOT	LDH	HBD
Rise Begins	4 hours	12-18 hours	12-18 hours	12-18 hours
Peak Activity	36 hours	24 hours	72 hours	72 hours
Return to Normal	4 days	5 days	7-10 days	10-14 days

CPK elevation due to recent electrical defibrillation has been observed. Fifty percent of patient may show an elevation due to powerful intercostal muscle contraction.[25,32] Prolonged arrhythmias may on occasion produce an elevation of serum enzymes, such as GOT especially with a cardiac rate over 160 beats per minute.[3] The GOT may be derived from the liver in this condition. In the pseudomyocardial infarction syndrome, serum GOT is elevated. After a morphine injection, spasm of the Sphincter of Oddi occurs, with elevation of the intra-biliary duct pressure and leakage of GOT from the liver.

One of the valuable reasons for determination of CPK is the clinical appraisal of patients suffering from muscular dystrophy (Fig. 6-1). A markedly elevated CPK as much as a 50 fold rise, may be present in a majority of these patients.[26,34] CPK is also present in increased amounts in the serum of patients with active muscle necrosis due to varied causes. Patients with active acute myositis, trauma to skeletal muscle, surgery involving skeletal muscle, injections of drugs into the skeletal muscle, severe muscular

Creatine Phosphokinase and Aldolase (Continued):

exercise, acute muscular necrosis in alcoholism, all show elevation of serum CPK (Fig. 6-2).[7,11,23,28,31,35,41] Serum CPK elevation is extremely helpful in detecting the preclinical muscular dystrophy state.[2] In addition, asymptomatic carriers usually show a rise in the enzyme.[24,33] In such patients it is extremely important to advise them not to exercise before a CPK determination is obtained. CPK is usually normal in the neurogenic atrophies of skeletal muscle and in myasthenia gravis.

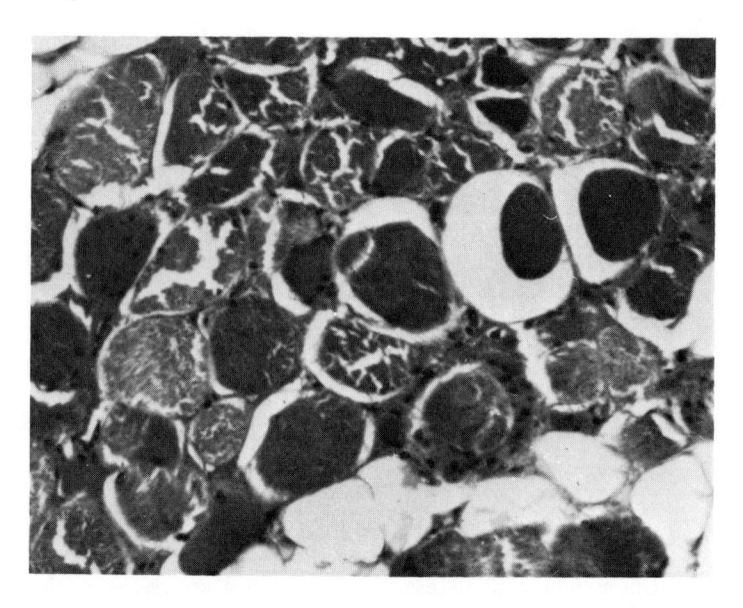

Fig. 6-1 Photomicrograph of skeletal muscle biopsy of a 2 year old male child who has muscular dystrophy. The CPK was 490 I.U.

Malignant hyperthermia is an abnormal response to certain anesthetics of neuromuscular blockers or both and is manifested as a rapid and alarming rise in temperature often to levels incompatible with life. Metabolic and respiratory acidosis, hypoxia, hyperkalemia, myoglobinemia and myoglobinuria, hemolysis, and coagulation difficulties are often associated with this complication. The disease is genetically determined and has been reported in a number of families. A number of investigators have reported elevated levels of creatine phosphokinase in patients surviving an episode of malignant hyperthermia.

Creatine Phosphokinase and Aldolase (Continued):

Similar findings have been made in their asymptomatic relatives. Thus, in any patient with a suggestive positive history, or when a possible episode has occurred in a relative, levels of CPK should be determined and, if elevated, halogenated anesthetics and depolarizing neuromuscular blockers should be avoided. However, the presence of an elevated CPK level does not necessarily indicate that the patient is predisposed to malignant hyperthermia.

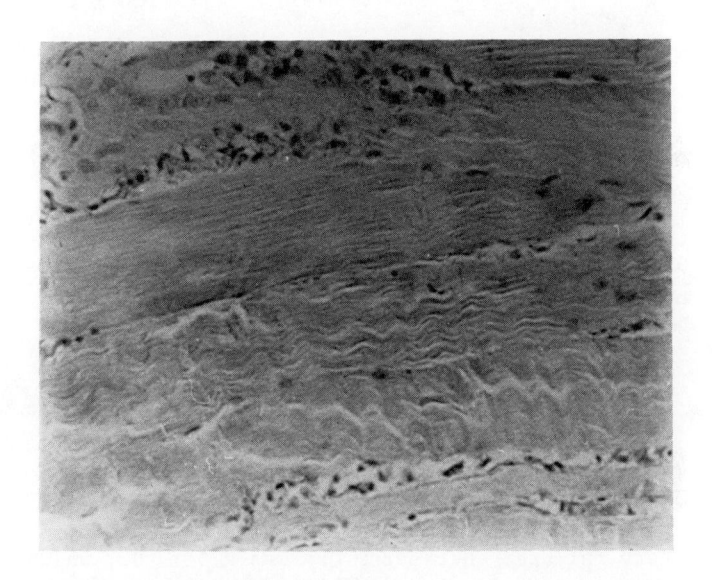

Fig. 6-2 Photomicrograph of a patient who developed diffuse necrosis of skeletal muscle secondary to alcoholism. The patient's CPK was 17,000 I. U. Transient renal acute tubular necrosis was present related to myoglobinuria from rhabdomyolysis. This lesion is known as alcoholic myopathy.

A majority of patients with hypothyroidism have been shown to have an increase in serum CPK, [20,30] the exact etiology of the increase being unknown. Some investigators feel that there may be a release of CPK from the damaged thyroid tissue; others that there is an induction and increase of CPK activity by excess thyroid-stimulating

Creatine Phosphokinase and Aldolase (Continued):

hormone (TSH) activity in patients with hypothyroidism. The isoenzyme is a skeletal muscle one. The assay of serum CPK activity is extremely valuable in screening for hypothyroidism (Fig. 6-3).

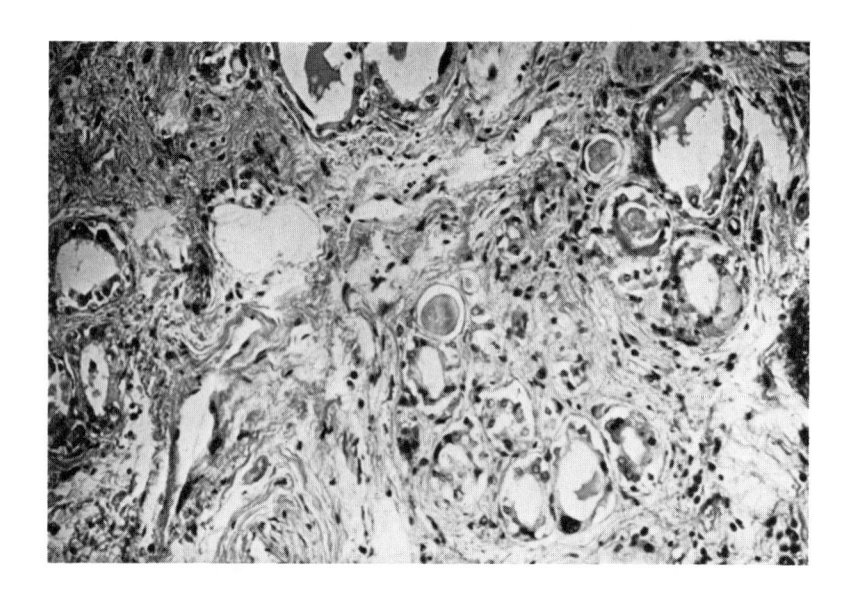

Fig. 6-3 Photomicrograph of atrophic thyroid gland from a patient who died with myxedema secondary to pituitary failure. Her serum CPK was 275 I. U., which was elevated due to hypothyroid myopathy.

CPK plays a role in the confirmation of diseases of the brain.[6,2?] CPK activity is elevated during the acute phase of cerebral damage and usually returns to normal within 2 weeks. It is usually elevated in patients with infarction of the cerebral cortex (Fig. 6-4),[1,8] and those with necrosis within a brain tumor.[14,37] It has also been reported to be elevated in patients with schizophrenia. There is a lack of correlation between elevation of serum CPK and the degree of central nervous system damage.

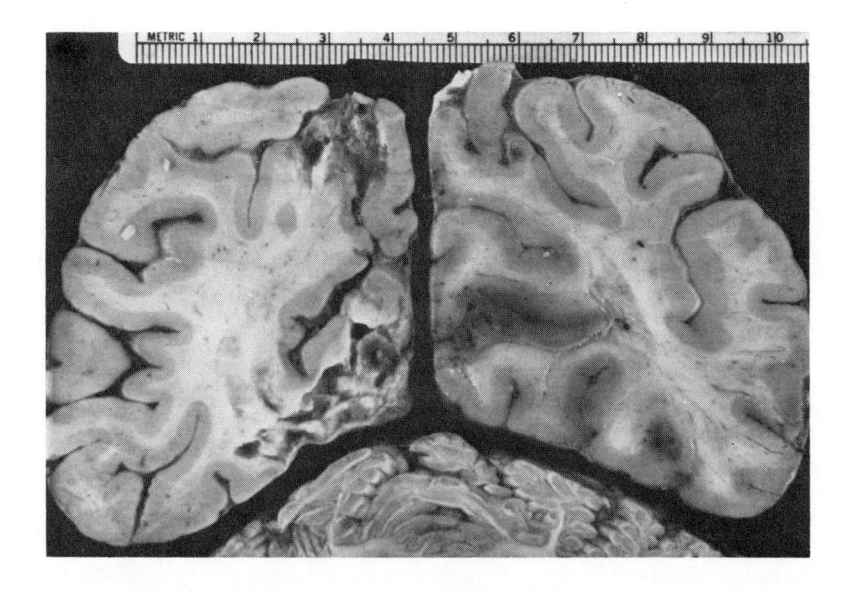

Creatine Phosphokinase and Aldolase (Continued):

Fig. 6-4 Gross photograph of brain demonstrating a recent cerebral infarct which caused an elevated serum CPK of 370 I.U.

Cerebrospinal fluid CPK is normal in transient cerebral ischemia without damage and in the demylinating diseases. The cerebellum has very little CPK activity and thus cerebellar lesions do not produce a rise in cerebrospinal fluid CPK activity.[17]

The level of CPK in umbilical cord blood is higher than that found in the serum during infancy and in adults,[19] as a result of intense muscular activity associated with labor. Higher values are also present in umbilical cord blood in infants born to mothers with obstetric difficulties. The elevated CPK levels in these conditions usually return to normal within 7 days after birth.

A minimal rise in CPK in maternal blood occurs within the first few days after delivery with a return to normal within a week.[16,27] The elevation probably results from involution of the uterus, and the intense muscular activity associated with labor. Extraction studies have shown a 15-fold rise in CPK activity in the pregnant myometrium over the non-pregnant myometrium.

Creatine Phosphokinase and Aldolase (Continued):

Causes of a decreased serum CPK level are:
1. Storage of serum specimen
2. Oxalates and utilization of plasma
3. Inhibitors such as drugs in serum

Aldolase

The adult serum aldolase level is 2 to 14 I.U. The newborn has five times the adult level, and children have twice the adult level.

Aldolase is present in many organs such as the liver and heart, but is in highest concentration in the skeletal muscle. The red blood cell contains approximately 150 times more aldolase than is present in the plasma, and large amounts have also been found in platelets.[12] When platelets disrupt during clotting, very little of the enzyme is released into the serum because it is bound by the fibrin clot.

Causes of an elevated serum aldolase level are:[15,39]
1. Skeletal muscle necrosis
2. Myocardial necrosis
3. Liver necrosis
4. Hemolysis

The main value for determining serum aldolase is in the assessment of skeletal muscle disease, such as muscular dystrophy and particularly in differentiating primary inflammatory necrotizing or dystrophic diseases of skeletal muscle from disorders in which there is neurogenic atrophy. Aldolase is elevated in muscular dystrophy and acute muscular necrosis, but not in myasthenia gravis or neurogenic atrophy.

Various explanations have been offered for the high aldolase activity in the neonatal period.[18] One group feels that the elevated levels are related to the increased activity of the adrenal gland where the enzyme is present.[5] Increased blood aldolase in rabbits occurs after ACTH is administered. Another group feels that aldolase is high because of the increased requirements of glycolytic enzymes in the newborn in response to their extrauterine life.[38]

The enzyme is elevated in patients with viral hepatitis.[15] Some authors believe that the measurement of aldolase is the most useful enzyme assay in the assessment of patients with viral hepatitis. The activity of the enzyme is elevated in patients with acute viral hepatitis and parallels the activity of SGPT.

Creatine Phosphokinase and Aldolase (Continued):

REFERENCES

1. Acheson, J., James, D. C., Hutchinson, E. C., Westhead, R.: "Serum Creatine Kinase Levels in Cerebral Vascular Disease". Lancet 1:1306, 1965.

2. Aebi, U., Richterich, R., Colombo, J. P., Rossi, E.: "Progressive Muscular Dystrophy. II. Biochemical Identification of the Carrier State in the Recessive Sex-linked Juvenile (Duchenne) Type by Serum Creatine Phosphokinase Determination". Enzymol. Biol. Clin. 1:61, 1961/62.

3. Batsakis, J. G., Preston, J. A., Briere, R. O., Giesen, P. C.: "Iatrogenic Aberrations of Serum Enzyme Activity". Clin. Biochem. 2:126, 1968.

4. Barnard, C. N.: "Human Heart Transplantation: The Diagnosis of Rejection". Amer. J. Cardiol. 22:811, 1968.

5. Brenner, M. D.: "Studies on the Involution of the Fetal Cortex of the Adrenal Gland". Amer. J. Path. 16:787, 1940.

6. Buckell, M.: "Enzymes in the Cerebrospinal Fluid". Proc. Assoc. Clin. Biochem. 5:33, 1968.

7. Cherington, M., Lewin, E., McCrimmon, A.: "Serum Creatine Phosphokinase Changes following Needle Electromyographic Studies". Neurology 18:271, 1968.

8. Cohen, G., Werner, M.: "Serum Creatine Kinase in Cerebral Vascular Accidents". Lancet 1:389, 1967.

9. Cooley, D. A., Bloodwell, R. D., Hallman, G. L., Nora, J. J.: "Transplantation of the Human Heart: Report of Four Cases". J. Amer. Med. Assoc. 205:479, 1968.

10. Cooley, D. A., Hallman, G. L., Bloodwell, R. D., Nora, J. J., Leachman, R. D.: "Human Heart Transplantation: Experience with Twelve Cases". Amer. J. Cardiol. 22:804, 1968.

11. Crowley, L. V.: "Creatine Phosphokinase Activity in Myocardial Infarction, Heart Failure, and Following Various Diagnostic and Therapeutic Procedures". Clin. Chem. 14:1185, 1968.

Creatine Phosphokinase and Aldolase References (Continued):

12. Dale, R. A.: "Demonstration of Aldolase in Human Platelets: The Relation to Plasma and Serum Aldolase". Clin. Chim. Acta. 5:652, 1960.

13. Dawson, D. M., Fine, I. H.: "Creatine Kinase in Human Tissues". Arch. Neurol. 16:175, 1967.

14. Dubo, H., Park, D. C., Pennington, R. J. T., Kalbag, R. M., Walton, J. M.: "Serum Creatine Kinase in Cases of Stroke, Head Injury, and Meningitis". Lancet 2:743, 1967.

15. Eismann, J.: "Die Aktivitat der Aldolase, ein Beitrag zur Differential Diagnose des Ikterus". Deutsch. Med. J. 7:204, 1956.

16. Emery, A. E. H., Pascasio, F. M.: "The Effect of Pregnancy on the Concentration of Creatine Kinase in Serum, Skeletal Muscle, and Myometrium". Amer. J. Obstet. & Gynec. 91:18, 1965.

17. Fowler, W. M., Jr., Pearson, C. M.: "Diagnostic and Prognostic Significance of Serum Enzymes. II. Neurologic Diseases other than Muscular Dystrophy". Arch. Phys. Med. 45:125, 1964.

18. Friedman, M. M., Lapan, B.: "Serum Aldolase in the Neonatal Period: Including a Colorimetric Determination of Aldolase by Standardization with Dehydroxyacetone". J. Lab. Clin. Med. 51:745, 1958.

19. Griffiths, P. D.: "Serum Levels of Creatine Phosphokinase". J. Clin. Path. 17:56, 1964.

20. Griffiths, P. D.: "Serum Enzymes in Disease of the Thyroid Gland". J. Clin. Path. 18:660, 1965.

21. Hess, J. W., MacDonald, R. P.: "Serum Creatine Phosphokinase Activity: A New Diagnostic Aid in Myocardial and Skeletal Muscle Disease". J. Mich. Med. Soc. 62:1095, 1963.

22. Hess, J. W., MacDonald, R. P., Frederick, R. J., Jones, R. N. Neely, J., Gross, D.: "Serum Creatine Phosphokinase (CPK) Activity in Disorders of Heart and Skeletal Muscle". Ann. Intern. Med. 61:1015, 1964.

Creatine Phosphokinase and Aldolase References (Continued):

23. Hudson, A. J., Strickland, K. P., Wilensky, A. J.: "Serum Enzyme Studies in Familial Hyperkalemic Periodic Paralysis". Clin. Chim. Acta. 17:331, 1967.

24. Hughes, B. P.: "Serum Enzymes in Carriers of Muscular Dystrophy". Brit. Med. J. 2:923, 1962.

25. Hunt, D., Bailie, M. J.: "Enzyme Changes following Direct Current Counter Shock". Amer. Heart J. 76:340, 1968.

26. Kar, N. C., Pearson, C. M.: "Creatine Phosphokinase Isoenzymes in Muscle in Human Myopathies". Amer. J. Clin. Path. 43:207, 1965.

27. Konttinen, A., Pyorala, T.: "Serum Enzyme Activity in Late Pregnancy, at Delivery, and during Puerperium". Scand. J. Clin. Lab. Invest. 15:429, 1963.

28. Lafair, J. S., Myerson, R. M.: "Alcoholic Myopathy: With Special Reference to the Significance of Creatine Phosphokinase". Arch. Intern. Med. 122:417, 1968.

29. Nathan, M. J.: "Creatine Phosphokinase in the Cerebrospinal Fluid". J. Neurol. Neurosurg. & Psychiat. 30:52, 1967.

30. Norris, F. H., Jr., Panner, B. J.: "Hypothyroid Myopathy". Arch. Neurol. 14:574, 1966.

31. Nuttall, F. Q., Jones, B.: "Creatine Kinase and Glutamic Oxalacetic Transaminase Activity in Serum: Kinetics of Change with Exercise and Effect of Physical Conditioning". J. Lab. Clin. Med. 71:847, 1968.

32. Oram, S., Davies, J. P. H., Weinbaum, I., Taggert, P., Kitchen, L. D.: "Conversion of Atrial Fibrillation to Sinus Rhythm by Direct Current Shock". Lancet 2:159, 1963.

33. Pearce, J. M. S., Pennington, R. J. T., Walton, J. N.: "Serum Enzyme Studies in Muscle Disease. III. Serum Creatine Kinase Activities in Relatives of Patients with the Duchenne Type of Muscular Dystrophy". J. Neurol. Neurosurg. & Psychiat. 27:181, 1964.

34. Pearson, C. M.: "Muscular Dystrophy: Review and Recent Observations". Amer. J. Med. 35:632, 1963.

Creatine Phosphokinase and Aldolase References (Continued):

35. Perkoff, G. T., Hardy, P., Velez-Garcia, E.: "Reversible Acute Muscular Syndrome in Chronic Alcoholism". New Eng. J. Med. 274:1277, 1966.

36. Rosalki, S. B.: "Creatine Phosphokinase Isoenzymes". Nature 207:414, 1965.

37. Rosalki, S. B.: "Creatine Kinase and Brain Damage". Lancet 2:722, 1967.

38. Schapira, F.: "Hyperaldolasemie Plasmatique Provoquee par la Cortisone et l'ACTH". Soc. Biol. 148:1997, 1954.

39. Sibley, J. A., Fleisher, G. A.: "Clinical Significance of Serum Aldolase". Mayo Clin. Proc. 29:591, 1954.

40. Stinson, E. B., Dong, E., Bieber, C. P., Schroeder, J. S., Shumway, N. E.: "Cardiac Transplantation in Man. I. Early Rejection". J. Amer. Med. Assoc. 207:2233, 1969.

41. Swaimann, K. F., Awad, E. A.: "Creatine Phosphokinase and other Serum Enzyme Activity after Controlled Exercise". Neurology 14:977, 1964.

42. Velez-Garcia, E., Hardy, P., Dioso, M., Perkoff, G. T.: "Cysteine-stimulated Serum 5-Creatine Phosphokinase: Unexpected Results". J. Lab. Clin. Med. 68:636, 1966.

43. Vincent, W. R., Rapaport, E.: "Serum Creatine Phosphokinase in the Diagnosis of Acute Myocardial Infarction". Amer. J. Cardiol. 15:17, 1965.

44. Wacker, W. E. C., Rosenthal, M., Snodgrass, P. J., Amador, E.: "A Triad for the Diagnosis of Pulmonary Embolism and Infarction". J. Amer. Med. Assoc. 178:8, 1961.

45. Wiesmann, U., Colombo, J. P., Adam, A., Richterich, R.: "Determination of Cysteine-activated Creatine Kinase in Serum". Enzymol. Biol. Clin. (Basel) 7:266, 1966.

CHAPTER 7

SORBITOL, GLUTAMIC, ISOCITRIC, GLUCOSE-6-PHOSPHATE
DEHYDROGENASES AND GLUCOSE-6-PHOSPHATASE

Sorbitol Dehydrogenase

Sorbitol is a polyhydric alcohol derived from glucose. Its conversion to glucose takes place in the liver and is catalyzed by the action of the reductase, sorbitol dehydrogenase (SDH), which is mainly a liver enzyme. No serum SDH activity is present in the healthy adult. Unlike the liver enzymes GPT, GOT, and LDH, sorbitol dehydrogenase is not elevated in the serum in organ disease other than liver damage.[1] Myocardial infarction, skeletal muscle disease or kidney necrosis does not lead to elevated serum levels. Assays for sorbitol dehydrogenase in various tissues are: Liver - 57 percent; prostate - 14 percent; kidney - 12 percent; spleen - 5 percent; testis - 2 percent; heart and skeletal muscle - 1 percent each; all other tissues - 7 percent.

Erythrocytes and heart and skeletal muscle contain minimal amounts of the enzyme. Spurious elevations are not seen in hemolyzed blood, nor is there an elevated level after intramuscular injections as is the case for GOT and CPK.

SDH is usually elevated to the greatest degree in acute hepatitis.[9] Biliary tract obstruction does not cause elevated levels unless it is longstanding and then only if it leads to parenchymal cell damage. The liver specificity of this enzyme is emphasized by the normal levels of SDH in myocardial infarction. Furthermore in disease of the prostate or kidney which contain an appreciable amount of tissue SDH, serum levels are not increased.

Sorbitol dehydrogenase is easy to assay and shows promise as a valuable enzyme determination in the diagnosis of liver disease.

Glutamic Dehydrogenase

Glutamic dehydrogenase (GLDH) is found in abundant amount in the liver.[6] The heart and spleen contain small amounts, and only trace levels are present in the serum of healthy persons. GLDH is not present in erythrocytes; thus, hemolysis will not influence the assay.

Glutamic Dehydrogenase (Continued):

The enzyme contains zinc and has a high molecular weight of 1,000,000. It has been differentiated into six isoenzymes which have not been categorized into a pattern useful for diagnostic purposes. Elevated serum GLDH activity has not been consistently found in liver disease and is not as reliable as the more commonly used GOT, GPT, or LDH. Henley, Schmidt, and Schmidt have reported a significant elevation of GLDH in hepatic coma.[12]

Isocitric Dehydrogenase

Isocitric dehydrogenase (ICD) is an ubiquitous enzyme present in cell mitochondria.[4] It occurs in two isoenzyme forms: a fast and a slow electrophoretic fraction. The liver contains the fast component, whereas, the slow component is present in heart muscle. The slow portion is not as heat stable as the fast component. The enzyme is also present in certain neoplasms. Activity has also been demonstrated in platelets and in erythrocytes; very little activity is demonstrable in normal plasma.

Heart ICD activity is heat labile and is inactivated by 56°C.; liver ICD is heat stable at 56°C. After an acute myocardial infarct, the heat-labile heart isoenzyme is found for only a very short time in the blood. With liver damage, however, the activity of the serum ICD persists.[21] ICD elevation is a sensitive indicator of parenchymal hepatic disease, but increased enzyme activity cannot be used to differentiate various liver diseases.[26] Normal ICD activity is usually found after an acute myocardial infarct; however, if heart failure occurs as a complication, there may be an elevation due to the intense hepatic congestion with necrosis of centrilobular cells.[17] Various lesions such as viral hepatitis, metastatic carcinoma, hepatoma, and severe congestive heart failure have caused ICD elevation.[7] Enzyme activity is elevated early in the course of infectious hepatitis, persists for approximately 3 weeks, and then returns to normal with recovery from the illness. When the level remains elevated, one may assume that there is persistence of the viral hepatitis.[8,11]

Isocitric dehydrogenase is increased in patients with megaloblastic anemia; presumably the enzyme is produced in large amounts by the megaloblasts. This phenomenon is similar to the one that exists in pernicious anemia with serum elevation of LDH. Both ICD and LDH are mitochondrial enzymes, and if one stains the megaloblast with the tetrazolium-formazan technique, a large amount of ICD and LDH is found in the megaloblasts. With proliferation and death of the megaloblasts, a large amount of enzyme will be contributed to

Isocitric Dehydrogenase (Continued):

the serum. Rare reports of ICD elevations are present where patients with carcinoma of the pancreas, carcinoma of the prostate, infarction of the placenta, or myeloid leukemia have elevated serum ICD.[19] Isocitric dehydrogenase activity may be increased in the cerebrospinal fluid with a primary or metastatic carcinoma to the central nervous system, and in patients who have had cerebral infarction or acute bacterial meningitis.[24]

The normal level of isocitric dehydrogenase in the blood is up to 11 I.U. Isocitric dehydrogenase activity in the umbilical cord blood is 40 to 250 percent higher than that found in the serum of adults. These high levels are maintained in the infant during the first 3 months of life and may be elevated throughout the first year.

Glucose-6-Phosphate Dehydrogenase

During the last few years a number of hemolytic anemias have been investigated which are caused by inherited errors in the metabolism of glucose. The erythrocyte derives all its energy from the metabolism of glucose via the Embden-Meyerhof cycle associated with the pentose-monophosphate shunt. Red blood cells do not contain all of the components of the tricarboxylic acid cycle or cytochrome oxidase system, but the mature red cell does possess a functioning Krebs cycle. Over 90 percent of glucose consumption occurs through the Embden-Meyerhof anaerobic pathway. Glucose must be phosphorylated before it can be metabolized by the erythrocytes. A direct oxidative pathway involving glucose metabolism is also present within the red cell; this is known as the pentose-monophosphate shunt. Usually less than 10 percent of glucose undergoes direct oxidation by this shunt. The primary function of the anaerobic cycle is to provide sufficient adenosine triphosphate for the necessary metabolism of the red cell.[18] Many of the recognized abnormalities of red cells are associated with enzyme deficiencies which occur with increasing age of the cells.[16] These conditions can be considered as metabolic abnormalities that are imposed on the normal aging process.

Young red blood cells such as reticulocytes, utilize more glucose and possess higher levels of glycolytic enzymes than do older red cells. Aging red cells have a very rapid decrease in activity in glucose-6-phosphate dehydrogenase (G-6-PD) activity.[25] Along with this decrease is a drop in adenosine triphosphate (ATP) as well as in diphosphopyridine nucleotide or nicotinamide adenine dinucleotide (NAD). One

Glucose-6-Phosphate Dehydrogenase (Continued):

group of investigators has demonstrated that G-6-PD activity in reticulocytes is ten times greater than in older red cells.

G-6-PD deficiency may exist as a number of genetic variants related to different changes in enzyme structure. It may be cause for neonatal jaundice and kernicterus in highly affected groups such as in blacks. It has been more commonly recognized as the cause for hemolytic anemia in patients who have taken various drugs (Fig. 7-1).

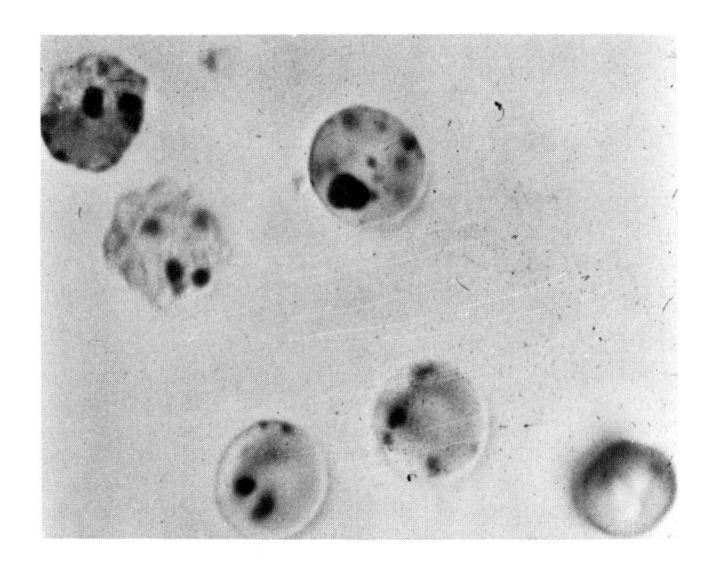

Fig. 7-1 Positive Heinz-Ehrlich body peripheral blood preparation in a patient who has a G-6-PD deficiency and developed a severe hemolytic anemia when he took Gantrisin for a urinary tract infection.

G-6-PD is an essential enzyme in the pentose monophosphate shun which produces reduced triphosphopyridine nucleotide or nicotinamide adenine dinucleotide (NADH). In the absence of sufficient NADH ther is a marked reduction in available reduced glutathione. Although th role of the latter compound in preserving red blood cell integrity i not entirely understood, it appears to be an important one. The glutathione stability test was originally proposed by Beutler.[2]

Glucose-6-Phosphate Dehydrogenase (Continued):

Reduced glutathione-sensitive cells are extremely sensitive to oxidation in the presence of drugs such as acetylphenylhydrazine. The test is not as reliable in detecting heterozygous females; in one series 30 to 50 percent negative results occurred. Another test is based on the inability of sensitive red cells to reduce the dye, brilliant cresyl blue.[3] This test is also unsatisfactory for the detection of the female heterozygote with the disease. In this test, decoloration should be complete within 100 minutes; partial decoloration at 100 minutes indicates partial enzyme deficiency. Because of the qualitative nature of the test, intermediate results are sometimes obtained and are difficult to interpret; those blood samples showing no decoloration at 3 to 6 hours have virtually no enzyme activity. The direct estimation of G-6-PD activity is the most specific method, particularly with respect to the detection of the heterozygous patient.[14] Normal values are 120 - 240 $mu/10^9$ RBC. Other tests for this condition are discussed in Chapter 12 on enzyme abnormalities in the reticuloendothelial system.[15]

A deficiency of G-6-PD was recognized many years ago to be present throughout the world. It was recognized in the black races and in the Sephardic Jews. A hemolytic anemia may arise in patients with this enzyme deficiency; these patients usually become anemic when exposed to certain drugs or develop various diseases. Patients with G-6-PD deficiency may have a hemolytic crisis when drugs of the following types are taken:[23]

1. Analgesics or antipyretics such as phenacetin
2. Sulfa drugs such as sulfapyridine or sulfaoxazole (Gantrisin)
3. Antimalarials such as primaquine, quinacrine (atabrine), or quinine
4. Nitrofurantoin or nitrofurazone
5. Other drugs such as chloramphenicol (chloromycetin), p-aminosalicylic acid (PAS), quinidine or vitamin K

Recently, conditions such as viral hepatitis and diabetic acidosis have been shown to induce hemolytic anemia in patients with G-6-PD deficiency. It is thought that certain oxidants increase in the blood in these patients and this leads to an acute hemolytic crisis.[5]

The deficiency is more severe in the Middle East, as in the Sephardic Jew than in blacks. It is unwise to use for blood transfusion purposes, donor blood from patients who have severe G-6-PD deficiency from the Middle East. However, individuals in the United States who have G-6-PD deficiency may serve as blood donors.

Glucose-6-Phosphate Dehydrogenase (Continued):

Patients with an acute hemolytic anemia progress through various phases; the first phase may last 2 weeks, a 7 to 10 day recovery phase follows, and this is followed by an equilibrium phase in which the anemia disappears.[10] The exact mechanism for the hemolysis is not clear; however, it has been suggested that because of the G-6-PD deficiency, there is an associated NADH deficiency and a defect in the sodium pump of the red blood cell membrane. Associated with this defect is a defective generation of ATP because of glutathione deficiency. The potassium loss of G-6-PD deficient red cells has been reported to be greater than that of normal red cells following incubation with drugs such as primaquine. However, no significant differences in potassium content of G-6-PD deficient or normal red cells after incubation with acetylphenylhydrazine has been demonstrated. These investigations emphasize that differences exist in the type of injury induced by different drugs and that no single explanation suffices to explain the mechanism of hemolysis. More and more drugs are being implicated in the causation of hemolysis in patients who have G-6-PD deficiency.

Usually spontaneous hemolysis does not occur in the newborn with this enzyme deficiency. Administration of an offending drug to the mother with a G-6-PD deficient fetus prior to delivery may be hazardo

Glucose-6-Phosphatase

Glucose-6-phosphatase is present in the serum; this enzyme should not be confused with glucose-6-phosphate dehydrogenase. The enzyme is normally present in the liver and is normally present in serum in very small amounts.[13] The normal serum level is 0.12 I.U.

Glucose-6-phosphatase hydrolyzes glucose-6-phosphate to glucose and phosphate. Increases in the serum are found in:

1. Hepatitis - Maximal elevations are seen approximately 6 hour after contact with toxic agents, such as carbon tetrachloride or chloroform.
2. Cirrhosis - Less marked increases are seen in cirrhosis than in hepatitis.
3. Renal disease - This is probably due to reduced clearance of the enzyme.
4. Hemolysis - Free phosphates from hemolyzed red blood cells will interfere with the assay for glucose-6-phosphatase; thus hemolyzed blood is not suitab for the enzyme assay.

Dehydrogenases (Continued):

REFERENCES

1. Asada, M., Galamos, J. T.: "Sorbitol Dehydrogenase and Hepatocellular Injury: An Experimental and Clinical Study". Gastroenterology 44:518, 1963.

2. Beutler, E.: "The Glutathione Instability of Drug-sensitive Red Cells. A New Method for the In Vitro Detection of Drug Sensitivity". J. Lab. Clin. Med. 49:84, 1957.

3. Beutler, E.: "Glucose-6-Phosphate Dehydrogenase Deficiency and Nonspherocytic Congenital Hemolytic Anemia". Seminars Hemat. 2:91, 1965.

4. Bowers, G. N., Jr.: "Measurement of Isocitric Dehydrogenase Activity in Body Fluids". Clin. Chem. 5:509, 1959.

5. Burka, E. R., Weaver, Z. III, Marks, P. A.: "Clinical Spectrum of Hemolytic Anemia associated with Glucose-6-Phosphate Deficiency". Ann. Intern. Med. 64:817, 1966.

6. Carlson, A. S., Siegelman, A. M., Robertson, T.: "Glutamic Dehydrogenase. II. Activity in Human Serums, contrasted with that of Lactic Dehydrogenase and Glutamic Oxalacetic Transaminase". Amer. J. Clin. Path. 38:260, 1962.

7. Cohen, N. N., Potter, H. P., Jr., Bowers, G. N., Jr.: "An Evaluation of Isocitric Dehydrogenase in Liver Disease". Ann. Intern. Med. 55:604, 1961.

8. Cohn, E. M., Winsten, S., Abramson, E. B.: "A Study of the Serum Isocitric Dehydrogenase Test in Patients with Jaundice". Amer. J. Gastroenterology 44:45, 1965.

9. DeRitis, F., Giusti, G., Piccinino, F., Cacciatore, L.: "Biochemical Laboratory Tests in Viral Hepatitis and Other Hepatic Diseases". Bull. WHO 32:59, 1965.

10. Dern, R. J., Beutler, E., Alving, A. S.: "The Hemolytic Effect of Primaquine. II. The Natural Course of the Hemolytic Anemia and the Mechanism of its Self-limited Character".

11. Emmanuel, B., West, M., Zimmerman, H. J.: "Serum Enzymes in Disease. XII. Transaminases, Glycolytic and Oxidative Enzymes in Normal Infants and Children". Amer. J. Dis. Child. 105:77, 1963.

Dehydrogenases References (Continued):

12. Henley, K. S., Schmidt, E., Schmidt, F. W.: "Serum Enzymes". J. Amer. Med. Assoc. 174:977, 1960.

13. Koide, H., Oda, T.: "Pathological Occurrence of Glucose-6-Phosphatase in Serum in Liver Disease". Clin. Chim. Acta. 4:55, 1959.

14. Levy, L. M., Walter, H., Sass, M. D.: "Enzymes and Radioactivity in Erythrocytes of Different Ages". Nature 184:643, 1959.

15. Marks, P. A.: "Glucose-6-Phosphate Dehydrogenase - Clinical Aspects". In W. A. Wood (Ed.) ENZYMES OF CARBOHYDRATE METABOLISM, Vol. IX. Academic Press, pg. 131 - 137, 1966.

16. Marks, P. A., Banks, J.: "Studies on the Mechanism of Aging of Human Red Cells". Ann. N. Y. Acad. Sci. 75:95, 1958.

17. Pojen, J., Ninger, E., Tovarek, J.: "Changes in Serum Isocitrate Dehydrogenase in Myocardial Infarction". Enzym. Biol. Clin. (Basel) 3:184, 1963.

18. Rose, I. A., O'Connell, E. L.: "The Role of Glucose-6-Phosphate in the Regulation of Glucose Metabolism in Human Erythrocytes". J. Biol. Chem. 239:12, 1964.

19. Schwartz, M. K., Greenberg, E., Bodansky, O.: "Comparative Values of Phosphatases and Other Serum Enzymes in Following Patients with Prostatic Carcinoma". Cancer 16:583, 1963.

20. Sterkel, R. L., Spencer, J. A., Wolfson, S. K., Jr., William Ashman, H. G.: "Serum Isocitric Dehydrogenase Activity with Particular Reference to Liver Disease". J. Lab. Clin. Med. 52:176, 1958.

21. Strandjord, P. E., Thomas, K. E., White, L. P.: "Studies on Isocitric and Lactic Dehydrogenase in Experimental Myocardial Infarction". J. Clin. Invest. 38:2111, 1959.

22. Stuckey, W. J., Jr.: "Hemolytic Anemia and Erythrocyte Glucose-6-Phosphate Dehydrogenase Deficiency". Amer. J. Med. Sci. 251:104, 1966.

23. Tarlow, A. R., Brewer, G. J., Carson, P. E., Alving, A. S.: "Primaquine Sensitivity. Glucose-6-Phosphate Dehydrogenase Deficiency. An Inborn Error of Metabolism of Medical and Biological Significance". Arch. Intern. Med. 109:209, 1962

Dehydrogenases References (Continued):

24. Van Rymentant, M., Robert, J.: "Enzymes in Cancer. II. The Isocitric Dehydrogenase of the Cerebrospinal Fluid in Various Cancerous and Noncancerous Conditions". Cancer 13:877, 1960.

25. Vuopio, R.: "Red Cell Enzymes in Anemia". Scand. J. Clin. Lab. Invest. (Suppl. 72) 15:99, 1963.

26. Zimmerman, H. J.: "Serum Enzymes in the Diagnosis of Hepatic Disease". Gastroenterology 46:613, 1964.

CHAPTER 8

ENZYME ABNORMALITIES IN NEOPLASTIC DISEASE

Various neoplasms are associated with elevations of specific serum enzymes, although at the present time there is no general enzyme test for the detection of cancer. A well established enzyme abnormality in neoplastic disease is the elevation of serum acid phosphatase in cancer of the prostate.[29] In this chapter we will specifically refer to enzyme abnormalities associated with specific cancers. The determination of serum acid phosphatase is a useful procedure in the assessment of metastatic carcinoma of the prostate. Patients with metastatic carcinoma of the prostate may not develop an elevation of serum acid phosphatase.[12] In order to stimulate production of the enzyme by these metastases, the Sullivan Test (described in Chapter 2) is employed. A slight increase in the serum acid phosphatase after such stimulation may be of diagnostic value. In this test 25 mg. of testosterone is injected for five days to stimulate the production of acid phosphatase by the anaplasti carcinoma cells. Ninety percent of prostatic cancer patients have acid phosphatase elevation when there are metastases to distant sites.[10,12,26,31]

It is important to follow the effectiveness of treatment in patients with carcinoma of the prostate by serial determinations of serum acid phosphatase. When orchiectomy is performed, and the serum acid phosphatase was elevated prior to the orchiectomy, there should be a marked decrease in the enzyme level approximately 4 to 6 weeks after the surgery.[56] Estrogen therapy will also induce a decrease. If acid phosphatase levels remain elevated following castration, estrogen therapy or both, the therapy has not been successful. If therapy is successful and there is an exacerbation of the carcinoma, acid phosphatase may become elevated again.

With hyperplasia or neoplasia of the prostate gland, simple palpation of the gland may cause an elevation of the serum acid phosphatase.[13,18,24] However, elevation following a rectal examination is more frequent in patients with cancer of the prostate than in those without malignancies.[34] Rectal examination of a nondiseased prostate gland may produce a slight increase in the tartrate-inhibitable fraction of the acid phosphatase. Infarction of the prostate has been associated with a rise in the acid phosphatase.[50] Furthermore, trauma of the prostate by either surgery or catheterization may produce a transient elevation of acid phosphatase.

Enzyme Abnormalities in Neoplastic Disease (Continued):

In summary, acid phosphatase is elevated in patients with cancer of the prostate when: (1) the cancer has spread beyond the capsule of the prostate, and (2) it may provide initial evidence of metastasis from a prostatic carcinoma. It may indicate that the origin of a metastatic lesion is the prostate when the primary site of the malignancy is not clinically known. Serum acid phosphatase should be elevated and should regress after successful therapy has been instituted. It is a useful determination to follow the results of treatment in a patient with metastatic carcinoma of the prostate.[10]

The enzyme is relatively unstable in the presence of heat. An interesting association of fever and depression of serum acid phosphatase has been observed. In patients with elevated serum acid phosphatase from metastatic carcinoma of the prostate, onset of fever will cause a fall in the level.[35] When the fever regresses, the level will rise again.[55]

A new substrate has recently been described termed the thymolphthalein monophosphate which has specificity for only the prostatic isoenzyme. This substrate is currently being evaluated for detection of elevation of the prostatic isoenzyme associated with various prostatic diseases.[43]

It is recommended that the L-tartrate inhibition test be performed in the work-up of a patient who has metastatic carcinoma of the prostate (Table 8-1).

TABLE 8-1

INHIBITION VS. NON-INHIBITION OF ISOENZYMES
OF
ACID PHOSPHATASE [1]

	L-Tartrate	Formaldehyde	Ethyl Alcohol
Prostate	Inhibited	Not Inhibited	Inhibited
Liver	Inhibited		Not Inhibited
Kidney	Inhibited		
Spleen	Inhibited		Not Inhibited
Gaucher's disease (Fig. 8-1)	Not Inhibited		
Red Blood Cells (Fig. 8-2)	Not Inhibited	Inhibited	Inhibited

192

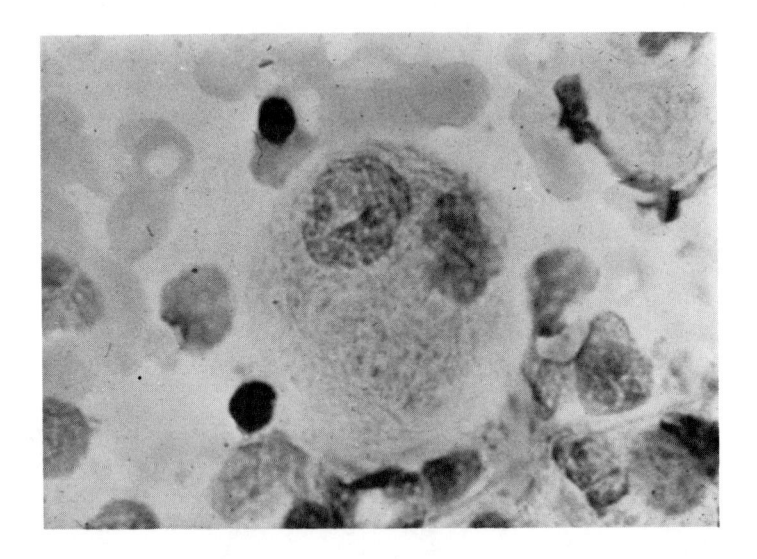

Enzyme Abnormalities in Neoplastic Disease (Continued):

Fig. 8-1 Photomicrograph of smear of bone marrow from a patient with Gaucher's disease. A Gaucher cell is present. A pancytopenia was present and the serum acid phosphatase was 2.6 I.U. and not inhibitable by L-tartrate.

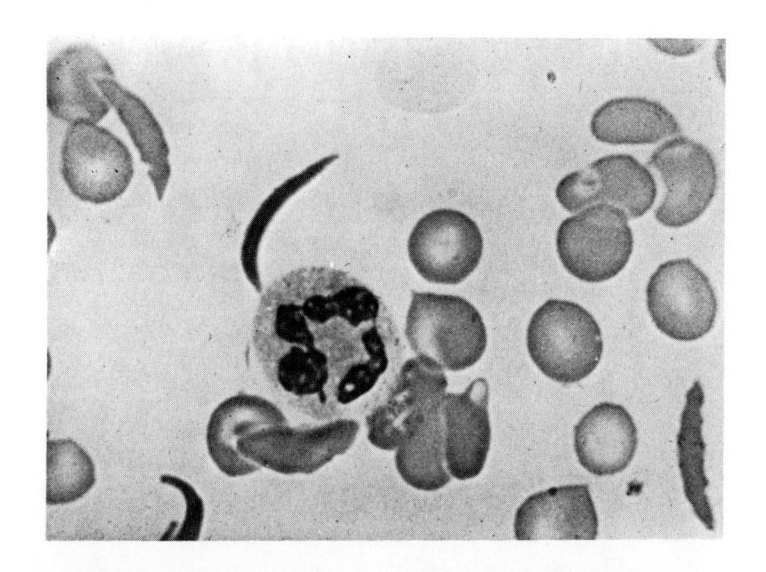

Enzyme Abnormalities in Neoplastic Disease (Continued):

Fig. 8-2 Photomicrograph of smear of peripheral blood
from a patient with an acute sickle cell crisis
hemolytic anemia. Note the numerous sickle
cells. The serum acid phosphatase was 1.96 I.U.

Various types of neoplastic bone disease such as osteogenic sarcoma and Paget's disease will cause an elevated serum acid phosphatase. Furthermore, extensive osteolytic bone disease such as is found in multiple myeloma may cause an elevated serum acid phosphatase (Fig. 8-3).[37]

Other causes for an elevated acid phosphatase are: 1). Gaucher's disease, L-tartrate will not inhibit the elevation of serum acid phosphatase in this disease, nor will formaldehyde;[48] 2). Necrosis of tissue such as renal or liver infarction, or metastatic cancer to liver;[19,23] 3). Hemolytic anemia; 4). Thrombocytopenia[41] with normal numbers of bone marrow megakaryocytes.

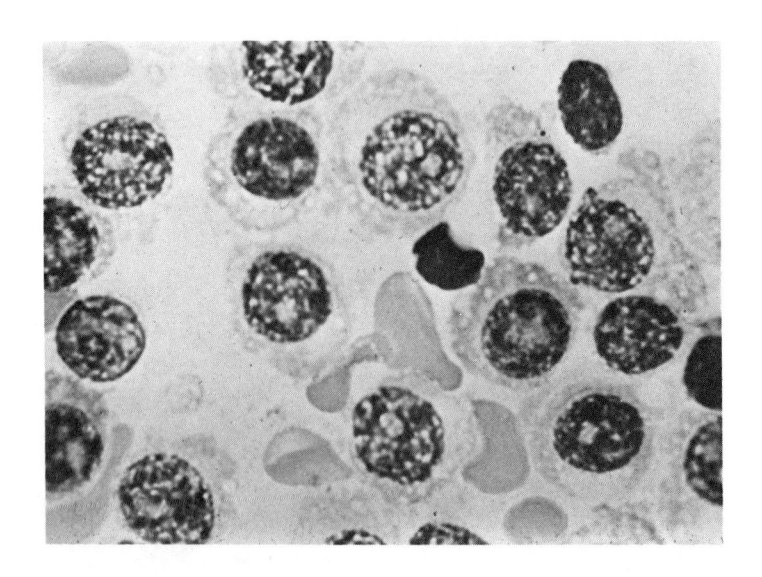

Enzyme Abnormalities in Neoplastic Disease (Continued):

Fig. 8-3 Photomicrograph of bone marrow of patient who died from multiple myeloma. The serum acid phosphatase was 2.3 I.U.

Lactic Dehydrogenase

Neoplastic involvement of pleural, pericardial, or peritoneal surfaces will produce a collection of fluid termed an exudate. This fluid will have a specific gravity greater than 1.018, a total protein greater than 2.5 gm. percent, and a glucose content less than that of plasma. The lipid content of an exudate is greater than that of plasma. Many malignant cells may be present in the exudate. Exudates may be milky and will have an opalescence that resembles chylous fluid. This type of fluid is called pseudochylous since it does not arise from rupture of the thoracic duct[10] from which chylous fluid originates.

The preferred biochemical method of differentiating a transudate from an exudate is by determining the LDH content.[11] Transudates are characterized by an LDH content similar to that of plasma; that is, 200 I.U. or less. In contrast, exudates have a greatly increased

Enzyme Abnormalities in Neoplastic Disease (Continued):

LDH content; that is above 200 I.U. and usually above 500 I.U. One must bear in mind that when the LDH content of plasma is elevated, it will also be elevated in pleural, pericardial, or peritoneal fluid as a reflection of the plasma LDH content. Thus, one might err in classifying a transudate as an exudate if the patient has a high blood LDH content. The increased LDH content of the neoplastic exudate results from the presence of great numbers of malignant cells that contribute their LDH to the fluid. Effusion LDH should be compared to serum LDH. With malignant effusions, the effusion:serum ratio is greater than one; with benign effusions, less than one.

Amylase and Lipase

Patients with carcinoma of the pancreas may present with elevated serum amylase and lipase if a secondary pancreatitis results from the carcinoma. When a middle-aged or elderly individual presents with acute pancreatitis without a history of alcoholism or biliary tract disease, the clinician should consider the possible presence of carcinoma of the pancreas causing pancreatic duct obstruction with subsequent acute pancreatitis.[17,38] However, serum amylase and lipase may be markedly decreased if the pancreas has been destroyed by the carcinoma. In addition to elevations of serum amylase and lipase in carcinoma of the pancreas, one should bear in mind that serum alkaline phosphatase, 5' nucleotidase, and leucine aminopeptidase may be elevated with carcinoma of the head of the pancreas.[5,9,25,27,44]

Alkaline Phosphatase

Patients with cancer metastatic to the liver from various primary sites, such as lung or colon, may present with minimal retention of Bromsulphalein (BSP), slightly elevated serum bilirubin, and markedly elevated serum alkaline phosphatase. The alkaline phosphatase elevation may be greater than that seen with biliary tract obstruction.[22,39,46] GOT and LDH, especially LDH_4 and LDH_5 may be slightly increased.[57]

The elevated serum alkaline phosphatase is usually a biliary tract type in that it is moderately heat stable and is not inhibited by phenylalanine. It is derived from proliferating bile ducts associated with the carcinoma metastatic to the liver or from liver parenchymal cells associated with intrahepatic obstruction.[42] Some observers feel that the alkaline phosphatase associated with liver metastases

Enzyme Abnormalities in Neoplastic Disease (Continued):

is derived from vascular endothelium which is increased surrounding the metastatic carcinoma.

Various types of carcinoma metastasize to bone, causing both osteolytic and osteoblastic lesions. Carcinoma of the prostate is a leading candidate for causing osteoblastic metastases. Both serum acid phosphatase and serum alkaline phosphatase are elevated in these patients.[21] Carcinomas that metastasize to bone causing an osteolytic type of lesion are from lung, thyroid, breast, kidney, and testis.

Although the primary pathologic process is osteolysis, osteoblastic activity is present and serum alkaline phosphatase will be elevated. Alkaline phosphatase from metastatic disease to bone is heat labile and is not inhibited by phenylalanine.[51] Furthermore, primary malignancies of bone such as osteogenic sarcoma may cause an elevated serum alkaline phosphatase. In addition, serum acid phosphatase may also be elevated in various malignancies involving bone. Elevation of serum alkaline phosphatase may be seen in carcinomas which metastasize to bone and concomitantly to the liver; these cases may present problems in differential diagnosis. The serum heat test and phenylalanine inhibition are helpful to determine whether either bone, liver, or both are involved. The differential diagnosis may also be aided by determining 5' nucleotidase, serum leucine aminopeptidase, or gamma glutamyl transpeptidase. Alkaline phosphatase may be elevated in patients with lymphoma such as Hodgkin disease or other malignancies possibly because the malignant neoplasm is producing alkaline phosphatase itself; this may be the Regan type of enzyme which is extremely heat stable and is inhibited by phenylalanine. Another explanation for elevation of alkaline phosphatase in Hodgkin's disease is liver or bone involvement or presence of fever (Fig. 8-4).[2]

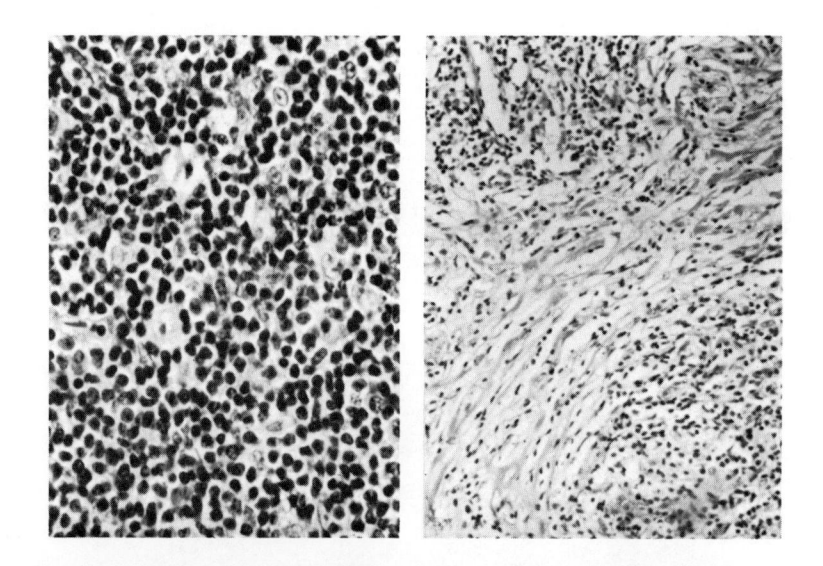

Enzyme Abnormalities in Neoplastic Disease (Continued):

Fig. 8-4 Photomicrographs of section of left supraclavicular lymph node from a patient with Hodgkin's disease. The lesion was classified as nodular sclerosing. Note the typical Reed-Sternberg cells present as lacunar cells. The serum alkaline phosphatase was 755 I.U.

Lactic Dehydrogenase

Lactic dehydrogenase is significantly elevated in patients with various types of malignant neoplasms.[16] The usual isoenzyme pattern in malignancy is an increase in isoenzymes 2, 3, 4, and sometimes 5. However, this similar pattern may also be found in infectious mononucleosis, shock, surgery, trauma, and immunologic disease. The serum LDH may be elevated due to metastases to various organs, including liver, lung, kidney, heart, or brain. Even without metastatic disease, a primary tumor may cause elevation of serum LDH. Elevations of LDH primarily, LDH_1, LDH_2, and LDH_3 occur with acute leukemia.[54] Those with chronic leukemia may also show elevations. Approximately 50 percent of patients with various lymphomas will exhibit an increase in serum LDH. The LDH is derived from the

Enzyme Abnormalities in Neoplastic Disease (Continued):

proliferating and dying leukemic or neoplastic lymphoma cells.[47]

Carcinoma of the colon and testicular carcinoma may produce elevation of LDH_1 and LDH_2.[58] LDH_4 and LDH_5 are elevated in patients with carcinoma of the prostate, lung, kidney, testes, and multiple myeloma.[3,14,49,53]

Patients who have choriocarcinoma or hydatidiform mole may develop elevated serum LDH, isocitric dehydrogenase, and GOT (Fig. 8-5).[52]

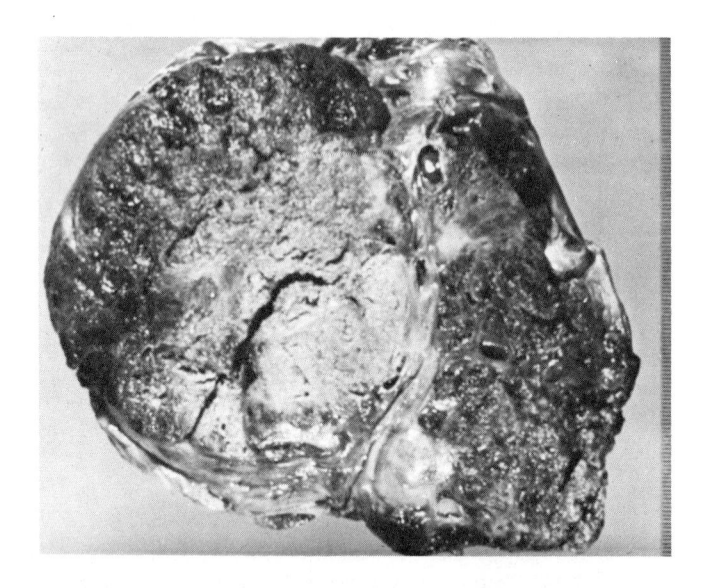

Fig. 8-5 Gross photograph of choriocarcinoma involving right ovary which metastasized widely. The serum LDH was 485 I.U.

Phosphoglucose Isomerase

Elevations in serum phosphoglucose isomerase result from various malignant neoplasms such as carcinoma of the lung, reproductive organs breast, stomach, pancreas, and colon.[32]   Patients with myeloid

Enzyme Abnormalities in Neoplastic Disease (Continued):

leukemia may have elevated phosphoglucose isomerase, perhaps resulting from destruction of myeloid cells.[30] Decrease in enzyme levels may result when the tumor is successfully treated. Elevation of this enzyme in the serum does not necessarily point to a malignancy; it may be associated with arteriosclerotic heart disease with myocardial infarction, acute hepatitis, hemolytic anemia, and Duchenne muscular dystrophy.[7]

Aldolase

Increase in serum aldolase may occur in certain types of cancer such as advanced cancer of the prostate. The elevation may arise from increased muscular wasting since the tumor may use skeletal muscle as its prime source of protein.[4]

6-Phosphogluconic Acid Dehydrogenase

The determination of 6-phosphogluconic acid dehydrogenase (6-PGD) in the vaginal fluid was proposed as a diagnostic test for carcinoma of the uterine cervix.[8] However, it has been shown that inflammatory conditions of the cervix and vagina as well as cervical carcinoma, will cause an elevation of this enzyme in the vaginal fluid.[6] The elevated 6-PGD in carcinoma of the cervix arises from the malignant cells, present and that associated with inflammation of the cervix or vagina, is from leukocytes present in the vagina. In addition, 6-PGD is present in the red blood cell, and bleeding lesions will increase the 6-PGD activity in the vaginal fluid.[28,36,40]

Beta-Glucuronidase

Beta-glucuronidase is also elevated in the vaginal fluid in carcinoma of the cervix, but the rise may also be due to other conditions such as inflammation, so the determination of this enzyme is not useful for the diagnosis of cancer of the cervix.[33]

Neoplastic Disease References (Continued):

REFERENCES

1. Abdul-Fadl, M. A., King, E. J.: "Properties of Acid Phosphatase of Erythrocytes and of the Human Prostate Gland". Bio. Chem. J. 45:51, 1949.

2. Aisenberg, A. C., Kaplan, M. M., Rieder, S. V., Goldman, J. M.: "Serum Alkaline Phosphatase at the Onset of Hodgkin' Disease". Cancer 26:2, 1970.

3. Ari, T.: "Diagnostic and Prognostic Evaluation of Serum Lactic Dehydrogenase Isozyme in Lung Cancer". Sci. Rep. Res. Inst. Tohoku Univ. Med. 14:28, 1967.

4. Baker, R., Gowan, D.: "The Effect of Hormonal Therapy of Prostatic Cancer on Serum Aldolase". Cancer Res. 13:141, 1953.

5. Banks, B. M., Pineda, E. P., Goldberg, J. A., Rutenberg, A. M.: "The Colorimetric Determination of Leucine Aminopeptidase Activity with L-Leucyl-beta-naphthylamide Hydrochloride". Arch. Biochem. 57:458, 1955.

6. Bell, J. L., Egerton, M. E.: "6-Phosphogluconate Dehydrogenase Estimation in Vaginal Fluid in the Diagnosis of Cervical Cancer". J. Obstet. & Gynec. Brit. Comm. 72:603, 1965.

7. Bodansky, O, Calitri, D.: "Serum Phosphohexose Isomerase in Cancer. I. Method of Determination and Establishment of Range of Normal". Cancer 7:1191, 1954.

8. Bonham, D. G., Gibbs, D. F.: "A New Enzyme Test for Gynaecological Cancer". Brit. Med. J. 11:823, 1962.

9. Bressler, R., Forsyth, B. R., Klatskin, G.: "Serum Leucine Aminopeptidase Activity in Hepatobiliary and Pancreatic Disease". J. Lab. Clin. Med. 56:417, 1960.

10. Cantarow, A., Trumper, M.: CLINICAL BIOCHEMISTRY Saunders Co., pg. 312, 1962.

11. Chandrasekhr, A. J., Palatao, A., Dubin A., Levin H.: Pleural Fluid Lactic Acid Dehydrogenase Activity and Protein Content". Arch. Intern. Med. 128:48, 1969.

Neoplastic Disease References (Continued):

12. Cook, W. B., Fishman, W. H., Clarke, B. G.: "Serum Acid Phosphatase of Prostatic Origin in the Diagnosis of Prostatic Cancer: Clinical Evaluation of 2408 Tests by the Fishman - Lerner Method". J. Urol. 88:281, 1962.

13. Daniel, O., Van Zyl, J. S.: "Rise of Serum Acid Phosphatase Level following Palpation of Prostate". Lancet 1:998, 1952.

14. Denis, L. J., Prout, G. R.: "Lactic Dehydrogenase in Prostatic Cancer". Invest. Urol. 1:101, 1963.

15. Doe, R. P., Seal, U. S.: "Acid Phosphatase in Urology". Surg. Clin. N. Amer. 45:1455, 1965.

16. Douglas, W. R.: "Relationships of Enzymology to Cancer: A Review. Brit. J. Cancer 17:415, 1963.

17. Dreiling, D. A., Richman, A.: "Evaluation of Provocative Blood Enzyme Tests employed in the Diagnosis of Pancreatic Disease". Arch. Intern. Med. 94:197, 1954.

18. Dybkaer, R., Jensen, G.: "Acid Phosphatase Levels following Massage of the Prostate". Scand. J. Clin. Lab. Invest. 10:349, 1958.

19. Fernandez, R., Seal, U. S., Mellinge, G. T., Doe, R. P.: "Renal Origin of Urinary Acid Phosphatase in Normal and in Patients with Cancer of the Prostate". Invest. Urol. 2:328, 1965.

20. Filippini, L., Ammann, R.: "Functional Clinical Diagnostic Tests for Carcinoma of the Pancreas (with Special Reference to the Pancreozymin Secretin Test and Proteolytic Activity of the Stool)". Schweiz Med. Wschr. 97:803, 1967.

21. Fishman, W. H., Lerner, F.: "Method for Estimating Serum Acid Phosphatase of Prostatic Origin". J. Biol. Chem. 200:89, 1953.

22. Galluzzi, N. J., Weingaarten, W. Regan, F. D., Doerner, A.: "Evolution of Hepatic Tests and Clinical Findings in Primary Hepatic Carcinoma". J. Amer. Med. Assoc. 15:152, 1953.

Neoplastic Disease References (Continued):

23. Gault, M. H., Steiner, G.: "Serum and Urinary Enzyme Activity after Renal Infarction". Canad. Med. Assoc. J. 93:1101, 1965.

24. Glenn, J. F., Spanel, D. L.: "Serum Acid Phosphatase and Effect of Prostatic Massage". J. Urol. 82:240, 1959.

25. Harkness, J., Roper, B. W., Durant, J. A., Miller, H.: "The Serum Leucine Aminopeptidase Test. An Appraisal of Its Value in Diagnosis of Carcinoma of Pancreas". Brit. Med. J. 1:1787, 1960.

26. Herbert, F. H.: "Estimation of Prostatic Phosphatase in Serum and Its Use in the Diagnosis of Prostatic Carcinoma". Quart. J. Med. 15:221, 1946.

27. Hoffman, E., Nachlas, M. M., Gaby, S. D., Abrams, S. J., Seligman, A. M.: "Limitations in the Diagnostic Value of Serum Leucine Aminopeptidase". New Eng. J. Med. 263:541, 19 1960.

28. Hoffman, R. L., Merritt, J. W.: "6-Phosphogluconate Dehydrogenase in Uterine Cancer Detection". Amer. J. Obstet. & Gynec. 92:650, 1965.

29. Huggins, C., Hodges, C. V.: "Studies on Prostatic Cancer. Effect of Castration, of Estrogen, and of Androgen Injection on Serum Phosphatase, in Metastatic Carcinoma of Prostate". Cancer Res. 1:293, 1941.

30. Israels, L. J., Delory, G. E.: "Serum Isomerase in Leukemia". Brit. J. Cancer 10:318, 1956.

31. Jacobsson, K.: "Determination of Tartrate-inhibited Phosphatase in Serum". Scand. J. Clin. Lab. Invest. 12:367, 1960.

32. Joplin, G. E., Jagatheesan, K. A.: "Serum Glycolytic Enzymes and Acid Phosphatases in Mammary Carcinomatosis". Brit. Med. J. 1:827, 1962.

33. Kasdon, S. C., Homberger, F., Yorshis, E., Fishman, W. H.: "Beta-Glucuronidase Studies in Women. VI. Premenopausal Vaginal Fluid Values in Relation to Invasion Cervical Cancer". Surg. Gynec. & Obstet. 97:579, 1953.

Neoplastic Disease References (Continued):

34. Kendall, A. R.: "Acid Phosphatase Elevation following Prostatic Examination in the Earlier Diagnosis of Prostatic Carcinoma". J. Urol. 86:442, 1961.

35. Kutscher, W., Worner, A.: "Prostatic Phosphatase. Mitteilung Hoppe-Seyler". Physiol. Chem. 239:109, 1936.

36. Lawson, J. G., Watkins, D. K.: "Vaginal Fluid Enzymes in Relation to Cervical Cancer". J. Obstet. & Gynec. Brit. Comm. 72:1, 1965.

37. Lepow, H., Schoenfeld, M. R., Messeloff, C. R., Chu, F.: "Nonprostatic Causes of Acid Hyperphosphatasemia: Report of a Case due to Multiple Myeloma". J. Urol. 87:991, 1962.

38. Lundh, G.: "Pancreatic Exocrine Function in Neoplastic and Inflammatory Disease. A Simple Reliable New Test". Gastroenterology 42:275, 1962.

39. Mendelsohn, M. L., Bodansky, O.: "The Value of Liver Function Tests in the Diagnosis of Intrahepatic Metastases in the Nonicteric Cancer Patient". Cancer 5:1, 1952.

40. Muir, G. G., Canti, G., Williams, D. K.: "Use of 6-Phosphogluconate Dehydrogenase as a Screen Test for Cervical Carcinoma in Normal Women". Brit. Med. J. 11:1563, 1964.

41. Oski, F. A., Naiman, J. L., Diamond, L. K.: "Use of the Plasma Acid Phosphatase Value in the Differentiation of Thrombocytopenic States". New Eng. J. Med. 268:1428, 1963.

42. Posen, S., Neale, F. C., Clubb, J. S.: "Heat Inactivation in the Study of Human Alkaline Phosphatases". Ann. Intern. Med. 62:1234, 1965.

43. Roy, A. V., Brower, M. E., Hayden, J. E.: "Sodium Thymolphthalein Monophosphate: A New Acid Phosphatase Substrate with Greater Specificity for the Prostatic Enzyme in Serum". Clin. Chem. 17:1093, 1971.

44. Schwartz, M. D., Bodansky, O.: "Serum 5 Nucleotidase in Patients with Cancer". Cancer 18:886, 1965.

Neoplastic Disease References (Continued):

45. Schwartz, M. K., Greenberg, E., Bodansky, O.: "Comparative values of Phosphatases and Other Serum Enzymes in following Patients with Prostatic Carcinoma. Consideration of Phosphohexose Isomerase, Glutamic Oxalacetic Transaminase, Isocitric Dehydrogenase, and Acid and Alkaline Phosphatase". Cancer 16:583, 1963.

46. Shay, H., Siplet, H.: "The Value of Serum Alkaline Phosphatase Determination and Bromsulphalein Test in the Diagnosis of Metastatic Cancer of the Liver". J. Lab. Clin. Med. 43:741, 1954.

47. Spector, I., McFarland, W., Trujillo, N. P., Ticktin, H. E.: "Bone Marrow Lactic Dehydrogenase in Hematologic and Neoplastic Disease". Enzym. Biol. Clin. (Basel) 7:78, 1966.

48. Stanbury, J. B., Wyngaarden, J. B., Fredrickson, D. S.: THE METABOLIC BASIS OF INHERITED DISEASE McGraw-Hill, pg. 576, 1966.

49. Starkweather, W. H., Green, R. A., Spencer, H. H., Schoch, H. K.: "Alterations of Serum Lactic Dehydrogenase Isoenzymes during Therapy directed at Lung Cancer". J. Lab. Clin. Med. 68:314, 1966.

50. Stewart, C., Sweetser, T. H., Delory, G. E.: "A Case of Benign Prostatic Hypertrophy with Recent Infarcts and Associated High Serum Acid Phosphatase". J. Urol. 63:128, 1950.

51. Stolbach, L. L., Krant, M. J., Fishman, W. H.: "Ectopic Production of Alkaline Phosphatase Isoenzyme in Patients with Cancer". New Eng. J. Med. 281:14, 1969.

52. Tobin, S. M.: "A Further Aid in the Diagnosis of Hydatidiform Mole. The Serum Glutamic Oxalacetic Transaminase". Amer. J. Obstet. & Gynec. 87:213, 1963.

53. Wacker, W. E. C., Dorfman, L. E.: "Urinary Lactic Dehydrogenase Activity. I. Screening Method for Detection of Cancer of Kidneys and Bladder". J. Amer. Med. Assoc. 181:972, 1962.

54. West, M., Heller, P., Zimmerman, H. J.: "Serum Enzymes in Disease. II. Lactic Dehydrogenase and Glutamic Oxalacetic Transaminase in Patients with Leukemia and Lymphoma". Amer. J. Med. Sci. 235:689, 1958.

Neoplastic Disease References (Continued):

55. Woodard, H. W.: "A Note on the Inactivation by Heat of Acid Glycerophosphatase in Alkaline Solution". J. Urol. 65:688, 1951.

56. Woodard, H. Q.: "The Clinical Significance of Serum Acid Phosphatase". Amer. J. Med. 27:902, 1959.

57. Zimmerman, H. J.: "Serum Enzymes in the Diagnosis of Hepatic Disease". Gastroenterology 46:613, 1964.

58. Zondag, H. A.: "Enzyme Activity in Dysgerminoma and Seminoma. A Study of Lactic Dehydrogenase Isoenzymes in Malignant Diseases". Rhode Island Med. J. 47:273, 1964.

CHAPTER 9

ENZYME PATTERNS IN OBSTETRIC AND GYNECOLOGIC PRACTICE

Pregnancy is associated with an elevation of serum alkaline phosphatase. The enzyme begins to rise in the mother's serum at approximately the sixteenth week of gestation and remains elevated until 4 weeks after delivery.[20] Serial serum alkaline phosphatase determinations may be used to assess fetal maturity. The alkaline phosphatase is derived from the placenta.[25,36] It is extremely heat stable and is inhibited by phenylalanine; placental alkaline phosphatase is the most heat stable alkaline phosphatase isoenzyme. Heat stability is determined by measuring the enzyme activity after serum is heated in a water bath or in a heat block at 56°C. for 10 to 15 minutes. Eighty percent of placental alkaline phosphatase will be inhibited by incubation with 0.005 M phenylalanine (Tables 9-1 and 9-2).

TABLE 9-1

	Heat Test	Phenylalanine
Placental Isoenzyme	Heat Stable	Inhibited
Regan Isoenzyme (Fig. 9-1)	Heat Stable	Inhibited

TABLE 9-2

PERCENT HEAT STABLE OF TOTAL SERUM ALKALINE PHOSPHATASE DURING NORMAL PREGNANCY [17]

First Trimester	Second Trimester	Third Trimester
10 Percent	30 Percent	50 Percent

A majority of the placental alkaline phosphatase activity remains active and thus, this is an indication of heat stability. However, incubation of the serum with phenylalanine will result in inactivation of the placental isoenzyme.[23]

Enzyme Patterns in Obstetric and Gynecologic Practice (Continued):

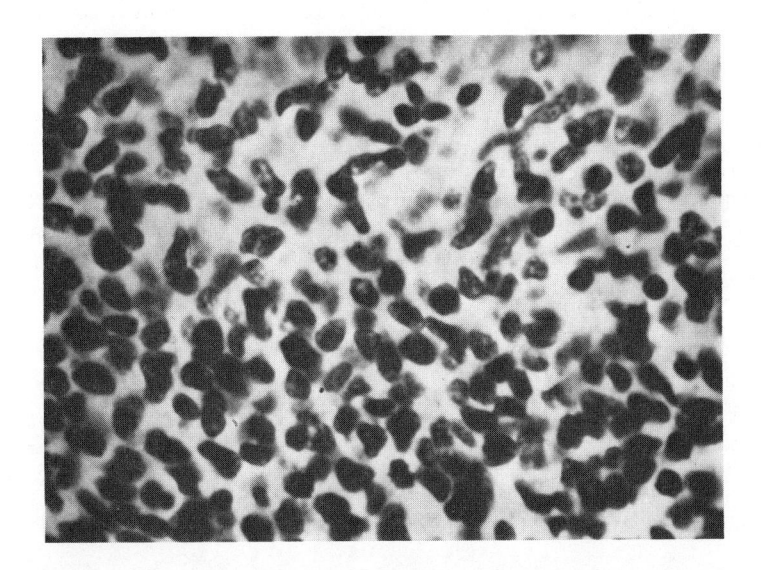

Fig. 9-1 Photomicrograph of an undifferentiated bronchogenic carcinoma which produced alkaline phosphatase and caused a serum alkaline phosphatase of 270 I.U. The alkaline phosphatase was the Regan type of isoenzyme.

Plasma expanders from human placenta are available commercially. These plasma proteins substitutes cause a dramatic elevation in serum alkaline phosphatase in patients who have received them for treatment of diminished plasma albumin or volume.[2] The prominent elevation in alkaline phosphatase is related to the alkaline phosphatase derived from the human placenta in the commercial product.

A number of reports have suggested that the elevation of alkaline phosphatase in maternal serum during pregnancy is related to mobilization of calcium from maternal bone with osteomalacia for the benefit of the fetus.[31] However, this speculation has been shown to be incorrect; bone alkaline phosphatase is extremely heat labile whereas elevated alkaline phosphatase in maternal serum during pregnancy is heat stable. Furthermore, techniques such as electrophoresis and immunologic methods have demonstrated that maternal alkaline phosphatase is not derived from a bone source but from the placenta.[6]

Enzyme Patterns in Obstetric and Gynecologic Practice (Continued):

During a normal pregnancy, the placenta produces acid phosphatase up to the thirty-second week of gestation with a decrease towards term. Acid phosphatase increases in anoxic placentas associated with pre-eclampsia and eclampsia.[41]

Beta-Glucuronidase

Beta-glucuronidase activity in the serum rises with the rise of serum estrogen in pregnancy. Thus, increased urinary excretion of pregnanediol and estriol is accompanied by an increase in beta-glucuronidase. Serum enzyme activity reaches normal levels 1 week postpartum.

Creatine Phosphokinase

Serum creatine phosphokinase (CPK) rises slowly during pregnancy and is three times normal in the third trimester. Because of the strenuous muscular exertion during labor, CPK is elevated during labor and for a few days thereafter. The CPK elevation may also be due in part to the large amount of enzyme contained in the pregnant myometrium compared to the nonpregnant myometrium.[10,12]

Isocitric Dehydrogenase, Lactic Dehydrogenase, and Glutamic Oxalacetic Transaminase

During and following labor, there may be an elevation of isocitric dehydrogenase (ICD) and lactic dehydrogenase (LDH) in maternal blood. These elevations are related to the extreme muscular exertion and to necrosis of the placenta. Placental LDH has been characterized as isoenzymes LDH_3 and LDH_4.[18,26] Glutamic oxalacetic transaminase (GOT) activity may be diminished during pregnancy because of negative nitrogen balance or pyridoxine deficiency.[37,40]

Aminopeptidases

Leucine aminopeptidase (LAP) and cystine aminopeptidase (CAP) are elevated during pregnancy; the increased enzymes are derived from the placenta.[30,34] The elevations begin at approximately the end of the first month of pregnancy and reach a maximum at term. The elevations persist until approximately 1 month after delivery, similar to the elevation of alkaline phosphatase in maternal serum.[27] Two isoenzymes of cystine aminopeptidase also known as oxytocinase, are present in the serum during pregnancy. Isoenzyme one appears early in pregnancy and decreases in the later months. The second isoenzyme makes its

Enzyme Patterns in Obstetric and Gynecologic Practice (Continued):

appearance during the sixth month and is the commoner fraction during the third trimester. Following delivery, the level becomes normal in 3 to 4 weeks.

Cystine aminopeptidase in maternal serum is thought to relate to an attempt to decrease the muscular activity of the myometrium of the uterus during pregnancy.[15] A number of investigators have speculated that the decrease in enzyme noted at the onset of labor results in increased oxytocin and thus initiates labor.[13]

The presence of increased aminopeptidase can serve as a highly accurate diagnostic pregnancy test. This enzyme is increased in maternal serum as early as 4 weeks following conception. The determination of this enzyme has a high degree of accuracy as a pregnancy test; other conditions associated with elevations of leucine aminopeptidase in human sera are intra and extra-hepatic biliary obstruction and extra-uterine pregnancy.

Pregnancy with Complications

Complications of pregnancy may cause various enzyme alterations in maternal serum. Toxemia may cause elevated LDH and GOT, especially if renal or hepatic disease is present. Glutamic pyruvic transaminase (GPT) activity increases with increasing hepatic necrosis associated with toxemia of pregnancy.[1,8] An elevated LDH and more prominent elevation in beta-glucuronidase may be associated with toxemia of pregnancy.[26,38]

Increased LDH and isocitric dehydrogenase may occur when abruptio placenta is present. Because of the massive placental clot, LDH will be elevated,[9] and is related to the red cell destruction in the blood clot.[5,21,22]

Choriocarcinoma or hydatidiform mole may cause an elevated serum LDH, isocitric dehydrogenase or GOT level.[39] Cystine aminopeptidase may be absent or lower than normal.[35]

Other Gynecologic Conditions

Determinations of enzymes in vaginal fluids may be of diagnostic value. Determination of acid phosphatase in the vaginal fluid suggests that there has been recent sexual intercourse.[11] The medical examiner uses this enzyme determination to determine if rape has occurred. This test is more valuable than a search for spermatozoa

Enzyme Patterns in Obstetric and Gynecologic Practice (Continued):

in the vaginal fluid because spermatozoa degenerate rapidly whereas acid phosphatase persists for a longer period of time. Semen has a high acid phosphatase content which is of prostatic origin. Prostatic acid phosphatase is inhibited by L-tartrate but not by formaldehyde. This test is important in identifying acid phosphatase of prostatic origin as opposed to, for example, that from red blood cells which is not inhibited by L-tartrate. Prostatic acid phosphatase is present in vaginal fluid up to 12 hours after coitus. During menstruation, red blood cell acid phosphatase will be present in the vagina.[32,33]

6-Phosphogluconic Acid Dehydrogenase

6-phosphogluconic acid dehydrogenase (6-PGD) is elevated in the vaginal fluid in carcinoma of the uterine cervix; the enzyme is derived from the malignant cells (Fig. 9-2). This was at one time proposed as a test for cervical cancer, but it has since been found to be nonspecific (see the discussion of 6-PGD in Chapter 8 on Neoplastic Disease).[3,4,14,19,28,29]

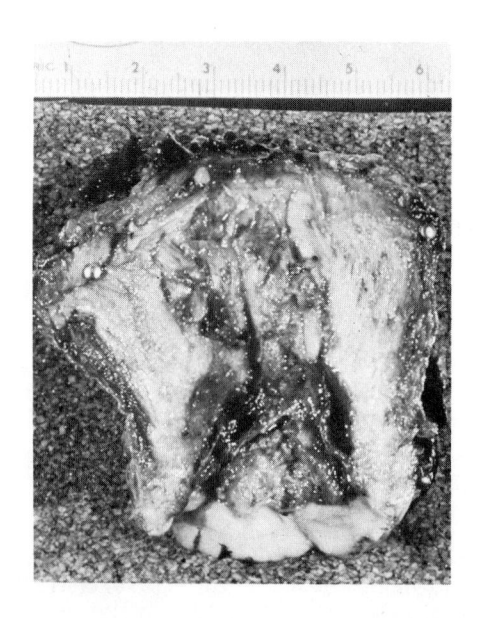

Fig. 9-2 Gross photograph of carcinoma of the uterine cervix which caused an elevation of 6-PGD in the vaginal fluid.

Enzyme Patterns in Obstetric and Gynecologic Practice (Continued):

Beta-Glucuronidase

Beta-glucuronidase is also elevated in the vaginal fluid in carcinoma of the cervix but again the rise is nonspecific and may be due to other conditions such as inflammation.[16]

REFERENCES

1. Atuk, N. O., Wax, S. H., Word, B. H., McGaughey, H. S., Corey, E. L., Wood, J. E.: "Observations of the Steady State of Lactic Dehydrogenase Activity Across the Human Placental Membrane". Amer. J. Obstet. & Gynec. 82:271, 1961.

2. Bark, C. J.: "Artifactual Serum Alkaline Phosphatase from Placental Albumin". J. Amer. Med. Assoc. 207:953, 1969.

3. Bell, J. L., Egerton, M. E.: "6-Phosphogluconate Dehydrogenase Estimation in Vaginal Fluid in the Diagnosis of Cervical Cancer". J. Obstet. & Gynec. Brit. Comm. 72:603, 1965.

4. Bonham, D. G., Gibbs, D. F.: "A New Enzyme Test for Gynaecological Cancer". Brit. Med. J. 11:823, 1962.

5. Boutselis, J. G., Wensinger, J. A., Sollarsk, J.: "Serum Lactic Dehydrogenase in Abruptio Placentae". Amer. J. Obstet. & Gynec. 86:762, 1963.

6. Boyer, S. H.: "Alkaline Phosphatase in Human Sera and Placentae". Science 134:1002, 1961.

7. Christian, D. G.: "Drug Interference with Laboratory Blood Chemistry Determinations". Amer. J. Clin. Path. 54:118, 1970.

8. Crisp, W. E., Miesfeld, R. L., Frajola, W. J.: "Serum Glutamic Oxalacetic Transaminase Levels in Toxemias of Pregnancy". Obstet. & Gynec. 13:487, 1959.

9. Dawkins, M. J. R., Wigglesworth, J. S.: "Serum Isocitric Dehydrogenase in Normal and Abnormal Pregnancy". J. Obstet. & Gynec. Brit. Comm. 68:264, 1961.

Obstetric and Gynecologic References (Continued):

10. Emery, A. E. H., Pascasio, F. M.: "The Effect of Pregnancy on the Concentration of Creatine Kinase in Serum, Skeletal Muscle, and Myometrium". Amer. J. Obstet. & Gynec. 91:18, 1965.

11. Fisher, R. S.: "Acid Phosphatase Tests as Evidence of Rape". New Eng. J. Med. 240:738, 1949.

12. Fleisher, G. A., McConahey, W. M., Pankow, M.: "Serum Creatine Kinase, Lactic Dehydrogenase, and Glutamic Oxalacetic Transaminase in Thyroid Diseases and Pregnancy". Mayo Clin. Proc. 40:300, 1965.

13. Hilton, J. G., Johnson, R. F.: "Changes in Blood Oxytocinase during Parturition". Amer. J. Obstet. & Gynec. 78:479, 1959.

14. Hoffman, R. L., Merritt, J. W.: "6-Phosphogluconate Dehydrogenase in Uterine Cancer Detection". Amer. J. Obstet. & Gynec. 92:650, 1965.

15. Ichaliotis, S. D., Lambrinopoulos, T. C.: "Serum Oxytocinase in Twin Pregnancies". Obstet. & Gynec. 25:270, 1965.

16. Kasdon, S. C., Homburger, F., Yorshis, E., Fishman, W. H.: "Beta-Glucuronidase Studies in Women. VI. Premenopausal Vaginal Fluid Values in Relation to Invasive Cervical Cancer". Surg. Gynec. & Obstet. 97:579, 1953.

17. Konttinen, A., Pyorala, T.: "Serum Enzyme Activity in Late Pregnancy, at Delivery, and during Puerperium". Scand. J. Clin. Lab. Invest. 15:429, 1963.

18. Lapan, B., Friedman, M. A.: "A Comparative Study of Fetal and Maternal Serum Enzyme Levels". J. Lab. Clin. Med. 54:417, 1959.

19. Lawson, J. G., Watkins, D. K.: "Vaginal Fluid Enzymes in Relation to Cervical Cancer". J. Obstet. & Gynec. Brit. Comm. 72:1, 1965.

20. Levine, G., Wood, W.: "Maternal Serum Alkaline Phosphatase and Placental Function". Amer. J. Obstet. & Gynec. 91:967, 1965.

Obstetric and Gynecologic References (Continued):

21. Little, W. A.: "Serum Enzyme Alterations in Abruptio Placentae". Surg. Forum 10:716, 1959.

22. Little, W. A.: "Serum Lactic Dehydrogenase in Pregnancy". Obstet. & Gynec. 13:152, 1959.

23. Luke, D., Wolf, P. L.: "Evaluation of the Heat Stability-Lability of Serum Alkaline Phosphatase utilizing a New Chemical Technique". Enzymologia 39:9, 1970.

24. McDonald, D. F., Odell, L. D.: "Serum Glucuronidase Activity during Normal and Toxemic Pregnancy". J. Clin. Endocr. 7:535, 1947.

25. McMaster, Y., Tennant, R., Clubb, J. S., Neal, F. C., Posen, S.: "The Mechanism of the Evaluation of Serum Alkaline Phosphatase in Pregnancy". J. Obstet. & Gynec. Brit. Comm. 71:735, 1964.

26. Meade, B. W., Rosalki, S. B.: "The Origin of Increased Maternal Serum Enzyme Activity in Pregnancy and Labour". J. Obstet. & Gynec. Brit. Comm. 70:862, 1963.

27. Melander, S.: "Plasma Oxtocinase Activity. A Methodological and Clinical Study with Special Reference to Pregnancy". Acta Endocr. (Kobenhavn) Suppl. 96:94, 1965.

28. Muir, G. G., Canti, G., Williams, D. K.: "Use of 6-Phosphogluconate Dehydrogenase as a Screen Test for Cervical Carcinoma in Normal Women". Brit. Med. J. 11:1563, 1964.

29. Nerdrum, H. J.: "The Activity of 6-Phosphogluconic Acid Dehydrogenase in Vaginal Secretions". Scand. J. Clin. Lab. Invest. 16:565, 1964.

30. Page, E. W., Titus, M. A., Mohun, G., Glending, M. B.: "The Origin and Distribution of Oxytocinase". Amer. J. Obstet. & Gynec. 82:1090, 1961.

31. Ramsey, J., Thierens, V. T., Magee, H. E.: "Composition of Blood in Pregnancy". Brit. Med. J. 1:1199, 1938.

32. Reinstein, H., Benotti, N., McBay, A. J.: "The Demonstration of Prostatic Origin of Seminal Stain Acid Phosphatases". Amer. J. Clin. Path. 39:583, 1963.

Obstetric and Gynecologic References (Continued):

33. Riisfeldt, O.: "Acid Phosphatase Employed as a New Method of Demonstrating Seminal Spots in Forensic Medicine". Acta. Path. Micro. Biol. Scand. Suppl 58:1, 1946.

34. Siegel, I. A.: "Leucine Aminopeptidase in Pregnancy". Obstet. & Gynec. 14:488, 1959.

35. Smith, E. E., Rutenburg, A.: "Studies of Serum Aminopeptidase in Pregnancy, Hydatidiform Mole, and Choriocarcinoma". Amer. J. Obstet. & Gynec. 96:301, 1966.

36. Stolbach, L. L., Krant, M. J., Fishman, W. H.: "Ectopic Production of Alkaline Phosphatase Isoenzyme in Patients with Cancer". New Eng. J. Med. 281:757, 1969.

37. Stone, M. L., Lending, M., Slobody, M.D., Mestern, J.: "Glutamic Oxalacetic Transaminase and Lactic Dehydrogenase in Pregnancy". Amer. J. Obstet. & Gynec. 80:104, 1960.

38. Stone, M. L., Lending, M., Slobody, L. B., Salzman, M., Mestern, J.: "Glutamic Oxalacetic Transaminase and Lactic Dehydrogenase in Pregnancy. II. Complications of Pregnancy" Amer. J. Obstet. & Gynec. 83:1342, 1962.

39. Tobin, S. M.: "A Further Aid in the Diagnosis of Hydatidiform Mole. The Serum Glutamic Oxalacetic Transaminase". Amer. J. Obstet. & Gynec. 87:213, 1963.

40. West, M., Zimmerman, H. J.: "Lactic Dehydrogenase and Glutamic Oxalacetic Transaminase in Normal Pregnant Women and Newborn Children". Amer. J. Med. Sci. 235:443, 1958.

41. Young, B. K., Beller, F. K.: "Plasma Acid Phosphatase in Normal and Pre-eclamptic Pregnancy". Amer. J. Obstet. & Gynec. 101:1068, 1968.

CHAPTER 10

ENZYME ABNORMALITIES IN LIVER DISEASE

Various enzymes may increase in the serum of patients with hepatic disease. The first type of serum enzyme abnormality to be discussed is that associated with intra-hepatic or extra-hepatic obstructive jaundice. The serum enzymes that are elevated in obstructive jaundice are: Alkaline phosphatase, leucine aminopeptidase, 5'-nucleotidase, and gamma-glutamyl-transpeptidase. Alkaline phosphatase also may be elevated in conditions such as pregnancy, osteoblastic lesions of bone, bone growth, malignant neoplasms producing alkaline phosphatase (the Regan enzyme), infarction of kidney or spleen, lung, or various lesions involving gastrointestinal tract. In contrast, leucine aminopeptidase, 5'-nucleotidase, and gamma-glutamyl-transpeptidase are normal in patients with diseases of bone, kidney, gastrointestinal tract and lung.

Alkaline Phosphatase

Alkaline phosphatase is elevated in patients with extra-hepatic obstructive jaundice secondary to causes such as carcinoma of the head of the pancreas, choledocholithiasis, or sclerosing cholangitis. If prominent obstruction of the common bile duct is present, alkaline phosphatase may be two to ten times normal. If metastases from the carcinoma of a pancreas are also present to the liver, the alkaline phosphatase elevation will be much greater.[44,67,92] Liver alkaline phosphatase is moderately heat stable and not inhibited by phenylalanine.[56] The alkaline phosphatase elevation in liver metastases may arise from either the intra-hepatic obstructive lesion or from proliferation of vascular endothelium due to vascularity of the metastases. Tumor angiogenesis factor elaborated from the metastasic lesion may stimulate the vascularity.

Patients with chronic liver disease may develop osteomalacia and an elevated serum alkaline phosphatase derived from bone. The cause is decreased intake of vitamin D.[52]

Alkaline phosphatase may be elevated in other infiltrative diseases of the liver such as miliary tuberculosis or sarcoidosis.[12]

Drug-induced cholestasis caused by chlorpromazine hydrochloride (Thorazine) or other drugs may cause an elevated alkaline phosphatase

Enzyme Abnormalities in Liver Disease (Continued):

(Fig. 10-1).[70,90] Furthermore, acute viral hepatitis may cause prominent ballooning of hepatic cells leading to swelling and compression of the bile canaliculi with cholestasis and elevated alkaline phosphatase (Fig. 10-2). Fatty metamorphosis of the liver usuall induces a minimal increase in the alkaline phosphatase.[5,11] If an alcoholic has a prominent alkaline phosphatase increase, it probably is secondary to steatonecrosis with intra-hepatic cholestasis induced by the alcoholism, but it may be due to acute pancreatitis, severe fatty liver, or to choledocholithiasis.[80]

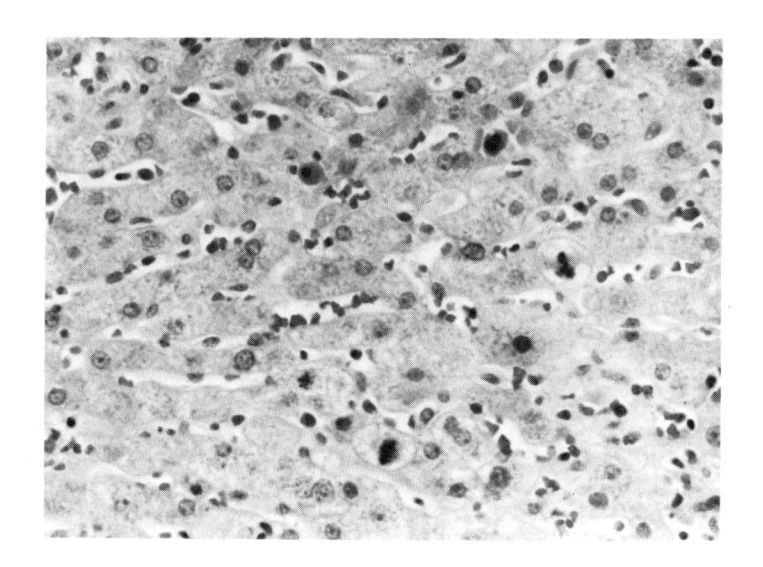

Fig. 10-1 Photomicrograph of section of liver in which a Thorazine-induced cholestasis is present. The serum alkaline phosphatase was 190 I.U.

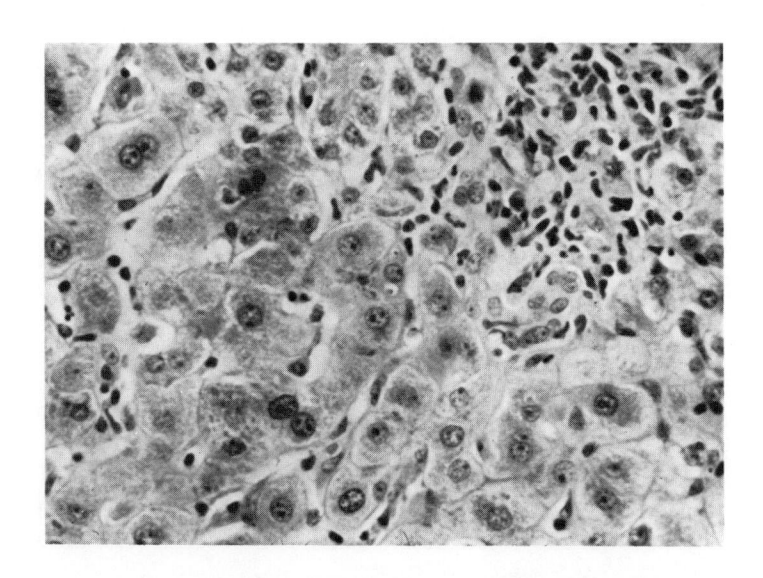

Enzyme Abnormalities in Liver Disease (Continued):

Fig. 10-2   Photomicrograph of liver in which viral hepatitis caused ballooning of cells with intrahepatic obstruction. The serum alkaline phosphatase was 155 I.U.

Portal cirrhosis is associated with minimal increase in alkaline phosphatase which is not as high as that with post-necrotic cirrhosis.[50,84] The increase in post-necrotic cirrhosis may simulate that of biliary cirrhosis, which produces a very prominent elevation.[91]

It has recently been recognized that infiltrative disease of the liver may have extreme elevations of alkaline phosphatase with the absence of jaundice. In addition, BSP excretion is impaired and LDH is prominently increased. Lesions that may cause this type of biochemical pattern are metastatic carcinoma to the liver, diffuse liver abscesses, lymphoma or granuloma secondary to tuberculosis, or sarcoidosis (Fig. 10-3).

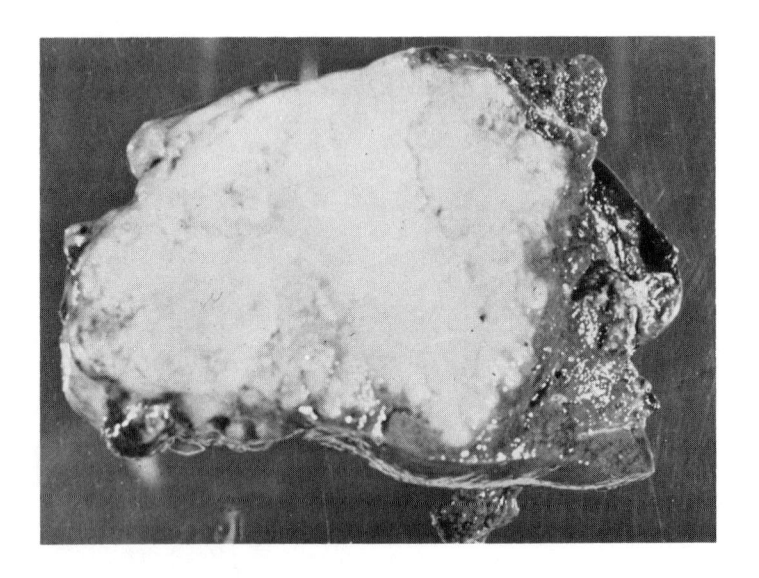

Enzyme Abnormalities in Liver Disease (Continued):

Fig. 10-3   Gross photograph of the liver to which a carcinoma of the pancreas has metastasized. The serum alkaline phosphatase was 580 I.U. secondary to the infiltrative lesion of the liver.

Increased serum levels of alkaline phosphatase in patients with hepatic disease probably reflect an increased formation of the enzyme by the liver and bile duct cells with impaired excretion by the damaged liver. Obstruction causes decreased excretion and increased synthesis. The half life of the enzyme is a number of weeks with a gradual decrease in the serum following relief of obstruction.[55]

A number of observers have pointed out the value of determining serum alkaline phosphatase to detect metastases to the liver. It has been shown that there is a close association between the number of metastases and the elevation of serum alkaline phosphatase.[30]

Congestive heart failure and long-standing chronic passive congestion of the liver may elevate serum alkaline phosphatase.[27] Infectious mononucleosis may cause prominent elevation of alkaline phosphatase without conspicuous jaundice.[9,29]

Enzyme Abnormalities in Liver Disease (Continued):

The detection of elevated serum alkaline phosphatase is a sensitive method for determining sensitivity to various drugs, such as phenothiazine, sulfonamides, diuretics, antidiabetic drugs, oral contraceptives, isoniazid (INH), PAS leading to cholestasis and intrahepatic obstructive disease.[23] Drug-induced liver disease may be cholestatic, cytotoxic, or mixed, and the biochemical picture is that of cholestasis with hepatocellular necrosis.

Leucine Aminopeptidase

Infiltrative lesions of the liver as described above will increase leucine aminopeptidase [61,76] as will extra-hepatic obstructive jaundice.[40] The greater the obstruction, the greater will be the LAP elevation. Thus, patients with carcinoma of the head of the pancreas and choledocholithiasis will have larger elevations of LAP than hepatitis. Carcinoma of the body or tail of the pancreas may not have an elevated LAP. Another important cause for an elevated serum LAP is pregnancy.[33,69,78]

Striking elevations of LAP are seen in biliary atresia. Also present are distinctive changes in the distribution of isoenzyme patterns. The post-albumin zone (LAP$_1$) which is demonstrated on electrophoresis as only one clear zone is present in neonatal hepatitis. With biliary atresia, an intensified zone of LAP$_1$ extends to the alpha-globulin zone along with a second band which is consistently found situated between the alpha-2 and globulin zones. This finding is consistent with the pattern demonstrated in patients with adult biliary cirrhosis.[13,37,58]

5'-Nucleotidase

Patients with obstructive jaundice usually develop elevated 5'-nucleotidase.[40] Individuals with infiltrative disease of the liver as well as those with extra-hepatic lesions causing obstructive jaundice will also show increased serum 5'-nucleotidase.[24] Inflammatory hepatic disease will cause elevations of this enzyme but not to the same degree as obstructive jaundice.[7] 5'-nucleotidase levels are normal in patients with bone disease.[64]

Enzyme Abnormalities in Liver Disease (Continued):

Glutamic Oxalacetic Transaminase and Glutamic Pyruvic Transaminase

GOT and GPT may increase in the serum with hepatobiliary tract disease. GOT is found in the brain, myocardium, skeletal muscle, kidney, skin, red blood cells, and pancreas. GOT values are markedly increased in acute viral hepatitis and in hepatocellular necrosis secondary to drugs or chemicals. With marked destruction of the liver, GOT may not be increased because of exhaustion of the enzyme from the markedly necrotic liver cells.[66] Shock may cause extensive liver necrosis with prominent increase of GOT. Extra-hepatic or intra-hepatic obstruction due to drugs, neoplasia, or calculi will cause elevated GOT values, but not to the extent seen in necrosis of the liver. The GOT increase with obstruction is due to hepatic injury from ascending cholangitis,[2,28,41,48] or the increased intra-biliary pressure may result in leakage of GOT from hepatic cells.

Fatty metamorphosis will result in only slight elevation of GOT, whereas cirrhosis will cause a moderate increase, with higher values seen in active alcoholic portal cirrhosis.[5,11] If high GOT levels are present, one should suspect the coexistence of alcoholic myopathy or pancreatitis. In active post-necrotic cirrhosis or chronic active hepatitis, GOT values may be low as is seen in alcoholic cirrhosis, or extremely high, as is seen in viral hepatitis.[16] Cancer of the liver or lymphoma involving the liver may cause moderate elevations of GOT due to hepatocellular necrosis or production by the neoplasm.

An elevated level of GOT occurs early in the course of acute viral hepatitis. It may become elevated 1 to 4 weeks prior to clinical jaundice. Peak levels occur during the first week of the disease with a progressive decrease until a normal value is reached within a month. If the level increases again, this may indicate a relapse. If there is persistent elevation of GOT, incomplete recovery has occurred. Some patients have a persistent elevation for months, indicating continuing activity of the disease. In these individuals a spurious elevation of GOT induced by drugs should be suspected. Most individuals with active acute hepatitis will have GOT values up to 3,000 I.U. Such high values are equaled at times by liver injury caused by drugs or chemicals.

GOT is elevated in the urine when SGOT is elevated in viral hepatitis patients.

The dissimilarity of enzyme relationships in an organ and in the serum provides diagnostic clues at times. The degree and type of damage may be determined by this distortion analysis. For example

Enzyme Abnormalities in Liver Disease (Continued):

in the liver, GOT is higher than GPT; however, one-half of the GOT is in the hepatic cell mitochondria. In acute hepatitis, usually only the cytoplasmic or cell sap enzymes are released and serum GPT is greater than serum GOT. The mitochondrial glutamic dehydrogenase (GLDH) elevation is only slight in the usual patient with viral hepatitis.

In severe toxic necrosis of the liver or severe viral hepatitis, release of GOT and GLDH from mitochondria occurs and GOT is much higher than GPT, and glutamic dehydrogenase is greatly elevated.

The $\frac{GOT + GPT}{GLDH}$ ratio is useful in the assessment of liver disease. Typical ratios are:

Acute hepatitis	Above 50
Chronic hepatitis	30 - 40
Obstructive jaundice	5 - 15
Cancer metastatic to liver	Below 10
Cholestatic drug-induced liver disease	40 - 50

The DeRitis GOT/GPT ratio may be clinically useful, but should not be utilized when the GOT is above 300 units. When the GOT value is below that level, the relative elevations of GOT and GPT are of differential value. Patients with infectious mononucleosis or ascending cholangitis secondary to obstructive jaundice may have a GOT/GPT ratio of one or below.[22] Patients with granulomatous or infiltrative disease of the liver such as tuberculosis, lymphoma, or alcoholic hepatitis may have a ratio above one. Viral hepatitis may be distinguished from alcoholic hepatitis by utilization of the ratio. The alcoholic with cirrhosis with a GOT/GPT ratio below one may have a complication such as acute pancreatitis.[21] The GOT/GPT ratio also is useful in cardiac patients. Acute myocardial infarction causes a ratio of greater than one. If the patient develops congestive heart failure, the ratio becomes less than one due to leakage of GPT from the congested liver.

GOT may be elevated prior to the onset of clinical jaundice in patients with viral hepatitis. Thus, it may be of value to determine GOT to detect anicteric viral hepatitis. However, the chemical screening of blood for GOT to detect viral hepatitis has not proven valuable. Some investigators have used the GOT test to screen blood

Enzyme Abnormalities in Liver Disease (Continued):

donors in an attempt to detect viral hepatitis carriers, but determination of GPT may be of greater value for this purpose.[1,6] The detection of the hepatitis associated antigen (HAA) is the preferred screening procedure for viral hepatitis in donor blood bank blood.[57]

Narcotics produce contraction of the Sphincter of Oddi with increase in the pressure in the bile duct system, abdominal and chest pain, and ECG changes. Serum amylase may increase after narcotics in normal individuals, but GOT may rise in individuals with an absent or nonfunctioning gallbladder. The GOT rise begins at 2 hours, peaks at 8 hours, and lasts 24 hours.

The serum enzymes do not increase with chronic cholecystitis and cholelithiasis in the absence of secondary liver disease. However, choledocholithiasis causes a high GOT. The elevated GOT may be due to ascending cholangitis or leakage of GOT from the hepatic cells or increased GOT synthesis induced by the increase in intrabiliary pressure with regurgitation of the enzyme into the blood.[49]

Thus, acute injury to the liver may cause a GOT value of over 1,000 units. Individuals with GOT values varying from a few hundred to 1,000 units may have moderate viral hepatitis, shock, or ascending cholangitis related to extra-hepatic obstructive jaundice. If the values are below a few hundred units, a variety of hepatic or biliary conditions may be the cause.

Infectious mononucleosis causes a moderate elevation in GOT and jaundice may be minimal. Alkaline phosphatase elevation may also occur. GOT elevation as high as several hundred units may be present

Portal cirrhosis causes GOT elevations of a few hundred units. Very little correlation exists between the severity of the liver disease and the enzyme level (Fig. 10-4).

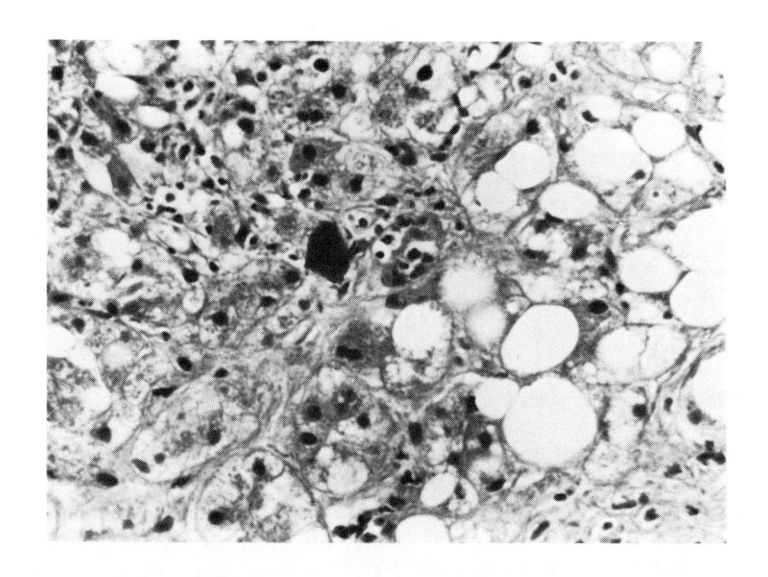

Enzyme Abnormalities in Liver Disease (Continued):

Fig. 10-4 Photomicrograph of portal cirrhosis with fatty metamorphosis from an alcoholic patient. The SGOT was 210 I.U.

Only moderate GOT elevations of a few hundred units are observed in patients with primary cancer of the liver, lymphoma involving the liver (such as Hodgkin's disease), tuberculosis, congestive heart failure and acute alcoholic hepatitis with jaundice.

The value of determining GPT in patients with liver disease has been established. The enzyme is widely distributed in various tissues but has its greatest concentration in the liver. Elevations of GPT reflect acute hepatic injury with more specificity than do GOT increases. With acute injury to the liver, GPT values are usually as elevated or higher than those of GOT. GPT values are somewhat lower than GOT in patients with cirrhosis and cancer of the liver.

Lactic Dehydrogenase

Lactic dehydrogenase is increased in the serum in patients with a wide variety of conditions, and this subject has been extensively

Enzyme Abnormalities in Liver Disease (Continued):

reviewed in previous chapters. Lactic dehydrogenase may be elevated in patients with hematologic conditions, such as pernicious anemia, hemolytic anemia, leukemia, in acute myocardial infarction, pulmonary infarction, renal infarction, in association with various malignant neoplasms, and with skeletal muscle diseases.

Lactic dehydrogenase is found within hepatic parenchymal cells with minimal activity in bile duct epithelium. It is practically absent in bile of normal individuals but is present in bile of patients with active liver disease. Its presence in bile signifies acute liver injury. It parallels the transaminases as to time of elevation and degradation following the liver insult.[91]

A jaundiced patient with marked elevation of lactic dehydrogenase may have a malignant neoplasm metastatic to the liver. The elevation is related to the malignant neoplasm and not the hepatic disease. The usual isoenzyme related to hepatic disease is LDH_5. This slow migrating isoenzyme is heat labile, in contrast to LDH_1 derived from a myocardial or hematologic source which is heat stable (Fig. 10-5). Unless there is prominent hepatic necrosis or infiltrative hepatic disease, LDH is considered an insensitive determination to ascertain hepatic disease.[74,75]

In acute hepatitis the increase in LDH_5 is an early manifestation of the disease. It may precede the onset of jaundice by a number of days. Once the patient with acute viral hepatitis becomes jaundiced, the LDH isoenzyme tends to return quickly to normal. The tissue supply of LDH_5 from the liver becomes exhausted and the serum level returns to normal. If chronic hepatitis ensues, the LDH_5 may again become elevated. Only minimal elevations of LDH_5 may be present and the total LDH may not be significantly elevated.

Congestive heart failure secondary to myocardial infarction may cause elevated LDH_1 from the myocardial infarct and elevated LDH_5 from the congestion and necrosis of the central part of the lobule.[74,77,81,82]

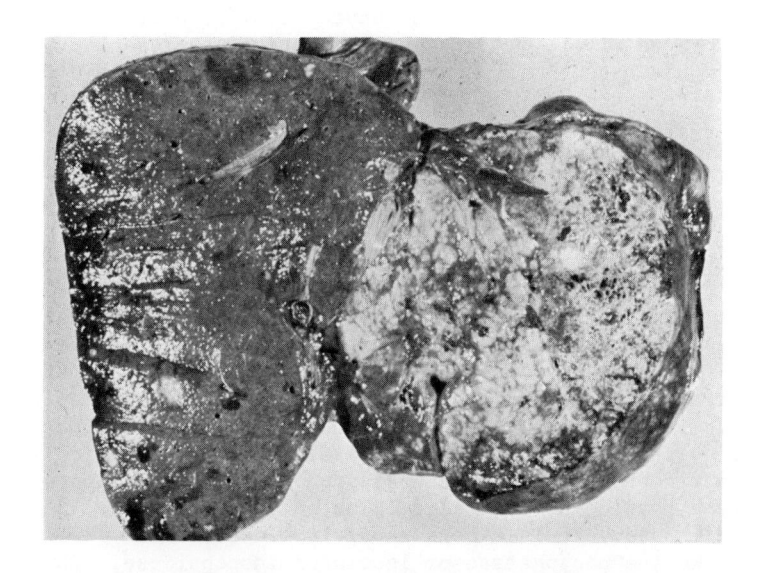

Enzyme Abnormalities in Liver Disease (Continued):

Fig. 10-5   Gross photograph of a cirrhotic liver in which a hepatoma developed. The serum LDH was 510 I.U. secondary to the malignancy involving the liver.

Glutamic Dehydrogenase

Glutamic dehydrogenase is present in organs such as liver, heart, and kidney. It is present in mitochondria within the cell cytoplasm. It is found normally in only small amounts in normal serum. Moderate elevations occur in serum secondary to acute viral hepatitis and toxic necrosis of the liver due to mitochondrial damage. Patients with portal cirrhosis develop minimal elevation. Individuals with extra-hepatic obstructive jaundice develop minimal serum elevation. A rise in the enzyme usually signifies cellular necrosis.[14]

Guanase

Liver guanase is needed to convert guanine to xanthine. This enzyme is also present in high amounts in kidney and brain, and in

Enzyme Abnormalities in Liver Disease (Continued):

moderate amounts in colon and small intestine. It has the potential of being a sensitive indicator of acute viral hepatitis or toxic necrosis of the liver.[35] Prominent elevation is present in patients with these diseases.[19] It has also been found to be elevated in patients with cancer of the pancreas and biliary cirrhosis. The level of guanase elevation is parallel to that of GOT in that it rises rapidly with damage to the liver and persists for approximately 3 weeks.[38]

Gamma-Glutamyl-Transpeptidase

This enzyme is involved in peptide metabolism; it catalyzes the hydrolysis of peptides and the simultaneous transfer of the glutamyl group. It is present in liver, pancreas, and kidneys; hepatobiliary and pancreatic disease is the usual cause for serum enzyme increase. A moderate increase occurs with parenchymal liver injury but a greater one results from obstructive disease.[62] The obstruction may be intra-hepatic or extra-hepatic.[63] Injured liver cells produce more of this enzyme than do normal cells. The enzyme is elevated in all types of liver disease. It is a more sensitive indicator of liver disease than is alkaline phosphatase or leucine aminopeptidase. Gambino has[42] found the enzyme to be normal in pregnancy, bone disease, and renal failure, and increased in pancreatitis, carcinoma of the pancreas, and obstructive hepatobiliary disease. Recently Ewen and Griffiths [26] reported that the enzyme was elevated in epileptic patients and in patients with highly vascular brain lesions. It rarely is increased in post-myocardial infarction patients. It is elevated in alcoholics without liver disease because of enzyme induction by alcohol.[3,51]

Monamine Oxidase

A soluble monamine oxidase which deaminates monamines is present in serum. The serum enzyme level is increased in post-necrotic and nutritional cirrhosis or metastatic cancer to the liver. It is not increased in acute hepatitis. The association of an increased serum monamine oxidase with hepatic fibrosis suggests the enzyme may catalyze maturation of connective tissue fibers.[43]

Aldolase

Acute viral hepatitis or hepatic necrosis due to chemicals or drugs will cause an elevation of serum aldolase.[25,68] Obstructive

Enzyme Abnormalities in Liver Disease (Continued):

jaundice or portal cirrhosis usually causes a minimal increase. Although the primary value of determination of this enzyme is in detecting primary myopathy, it is of value in the assessment of acute viral hepatitis or toxic necrosis of the liver.

The enzyme may also be increased in hematologic conditions, such as hemolysis, leukemia, or pernicious anemia, and in acute myocardial infarction.[20] The enzyme derived from skeletal muscle has specificity for the substrate fructose-1,6-diphosphate. The enzyme from the liver, especially in patients with infectious hepatitis, has specificity against both fructose-1,6-diphosphate, and fructose-1-phosphate. Elevations of serum aldolase may occur before the bilirubin rises in acute hepatitis. It usually remains high in the serum for approximately 3 weeks.[65]

Isocitric Dehydrogenase

Acute viral hepatitis or toxic necrosis of the liver, may cause an elevated isocitric dehydrogenase.[71] This enzyme may be minimally[83] elevated in patients with metastatic cancer of the liver, obstructive jaundice, or portal cirrhosis.[8] Marked serum elevations occur in patients with acute viral hepatitis or toxic necrosis.[17,18] The enzyme is widely distributed throughout the body, being found in the liver, heart, skeletal muscle, and malignant neoplasms.[54] The enzyme derived from cardiac muscle, is extremely heat labile at 56°C. for 15 minutes, whereas the liver enzyme is heat stable.[10] This pattern is distinctly opposite to that of lactic dehydrogenase, in which the heart enzyme is heat stable and the liver enzyme is heat labile. After acute hepatitis or toxic necrosis of the liver, the liver enzyme persists in the serum and is easily detected. Elevation of this enzyme occurs rapidly with acute viral hepatitis, infectious mononucleosis, or toxic necrosis of the liver. High levels of the enzyme are present until approximately 3 weeks after onset of the disease. Then the level of the enzyme returns to normal.

Sorbitol Dehydrogenase

Sorbitol dehydrogenase is an enzyme found primarily in the liver, but it is also present in prostate, kidney, lymphopoietic system, heart, and skeletal muscle.[86] It ordinarily is not present in human serum. Patients with acute viral hepatitis or toxic necrosis of the liver will develop a high serum activity.[4]

Enzyme Abnormalities in Liver Disease (Continued):

Ornithine Carbamoyl Transferase

Ornithine carbamoyl transferase is primarily a liver enzyme.[59] The enzyme is also found in the small intestine, colon, stomach, lung, lymphopoietic system, brain, and skeletal muscle.[73] A small amount is normally present in the serum.[60] With acute chemical necrosis of the liver or acute viral hepatitis, the enzyme is released into the blood stream. The level rises rapidly and persists for approximately 3 weeks. Increased serum levels have been found in patients with myocardial disease, immunologic conditions such as the collagen vascular diseases, and acute gastroenteritis. Elevations of 200 times the normal level have been described in patients with acute viral hepatitis or toxic necrosis of the liver. It is a sensitive and specific enzyme for hepatic injury. Slight elevations have occurred in patients with portal cirrhosis, cancer metastatic to the liver, and obstructive jaundice. Congenital deficiency of this liver enzyme may lead to an interesting clinical condition in children, in which high protein diets produce increased serum ammonia due to the inability to convert ammonia to urea with subsequent hepatic coma.

Cholinesterase

The cholinesterases are hydrolytic enzymes which act on choline esters. There are two types: Acetylcholinesterase, true, Type I, and cholinesterase, pseudo, Type II. Type I is present in the nervous system and the red blood cell while Type II is present in serum.[79,85] Thus, pseudocholinesterase is the term used to dintinguish the serum enzyme from true cholinesterase. Liver disease will cause decreased serum cholinesterase levels. Thus, viral hepatitis, cirrhosis, or metastatic liver cancer causes a decreased serum cholinesterase. The lowest level of enzyme is present in patients with acute viral hepatitis at the peak of the disease (Fig. 10-6).[46] Persistent depression of cholinesterase in patients with portal cirrhosis signifies a poor prognosis. Early obstructive jaundice is associated with normal serum cholinesterase, but with the onset of biliary cirrhosis, the cholinesterase may decrease. Other diseases that may cause low serum cholinesterase are: Myocardial infarction, the collagen diseases such as dermatomyositis, and conditions in which there is a low serum albumin such as malabsorption and exfoliative dermatitis (Fig. 10-7). However, some individuals with a low serum albumin may have increased cholinesterase levels as in the nephrotic syndrome.[47,89]

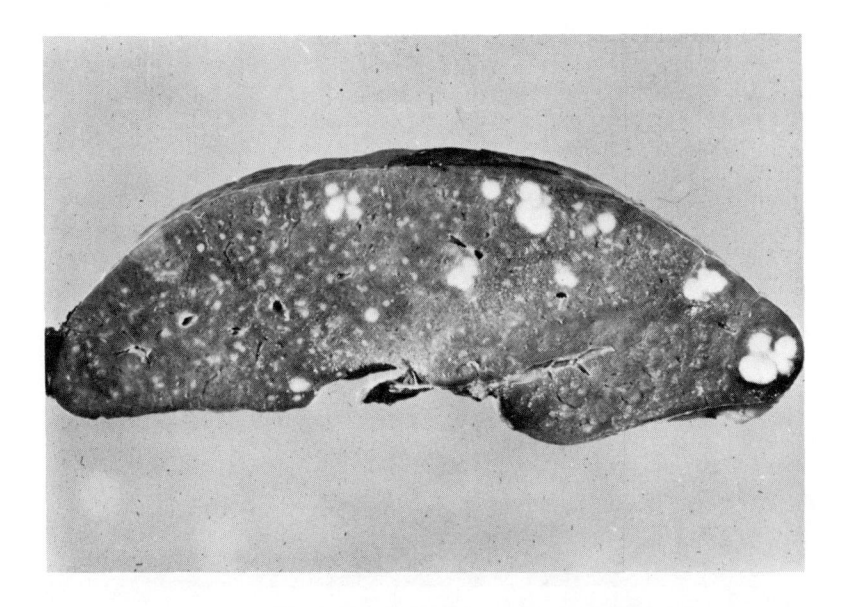

Enzyme Abnormalities in Liver Disease (Continued):

Fig. 10-6 Gross photograph of spleen with prominent Hodgkin's disease involvement. The ceruloplasmin was 1470 units before radiotherapy and returned to 310 units with remission of the disease.

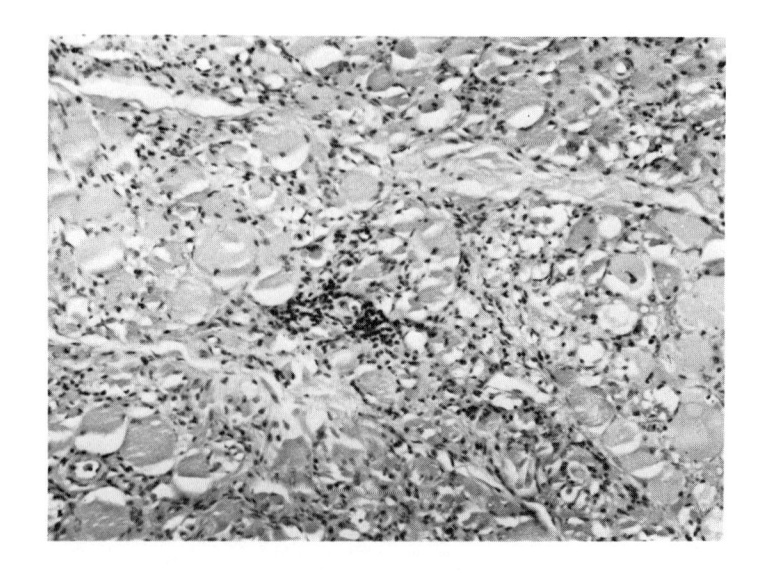

Enzyme Abnormalities in Liver Disease (Continued):

Fig. 10-7 Photomicrograph of section of skeletal muscle from a patient with dermatomyositis who had a prominent decrease in his serum pseudocholinesterase.

Measurement of serum or erythrocyte cholinesterase is valuable to determine exposure to organic phosphate insecticides. Serum cholinesterase activity may be reduced to 40 percent of the original level before clinical symptoms begin. Red blood cell cholinesterase is a better indicator of nervous system cholinesterase than serum cholinesterase. Clinical symptoms begin when erythrocyte cholinesterase activity falls to 20 percent of pre-exposure levels. A 40 percent decline is considered a dangerous level, while a 60 percent decline necessitates removal of the person from exposure.[87]

Fluoride and citrate inhibit cholinesterase, as do detergents such as quaternary ammonium compounds. Thus, special glassware must be used in the laboratory to determine this enzyme.[36]

Enzyme Abnormalities in Liver Disease (Continued):

A rapid test to qualitatively determine the plasma level of cholinesterase is a test paper known as Acholest Test Paper. It is utilized as a simple screening method for plasma cholinesterase. The best test, however, is a quantitative determination.

The succinylcholine compounds inhibit acetylcholinesterase. Prolonged paralysis is occasionally found in patients with low serum cholinesterase activity. This may be extremely important in patients with diseases associated with low cholinesterase activity or in apparently healthy individuals with an inherited deficiency of serum cholinesterase.

Since pseudocholinesterase hydrolyzes the succinylcholine, individuals with low levels or with a genetically altered cholinesterase (atypical cholinesterase variant) may develop prolonged apnea when the drug is administered in the course of anesthesia or during psychiatric therapy. Determination of this enzyme is important in suspected decrease due to poisoning by organic phosphate insecticides. Enzyme levels also are low in hepatic disease (Fig. 10-8). If succinylcholine is utilized before surgery in patients with hepatic disease where there are low enzyme levels, patients may suffer undesirable "overdose" type effect.

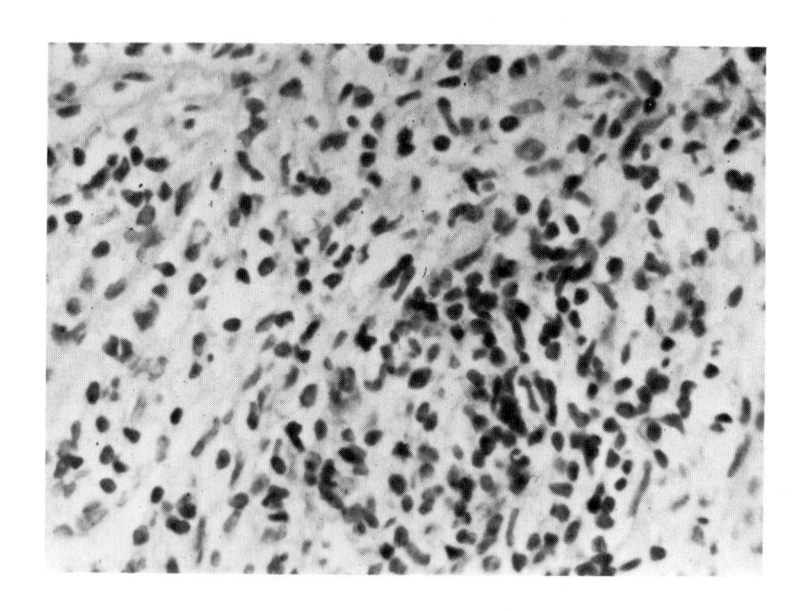

Enzyme Abnormalities in Liver Disease (Continued):

Fig. 10-8 Photomicrograph of liver biopsy from a patient with prominent viral hepatitis. Decreased serum pseudocholinesterase was synthesized and thus the serum level was low.

Thus the determination of plasma cholinesterase may be extremely useful in the surgical patient. The lab kit consists of a test strip impregnated with a special substrate, and a control strip is a measure of cholinesterase activity which is reported as normal or decrease The plasma cholinesterase acts on the substrate to liberate choline and acetic acid. The acetic acid changes the color of the test strip containing an acid base indicator.

Ceruloplasmin

Copper is present in two forms in the serum. Five percent is free or loosely bound to albumin. Ninety-five percent is firmly bound to alpha-two globulin. Ceruloplasmin is a blue protein and acts as a ferroxidase. The chief aspect of determining ceruloplasmin is in the evaluation of Wilson's disease in which the liver is cirrhotic (Fig. 10-9).[31]The serum ceruloplasmin is markedly decreased in Wilson's

Enzyme Abnormalities in Liver Disease (Continued):

disease. Loss of the alpha-two globulin in nephrosis and in sprue will also result in a low ceruloplasmin (Fig. 10-10).[15,72]

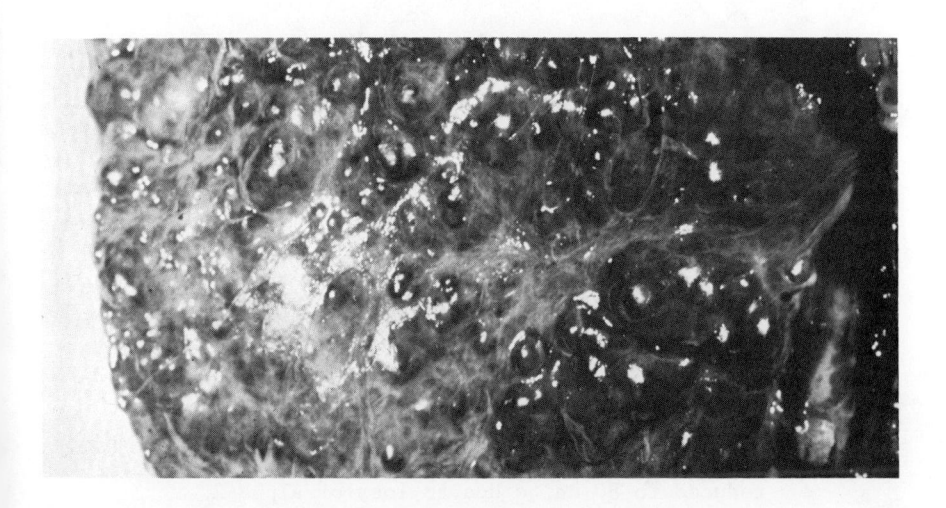

Fig. 10-9   Gross photograph of a cirrhotic liver from a patient with Wilson's disease. The ceruloplasmin was reduced to 60 units.

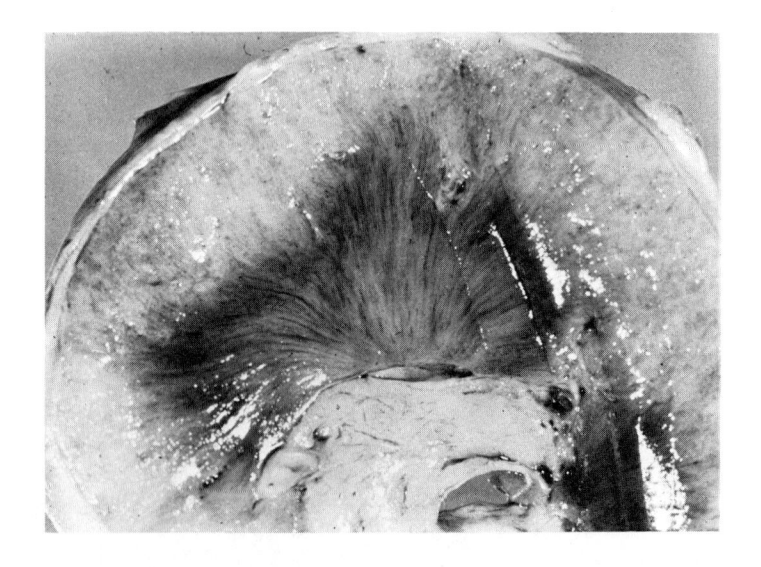

Enzyme Abnormalities in Liver Disease (Continued):

Fig. 10-10 Gross photograph of kidney from a patient who died from nephrosis. The ceruloplasmin was reduced to 80 units due to loss of alpha-2 globulin into the urine.

There are many causes for an elevated serum ceruloplasmin. These are pregnancy, utilization of oral contraceptives, Hodgkin's disease, hyperthyroidism, hepatic disease, tissue necrosis as is seen in myocardial infarction, and acute inflammatory states. Ceruloplasmin is an acute phase reactant and is elevated in active disease and returns to normal with successful treatment of the disease. The serum ceruloplasmin is a useful test to determine the activity of Hodgkin's disease.

The elevation of ceruloplasmin in pregnancy has been attributed to estrogens. When adminstered alone, estrogens can produce a markedly increased ceruloplasmin. The rise in plasma ceruloplasmin in women taking oral contraceptives is most probably due to the estrogen component.[88]

The green color of plasma secondary to elevated ceruloplasmin is more prominent when a large volume of plasma is present as in a blood

Enzyme Abnormalities in Liver Disease (Continued):

bank unit (Fig. 10-11). It is not as noticeable and infrequently seen in a clinical laboratory test tube or pilot tube attached to a blood bank unit.

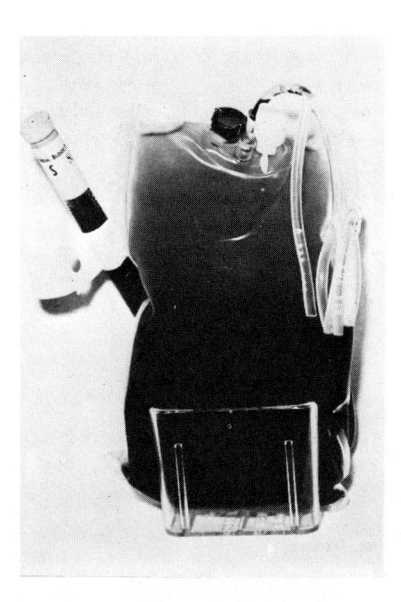

Fig. 10-11   Green plasma in a blood bank unit of blood from a female donor who was taking an oral contraceptive tablet. The ceruloplasmin was 1310 units and elevated causing the green plasma.

The presence of elevated levels of plasma ceruloplasmin in rheumatoid arthritis is unexplained. The green plasma color is easier to detect if there is a reduction in yellow plasma pigments, which are primarily carotenoids, bilirubin and heme. These yellow pigments are decreased in rheumatoid arthritis, and observation of the green color is easier.

Enzyme Abnormalities in Liver Disease (Continued):

TABLE 10-1

SERUM CERULOPLASMIN

Increased	Decreased
Pregnancy	Wilson's disease
Hyperthyroidism	Nephrosis
Oral Contraceptives	Malnutrition
Leukemia	Malabsorption
Lymphoma, active Hodgkin's	Newborn
Rheumatoid arthritis	
Acute and Chronic infection	
Neoplasia	
Tissue necrosis and myocardial infarction	
Collagen-vascular diseases	

Beta-Glucuronidase

Beta-glucuronidase is present in the liver. Increased serum levels may occur with metastatic cancer to the liver, viral hepatitis, or toxic necrosis of the liver.[32,53]

Glucose-6-Phosphatase

Glucose-6-phosphatase may increase markedly in patients with acute viral hepatitis or toxic necrosis of the liver.[39] The enzyme is normally present in the microsomes of the liver cells and is extremely heat labile.[93]

Alcohol Dehydrogenase

Alcohol dehydrogenase is present in the liver, gastrointestinal tract and retina. The enzyme is present primarily in the hepatic cells and minimally in bile duct epithelium.

Alcohol dehydrogenase activity in serum is detectable only in patients with active hepatic disease. With acute hepatitis, the level of the enzyme reflects the severity of the hepatic necrosis. The duration of serum enzyme elevation parallels GOT elevation. The height of the serum activity may also reflect presence of bile stasis

Enzyme Abnormalities in Liver Disease (Continued):

Uncomplicated obstructive jaundice does not cause elevation of the serum alcohol dehydrogenase.[45]

Summary of Enzyme Tests in Liver Disease Diagnosis

Thus, various enzyme tests are helpful in the diagnosis of hepatobiliary conditions. A jaundiced patient with an elevated GOT and GPT, alkaline phosphatase, and minor elevation of LDH most likely has acute viral hepatitis. Toxic necrosis will cause a high LDH. The jaundiced individual with high alkaline phosphatase, 5'-nucleotidase, leucine aminopeptidase, and only moderate elevations of GOT, GPT, and LDH most likely has obstructive jaundice or drug-induced intra-hepatic cholestasis. The mildly jaundiced individual with moderately elevated GOT, GPT, and alkaline phosphatase probably has portal cirrhosis. Alkaline phosphatase and LDH are prominently increased in metastatic cancer to the liver or other infiltrative lesions. Elevated enzyme levels may preceed other evidence of liver disease by several days. If enzymes such as GOT or GPT are normal, a diagnosis of acute hepatic injury can be excluded.

Alkaline phosphatase, GOT, and GPT are utilized to appraise the potential hepatotoxicity of new drugs. The best enzyme tests to monitor the use of immunosuppressive drugs are GOT and GPT. Infectious mononucleosis causes elevated alkaline phosphatase, GOT, GPT, and LDH.

Certain errors may occur in the diagnostic assessment of liver disease. One should not rely on any single enzyme test to specifically denote the cause of jaundice. At times, patients with obstructive jaundice may have very high GOT and GPT but these are exceptions. Patients with metastatic cancer to the liver may have normal GOT or only moderate elevation. Patients with severe necrosis of the liver or acute viral hepatitis with necrosis may develop low levels of GOT and GPT. Exhaustion of the liver cells may occur and these enzymes may decrease markedly. At times a delay occurs in a return to normal of GOT and GPT in patients with acute viral hepatitis. In most patients the levels return to normal within 3 weeks after onset of the hepatitis. It should not be assumed that patients have progressed into chronic active hepatitis if there is a late return of these enzymes to normal. Patients with portal cirrhosis may have a normal GOT level. One should not exclude the diagnosis of cirrhosis if the GOT level is normal.

Enzyme Abnormalities in Liver Disease (Continued):

REFERENCES

1. Alsever, J. B.: "The Blood Bank and Homologous Serum Jaundice. A Review of Medicolegal Considerations". New Eng. J. Med. 261:383, 1959.

2. Aronsen, K. F.: "Liver Function Studies during and after Complete Extrahepatic Biliary Obstruction in the Dog". Acta. Chir. Scand. (Suppl.) 257:1, 1961.

3. Aronsen, K. F., Hanson, A., Nosslini, B.: "The Value of Gamma Glutamyl Transpeptidase in Differentiating Viral Hepatitis from Obstructive Jaundice. A Statistical Comparison with Alkaline Phosphatase". Acta. Chir. Scand. 130:92, 1965

4. Asada, M., Galambos, J. T.: "Sorbitol Dehydrogenase and Hepatocellular Injury: An Experimental and Clinical Study". Gastroenterology 44:578, 1963.

5. Ballard, H., Bernstein, M., Farrar, J. T.: "Fatty Liver Presenting as Obstructive Jaundice". Amer. J. Med. 30:196, 1961.

6. Bang, N. V., Ruegsegger, P., Ley, A. B., LaDue, J. S.: "Detection of Hepatitis Carrier by Serum Glutamic Oxalacetic Transaminase Activity". J. Amer. Med. Assoc. 171:2303, 1959.

7. Bardauill, C., Chang, C.: "Serum Lactic Dehydrogenase, Leucine Aminopeptidase, and 5'-Nucleotidase Activities: Observations in Patients with Carcinoma of the Pancreas and Hepatobiliary Disease". Canad. Med. Assoc. J. 89:755, 1962.

8. Baron, D. N., Bell, J. L.: "Serum Isocitric Dehydrogenase in the Investigation of Liver Cell Damage". J. Clin. Path. 12:385, 1959.

9. Barondess, J. A., Erle, H.: "Serum Alkaline Phosphatase Activity in Hepatitis of Infectious Mononucleosis". Amer. J. Med. 29:43, 1960.

10. Bowers, G. N., Jr.: "Isocitric Dehydrogenase (ICD): The NADP-linked Enzyme". In Pre-Workshop Manual, J. B. Henry (Ed.) CLINICAL ENZYMOLOGY Amer. Soc. Clin. Path., pg. 118, 1964.

Liver Disease References (Continued):

11. Bradus, S., Korn, R. J., Chomet. B., West, M., Zimmerman, H. J.: "Hepatic Function and Serum Enzyme Levels in Assoc. iation with Fatty Metamorphosis of the Liver". Amer. J. Med. Sci. 246:69, 1963.

12. Brem, T. H.: "Use of Hepatic Function Tests in the Diagnosis of Amebic Abscess of the Liver". Amer. J. Med. Sci. 229:135, 1955.

13. Bressler, R., Forsyth, B. R., Klatskin, G.: "Serum Leucine Aminopeptidase Activity in Hepatobiliary and Pancreatic Disease". J. Lab. Clin. Med. 56:417, 1960.

14. Carlson, A. S., Siegelman, A. M., Robertson, T.: "Glutamic Dehydrogenase. II. Activity in Human Serums, Contrasted with that of Lactic Dehydrogenase and Glutamic Oxalacetic Transaminase". Amer. J. Clin. Path. 38:260, 1962.

15. Cartwright, G. E., Markowitz, H., Shields, G. S., Wintrobe, M. M.: "Studies on Copper Metabolism. XXIX. A Critical Analysis of Serum Copper and Ceruloplasmin Concentrations in Normal Subjects, Patients with Wilson's Disease and Relatives of Patients with Wilson's Disease". Amer. J. Med. 28:555, 1960.

16. Clermont, R. J., Chalmers, T. C.: "The Transaminase Tests in Liver Disease". Medicine 46:197, 1967.

17. Cohen, N. N., Potter, H. P., Jr., Bowers, G. N., Jr.: "An Evaluation of Isocitric Dehydrogenase in Liver Disease". Ann. Intern. Med. 55:604, 1961.

18. Cohn, E. M., Winsten, S., Abramson, E. B.: "A Study of the Serum Isocitric Dehydrogenase Test in Patients with Jaundice". Amer. J. Gastroenterology 44:45, 1965.

19. Coodley, E. L.: "Serum Guanase in the Diagnosis of Liver Disease". Amer. J. Gastroenterology 50:55, 1968.

20. Dale, R. A.: "Demonstration of Aldolase in Human Platelets: The Relation to Plasma and Serum Aldolase". Clin. Chim. Acta. 5:652, 1960.

21. DeRitis, F., Coltorti, M., Giusti, G.: "Diagnostic Value and Pathogenic Significance of Transaminase Activity Changes in Viral Hepatitis". Minerva Med. 47:167, 1956.

Liver Disease References (Continued):

22. DeRitis, F., Giusti, G., Piccinino, F., Cacciatore, L.: "Biochemical Laboratory Tests in Viral Hepatitis and other Hepatic Diseases". Bull. WHO 32:59, 1965.

23. Dickes, R., Schenker, V., Deutsch, L.: "Serial Liver Functi and Blood Studies in Patients receiving Chlorpromazine". New Eng. J. Med. 256:1, 1957.

24. Dixon, T. F., Purdom, M.: "Serum 5'-Nucleotidase". J. Clin. Path. 7:341, 1954.

25. Eismann, J.: "Die Aktivitat der Aldolase; ein Beitrag zur Differential Diagnose des Ikterus". Deutsch. Med. J. 7:204, 1956.

26. Ewen, L. M., Griffiths, J.: "Serum Gamma Glutamyl Transpeptidase: Elevated Levels in Certain Neurological Diseases" Clin. Chem. 17:642, 1971.

27. Felder, L., Mund, A., Parker, J. G.: "Liver Function Tests in Chronic Congestive Heart Failure". Circulation 2:286, 1950.

28. Frank, H. D., Merritt, J. H.: "Enzyme Activity in the Serum and Common Bile Duct of Dogs". Amer. J. Gastroenterology 31:166, 1959.

29. Gall, E. A.: "Serum Phosphatase and other Tests of Liver Function in Infectious Mononucleosis". Amer. J. Clin. Path 17:529, 1947.

30. Galluzzi, N. J., Weingarten, W., Regan, F. D., Doerner, A.: "Evolution of Hepatic Tests and Clinical Findings in Primar Hepatic Carcinoma". J. Amer. Med. Assoc. 15:152, 1953.

31. Gault, M. H., Stein, J., Aronoff, A.: "Serum Ceruloplasmir in Hepatobiliary and other Disorders: Significance of Abnormal Values". Gastroenterology 50:8, 1966.

32. Goldbarg, J. A., Pineda, E. P., Banks, B. M., Rutenberg, A. M.: "A Method for the Colorimetric Determination of Beta-Glucuronidase in Urine, Serum and Tissue: Assay of Enzymatic Activity in Health and Disease". Gastroenterology 36:193, 1959.

Liver Disease References (Continued):

33. Harkness, J., Roper, B. W., Durant, J. A., Miller, H.: "The Serum Leucine Aminopeptidase Test. An Appraisal of its Value in Diagnosis of Carcinoma of Pancreas". Brit. Med. J. 1:1787, 1960.

34. Hoffman, E., Nachlas, M. M., Gaby, S. D., Abrams, S. J., Seligman, A. M.: "Limitations in the Diagnostic Value of Serum Leucine Aminopeptidase". New Eng. J. Med. 263:541, 1960.

35. Hugh, A. C., Free, A. H.: "An Improved Method for the Determination of Guanase in Serum of Plasma". Clin. Chem. 11:708, 1965.

36. Johnston, D. G., Huff, W. C.: "Stability of Cholinesterase in Frozen Plasma". Clin. Chem. 11:729, 1965.

37. Jones, W. A., Tisdale, W. A.: "Posthepatic Cirrhosis Clinically Simulating Extrahepatic Biliary Obstruction (so-called 'Primary Biliary Cirrhosis')". New Eng. J. Med. 268:629, 1963.

38. Knights, E. M., Whitehouse, J. L., Hue, A. C., Santos, C. L.: "Serum Guanase Determination: A Liver Function Test". J. Lab. Clin. Med. 65:355, 1965.

39. Koide, H., Oda, T.: "Pathological Occurrence of Glucose-6-Phosphatase in Serum in Liver Disease". Clin. Chim. Acta. 4:554, 1959.

40. Kowlessar, O. D., Haeffner, L. J., Riley, E. M., Sleisnger, M. H.: "Comparative Study of Serum Leucine Aminopeptidase, 5'-Nucleotidase and Non-specific Alkaline Phosphatase in Diseases affecting the Pancreas, Hepatobiliary Tree and Bone". Amer. J. Med. 31:231, 1961.

41. Linde, S.: "On the Mechanism of the Elevation of Serum Glutamic Oxalacetic Transaminase in Obstructive Jaundice". Scand. J. Clin. Lab. Invest. 10:308, 1959.

42. Lum, G., Gambino, S. R.: "Serum Gamma Glutamyl Transpeptidase Activity as an Indicator of Disease of Liver, Pancreas, or Bone". Clin. Chem. 18:358, 1972.

Liver Disease References (Continued):

43. McEwen, C. M., Castell, D. O.: "Abnormalities of Serum Monoamine Oxidase in Chronic Liver Disease". J. Lab. Clin. Med. 70:36, 1967.

44. Mendelsohn, M. L., Bodansky, O.: "The Value of Liver Function Tests in the Diagnosis of Intrahepatic Metastases in the Nonicteric Cancer Patient". Cancer 5:1, 1952.

45. Mezey, E., Cherrick, G. R., Holt, P. R.: "Serum Alcohol Dehydrogenase, an Indicator of Intrahepatic Cholestasis". New Eng. J. Med. 279:241, 1968.

46. Michel, H. O.: "An Electrometric Method for the Determination of Red Cell and Plasma Cholinesterase Activity". J. Lab. Clin. Med. 34:1564, 1949.

47. Morrow, A. C., Motulsky, A. G.: "Rapid Screening Method for the Common Atypical Pseudocholinesterase Variant". J. Lab. Clin. Med. 71:350, 1968.

48. Mossberg, S. M.: "Myocardial Infarction-like Syndrome in Cholecystectomized Patients given Narcotics". Brit. Med. J. 1:948, 1964.

49. Mossberg, S. M., Ross, G.: "High Serum Transaminase Activity associated with Extrahepatic Biliary Disease". Gastroenterology 45:345, 1963.

50. Musser, A. W., Ortigoza, C., Vazquez, M., Riddick, J.: "Correlation of Serum Enzymes and Morphologic Alterations of the Liver". Amer. J. Clin. Path. 46:82, 1966.

51. Orlowski, M., Szezeklik, A.: "Heterogeneity of Serum Gamma Glutamyl Transpeptidase in Hepatobiliary Disease". Clin. Chim. Acta. 15:387, 1967.

52. Paterson, C. R., Losowsky, M. S.: "The Bones in Chronic Liver Disease". Scand. J. Gastroenterology 2:293, 1967.

53. Pineda, E. P., Goldbarg, J. A., Banks, B. M., Rutenberg, A. M.: "The Significance of Serum Beta-Glucuronidase Activity in Patients with Liver Disease. A Preliminary Report". Gastroenterology 36:202, 1959.

Liver Disease References (Continued):

54. Pojen, J., Ninger, E., Tovarek, J.: "Changes in Serum Isocitric Dehydrogenase in Myocardial Infarction". Enzym. Biol. Clin. (Basel) 3:184, 1963.

55. Polin, S. G., Spellberg, M. A., Teitelman, L., Okumura, M.: "The Origin of Elevation of Serum Alkaline Phosphatase in Hepatic Disease". Gastroenterology 42:431, 1962.

56. Posen, S.: "Alkaline Phosphatase". Ann. Intern. Med. 67:183, 1967.

57. Prince, A. M., Gersohn, R. K.: "The Use of Serum Enzyme Determinations to Detect Anicteric Hepatitis". Transfusion 5:120, 1965.

58. Pruzanski, W., Fischi, J.: "The Evaluation of Serum Leucine Aminopeptidase Estimation". Amer. J. Med. Sci. 248:581, 1964.

59. Reichard, H.: "Ornithine Carbamyl Transferase Activity in Human Tissue Homogenates". J. Lab. Clin. Med. 56:218, 1960.

60. Reichard H.: "Studies on Ornithine Carbamyl Transferase Activity in Blood Serum: Methodologic, Experimental, and Clinical Investigation". Acta. Med. Scand. 70:390, 1962.

61. Rutenberg, A. M., Goldbarg, J. A., Pineda, E. P.: "Leucine Aminopeptidase Activity. Observations in Patients with Cancer of the Pancreas and other Diseases". New Eng. J. Med. 259:469, 1958.

62. Rutenberg, A. M., Goldbarg, J. A., Pineda, E. P.: "Serum Gamma Glutamyl Transpeptidase Activity in Hepatobiliary Pancreatic Disease". Gastroenterology 46:43, 1963.

63. Rutenberg, A. M., Smith, E. E., Fischbein, J. W.: "Electrophoretic Mobilities of Serum Gamma Glutamyl Transpeptidase and its Clinical Application in Hepatobiliary Disease". J. Lab. Clin. Med. 69:504, 1967.

64. Schwartz, M. K., Bodansky, O.: "Serum 5'-Nucleotidase in Patients with Cancer". Cancer 18:886, 1965.

65. Schapira, F.: "Fructose-1-Phosphate Aldolase of the Serum". Path. Biol. (Paris) 9:63, 1961.

Liver Disease References (Continued):

66. Searcy, R.: DIAGNOSTIC BIOCHEMISTRY McGraw-Hill Co., pg. 511, 1969.

67. Shay, H., Siplet, H.: "The Value of Serum Alkaline Phosphatase Determination and Bromsulphalein Test in the Diagnosis of Metastatic Cancer of the Liver". J. Lab. Clin. Med. 43:741, 1954.

68. Sibley, J. A., Fleisher, G. A.: "Clinical Significance of Serum Aldolase". Mayo Clin. Proc. 29:591, 1954.

69. Siegel, I. A.: "Leucine Aminopeptidase in Pregnancy". Obstet. & Gynec. 14:488, 1959.

70. Smetana, H. F.: "The Histopathology of Drug-induced Liver Disease". Ann. N. Y. Acad. Sci. 104:821, 1963.

71. Sterkei, R. L., Spincer, J. A., Wolfson, S., Jr., Williams-Ashman, H. G.: "Serum Isocitric Dehydrogenase Activity with Particular Reference to Liver Disease". J. Lab. Clin. Med. 52:176, 1958.

72. Sternlieb, I., Scheinberg, I. H.: "The Diagnosis of Wilson Disease in Asymptomatic Patients". J. Amer. Med. Assoc. 183:747, 1963.

73. Strandjord, P. E., Clayson, K. J.: "An Automatic Method for the Determination of Ornithine Carbamyl Transferase Activity". J. Lab. Clin. Med. 67:154, 1966.

74. Strandjord, P. E., Clayson, K. J., Freier, E. F.: "Heat-Stable Lactate Dehydrogenase in the Diagnosis of Myocardial Infarction". J. Amer. Med. Assoc. 182:1099, 1962.

75. Strandjord, P. E., Thomas, K. E., White, L. P.: "Studies on Isocitric and Lactic Dehydrogenase in Experimental Myocardial Infarction". J. Clin. Invest. 38:2111, 1959.

76. Szasz, G.: "Serum Leucine Aminopeptidase Activity in Acute Lesions of the Liver Parenchyma". Lancet 1:441, 1964.

77. Tan, C. O., Cohen, J., West, M., Zimmerman, H. J.: "Serum Enzymes in Disease. XIV. Abnormality of Levels of Transaminases and Glycolytic and Oxidative Enzymes and of Liver Functions as related to the Extent of Metastatic Carcinoma of the Liver". Cancer 16:1373, 1963.

Liver Disease References (Continued):

78. Titus, M. A., Reynolds, D. A., Glending, M. B., Page, E. W.: "Plasma Aminopeptidase Activity (Oxytocinase) in Pregnancy and Labor". Amer. J. Obstet. & Gynec. 80:1124, 1960.

79. Vorhaus, L. J., Scudamore, H. H., Kark, R. M.: "Measurement of Serum Cholinesterase Activity; Useful Test in the Management of Acute Hepatitis". Amer. J. Med. Sci. 221:140, 1951.

80. Weinstein, B. R., Korn, R. J., Zimmerman, H. J.: "Obstructive Jaundice as a Complication of Pancreatitis". Ann. Intern. Med. 58:245, 1963.

81. West, M., Eshchar, J., Zimmerman, H. J.: "Serum Enzymology in the Diagnosis of Myocardial Infarction and Related Cardiovascular Conditions". Med. Clin. N. Amer. 50:171, 1966.

82. West, M., Gleb, D., Pilz, C. G., Zimmerman, H. J.: "Serum Enzymes in Disease. VII. Significance of Abnormal Serum Enzymes in Cardiac Failure". Amer. J. Med. Sci. 241:350, 1961.

83. West, M., Schwartz, M. A., Cohen, J., Zimmerman, H. J.: "Serum Enzymes in Disease. IX. Glycolytic and Oxidative Enzymes and Transaminases in Patients with Carcinoma of the Kidney, Prostate and Urinary Bladder". Cancer 17:432, 1964.

84. West, M., Zimmerman, H. J.: "Serum Enzymes in Hepatic Disease". Med. Clin. N. Amer. 43:371, 1959.

85. Wetstone, H. J., LaMotta, R. U.: "The Clinical Stability of Serum Cholinesterase Activity". Clin. Chem. 11:653, 1965.

86. Williams-Ashman, H. G., Banks, J., Wolfson, S. K., Jr.: "Oxidation of Polyhydric Alcohols by the Prostate Gland and Seminal Vesicle". Arch. Biochem. 72:485, 1957.

87. Witter, R. F.: "Measurement of Blood Cholinesterase. A Critical Account of Methods of Estimating Cholinesterase with Reference to their Usefulness and Limitations under Difficult Conditions". Arch. Environ. Health 6:537, 1963.

88. Wolf, P. L., Dalziel, J., Swanson, S., Enlander, D.: "Brief Recording: Green Plasma in Blood Donors". New Eng. J. Med. 281:205, 1969.

Liver Disease References (Continued):

89. Zavon, M. R.: "Blood Cholinesterase Levels in Organic Phosphate Intoxication". J. Amer. Med. Assoc. 192:51, 1965.

90. Zimmerman, H. J.: "Clinical and Laboratory Manifestations of Hepatotoxicity". Ann. N. Y. Acad. Sci. 104:954, 1963.

91. Zimmerman, H. J.: "Serum Enzymes in the Diagnosis of Hepati Disease". Gastroenterology 46:613, 1964.

92. Zimmerman, H. J., West, M.: "Serum Enzyme Levels in the Diagnosis of Hepatic Disease". Amer. J. Gastroenterology 40:387, 1963.

93. Zuppinger, K.: "Die Identitat der Serum Glukose-6-Phosphatase mit der Alkalischen Phosphatase". Clin. Chim. Acta. 6:759, 1961.

CHAPTER 11

ENZYME ABNORMALITIES IN PANCREATIC AND SALIVARY GLAND DISEASE

Serum enzyme determinations as aids in the diagnosis of pancreatic disease have been established. The tests that are customarily ordered are serum amylase and serum lipase.

Amylase

Amylase exists in different forms. Beta-amylase is usually present in plants in contrast to alpha-amylase which is an animal enzyme.[1] The highest concentration of alpha amylase in man is in the salivary glands and the pancreas. Oviduct mucosa, liver, skeletal muscle, gastric and small intestinal mucosa, and red blood cells also have amylase activity.[44] A majority of the serum amylase in humans arises from the liver, pancreas, and salivary glands.[7,46] When subjected to electrophoresis, serum amylase will migrate with the albumin and gamma-globulin.[29] In acute pancreatitis, it usually migrates with the gamma-globulin fraction.[36,69,70] When the isoenzymes of amylase are subjected to electrophoretic separation, pancreatic amylase is shown to be distinct from salivary amylase.

Serum amylase is low in early infancy, is first demonstrable at 2 months of age, and reaches adult levels at the age of 1 year.[61] Various conditions besides acute pancreatitis may cause an elevated serum amylase. These are: Parotitis; carcinoma of the pancreas causing an acute pancreatitis; choledocholithiasis obstructing the ampulla of Vater; perforation or penetration of a duodenal ulcer into the pancreas; intestinal obstruction; trauma to the abdomen inducing a pancreatitis; pancreatic pseudocyst; intra-abdominal surgery resulting in acute pancreatitis; rupture of an ectopic pregnancy with liberation of amylase from the oviduct mucosa; tubo-ovarian abscess; opiates causing spasm of the ampulla of Vater; cholecystography; renal insufficiency with decreased excretion of amylase, small intestinal lymphoma resulting in elevation of I_gA or IgG immunoglobulin which binds to amylase causing a macroamylasemia, hemolysis, and hepatic necrosis. The serum amylase may be prominently increased in patients who have had gastric surgery.

The commonest cause of a high serum amylase is acute pancreatitis.[14] Destruction of pancreatic tissue with or without obstruction

247

Enzyme Abnormalities in Pancreatic and Salivary Disease (Continued):

of flow of pancreatic secretion results in escape of pancreatic enzyme into the peritoneal cavity. Subsequently, absorption of pancreatic enzymes into the blood with elevation of the serum amylase occurs. Persistence of the elevated circulating amylase usually depends on continued secretion of the enzyme and increased amylase in the thoracic duct results. Constant drainage of thoracic duct fluid in acute pancreatitis may lead to considerable improvement in the clinical condition.[26,55]

In acute pancreatitis, the serum amylase invariably increases simultaneously with the onset of symptoms (Fig. 11-1). It may rise above 500 Somogyi units.[2] The peak level of the enzyme is reached within 24 hours, after which there usually is a rapid decrease with a return to normal within 2 to 4 days. Absence of increase in serum amylase in the first 24 hours after the onset of symptoms is evidence against a diagnosis of acute pancreatitis. In severe pancreatitis, the height of the serum amylase elevation is inversely related to the severity of the pancreatitis.

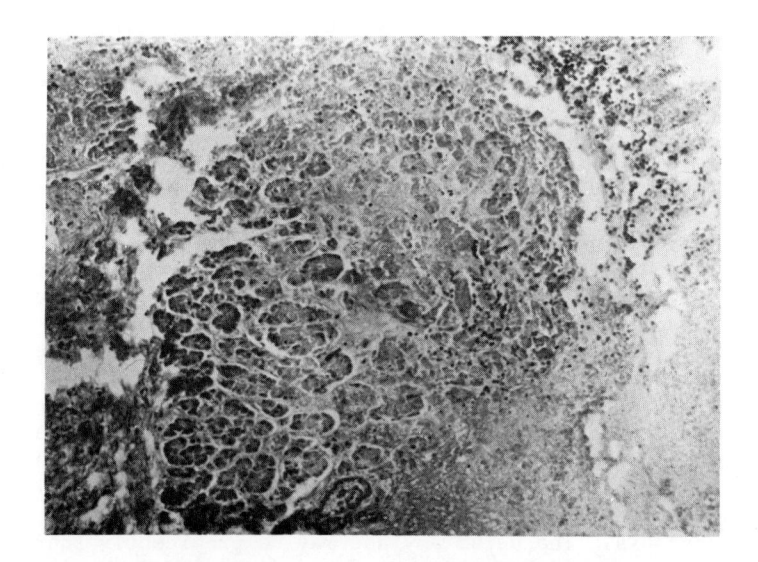

Fig. 11-1 Photomicrograph of section of pancreas from patient who died of acute hemorrhagic pancreatitis. The patient was an alcoholic who developed severe epigastric pain and had a serum amylase of 680 I.U. Note the extensive pancreatic necrosis with fat necrosis.

Enzyme Abnormalities in Pancreatic and Salivary Disease (Continued):

Abdominal paracentesis with determination of amylase in the peritoneal fluid is another useful technique to detect acute pancreatitis. Elevation of the enzyme in peritoneal fluid is strong evidence of acute pancreatitis.[13,31,37,47,51,53] However, other intra-abdominal pathologic conditions , such as rupture of an ectopic pregnancy, may also elevate peritoneal amylase activity.[57]

In the mild form of the disease, there will be a transient elevation of amylase and lipase if the pancreatitis is of moderate severity, and complete resolution may occur within 2 weeks. When acute pancreatitis is severe, there is marked pancreatic destruction and death may result. Patients with pancreatitis may have a frequent association with biliary tract disease; serum amylase should be determined in patients with choledocholithiasis. Pancreatitis may be induced by mumps, or coxsackie virus. Hyperlipemia and hyperparathyroidism are metabolic conditions that are at times associated with pancreatitis.

Patients with abdominal trauma in which there is an elevated serum amylase may have to be explored to exclude pancreatic trauma such as contusion with possible major injury or transsection of the pancreas.[35] Hematomas may occur in the pancreas. Patients with pseudocysts have an increased serum amylase which is prolonged.

Pseudocysts of the pancreas may result from acute pancreatitis.[63] A marked elevation of serum amylase may be the first indication of such a cyst. Elevations of serum amylase are never as great as those of the cyst fluid. The cyst may suddenly rupture into the peritoneal cavity or into the retroperitoneal space. The fluid within the pseudocyst usually has a large amount of amylase and lipase.[54] Up to 2000 Somogyi units or higher of amylase has been reported in some cysts, and some levels are as great as 25,000 Somogyi units.[24,25,43]

In acute pancreatitis, the peritoneal amylase reaches a higher level than the serum amylase, perhaps ten times as much, and the elevation persists longer.[42] In addition to acute pancreatitis causing a raised peritoneal fluid amylase, perforated peptic ulcer into the pancreas, infarcted obstructed small intestine, trauma to the pancreas, and ruptured ectopic pregnancy may result in increase in peritoneal fluid amylase. Some investigators have effectively utilized peritoneal fluid dialysis to decrease the amount of lipase and amylase in the peritoneal fluid as a therapeutic procedure.

Enzyme Abnormalities in Pancreatic and Salivary Disease (Continued):

Amylase is excreted in the urine as a threshhold substance; thus, elevation of urine amylase reflects an increase in serum levels.[15] Certain investigators have suggested that urinary amylase elevation is more diagnostic if it is determined quantitatively as a rate of secretion,[20,58] and that secretion rates are more closely correlated with the course of acute pancreatitis than are serum amylase levels. When patients with acute pancreatitis are monitored with urinary amylase determinations periodically, fluctuations in the abdominal condition can be assessed.[5,30] Urinary excretion also increases following operative injury to the pancreas. For diagnostic purposes, simultaneous determinations of amylase in blood and urine are superior to a single determination of the enzyme in the blood or urine alone.[38]

Extrapancreatic causes for elevation of serum amylase are usually associated with diseases involving the parotid, gastrointestinal tract bile duct system, and oviduct. Elevation of the enzyme following administration of drugs results from spasm of the Sphincter of Oddi. secretin stimulation of the pancreas after administration of morphine or codeine has been advocated as a test of pancreatic function.[17,23] High serum amylase values may persist from 24 to 48 hours after the injection of drugs. False positive elevations are always only moderat emphasizing the diagnostic reliability of high serum amylase in pancreatitis.[22]

Patients with hemolytic anemia may have a rise in amylase activi If the serum is allowed to stand at room temperature, amylase may increase. Anticoagulants such as oxalate and citrate will decrease amylase activity.[18]

Amylase is also present in hepatic cells. In massive hepatic necrosis, elevated serum amylase occurs. Some investigators believe that this elevation results usually from the renal failure which may accompany the hepatic necrosis.

It is common to observe prominent elevation of serum amylase in acute renal insufficiency, with high serum amylase secondary to decreased renal clearance of the enzyme.[8]

Individuals with malabsorption secondary to small intestinal lymphoma (Fig. 11-2), may develop a high serum amylase.[11] The lympho results in production of a large amount of I_gA or I_gG immunoglobulin which binds serum amylase, resulting in the macroamylase macromolecul The presence of this macroamylase in the serum causes a high serum amylase simulating the level in acute pancreatic disease. Unlike

Enzyme Abnormalities in Pancreatic and Salivary Disease (Continued):

elevations of ordinary amylase, macroamylase will not produce an elevated urinary amylase. This is because of the nonclearance of this macromolecule by the kidney. The macroamylase has a molecular weight of 100,000 and is tightly bound to the I_gA or I_gG immunoglobulin.[40]

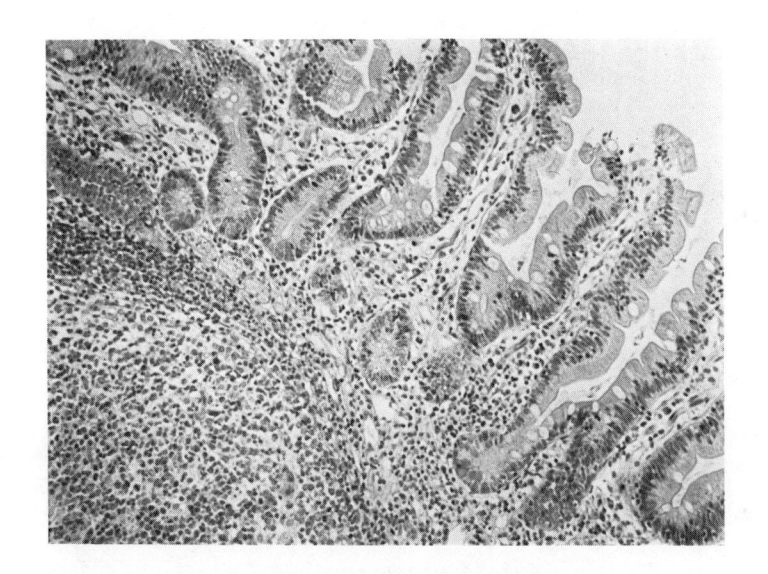

Fig. 11-2 Photomicrograph of small intestine from a patient with malabsorption due to small intestinal lymphoma. His serum amylase was 970 I.U. and it was characterized as a macroamylase. The serum I_gA was markedly elevated.

The parotid glands are a rich source of amylase. Infections of the parotids due to mumps, bacterial parotitis, and calculi in the parotid ducts will cause a high serum amylase.

Cancer of the pancreas is rarely associated with marked elevation of amylase or lipase in the serum, but it does produce alterations in these enzymes in the duodenal contents. Provocative enzyme studies with pancreozymin, secretin, or both may define cancer of the pancreas

Enzyme Abnormalities in Pancreatic and Salivary Disease (Continued):

by study of the enzyme levels in the duodenal contents.[27] Cancer of the pancreas may cause complete or partial obstruction of the pancreatic duct and with pancreozymin or secretin stimulation, a significant elevation of amylase or lipase may occur in the duodenal contents

Pancreatic fistula may occur after operation on the pancreas or following trauma to the abdomen with continuous drainage of the fistula. Major ones may persist for a month or longer. A minor or major fistula may be associated with increase in serum amylase or lipase. The presence of either enzyme in the fistula drainage indicates the presence of a pancreatic fistula.

Provocative enzyme tests aid in the diagnosis of chronic pancreatitis. They are based on the assumption that if a gland does not have ductal obstruction, neither stimulation of enzyme production, nor stimulation of flow will increase enzyme levels in the blood. Two factors, damage to the pancreas and the degree of obstruction, determine the amount of blood enzyme found in pancreatic disease. Some investigators have used a pancreozymin secretin test and claim to have increased over previous tenchiques, the percentage of positive results in patients with pancreatitis. Only serum amylase was elevated; serum lipase did not increase.[28,62]

Patients with chronic pancreatitis usually have no elevation of serum amylase or lipase. When secretin or pancreozymin testing is employed, however, significant decrease in the duodenal pancreatic secretion is demonstrated; bicarbonate secretion may be depressed.[71]

Prolonged serum amylase elevation is of diagnostic importance. The condition has been noted in patients with pancreatic pseudocysts. There is no quantative relationship between serum amylase elevation and amylase in the cyst fluid. Other conditions which may cause prolonged serum amylase elevation in the serum are: Relapsing pancreatitis; pancreatic abscess; and pancreatic fistula.

Lipase

Besides amylase, the other serum enzyme test for pancreatic disease diagnosis is lipase. Lipase activity is present in pancreas, gastrointestinal mucosa, especially gastric mucosa, and white blood cells.[68] Very little activity is present in liver, brain, kidney, or skeletal muscle; some is present in lung. Estimation of serum lipase activity to assess pancreatic disease is not as widely used as estimation of serum amylase. Serum lipase activity rises after 24 hours

Enzyme Abnormalities in Pancreatic and Salivary Disease (Continued):

following onset of acute pancreatitis.[60] Increases in lipase may be more pronounced than those of amylase and decreases may occur more slowly. Elevations of serum lipase may last as long as 2 weeks after the onset of acute pancreatitis.[16,21,32] Elevated serum lipase may result from other intra-abdominal pathologic lesions leading to acute pancreatitis such as pancreatic pseudocysts, carcinoma of the pancreas, or choledocholithiasis with resultant pancreatitis, perforation or penetration of a duodenal ulcer into the pancreas, intestinal obstruction with infarction of the intestine, peritonitis involving the pancreas, opiates causing spasm of the Sphincter of Oddi, and renal insufficiency with lack of excretion of lipase (Fig. 11-3).[9]

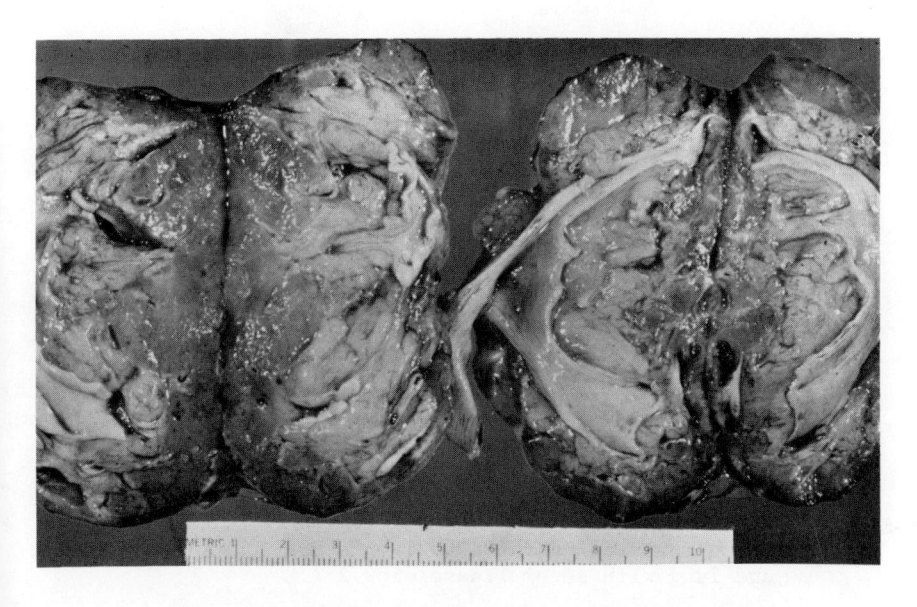

Fig. 11-3 Gross photograph of markedly atrophic kidneys from a patient who died of chronic pyelonephritis causing severe renal insufficiency. The serum amylase was 360 I.U. and the lipase 2.4 I.U. The excretion of these enzymes was reduced because of renal insufficiency.

Enzyme Abnormalities in Pancreatic and Salivary Disease (Continued):

Patients with fat embolism to the lungs will have an elevated serum lipase resulting from liberation of lung lipase into the serum (Fig. 11-4).[4,6,39] Decrease in the serum lipase is caused by use of anticoagulants such as oxalate or citrate and by the presence of a hemolyzed specimen,[4] since hemoglobin inhibits the action of lipase. Urinary lipase measurements are not advised for clinical diagnosis.[6,39,49,52]

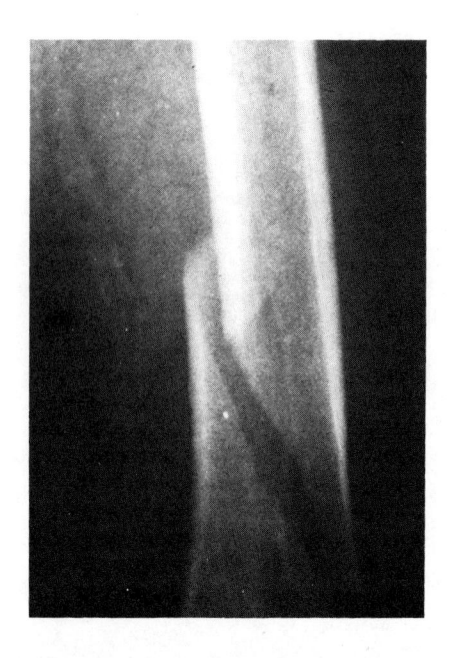

Fig. 11-4 Gross photograph of x-ray of fracture of femur which resulted in fat embolism to brain, kidneys and lung with serum lipase of 2.2 I.U.

Decreased lipase activity is present when there has been destruction of the pancreatic gland and diminished production of the enzyme.[64]

Lipase catalyzes the hydrolysis of triglycerides to fatty acids and glycerol.[10,33] The substrates that can be utilized to determine serum lipase activity are olive oil, coconut oil, and tributyrin. The coconut oil method is a more sensitive indicator of pancreatic disease than is the one using olive oil. Various recent studies

Enzyme Abnormalities in Pancreatic and Salivary Disease (Continued):

indicate that the lipases may be divided into different isolipases.[19,66]

Trypsin

The estimation of trypsin in pancreatic juice and serum is more difficult than the estimation of amylase or lipase.[48] This is because antitryptic factors are present in the body fluids. The body protects itself against autodigestion by these antitryptic factors. In active pancreatitis, a great increase in the antitryptic factors occurs. This is also true of patients with pancreatic pseudocysts and cancer of the pancreas.[3,67]

Trypsinogen is transformed to trypsin by enterokinase. The assay for trypsin in duodenal juice was a definitive test for cystic fibrosis until the sweat chloride test was recognized.[56] Pancreatic secretions may be decreased in a variety of diseases such as cystic fibrosis. Lipase deficiency and trypsinogen deficiency may occur.[59] The normal trypsin content in the duodenal juice is 160 to 180 micrograms per milliliter.[50] The determination of this enzyme may be valuable to assess cystic fibrosis. Under the age of 1 year, failure of the stool to digest gelatin in a dilution of 1:100 is suggestive of fibrocystic disease. Lower values are found, however, in meconium stools and in fecal stools in the first 10 days of life. Above 2 years many normal stools from children will digest in a dilution of 1:12.5. Failure of digestion at this dilution is suggestive of fibrocystic disease.[51]

A majority of children with fibrocystic disease show no digestion of gelatin with duodenal juice. Occasional titers up to 1:50 are recorded. Lower titers of 1:12.5 are however, normally seen in normal children; especially over the age of two years.[34,65]

Lactic Dehydrogenase

Recently a number of reports have indicated the usefulness of determining lactic dehydrogenase (LDH) in saliva. Elevated LDH has been found in the saliva in diseases such as carcinoma and acute and chronic inflammations of the salivary glands.[72]

Enzyme Abnormalities in Pancreatic and Salivary Disease (Continued):

REFERENCES

1. Abderhalden, R.: ENZYMOLOGY: ENZYMES IN PATHOGENESIS, DIAGNOSIS AND THERAPY Princeton: Van Nostrand, pg. 30, 1961.

2. Adams, J. T., Libertino, J. A., Schwartz, S. I.: "Significance of an Elevated Serum Amylase". Surgery 63:877, 1968.

3. Adham, N. F., Dyce, B., Haverback, J. B.: "Elevated Serum Trypsin Binding Activity in Patients with Hereditary Pancreatitis". Amer. J. Dig. Dis. 13:8, 1968.

4. Adler, F., Peltier, L. F.: "The Laboratory Diagnosis of Fat Embolism". Clin. Orthop. 21:226, 1961.

5. Ambromovage, A. M., Howard, J. M., Pairent, F. N.: "The Twenty-four Hour Excretion of Amylase and Lipase in the Urine". Ann. Surg. 167:539, 1968.

6. Armstrong, H. J., Kuenzig, M. C., Peltier, L. F.: "Lung Lipase Levels in Normal Rats and Rats with Experimentally Produced Fat Embolism". Proc. Soc. Exp. Biol. Med. 124:959, 1967.

7. Arnold, M., Ritter, W. J.: "Liver Amylase. III. Synthesis by the Liver and Secretion into the Perfusion Medium". J. Biol. Chem. 238:2760, 1963.

8. Baily, G. L., Katz, A. I., Hampers, C. L., Merrill, J. P.: "Alterations in Serum Enzymes in Chronic Renal Failure". J. Amer. Med. Assoc. 213:2263, 1970.

9. Barowman, J. A., Borgstrom, B.: "Specifity of Certain Methods for the Determination of Pancreatic Lipase". Gastroenterology 55:601, 1968.

10. Berk, J. E.: "Serum Amylase and Lipase". J. Amer. Med. Assoc. 199:98, 1967.

11. Berk, J. E., Kizu, H., Wilding, P., Searcy, R. L.: "Macroamylasemia: A Newly Recognized Cause for Elevated Serum Amylase Activity". New Eng. J. Med. 277:941, 1967.

Pancreatic and Salivary Disease References (Continued):

12. Berk, J. E., Searcy, R. L., Wilding, P., Kizu, H., Svoboda, A. C.: "Macroamylase: A New Cause for Elevated Serum Amylase Activity". J. Amer. Med. Assoc. 200:545, 1967.

13. Bolooki, H., Gliedman, M. L.: "Peritoneal Dialysis in the Treatment of Acute Pancreatitis". Surgery 64:466, 1968.

14. Burnett, N., Ness, T. D.: "Serum Amylase and Acute Abdominal Disease". Brit. Med. J. 2:770, 1955.

15. Calkins, W. G.: "Study of Urinary Amylase Excretion in Normal Persons". Amer. J. Gastroenterology 46:407, 1966.

16. Cherry, I. S., Crandall, L. A., Jr.: "Specificity of Pancreatic Lipase: Its Appearance in Blood after Pancreatic Injury". Amer. J. Physiol. 100:266, 1932.

17. Chey, W. Y., Shary, H., Nielsen, O. F., Lorber, S. H.: "Evaluation of Tests of Pancreatic Function in Chronic Pancreatic Disease". J. Amer. Med. Assoc. 201:347, 1967.

18. Christian, D. G.: "Drug Interference with Laboratory Blood Chemistry Determinations". Amer. J. Clin. Path. 54:118, 1970.

19. Comfort, M. W.: "Serum Lipase: Its Diagnostic Value". Amer. J. Dig. Dis. 3:817, 1937.

20. Danker, A., Heifetz, C. J.: "Interrelationship of Blood and Urine Diastase during Transient Acute Pancreatitis". Gastroenterology 18:207, 1951.

21. Desnuelle, P., Savary, P.: "Specificities of Lipase". J. Lipid Res. 4:369, 1963.

22. Dreiling, D. A., Hollander, F.: "Studies in Pancreatic Function. II. Statistical Study of Pancreatic Secretion following Secretin in Patients without Pancreatic Disease". Gastroenterology 15:620, 1950.

23. Dreiling, D. A., Richman, A.: "Evaluation of Provocative Blood Enzyme Tests Employed in the Diagnosis of Pancreatic Disease". Arch. Intern. Med. 94:197, 1954.

Pancreatic and Salivary Disease References (Continued):

24. Ebbesen, K. E., Schoenebeck, J.: "Prolonged Amylase Elevation". Acta. Chir. Scand. 133:61, 1967.

25. Edlin, P.: "Mediastinal Pseudocyst of Pancreas: Case Report and Discussion". Gastroenterology 17:96, 1951.

26. Egdahl, R. H.: "Mechanism of Blood Enzyme Changes following the Production of Experimental Pancreatitis". Ann. Surg. 148:389, 1958.

27. Fillippini, L., Ammann, R.: "Functional Clinical Diagnostic Tests for Carcinoma of the Pancreas (with Special Reference to the Pancreozymin-Secretin Test and Proteolytic Activity of the Stool)". Schweiz. Med. Wschr. 97:803, 1967.

28. Fitzgerald, O., Fitzgerald, P., Finnelly, J., McMullin, J. P., Boland, S. J.: "A Clinical Study of Chronic Pancreatitis". Gut 4:193, 1963.

29. Franzini, C.: "Electrophoretic Behavior of Human Urinary Amylase". J. Clin. Path. 18:664, 1965.

30. Gambill, E. E., Mason, H. L.: "One-Hour Value for Urinary Amylase in 96 Patients with Pancreatitis: Comparative Diagnostic Value of Tests of Urinary and Serum Amylase and Serum Lipase". J. Amer. Med. Assoc. 186:24, 1963.

31. Gjessing, J., Dencher, H.: "Abdominal Paracentesis with Dialysis Catheter and Peritoneal Lavage. A Diagnostic Technique in Acute Abdominal Conditions". Acta. Chir. Scand. 134:351, 1968.

32. Gomori, G.: "Assay of Pancreatic Lipase in Serum". Amer. J. Clin. Path. 27:170, 1957.

33. Henry, R. J., Sobel, C., Berkman, S: "On the Determination of 'Pancreatic Lipase' in Serum". Clin. Chem. 3:77, 1957.

34. Horsfield, A.: "Proteolytic Activity of Stools". J. Med. Lab. Tech. 10:18, 1952.

35. Jones, R. C., Shires, G. T.: "The Management of Pancreatic Injuries". Arch. Surg. 90:502, 1965.

Pancreatic and Salivary Disease References (Continued):

36. Joseph, R. R., Olivero, E., Ressler, N.: "Electrophoretic Study of Human Isoamylases: A New Saccharogenic Staining Method and Preliminary Results". Gastroenterology 51:377, 1966.

37. Keith, L. M., Jr., Zollinger, R. M., McCleary, R. S.: "Peritoneal Fluid Amylase Determinations as an Aid in the Diagnosis of Acute Pancreatitis". Arch. Surg. 61:930, 1950.

38. Kirshen, R., Gambill, E. E., Mason, H. L.: "Comparison of Urinary and Serum Amylase Values following Pancreatic Stimulation in Patients with and without Pancreatic Disease". Gastroenterology 48:579, 1965.

39. Korn, E. D.: "Clearing Factor, a Heparin-activated Lipoprotein Lipase". J. Biol. Chem. 215:1, 1955.

40. Levitt, M. D., Cooperband, S. R.: "Hyperamylasemia from the Binding of Serum Amylase by an 11s I_gA Globulin". New Eng. J. Med. 278:474, 1968.

41. Lundh, G.: "Determination of Trypsin and Chymotrypsin in Human Intestinal Content". Scand. J. Clin. Lab. Invest. 9:229, 1957.

42. MacFate, R. P.: "Amylase in Biological Fluids". In F. W. Sunderman & F. W. Sunderman, Jr. (Eds.) MEASUREMENTS OF EXOCRINE AND ENDOCRINE FUNCTIONS OF THE PANCREAS Lippincott, pg. 14, 1961.

43. McClintock, J. T., McFee, J. L., Quimby, R. L.: "Pancreatic Pseudocyst presenting as a Mediastinal Tumor". J. Amer. Med. Assoc. 192:573, 1965.

44. McGeachin, R. L., Gleason, J. R., Adams, M. R.: "Amylase Distribution in Extrapancreatic, Extrasalivary Tissues". Arch. Biochem. 104:314, 1964.

45. McGeachin, R. L., Lewis, J. P.: "Electrophoretic Behavior of Serum Amylase". J. Biol. Chem. 234:795, 1959.

46. McGeachin, R. L., Potter, B. A., Lindsey, A. C.: "Puromycin Inhibition of Amylase Synthesis in the Perfused Rat Liver". Arch. Biochem. 104:314, 1964.

Pancreatic and Salivary Disease References (Continued):

47. Maingot, R.: ABDOMINAL OPERATIONS Lewis & Co., Ltd., 4th Edition, pg. 524, 1964.

48. Nardi, G. L., Lees, C. W.: "Serum Trypsin: A New Diagnostic Test for Pancreatic Disease". New Eng. J. Med. 258:797, 1958

49. Notham, M. M., Pratt, J. H., Callow, A. D.: "Studies on Urinary Lipase. II. Urinary Lipase in Man". Arch. Intern. Med. 96:188, 1955.

50. O'Brien, D., Ibbott, F. A., Rodgerson, D. O.: LABORATORY MANUAL OF PEDIATRIC MICROBIOCHEMICAL TECHNIQUES Harper & Row, 4th Edition, pg. 319, 1968.

51. O'Brien, D., Ibbott, F. A., Rodgerson, D. O.: LABORATORY MANUAL OF PEDIATRIC MICROBIOCHEMICAL TECHNIQUES Harper & Row, 4th Edition, pg. 321, 1968.

52. Pfeffer, R. B., Dishman, A., Cohen, T., Tesler, M., Aronson, A. R.: "Urinary Lipase Excretion in Pancreatic and Hepatobiliary Disease". Surg. Gynec. & Obstet. 124:1071, 1967.

53. Pfeffer, R. B., Mixter, G., Jr., Hinton, W. J.: "Acute Haemorrhagic Pancreatitis. A Safe Effective Technique for Diagnostic Paracentesis". Surgery 43:550, 1958.

54. Pinkham, R. D.: "Pancreatic Collections (Pseudocysts) following Pancreatic Pancreatitis and Pancreatic Necrosis. Reviews and Analysis of 10 Cases". Surg. Gynec. & Obstet. 3:227, 1945.

55. Popper, H. L., Necheles, H.: "Pathways of Enzymes into the Blood in Acute Damage of the Pancreas". Proc. Soc. Exp. Biol. Med. 43:220, 1940.

56. Richmond, R. C., Schwachmann, H.: "Studies of Fibrocystic Disease of the Pancreas: Mucoviscidosis Chymotrypsin Activity in Duodenal Fluid". Pediatrics 16:207, 1955.

57. Rott, H. D., Hauser, C. W., McKinley, C. R., LaFave, J. W., Mendiola, R. P.: "Diagnostic Peritoneal Lavage". Surgery 57:633, 1965.

Pancreatic and Salivary Disease References (Continued):

58. Saxon, E. I., Hinklay, W. C., Vogel, W. C., Zieve, L.: "Comparative Value of Serum and Urinary Amylase in the Diagnosis of Acute Pancreatitis". Arch. Intern. Med. 99:607, 1957.

59. Sheldon, W.: "Congenital Pancreatic Lipase Deficiency". Arch. Dis. Child. 39:268, 1964.

60. Sherrick, J. C., DeLa Huerga, J.: "Lipases in Biologic Fluids". In F. W. Sunderman & F. W. Sunderman, Jr. (Eds.) MEASUREMENTS OF EXOCRINE AND ENDOCRINE FUNCTIONS OF THE PANCREAS Lippincott, pg. 36, 1961.

61. Somogyi, M.: "Amylase Activity in Human Blood". Arch. Intern. Med. 67:678, 1941.

62. Sun, D. C. H.: "Diagnostic Tests for Chronic Pancreatic Disease". Arch. Intern. Med. 115:57, 1965.

63. Sybers, H. D., Shelp, W. D., Morrissey, J. F.: "Pseudocyst of the Pancreas with Fistulous Extension into the Neck". New Eng. J. Med. 278:1058, 1968.

64. Tauber, H.: "New Olive Oil Emulsion for Lipase and New Observation concerning 'Serum Lipase'". Proc. Soc. Exp. Biol. Med. 90:375, 1955.

65. Townes, P. L.: "Trypsinogen Deficiency Disease". J. Pediat. 66:275, 1965.

66. Vogel, W. C., Zieve, L.: "A Rapid and Sensitive Turbidimetric Method for Serum Lipase based upon Differences between the Lipase of Normal and Pancreatitis Serum". Clin. Chem. 9:168, 1963.

67. Warter, J., Metais, P., Bieth, S.: "Tryptic and Chymotryptic Inhibitory Capacity of Serum and Serous Effusions in Pancreatic Disease". Gut 9:258, 1968.

68. Webster, P. D., Zieve, L.: "Alterations in Serum Content of Pancreatic Enzymes". New Eng. J. Med. 267:604, 1962.

69. Wilding, P.: "Use of Gel Filtration in Study of Human Amylase". Clin. Chim. Acta. 8:918, 1963.

Pancreatic and Salivary Disease References (Continued):

70. Wilding, P.: "The Electrophoretic Nature of Human Amylase and the Effect of Protein on the Starch-Iodine Reaction". Clin. Chim. Acta. 12:97, 1965.

71. Wormsley, K. G.: "Maximal Bicarbonate Secretion in Man". Gut 9:257, 1967.

72. Wroblewski, F., Levine, I., Augsth, F., Sansur, M.: "Clinical Implications of Lactic Dehydrogenase Activity in Sputum". J. Amer. Med. Assoc. 207:2436, 1969.

CHAPTER 12

ENZYME ABNORMALITIES IN RETICULOENDOTHELIAL DISEASES

This chapter will be concerned first of all with the serum enzyme abnormalities in diseases of the reticuloendothelial system and will in addition, analyze enzyme abnormalities of the red cell and the white cell.

Lactic Dehydrogenase

A prominent increase in LDH is always observed in patients who have megaloblastic anemia.[22] Thus, pernicious anemia, sprue, megaloblastic anemia of pregnancy, drug-induced megaloblastic anemia, caused by such drugs such as dilantin or chemotherapeutic agents, megaloblastic anemia following resection of the stomach, and megaloblastic anemia associated with fish tapeworm Diphyllobothrium latum infection, are associated with an elevated serum LDH. The increased LDH activity is related to the megaloblastic erythropoiesis in the bone marrow.[1] Tetrazolium histochemical staining of the bone marrow in megaloblastic anemia demonstrates that there is prominent megaloblastic mitochondrial activity producing LDH. The short survival of the megaloblastic cells in the marrow also contributes a prominent amount of LDH to the serum.[33] The hemolytic anemia peripherally found in megaloblastic anemias also accounts for some increase in the LDH activity in the peripheral blood. LDH activity falls promptly once the megaloblastic anemia is successfully treated. When reticulocytosis begins, the LDH activity decreases. The heat stable LDH_1 and LDH_2 isoenzymes are the ones involved in megaloblastic anemia; electrophoretically they simulate the isoenzymes associated with myocardial infarction.[32]

The isoenzymes LDH_1, LDH_2 and LDH_3 are increased in patients with leukemia or lymphoma. When stained with the tetrazolium formazan technique, peripheral blood smears and sections of lymphoreticular tissue involved with malignant lymphoma demonstrate that prominent LDH activity is present in the leukemic cells and malignant lymphoma cells in the lymphoreticular system. Multiple myeloma may cause increased LDH_5 isoenzyme in the peripheral blood. Here again, the malignant plasmablasts produce the isoenzyme.

In hemolytic anemia, there is a prominent increase in the LDH and an elevation of GOT in the peripheral blood. Hemolytic anemias associated with calcific cardiac valve disease or with prosthetic

Enzyme Abnormalities in Reticuloendothelial Diseases (Continued):

cardiac valves show the increases (Fig. 12-1).[47] Both enzymes may be elevated in patients with disseminated intravascular coagulation or consumptive coagulopathy in which there is agglutination and then hemolysis of red cells. When hemolytic anemia develops associated with abnormal prosthetic cardiac valves, the red cells appear as schistocytes and acanthrocytes.[65] Thus, this hemolytic syndrome is secondary to mechanical damage of the erythrocytes.[2,48] Recently Gerlach's LDH/GOT ratio has been proposed as useful for the diagnosis of acute hemolytic anemia. Normally the ratio is about 5:1; in acute hemolytic anemia a prominent elevation in the ratio occurs, in some instances to approximately 20:1.

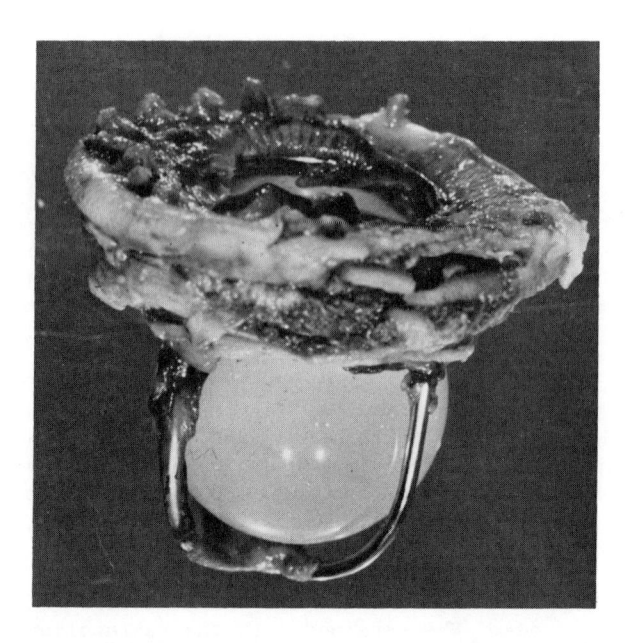

Fig. 12-1 Gross photograph of heart from patient who had rheumatic aortic stenosis. A Starr-Edwards valve was inserted to correct the valvular deformity. The patient developed a hemolytic anemia with an LDH of 380 I.U. and a GOT of 70 I.U.

Enzyme Abnormalities in Reticuloendothelial Diseases (Continued):

Muramidase

Serum and urine muramidase is elevated in individuals who have monocytic leukemia (Fig. 12-2) and myelomonocytic anemia.[44] This subject will be discussed more fully later in this Chapter when enzyme abnormalities of white cells are evaluated.

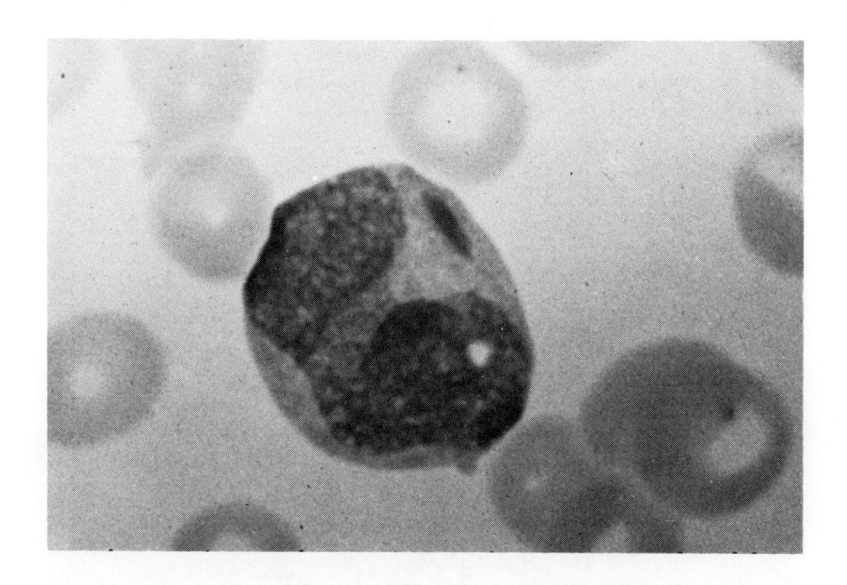

Fig. 12-2 Photomicrograph of monoblast containing an auer rod in peripheral blood of a patient with acute monocytic leukemia. The serum muramidase was elevated to 190 micrograms/100 ml.

Acid Phosphatase

Blood platelets contain acid phosphatase. Patients with the myeloproliferative syndrome will have increased serum acid phosphatase; the enzyme is derived from destroyed platelets. Estimation of the serum acid phosphatase in patients with thrombocytopenia is useful in indicating whether there is aplasia of bone marrow or if megakaryocytes still are present in the bone marrow. If there is

Enzyme Abnormalities in Reticuloendothelial Diseases (Continued):

thrombocytopenia with normal or increased numbers of megakaryocytes in the bone marrow, the serum acid phosphatase will be elevated (Fig. 12-3). If there is thrombocytopenia and decreased megakaryocytes in the bone marrow, acid phosphatase will not be elevated. Thus, increased serum acid phosphatase may signify the presence of increased platelet destruction with the presence of a normal or increased number of megakaryocytes in the bone marrow. This condition suggests Werlhof's disease, also known as idiopathic thrombocytopenic purpura. [42]

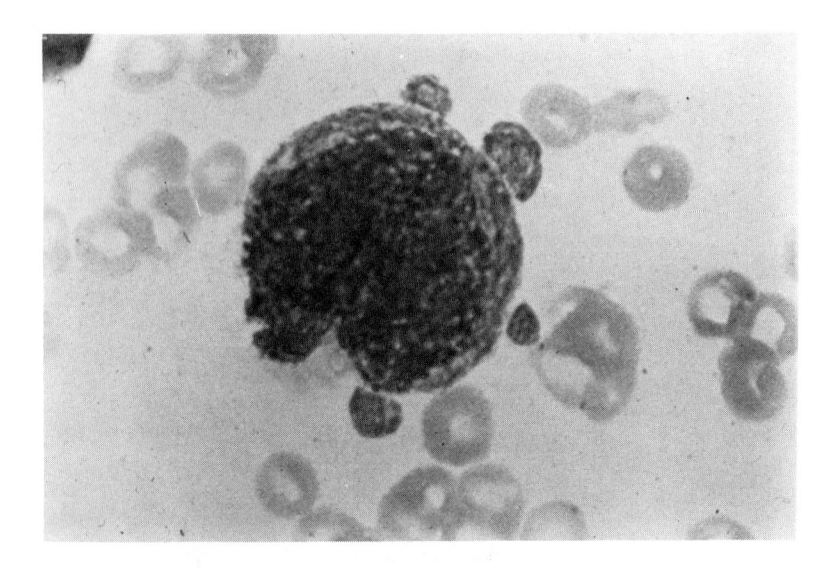

Fig. 12-3 Photomicrograph of bone marrow smear from a patient with idiopathic thrombocytopenic purpura. The platelet count was 5,000 per cu. mm. Increased numbers of megakaryocytes were present in the marrow. The megakaryocytes had serrated cytoplasm which is associated with ITP.

Red blood cells normally contain acid phosphatase, and in hemolytic anemia the enzyme increases in the serum (Fig. 12-4). Hemolyzed

Enzyme Abnormalities in Reticuloendothelial Diseases (Continued):

specimens should not be used for acid phosphatase determination because the hemolyzed red cells will produce a spurious elevation.

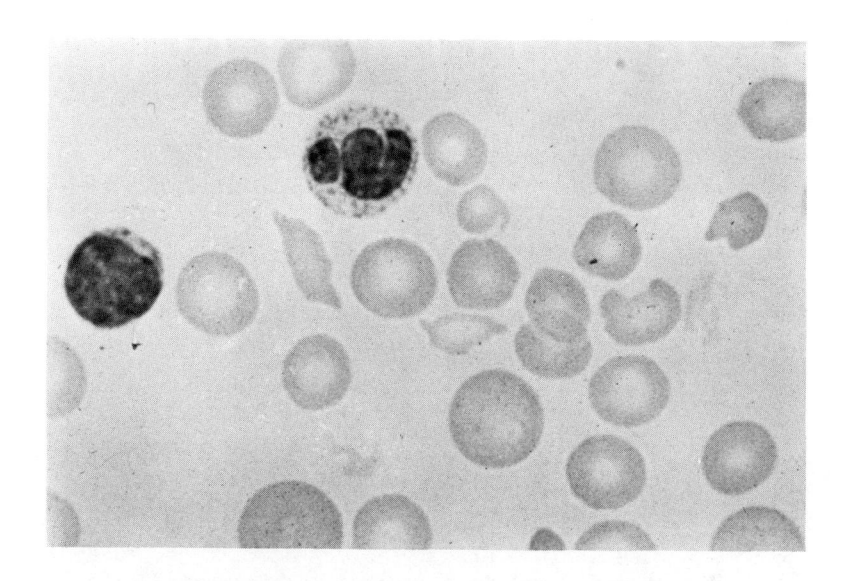

Fig. 12-4 Photomicrograph of peripheral blood smear from a patient with hemolytic anemia secondary to renal failure. The serum acid phosphatase was 1.85 I.U. due to the RBC destruction.

Leukocyte Alkaline Phosphatase

Serum enzymes are of great value in the diagnosis of hematologic conditions relevant to diseases of the white blood cells. Two important enzymes are alkaline phosphatase of the leukocytes and serum muramidase, also known as lysozyme.

The myeloid cell series contains alkaline phosphatase.[58] It first is detectable in the myelocyte; it is not present in the myeloblast or the promyelocyte.[26] It is present in the metamyelocyte, the band neutrophil, and the segmented neutrophil. Various methods

Enzyme Abnormalities in Reticuloendothelial Diseases (Continued):

are available to assess the alkaline phosphatase in neutrophils.[59]

TABLE 12-1

LEUKOCYTE ALKALINE PHOSPHATASE

Elevated	Normal
Leukocytosis	Chronic lymphocytic leukemia
Leukemoid reaction	Secondary polycythemia
Acute lymphocytic leukemia	Multiple myeloma
Polycythemia vera	
Myeloproliferative disorders	
Hodgkin's disease, active	Low
Pregnancy	Chronic myelocytic leukemia
Newborn infants	Acute myelocytic leukemia
Acute hemorrhage	Acute monocytic leukemia
	Infectious mononucleosis
	Paroxysmal nocturnal hemoglobinuria
	Congenital hypophosphatasia
	Sarcoidosis

Hematologic conditions that are associated with an elevated alkaline phosphatase include primary polycythemia (Fig. 12-5) in contrast to the normal level in secondary polycythemia. The myeloproliferative state [27,38] and Hodgkin's disease in an active state will also have an elevated leukocyte alkaline phosphatase; with reversion to inactivity, the enzyme level becomes normal.[4] Acute lymphocytic leukemia will cause an increased nutrophilic alkaline phosphatase as well. Other causes for increased leukocyte alkaline phosphatase are: (1) During the first week of life newborn infants have an increased leukocyte alkaline phosphatase which is related to the stress of birth, increased estrogens from mother, or conditions causing anoxia at the time of birth;[67] jaundice will cause a decrease in the leukocyte alkaline phosphatase in newborns. (2) Pregnancy will cause an increased leukocyte alkaline phosphatase (the cause for this is the high estrogen state; this is also evident in females taking oral contraceptive tablets containing estrogens)[66] (Fig. 12-6) Furthermore, leukocyte alkaline phosphatase may be used as a pregnanc test since it becomes elevated with pregnancy.[13]   (3) It may be used to diagnose ectopic pregnancy. (4) Acute hemorrhage will also cause an increase in leukocyte alkaline phosphatase. (5) The various

Enzyme Abnormalities in Reticuloendothelial Diseases (Continued):

causes for a myeloid leukemoid reaction will induce an elevated leukocyte alkaline phosphatase. In fact the latter condition is the main reason for performing this test. It is difficult at times to distinguish between myelogenous leukemia and a leukemoid state which may be secondary to bacterial infection, tissue necrosis, such as myocardial infarction, metabolic abnormality such as diabetic acidosis or uremia, or acute hemorrhage. The leukocyte alkaline phosphatase in leukemoid reactions will be markedly elevated but markedly depressed or absent in myelogenous leukemia. (6) In addition to estrogens causing a high leukocyte alkaline phosphatase, ACTH or cortisone will also cause an elevation of leukocyte alkaline phosphatase.

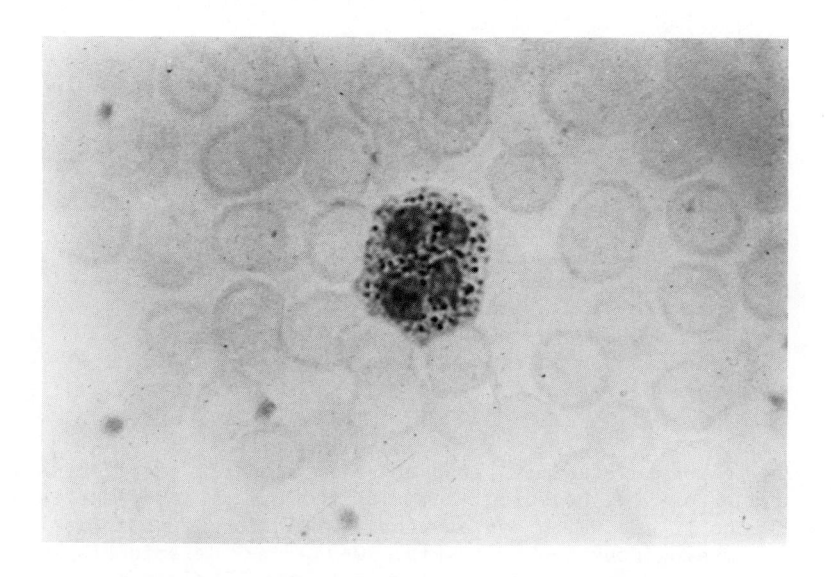

Fig. 12-5 Photomicrograph of peripheral blood leukocyte alkaline phosphatase from a patient with primary polycythemia. The leukocyte alkaline phosphatase score was significantly elevated to 280.

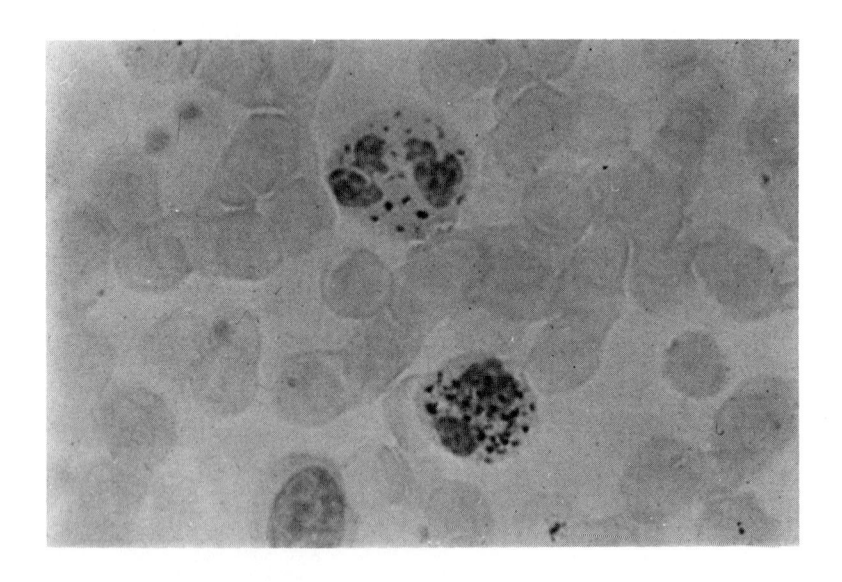

Enzyme Abnormalities in Reticuloendothelial Diseases (Continued):

Fig. 12-6 Photomicrograph of leukocyte alkaline phosphatase from a pregnant female. The score was 295 and the elevation was caused by the increase in maternal estrogen and progesterone during pregnancy.

Low leukocyte alkaline phosphatase is found in myelogenous leukemia and paroxysmal nocturnal hemoglobinuria in which there is a low white count and which may eventually transform into aplastic anemia or myelogenous leukemia (Fig. 12-7).[31,55] Sarcoidosis and infectious mononucleosis cause a low leukocyte alkaline phosphatase. Leukocyte alkaline phosphatase may be low in magnesium deficiency since the enzyme is dependent on magnesium for its function. Infants who appear to have rickets should have a leukocyte alkaline phosphatase determination to distinguish them from infants who have hypophosphatasia. Rickets will cause a normal or high leukocyte alkaline phosphatase in contrast to infants with hypophosphatasia who have a decreased or absent leukocyte alkaline phosphatase.[16] In addition, the serum alkaline phosphatase and tissue alkaline phosphatase are diminished or absent in hypophosphatasia.[28]

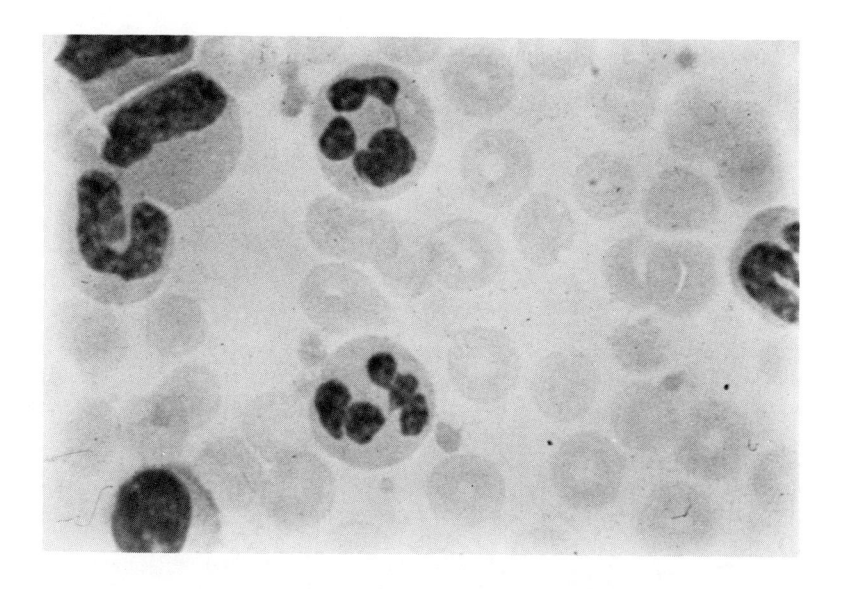

Enzyme Abnormalities in Reticuloendothelial Diseases (Continued):

Fig. 12-7 Photomicrograph of leukocyte alkaline phosphatase smear from a patient with chronic myelogenous leukemia. Note the absence of cytoplasmic leukocyte alkaline phosphatase.

A patient with chronic granulocytic leukemia with a low leukocyte alkaline phosphatase may develop a normal alkaline phosphatase when the disease goes into remission following myeleran therapy.[68]

Muramidase in Leukocyte Disease

Muramidase is a lysosomal enzyme found normally in cells such as the monocyte and the myeloid cell series.[18] The major part of muramidase in the serum is derived from granulocytes.[15] Patients with acute monocytic leukemia and other diseases in which monocytosis occurs may develop elevated muramidase activity in the serum and the urine. The serum enzyme is markedly elevated in acute monocytic leukemia and moderately elevated in acute and chronic myeloid leukemia.[46] It is usually normal or low in lymphocytic leukemia; thus, it is easy to

Enzyme Abnormalities in Reticuloendothelial Diseases (Continued):

exclude acute lymphocytic leukemia if the muramidase level is elevated. The muramidase activity in the serum varies with the total white blood cell count. In monocytic or myeloid leukemia, the elevated enzyme values return to normal when a remission occurs.[44] With a relapse, the serum muramidase will increase again and this rise may precede development of overt symptoms;[46] renal disease, tuberculosis, and sarcoidosis will cause an increased serum and urine muramidase level. Muramidase may cause damage to renal tubules with renal insufficiency.[21]

Hemolytic Anemias

Hereditary hemolytic anemias may be secondary to red blood cell enzyme abnormalities. Type I anemias are characterized by slight autohemolysis which is corrected in vitro by adding glucose to the defibrinated blood.[53] Type II anemias are characterized by marked autohemolysis which is not corrected in vitro by glucose but is corrected by adding adenosine triphosphate (ATP). These are the types I and II anemias of Dacie and Selwyn.

The remainder of this Chapter will be devoted to erythrocyte enzyme deficiencies.

TABLE 12-2

CHRONIC NONSPHEROCYTIC HEMOLYTIC ANEMIA
ERYTHROCYTE ENZYME DEFICIENCIES

Pentose Phosphate Pathway
1. Glucose-6-Phosphate Dehydrogenase
2. 6-Phosphogluconate Dehydrogenase

Nonglycolytic Pathway
1. Glutathione Reductase
2. Glutathione Peroxidase
3. Glutathione
4. Adenosine Triphosphatase

Anaerobic Glycolytic Pathway
1. Pyruvate Kinase
2. Triosephosphate Isomerase
3. Hexokinase

Enzyme Abnormalities in Reticuloendothelial Diseases (Continued):

4. Glucose Phosphate Isomerase
5. Phosphoglycerate Kinase
6. Glyceraldehyde-3-Phosphate Dehydrogenase
7. 2,3-Dephosphoglyceromutase

Glucose-6-Phosphate Dehydrogenase

One of the major enzyme abnormalities in hemolytic anemia is a genetic glucose-6-phosphate dehydrogenase (G-6-PD) deficiency.[40] Deficiency of G-6-PD may account for one-third of the persons with nonspherocytic hemolytic anemia. The anemia is manifest when the individual is exposed to drugs or to conditions in which oxidants are increased in the blood.[57] The trait is sex-linked, with males carrying only a single abnormal gene and females having two abnormal genes. The abnormality has been found to be more severe in Caucasians than in Negroes. The American Negro has a deficiency whereas G-6-PD deficient Caucasians of Mediterranean origin may have absence of the enzyme (Fig. 12-8). They have no erythrocyte G-6-PD activity.[9] Patients with G-6-PD deficiency are classified in the Dacie-Selwyn Type I group. Negroes with a mild G-6-PD deficiency usually have a self-limiting hemolytic crisis. Old red blood cells are more sensitive to hemolysis than young ones.[36] With destruction of the old cells, a self-limiting process occurs, since the young erythrocytes and reticulocytes that form contain sufficient enzyme and are not sensitive to the drug. When an affected individual is detected, family members should be studied especially mothers of male patients. Affected individuals should be counseled not to take drugs that will cause hemolysis.[30]

Enzyme Abnormalities in Reticuloendothelial Diseases (Continued):

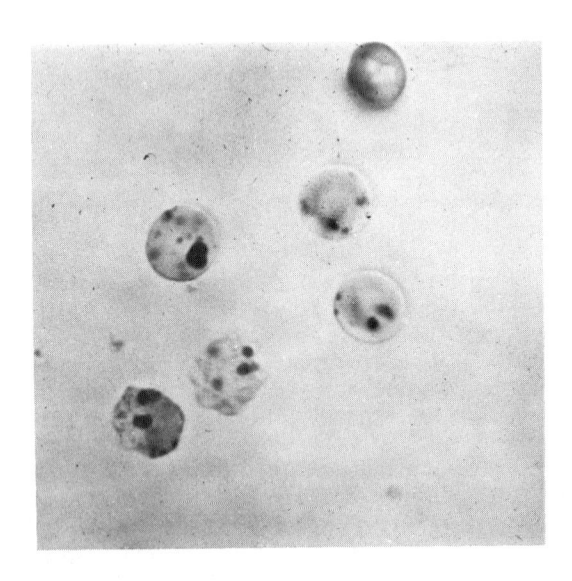

Fig. 12-8 Photomicrograph of a Heinz-Ehrlich preparation demonstrating the Heinz-Ehrlich bodies in the red blood cell representing precipitated hemoglobin.

As the red blood cell becomes older, there is a tendency for a decrease in glycolysis and energy production. Thus, the old red cell becomes more susceptible to oxidative processes.[63]

The age of red blood cells can be estimated by a study of its enzymes. G-6-PD activity in young red cells measures approximately ten times that of old cells. Aldolase and catalase may also be decreased in old cells. These red cells are more prone to formation of methemoglobin when exposed to various agents such as nitrites. In red cells with a reduced G-6-PD level reduced glutathione is also deficient. If a woman carrying a G-6-PD deficient fetus receives an oxidant drug, the fetus may develop a hemolytic anemia. Caucasians with G-6-PD deficiency should not serve as blood donors; transfusion of their blood into a normal recipient may be hazardous, especially if the recipient receives a drug which acts as an oxidant.[39]

Enzyme Abnormalities in Reticuloendothelial Diseases (Continued):

Assay Methods

Various methods are employed in the clinical laboratory to determine G-6-PD activity.

1. Glutathione Stability Test
   Blood is incubated with acetylphenylhydrazine causing a marked decrease in glutathione levels in G-6-PD deficient individuals.[5]

2. Dye Reduction Test
   Measurement is made of a change in color of a dye that is proportional to the amount of oxidation of a dye substance and the rate of generation of reduced nicotinamide adenine nucleotide (NADPH). Excess glucose-6-phosphate and NADP is added and the rate at which dye reduction occurs is observed as a measurement of the concentration of the enzyme. Decolorization of dye is delayed when there is marked decrease in G-6-PD.[6]

3. Direct Enzyme Measurement
   The rate of reduction of NADP as it is catalyzed by G-6-PD is measured.[34]

4. Methemoglobin Reduction Test
   This test is based on the fact that methemoglobin is reduced to hemoglobin if NADPH generation is sufficient.[11]

5. Cyanide and Ascorbate Test
   This test demonstrates that G-6-PD deficient red cells are different from normal red cells when their hemoglobin is rapidly oxidized to sulfhemoglobin by H_2O_2.[25]

6. Fluorescence Screening Test
   Ultraviolet light is utilized to detect the reduction of pyridine nucleotides.[7]

7. Tetrazolium Cytochemical Method
   The amount of G-6-PD is measured in red blood cells utilizing a tetrazolium formazan procedure.[14]

8. Heinz Body Test
   Acetylphenylhydrazine is added to blood and incubated. 50 percent of G-6-PD erythrocytes may contain five or more Heinz-Ehrlich bodies in the red cell. These bodies are denatured hemoglobin.[8] Normal red blood cells only contain one body.

Enzyme Abnormalities in Reticuloendothelial Diseases (Continued):

6-Phosphogluconate Dehydrogenase

6-phosphogluconate dehydrogenase (6-PGD) deficiency relates to the pentose phosphate pathway. Such deficient individuals may have minimal hemolysis. An assay for this enzyme is based on the reduction of NADP in the presence of excess 6-phosphogluconate.[35]

Glutathione Reductase

Several enzyme deficiencies affecting the nonglycolytic pathway occur in red cells. Included in this group is glutathione reductase.[6] These deficient individuals cannot be corrected by in vitro glucose but may be corrected by ATP, and thus fall in the Dacie-Selwyn Type II catagory. They may have a severe type of hemolytic anemia which frequently is drug induced.[10] The deficiency is diagnosed by tests similar to those for G-6-PD deficiency. In one assay, oxidized glutathione is mixed with NADPH and a hemolysate of the red cell. The disappearance of NADPH is measured spectrophotometrically at 340 nm.

Glutathione Peroxidase

Glutathione peroxidase protects the red cell from peroxidation. Deficiency of this enzyme may result in a hemolytic anemia.[45]

Glutathione Deficiency

Glutathione deficiency may present as a hemolytic anemia when th patient is exposed to an oxidant drug. A defect in glutathione synthetase may also be present.[5,50]

Adenosine Triphosphatase

Individuals with congenital nonspherocytic hemolytic anemia may have deficiency of ATPase. This enzyme plays an important part in the transport of potassium and sodium across the red blood cell membrane.[20] The deficiency is inherited as a dominant trait.

Pyruvate Kinase

One of the important enzyme deficiencies of the anaerobic glycolytic pathway causing hemolytic anemia is a deficiency of pyruvate

Enzyme Abnormalities in Reticuloendothelial Diseases (Continued):

kinase (PK). The most common cause next to G-6-PD deficiency of congenital and nonspherocytic hemolytic anemia is PK deficiency.[63] Individuals with heterozygous deficiency may have two to three times the levels found in the homozygous type.[54] The red cells may appear bizarre with marked acanthrocytosis.[43] Splenectomy affords some relief in this condition. Lack of the enzyme causes an inhibition of glycolysis with poor ATP synthesis. Individuals with PK deficiency have a prominent increase in the 2,3-diphosphoglycerate levels of the red cell. If total organic phosphate is present in high amounts in red cells, the possibility of PK deficiency should be considered. In diagnosis of the deficiency, the chemistry is related to the conversion of phosphoenolpyruvate to pyruvate in the presence of ATP. This reaction is catalyzed by PK in red cell homogenates. The white blood cells should be removed since PK activity is present in the white cells and will give false results.[56]

Another simple assay measures the disappearance of phosphoenolpyruvate; a third test is a screening test which shows that the PK reaction, in the absence of a buffer, is accompanied by a displacement of pH towards an alkaline one. A color indicator is used. Another screening test is a fluorescent one similar to that for G-6-PD deficiency. A decrease in fluorescence during the pyruvate kinase reaction occurs because NADH fluoresces and NAD does not.[12]

Hexokinase

Hexokinase deficiency causes a decrease in glucose utilization in the Embden-Meyerhof pathway but not the pentose phosphate pathway.[60] Autohemolysis may occur which is corrected by glucose and adenosine and belongs to Dacie-Selwyn Type I anemia. Splenectomy may improve the patient.

Phosphoglycerate Kinase

This enzyme is deficient in small amounts in affected individuals. Splenectomy is of value. The severity of this type of hemolytic anemia varies. Some deficient persons respond to correction in vitro by addition of glucose; others require the addition of ATP.[29]

Triphosphate Isomerase

Homozygous deficient individuals have a severe hemolytic anemia and severe neurologic disease which affects peripheral nerves and

Enzyme Abnormalities in Reticuloendothelial Diseases (Continued):

the central nervous system.[51] The enzyme deficiency occurs in the tissues and in the red and white cells.[61] There is recurrent bacteria infection and the red cells are acanthrocytic.

Glucose Phosphate Isomerase

Deficient individuals have a severe hemolytic anemia, with splenectomy giving some benefit. The white cell levels of the enzyme are low. The red cell deficiency is corrected by the addition of glucose in vitro,[3] and is thus a Type I anemia.

Glyceraldehyde-3-Phosphate Dehydrogenase

The autohemolysis caused by the enzyme deficiency is corrected by addition of glucose or ATP. Burr cells are present.[19]

Increased Methemoglobin

Increased erhthrocytic methemoglobin may be present in the red blood cell on the basis of a congenital enzyme deficiency or an acquired condition.[52] An increase in methemoglobin usually results from exposure to a chemical or drug which is an oxidant.[23] If the methemoglobinemia is secondary to a congenital cause, there may be an abnormal hemoglobin or deficiency of the enzyme NADH methemoglobin reductase. These individuals are usually homozygous and because of the enzyme deficiency there may be up to 50 percent methemoglobin in the red cell.[24]

Paroxysmal Nocturnal Hemoglobinuria

Patients with paroxysmal nocturnal hemoglobinuria (PNH) develop a hemolytic anemia which may eventually progress to aplastic anemia or myelogenous leukemia. The hemolysis is identified in vitro by acidifying the serum. The red cells may also hemolyze more rapidly in the presence of hypertonic sucrose. The PNH red blood cell is deficient in acetylcholinesterase.[37] If this enzyme is present in normal amounts, the red cells may not be lyzed by acidifying the serum. In view of the fact that this enzyme may also be deficient in individuals who do not have PNH and do not have hemolytic anemia, the acetylcholinesterase deficiency probably is not the major reason for hemolysis in PNH.[17]

Enzyme Abnormalities in Reticuloendothelial Diseases (Continued):

REFERENCES

1. Anderssen, N.: "The Activity of Lactic Dehydrogenase in Megaloblastic Anaemia". Scand. J. Haemat. 1:212, 1964.

2. Andersen, M. N., Gabrieli, E., Zizzi, J.: "Chronic Hemolysis in Patients with Ball-Valve Prostheses". J. Thorac. Cardiovasc. Surg. 50:501, 1965.

3. Baughan, M. A., Valentine, W. N., Paglia, D. E., Ways, P. O., Simon, E. R., DeMarsh, Q. B.: "Hereditary Hemolytic Anemia associated with Glucosephosphate Isomerase (GPI) Deficiency. A New Enzyme Defect of Human Erythrocytes". Blood 32:236, 1968.

4. Bennet, J. M., Nathanson, L., Rutenburg, A. M.: "Significance of Leukocyte Alkaline Phosphatase in Hodgkin's Disease". Arch. Intern. Med. 121:338, 1968.

5. Beutler, E.: "The Glutathione Instability of Drug-Sensitive Red Cells. A New Method for the In Vitro Detection of Drug Sensitivity". J. Lab. Clin. Med. 49:84, 1957.

6. Beutler, E.: "Glucose-6-Phosphate Dehydrogenase Deficiency and Nonspherocytic Congenital Hemolytic Anemia". Seminars Hemat. 2:91, 1965.

7. Beutler, E.: "A Series of New Screening Procedures for Pyruvate Kinase Deficiency, Glucose-6-Phosphate Dehydrogenase Deficiency and Glutathione Reductase Deficiency". Blood 28:553, 1966.

8. Beutler, E., Dern, R. J., Alving, A. S.: "The Hemolytic Effect of Primaquine. VI. An In Vitro Test for Sensitivity of Erythrocytes to Primaquine". J. Lab. Clin. Med. 45:40, 1955.

9. Beutler, E., Mathai, C. K., Smith, J. E.: "Biochemical Variants of Glucose-6-Phosphate Dehydrogenase giving rise to Congenital Nonspherocytic Hemolytic Disease". Blood 31:131, 1968.

10. Beutler, E., Yeh, M. K. H.: "Erythrocyte Glutathione Reductase". Blood 21:573, 1963.

Reticuloendothelial Disease References (Continued):

11. Brewer, G. J., Tarlov, A. R., Alving, A. S.: "Methemoglobin Reduction Test. A New Simple In Vitro Test for Identifying Primaquine Sensitivity". Bull WHO 22:633, 1960.

12. Brunetti, P., Nenci, G.: "A Screening Method for the Detection of Erythrocyte Pyruvate Kinase Deficiency". Enzymol. Biol. Clin. 4:51, 1964.

13. Climie, A. R. W.: "Neutrophilic Alkaline Phosphatase Test: A Review with Emphasis on Findings in Pregnancy". Amer. J. Clin. Path. 38:95, 1962.

14. Fairbanks, V. F., Lampe, L. T.: "Tetrazolium Cytochemical Method for G-6-PD in Erythrocytes". Blood 31:589, 1968.

15. Finch, S. C., Lamphere, J. P., Jablon, S.: "The Relationship of Serum Lysozyme to Leukocytes and other Constitutiona Factors". Yale J. Biol. Med. 36:350, 1964.

16. Fraser, D.: "Hypophosphatasia". Amer. J. Med. 22:730, 1957

17. Goldin, A. R., Rubenstein, A. H., Bradlow, B. A., Elliott, G. A.: "Malathion Poisoning with Special Reference to Effect of Cholinesterase Inhibition on Erythrocyte Survival" New Eng. J. Med. 271:1289, 1964.

18. Hammonds, F., Quaglino, D., Hayhoe, F. G. J.: "Blastic Crisis of Chronic Granulocytic Leukemia. Cytochemical, Cytogenetic and Autoradiographic Studies in Four Cases". Brit. Med. J. 1:1275, 1964.

19. Harkness, D. R.: "A New Erythrocyte Defect with Hemolytic Anemia: Glyceraldehyde-3-Phosphate Dehydrogenase Deficiency. Abstract. J. Lab. Clin. Med. 68:879, 1966.

20. Harvald, B., Hanel, K. H., Squire, R., Trap-Jensen, J.: "Adenosinetriphosphatase Deficiency in Patients with Nonspherocytic Hemolytic Anemia". Lancet 2:18, 1964.

21. Hayslett, J. P., Perillie, P. E., Finch, S. C.: "Urinary Muramidase and Renal Disease". New Eng. J. Med. 279:506,

22. Heller, P., Venger, N.: "Problems in the Differentiation of the Megaloblastic Anemias". Med. Clin. N. Amer. 46:121,

Reticuloendothelial Disease References (Continued):

23. Jaffe, E. R.: "The Reduction of Methemoglobin in Erythrocytes of a Patient with Congenital Methemoglobinemia, subjects with Erythrocyte Glucose-6-Phosphate Dehydrogenase Deficiency and Normal Individuals". Blood 21:561, 1963.

24. Jaffe, E. R.: "Hereditary Methemoglobinemias Associated with Abnormalities in the Metabolism of Erythrocytes". Amer. J. Med. 41:786, 1966.

25. Jacob, H. S., Jandl, J. H.: "A Simple Visual Screening Test for G-6-PD Deficiency Employing Ascorbate and Cyanide". New Eng. J. Med. 274:1162, 1966.

26. Kaplow, L. S.: "A Histochemical Procedure for Localizing and Evaluating Alkaline Phosphatase Activity in Smears of Blood and Marrow". Blood 10:1023, 1955.

27. Koler, R. D., Seaman, A. J., Osgood, E. E., Vanbellinghen, P.: "Myeloproliferative Diseases: Diagnostic Value of the Leukocyte Alkaline Phosphatase Test". Amer. J. Clin. Path. 30:295, 1958.

28. Korner, H. N.: "Distribution of Alkaline Phosphatase in the Serum Proteins in Hypophosphatasia". J. Clin. Path. 15:200, 1962.

29. Kraus, A. P., Langston, M. F., Jr., Lynch, B. L.: "Red Cell Phosphoglycerate Kinase Deficiency. A New Cause of Nonspherocytic Hemolytic Anemia". Biochem. Biophys. Res. Commun. 30, 173, 1968.

30. Levy, L. M., Walter, H., Sass, M. D.: "Enzymes and Radioactivity in Erythrocytes of Different Ages". Nature 184:643, 1959.

31. Lewis, S. M., Dacie, J. V.: "Neutrophil Alkaline Phosphatase in Paroxysmal Nocturnal Hemoglobinuria". Brit. J. Hemat. 11:549, 1965.

32. Libnoch, J. A., Yakulis, V. J., Heller, P.: "Lactate Dehydrogenase in Megaloblastic Bone Marrow". Amer. J. Clin. Path. 45:302, 1966.

33. McCarthy, C. F., Fraser, I. D., Read, A. E.: "Plasma Lactate Dehydrogenase in Megaloblastic Anemia". J. Clin. Path. 17:51, 1966.

Reticuloendothelial Disease References (Continued):

34. Marks, P. A.: "Glucose-6-Phosphate Dehydrogenase: Clinical Aspects". In W. A. Wood (Ed.) METHODS IN ENZYMOLOGY Vol. IX, Chapter on Carbohydrate Metabolism. Academic Press, pg. 131, 1966.

35. Marks, P. A.: "6-Phosphogluconate Dehydrogenase: Clinical Aspects". In W. A. Wood (Ed.) METHODS IN ENZYMOLOGY Vol. IX, Chapter on Enzymes of Carbohydrate Metabolism. Academic Press, pg. 141, 1966.

36. Marks, P. A., Banks, J.: "Studies on the Mechanism of Aging of Human Red Cells". Ann. N. Y. Acad. Sci. 75:95, 1958.

37. Metz, J., Bradlow, B. A., Lewis, S. M., Dacie, J. V.: "The Acetylcholinesterase Activity of Erythrocytes in Paroxysmal Nocturnal Haemoglobinuria in Relation to the Severity of the Disease". Brit. J. Haemat. 6:372, 1960.

38. Mitus, W. J., Bergna, L. J., Mednicoff, I. B., Dameshek, W.: "Alkaline Phosphatase of Mature Neutrophils in Chronic Forms of the Myeloproliferative Syndrome". Amer. J. Clin. Path. 30:285, 1958.

39. Motulsky, A. G., Campbell-Kraut, J. M.: "Population Genetics of Glucose-6-Phosphate Dehydrogenase Deficiency of the Red Cell". In B. S. Blumberg (Ed.) PROCEEDINGS OF THE CONFERENCE ON GENETIC POLYMORPHISM AND GEOGRAPHIC VARIATIONS IN DISEASE Grune & Stratton, pg. 159, 1961.

40. Motulsky, A. G., Stamatoyannopoulos, G.: "Clinical Implications of Glucose-6-Phosphate Dehydrogenase Deficiency". Ann. Intern. Med. 65:1329, 1966.

41. Noble, R. E., Fudenberg, H. H.: "Leukocyte Lysozyme Activity in Myelocytic Leukemia". Blood 30:465, 1967.

42. Oski, F. A., Naiman, J. L., Diamond, L. K.: "Use of the Plasma Acid Phosphatase Value in the Differentiation of Thrombocytopenic States". New Eng. J. Med. 268:1423, 1963.

43. Oski, F. A., Nathan, D. G., Sidel, V. W., Diamond, L. K.: "Extreme Hemolysis and Red Cell Distortion in Erythrocyte Pyruvate Kinase Deficiency". New Eng. J. Med. 270:1023,

Reticuloendothelial Disease References (Continued):

44. Osserman, E. F., Lawlor, D. P.: "Serum and Urinary Lysozyme (Muramidase) in Monocytic and Myelomonocytic Leukemia". J. Exp. Med. 124:921, 1966.

45. Paglia, D. E., Valentine, W. N.: "Studies on the Quantitative and Qualitative Characterization of Erythrocyte Glutathione Peroxidase". J. Lab. Clin. Med. 70:158, 1967.

46. Perillie, P. E., Kaplan, S. S., Lekowitz, E., Rogoway, W., Finch, S. C.: "Studies of Muramidase (Lysozyme) in Leukemia". J. Amer. Med. Assoc. 203:317, 1968.

47. Pirofsky, B.: "Hemolysis in Valvular Heart Disease". Ann. Intern. Med. 65:373, 1966.

48. Pirofsky, B., Sutherland, D. W., Starr, A., Griswold, H. E.: "Hemolytic Anemia Complicating Aortic Valve Surgery". New Eng. J. Med. 272:235, 1965.

49. Pon, N. G., Bondar, R. J. L.: "A Direct Spectrophotometric Assay for Pyruvate Kinase". Anal. Biochem. 19:272, 1967.

50. Prins, H. K., Oort, M., Loos, J. A., Zurcher, C., Beckers, T.: "Congenital Nonspherocytic Hemolytic Anemia Associated with Glutathione Deficiency of the Erythrocytes, Hematologic, Biochemical, and Genetic Studies". Blood 27:145, 1966.

51. Schneider, A. S., Valentine, W. N., Hattori, M., Heins, H. L., Jr.: "Hereditary Hemolytic Anemia with Triosephosphate Isomerase Deficiency". New Eng. J. Med. 272:229, 1965.

52. The Relation of Diaphorase of Human Erythrocytes to Inheritance of Methemoglobinemia". J. Clin. Invest. 39:1176, 1960.

53. Selwyn, J. G., Dacie, J. V.: "Autohemolysis and other Changes resulting from the Incubation In Vitro of Red Cells from Patients with Congenital Hemolytic Anemia". Blood 9:414, 1954.

54. Tanaka, K. R., Valentine, W. N.: "Pyruvate Kinase Deficiency". In E. Beutler (Ed.) HEREDITARY DISORDERS OF ERYTHROCYTE METABOLISM Grune & Stratton, pg. 229, 1968.

Reticuloendothelial Disease References (Continued):

55. Tanaka, K. R., Valentine, W. N., Fredricks, R. E.: "Diseases or Clinical Conditions Associated with Low Leukocyte Alkaline Phosphatase". New Eng. J. Med. 262:912, 1960.

56. Tanaka, K. R., Valentine, W. N., Miwa, S.: "Pyruvate Kinase (PK) Deficiency; Hereditary Nonspherocytic Hemolytic Anemia" Blood 19:267, 1962.

57. Tarlov, A. R., Brewer, G. J., Carson, P. E., Alving, A. S.: "Primaquine Sensitivity. Glucose-6-Phosphate Dehydrogenase Deficiency. An Inborn Error of Metabolism of Medical and Biological Significance". Arch. Intern. Med. 109:209, 1962.

58. Valentine, W. N., Beck, W. S.: "Biochemical Studies on Leukocytes. I. Phosphatase Activity in Health, Leukocytosis, and Myelocytic Leukemia". J. Lab. Clin. Med. 38:39, 1951.

59. Valentine, W. N., Follette, J. H., Solomon, D. H., Reynolds, J.: "The Relationship of Leukocyte Alkaline Phosphatase to 'Stress', to ACTH, and to 17-OH-Corticosteroids". J. Lab. Clin. Med. 49:723, 1957.

60. Valentine, W. N., Oski, F. A., Paglia, D. E., Baughan, M. A., Schneider, A. S., Naiman, J. L.: "Hereditary Hemolytic Anemia with Hexokinase Deficiency. Role of Hexokinase in Erythrocyte Aging". New Eng. J. Med. 276:1, 1967.

61. Valentine, W. N., Schneider, A. S., Baughan, M. A., Paglia, D. E., Hens, H. L., Jr.: "Hereditary Hemolytic Anemia with Triosephosphate Isomerase Deficiency. Studies in Kindreds with Coexistent Sickle Cell Trait and Erythrocyte Glucose-6-Phosphate Dehydrogenase Deficiency". Amer. J. Med. 41:27, 1966.

62. Valentine, W. N., Tanaka, K. R., Miwa, S.: "A Specific Erythrocyte Glycolytic Enzyme Defect (Pyruvate Kinase) in Three Subjects with Congenital Nonspherocytic Hemolytic Anemia". Trans. Assoc. Amer. Physicians 74:100, 1961.

63. Vuopio, R.: "Red Cell Enzymes in Anemia". Scand. J. Clin. Lab. Invest. (Suppl. 72) 15:99, 1963.

64. Waller, H. D.: "Glutathione Reductase Deficiency". In E. Beutler (Ed.) HEREDITARY DISORDERS OF ERYTHROCYTE METABOLISM Grune & Stratton, pg. 185, 1968.

Reticuloendothelial Disease References (Continued):

65. Westring, D. W.: "Aortic Valve Disease and Hemolytic Anemia". Ann. Intern. Med. 65:203, 1966.

66. Wolf, P. L., Enlander, D., Dalziel, J., Swanson, J.: "Green Plasma in Blood Donors". New Eng. J. Med. 281:205, 1969.

67. Wolf, P. L., Silberberg, B., Albert, S., Horwitz, J. P., Von der Muehll, E.: "Histochemical Similarities between Newborn and Leukemic Myeloid Cells and their Possible Significance". J. Clin. Path. 22:458, 1969.

68. Xefteris, E., Mitus, W. J., Mednicoff, E. B., Dameshek, W.: "Leukocytic Alkaline Phosphatase in Busulfan-induced Remissions of Chronic Granulocytic Leukemia". Blood 18:202, 1961.

CHAPTER 13

ENZYME ABNORMALITIES IN LUNG AND HEART DISEASE

Pulmonary Embolism

Pulmonary embolism often is clinically unsuspected and may result in death. The signs and symptoms are frequently nonspecific. Serum enzymology may be useful in aiding the clinician to make a specific diagnosis. The electrocardiogram is frequently normal and in approximately 25 percent of the cases, the chest x-ray is abnormal early in the course of the disease. Pulmonary scanning with nuclear medical techniques has improved the ability of the clinician to specifically diagnose pulmonary embolism, although it may give false negative results with infarcts less than 3.0 cm. in diameter. Angiography of the pulmonary vascular tree is an excellent method of detecting pulmonary emboli. A newer technique involving reflected ultrasound appears to be valuable for detecting pulmonary emboli but suffers from the major limitation of confusion when a subpleural density is present.[18]

When a pulmonary embolism is suspected, it is essential a battery of enzyme tests be determined. In the early 1950s, it was reported that glutamic oxalacetic transaminase (GOT) is elevated in approximately 50 percent of patients with pulmonary embolism,[21,30] and that lactic dehydrogenase (LDH) may be elevated along with elevation in unconjugated bilirubin.[28] Since these early reports, it has been shown that in patients with pulmonary infarction, the LDH is elevated and the GOT may be elevated or normal.[23,29] The elevation of GOT in pulmonary embolism may result from any of the following: (1) Myocardial ischemia associated with the pulmonary embolism, (2) hemolysis of red cells in the hemorrhagic pulmonary infarct, (3) congestive heart failure with central necrosis of the liver, or (4) shock, resulting in skeletal muscle ischemia. It is commoner to obtain a normal or only slightly increased GOT in pulmonary embolism. Associated with the normal GOT, there is an elevation of LDH, and there may be an increase in unconjugated bilirubin (Fig. 13-1). A slight elevation of CPK may occur.[25]

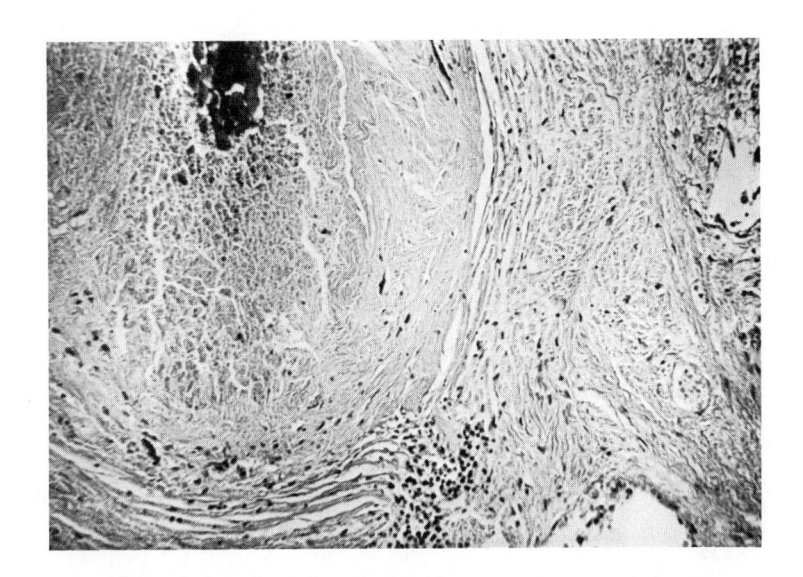

Enzyme Abnormalities in Lung and Heart Disease (Continued):

Fig. 13-1 Photomicrograph of an acute pulmonary embolus with a pulmonary infarct which occurred in the postoperative period following abdominal surgery in an 83 year old male patient. He developed a serum LDH of 425 I.U. which was primarily an elevation of LDH_3

LDH_3 is the commonest isoenzyme elevated in pulmonary embolism. However, the other isoenzymes may be increased;[5,27] for example, LDH_1 may be elevated if there is myocardial injury, from hemolyzed red cells in the infarct, or ischemic renal cortex from shock.[6] LDH_5 may be increased if heart failure complicates the pulmonary embolism with ischemia and congestion of the liver.[33] If shock ensues, LDH_4 and LDH_5 from necrotic skeletal muscle may rise in the patient's serum.[25] An elevated LDH, especially LDH_3, supports the clinical impression of pulmonary infarction, but a normal LDH may be present with pulmonary embolism.

An excellent battery of serum enzymes to be determined in suspected pulmonary embolism is LDH, creatine phosphokinase (CPK), and alpha-hydroxybutyrate dehydrogenase (α-HBDH).[7,22] This group is better than the originally suggested one of LDH and GOT. Patients

Enzyme Abnormalities in Lung and Heart Disease (Continued):

who complain of chest pain caused perhaps by either a pulmonary embolism or myocardial infarct are difficult to differentiate clinically. Differential diagnosis is usually possible on the basis of electrocardiograms, pulmonary angiograms, and serum enzymology.

Pulmonary Infarct

Patients with acute pulmonary infarction may have elevation of LDH_3, normal GOT levels and CPK levels. If the CPK level is prominently elevated with LDH, there is a strong possibility that the patient has sustained myocardial infarction. In some series, all patients with pulmonary infarcts have had increased LDH and normal CPK, but in others only half the patients with pulmonary infarcts have shown this enzyme picture.

The lung is a source of alkaline phosphatase. The enzyme is derived from pulmonary vascular endothelium. When a pulmonary infarct occurs, a release of alkaline phosphatase from the vascular endothelium occurs, resulting in a serum elevation[10] especially during the organization phase.

Carcinoma of the Lung

Serum enzymology has been utilized to aid in the detection of carcinoma of the lung. Serum LDH activity increases frequently in patients with carcinoma of the lung with or without metastatic disease (Fig. 13-2).[11,12] Carcinoma of the lung may cause thrombocytosis and this may be the source of the increased serum LDH. GOT activity is of little diagnostic value. At times alkaline phosphatase is elevated, especially if there are metastases to bones or liver. The Regan isoenzyme of alkaline phosphatase is similar to the placental isoenzyme in its marked heat stability at 56°C. and its inhibition by phenylalanine.[26] The Regan isoenzyme was originally described in Mr. Regan, a patient, who had a bronchogenic carcinoma. The bronchogenic carcinoma produced an alkaline phosphatase isoenzyme similar to the placental isoenzyme.

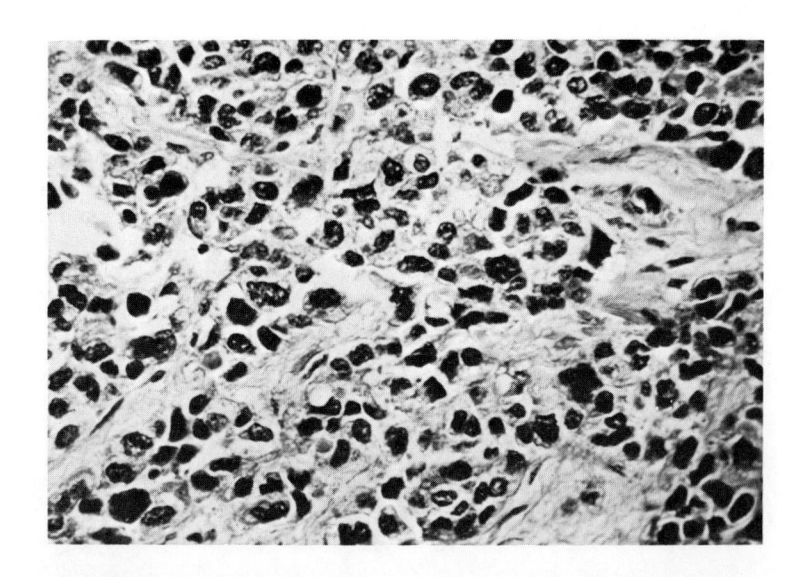

Enzyme Abnormalities in Lung and Heart Disease (Continued):

Fig. 13-2 Photomicrograph of section of undifferentiated carcinoma of lung which developed in a 65 year old male who was a heavy smoker. The serum LDH was 365 I.U.

When a patient has an elevated serum LDH associated with bronchogenic carcinoma, the elevation may regress when therapy has been successful.[12] The LDH isoenzymes associated with bronchogenic carcinoma are LDH_1, LDH_2, and LDH_3.

Fat Embolism of the Lung

Fat embolism of the lung may increase the serum lipase. Lipase activity of the lung has been increased experimentally with induced fat embolism of the lung; the increased serum lipase of lung fat embolism may be secondary to release of lung lipase.[1,2,9]

Enzyme Abnormalities in Lung and Heart Disease (Continued):

Emphysema

Patients with emphysema may have a deficiency of a specific globulin which represents a deficiency of alpha$_1$ anti-trypsin (Fig. 13-3). A majority of the anti-trypsin activity of serum is found with the alpha$_1$ globulin fraction, a minor part with alpha$_2$ globulin fraction. Alpha$_1$ anti-trypsin deficiency is recognized by performing electrophoresis or immunoelectrophoresis of the serum. An absence of alpha$_1$ globulin is detected with paper or agar gel electrophoresis.[13]

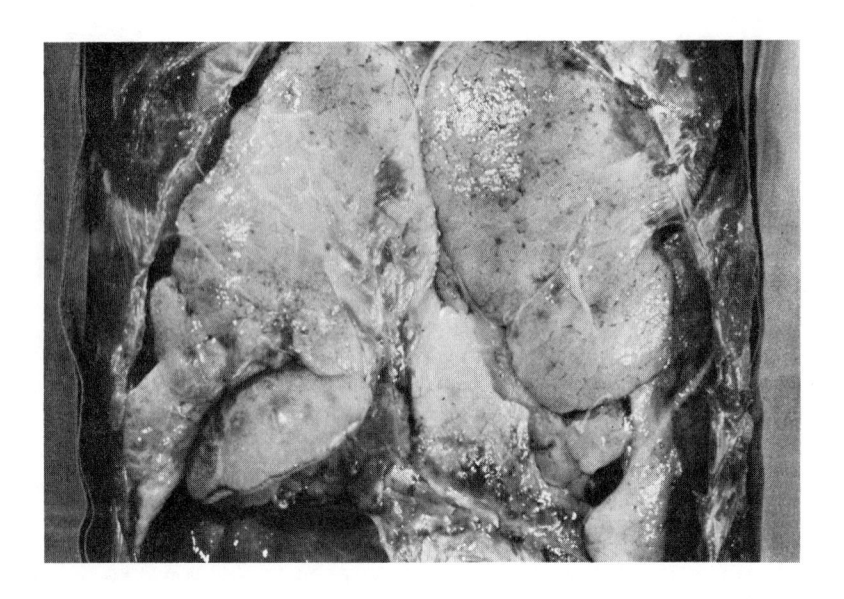

Fig. 13-3 Gross photograph of lung from a patient who died from severe respiratory insufficiency resulting from emphysema. His serum alpha$_1$ anti-trypsin was markedly deficient.

Alpha$_1$ anti-trypsin deficiency is familial and is thought to be transmitted through a recessive autosomal genetic process. Some investigators have found that even though the deficiency may exist, pulmonary disease may not occur; thus, the deficiency of this anti-enzyme associated with pulmonary disease may be coincidental. Severa investigators, however, have demonstrated that when anti-tryptic

Enzyme Abnormalities in Lung and Heart Disease (Continued):

activity is decreased, chronic obstructive lung disease is present and arterial hypoxia may be evident. Deficiency of this enzyme may be associated with liver cirrhosis. The function of anti-trypsin is to inhibit leukocyte tryptic activity.[14]

Myocardial Infarction and Other Heart Disease

The enzymes available for diagnosis of cardiac disease are GOT, LDH, α-HBD, and CPK. The clinician relies heavily on the laboratory to aid him in the diagnosis of cardiac disease and to follow its progression or regression.

Although the electrocardiogram is the most dependable aid in diagnosis of myocardial infarction, it may not be helpful if previous myocardial abnormalities are present or if there is a prior left bundle branch block. Furthermore, intramural infarction may not alter the ECG pattern. Serum enzymology has been extremely valuable in helping in the assessment of myocardial infarction. The ability to confirm the presence of an acute myocardial infarct by enzymes is close to 100 percent.

Glutamic Oxalacetic Transaminase

GOT rises abruptly after a myocardial infarction has occurred due to leakage of the enzyme into the serum from the necrotic myocardium. Patients with SGOT elevations over 1500 I.U. have a poor prognosis, whereas those with elevations between 50 and 300 I.U. have a better prognosis. Abnormal levels usually occur on the first day after infarction and may persist for 5 days. A direct relationship exists between the size of the myocardial infarct and the level of SGOT elevation (Fig. 13-4). An elevation of SGOT is particularly helpful to the clinician in the assessment of chest pain when the electrocardiogram does not indicate the presence of a myocardial infarct. Patients with angina pectoris do not have an increase in GOT.[3,19]

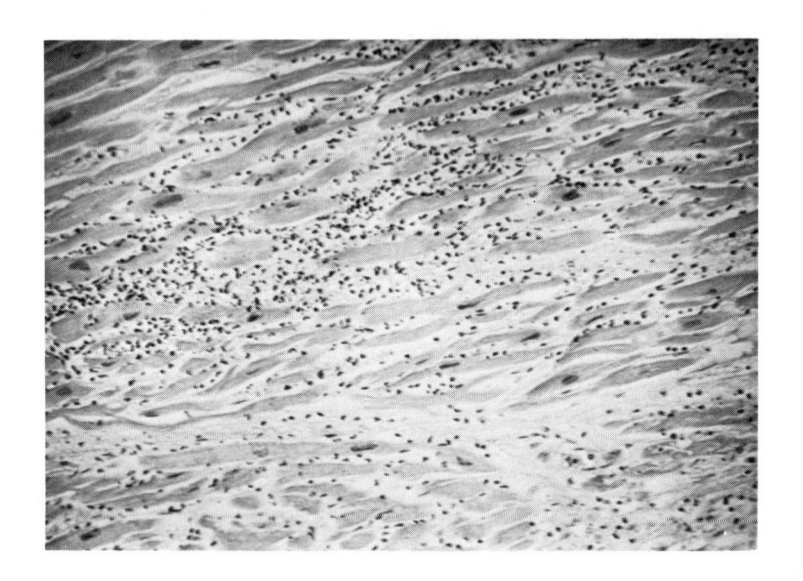

Enzyme Abnormalities in Lung and Heart Disease (Continued):

Fig. 13-4   Photomicrograph of section of heart demonstrating an extensive acute myocardial infarction which caused the death of a 60 year old white female. The GOT was 435 I.U.

Extensions of the acute myocardial infarct may cause persistent elevation in SGOT. Occasionally elevated SGOT may be seen in patients with severe tachycardia or arrythmia with a cardiac rate greater than 160 beats per minute. Active myocarditis secondary to a viral, metabolic, rheumatic, or bacterial etiology is associated with elevated transaminase levels. Here again, the severity of myocarditis may be assessed by the level of SGOT.[17] Patients who undergo cardiac catheterization, angiography or electrical cardioversion may have alterations in SGOT levels; presumably during the procedure the cardiac muscle is injured with release of GOT.[20,31] Patients who develop congestive heart failure have SGOT and SGPT elevations on the basis of hepatic congestion and hepatic central lobular necrosis.[15,3] One-half of patients who have had recent electrical defibrillation have a rise in their cardiac serum enzymes attributed to intercostal muscle contraction.[8]

Enzyme Abnormalities in Lung and Heart Disease (Continued):

Elevated levels of GOT in the urine in patients with myocardial infarction are usually correlated with GOT levels in the blood. In some patients, the concentrations in the urine remain elevated for a longer period of time than in the serum simulating the situation in acute pancreatitis where the urinary amylase levels remain elevated longer than that of the blood.[16]

In patients who have had an acute myocardial infarction, both GOT and GPT should be determined. A significantly higher activity of GOT compared to GPT at the onset is characteristic of acute myocardial infarction. If the activity of GPT increases above that of GOT in subsequent days, this is an indication of heart failure with hepatic congestion.

When infarction of the myocardium is suspected, daily measurements of GOT activity are more useful than a single determination, which may not detect the transient peak elevation. Pericarditis usually does not elevate the GOT. If the GOT is elevated in patients with pericarditis, there likely is subepicardial injury of the heart or congestive heart failure secondary to the pericarditis.[3,19]

The value of determining GOT in heart disease can be summarized as follows:

1. Differentiation between acute myocardial infarction and coronary insufficiency.

2. Diagnosis of acute myocardial infarction when the electrocardiogram does not indicate the infarct or when it is difficult to interpret because of the presence of a previous myocardial infarct, left bundle branch block, or digitalis effect.

3. Diagnosis of an extension of a myocardial infarct.

A large number of drugs may give a spurious elevation of GOT. This should be kept in mind in patients who have chest pain and who are receiving drugs. An incorrect diagnosis of myocardial infarction may result if the clinician is unaware that GOT may be elevated because of interaction of drugs with the chemical method for GOT assay, or because of liver disease induced by drugs, which then causes GOT elevation. Injections of drugs into skeletal muscle will elevate SGOT. Narcotics may increase the intra-biliary pressure with a rise in GOT from the liver.

Enzyme Abnormalities in Lung and Heart Disease (Continued):

Surgery on the myocardium will produce an elevated GOT. The GOT elevation after cardiac surgery is related to the surgical trauma to skeletal muscle of the chest wall, to the heart, and possibly anesthesia effects on the liver, with resultant GOT liberation from injured hepatic cells.

Lactic Dehydrogenase

LDH is usually elevated within 12 - 18 hours after acute myocardial infarction. Peak elevation occurs within 2 to 3 days with a return to normal within 7 to 10 days. The elevation correlates with the amount of myocardial injury. When LDH is elevated above 3,000 I.U., a poor prognosis exists. It is important to determine the LDH isoenzyme when one suspects a myocardial infarct. LDH_1 is the isoenzyme associated with myocardial injury (Fig. 13-5).

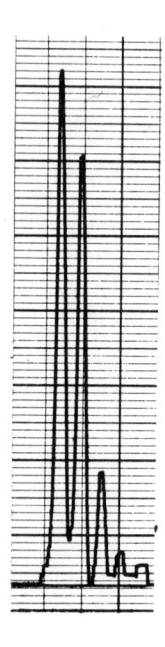

Fig. 13-5 Electrophoretic isoenzyme pattern of serum of a patient with acute myocardial infarction. A prominent elevation of LDH_1 is present.

Enzyme Abnormalities in Lung and Heart Disease (Continued):

The type of isoenzymes may be determined by electrophoresis or by chemical methods such as heat inactivation. LDH_1 elevated in myocardial infarction is heat stable. It is a rapid-migrating LDH isoenzyme and persists after a serum is incubated at 65°C. for 30 to 45 minutes. The liver isoenzyme is heat labile. The heat stable percentage of LDH_1 is not greater than 30 percent of total LDH in normal serum, whereas in patients with an acute myocardial infarction, the heat stable LDH_1 is between 50 and 90 percent of total LDH activity. The ratio of LDH_1/LDH_2 is usually greater than one in acute myocardial infarction. Urea in high concentration inactivates LDH_1 from a myocardial source, but it inactivates in low concentrations the isoenzymes derived from skeletal muscle and liver. This is a chemical method to differentiate the isoenzymes of heart, liver, and skeletal muscle.

Alpha-Hydroxybutyrate Dehydrogenase

α-HBD is another serum enzyme the clinician should request when there is a suspicion of acute myocardial infarction. The dehydrogenase activity of the alpha fractions of LDH are determined by using 2-oxybutyrate as a substrate. The greater sensitivity of α-HBD compared to LDH in detecting myocardial injury stems from the fact that myocardial enzymes rather than total LDH enzymes are being measured. Elevations of α-HBD associated with myocardial infarction peak at 12 - 18 hours and may persist for 2 weeks after the onset of the infarct. α-HBD elevation has been shown to be comparable to GOT elevation but outlasts the elevation of both GOT and LDH. Other conditions which will cause an elevated α-HBD are hemolytic anemia, megaloblastic anemia, or metastatic melanoma.

Creatine Phosphokinase

An increase in the serum CPK occurs within 4 hours after acute myocardial infarction with the peak value reached within 24 to 36 hours (Fig. 13-6). CPK usually returns to normal within 4 days after the onset of the infarction. CPK determination can be helpful in diagnosis of other cardiac or pulmonary conditions. SGOT may be elevated in congestive heart failure with liver congestion, but CPK will not be because the enzyme is not present in the liver. CPK will be elevated only when myocardial damage has occurred. To distinguish between myocardial and pulmonary infarction, it is important to request LDH and CPK. Both are increased in myocardial infarction but only LDH in pulmonary infarction. Only slight CPK elevation may occur in pulmonary infarction. Thus, CPK may be a more sensitive indicator of myocardial ischemia than any other enzyme. However,

Enzyme Abnormalities in Lung and Heart Disease (Continued):

injections of drugs into skeletal muscle, severe exercise, myopathy and brain injury will also cause CPK elevation.[4]

It has been reported that daily determination of a refrigerated sample of CPK shows a marked loss of CPK as high as 50 percent per day when specimens are stored in the refrigerator. Rapid loss of CPK activity in serum with time thus occurs. Probably a proteolytic enzyme in the serum causes degradation of CPK; thus, determinations of CPK should be done on the same day that the blood sample is obtained. In contrast, GOT activity remains stable in refrigerated sera for approximately 7 days. One may prevent the loss of CPK in the refrigerator by adding cysteine to the reaction mixture and thus, enzyme activity is restored to its previous level after the specimen has been stored. A deblocking of enzyme thiol groups may occur. When cysteine-stimulated CPK techniques are employed in the clinical laboratory, it has been shown that CPK is elevated in patients suffering from myopathy, acute brain or myocardial infarction, and hypothyroidism. Abnormalities are also found in patients who are being treated for diabetic acidosis. The mechanisms responsible for the elevation in patients who have diabetic acidosis are unknown. Lung tissue may contain CPK inhibitors which become inactivated when there is pulmonary congestion and edema, thus, CPK activity may be unmasked. A slight amount of CPK is present in the lung. The increase of CPK in patients recovering from diabetic acidosis may result from a direct effect of insulin on the cell membrane. It is unlikely that acidosis itself is responsible for CPK elevation. Low serum potassium may cause elevation of CPK and Clofibrate may also be associated with elevated CPK. Hypoparathyroidism may cause a myopathy with elevation of CPK.

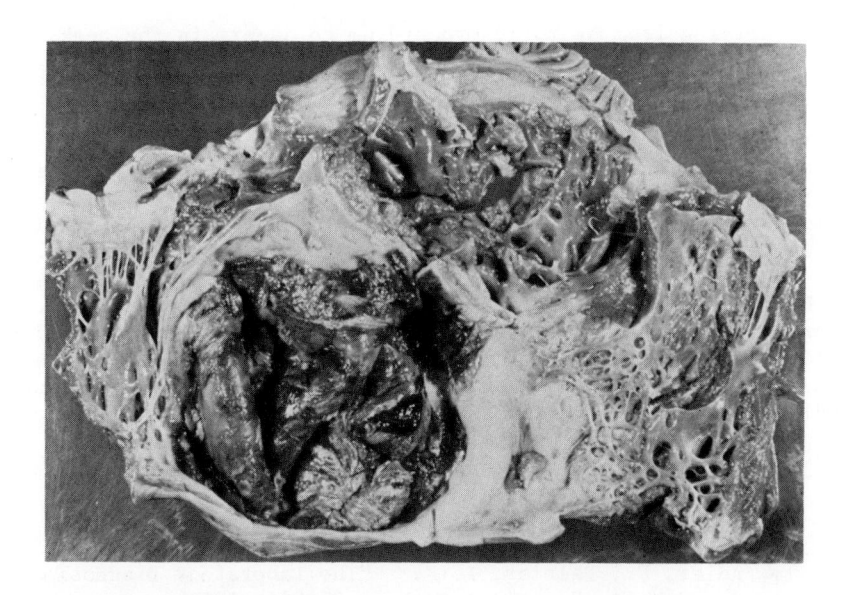

Enzyme Abnormalities in Lung and Heart Disease (Continued):

Fig. 13-6   Gross photograph of heart from patient who
            died 3 days after onset of acute myocardial
            infarction. The cause of death was rupture
            of the heart with cardiac tamponade. The
            serum CPK was 495 I.U.

Aldolase

The adult level of aldolase is 2 to 14 I.U. The newborn has 5 times the level as the adult, and children have twice the level.

Although the main value of determining serum aldolase is assessment of skeletal muscle disease, patients with acute myocardial infarction may have an elevation of the enzyme because this enzyme is also present in the liver and heart. Patients with congestive heart failure and congestion of the liver may also have higher levels.

Enzyme Tests in Myocardial Infarction Diagnosis

In utilizing enzyme levels as an aid in the diagnosis of myocardial infarction, one should perform frequent assays of different

Enzyme Abnormalities in Lung and Heart Disease (Continued):

enzymes arising from necrotic myocardium to ascertain the peak of the enzyme elevation and its regression. Levels of CPK, GOT, LDH, and α-HBD should be followed.

In patients with acute myocardial infarction, the GOT rises in 12 hours, peaks at 24 hours, and lasts 4 to 5 days. CPK rises in 4 hours, peaks at 36 hours, and lasts 4 days. LDH rises in 12 - 18 hours, peaks at 72 hours, and lasts 7 to 10 days. α-HBD rises in 12 - 18 hours, peaks at 72 hours, and lasts 10 to 14 days.

Enzyme levels may be correlated with the amount of necrotic myocardium and may serve as a prognostic indicator. The clinician should be aware that co-existing conditions such as abnormalities of lung, liver, kidney, and skeletal muscle may elevate these enzymes as well as usage of drugs. The enzyme values should be correlated with the clinical features and other laboratory data.

REFERENCES

1. Adler, F., Peltier, L. F.: "The Laboratory Diagnosis of Fat Embolism". Clin. Orthop. 21:226, 1961.

2. Armstrong, H. J., Kuenzig, M. C., Peltier, L. F.: "Lung Lipase Levels in Normal Rats and Rats with Experimentally Produced Fat Embolism". Proc. Soc. Exp. Biol. Med. 124:959 1967.

3. Brereton, W. F., Sherlock, P., Cameron, D. J.: "Pericardial Effusion: Marked Serum Transaminase Elevations. A Report of Four Cases". Arch. Intern. Med. 115:311, 1965.

4. Cherington, M., Lewin, E., McCrimmon, A.: "Serum Creatine Phosphokinase Changes following Needle Electromyographic Studies". Neurology 18:271, 1968.

5. Cohen, L., Djordjevich, J., Jacobsen, S.: "The Contribution of Isoenzymes of Serum Lactic Dehydrogenase to the Diagnosis of Specific Organ Injury". Med. Clin. N. Amer. 50:193, 1966.

6. Cohen, L., Djordjevich, J., Ormiste, V.: "Serum Lactic Dehydrogenase Isozyme Patterns in Cardiovascular and other Diseases with Particular Reference to Acute Myocardial Infarction". J. Lab. Clin. Med. 64:355, 1964.

Lung and Heart Disease References (Continued):

7. Coodley, E. L.: "Enzyme Profiles in the Evaluation of Pulmonary Infarction". J. Amer. Med. Assoc. 207:1307, 1969.

8. Crowley, L. V.: "Creatine Phosphokinase Activity in Myocardial Infarction, Heart Failure, and Following Various Diagnostic and Therapeutic Procedures". Clin. Chem. 14:1185, 1968.

9. Fuchsig, P., Brucke, P., Blumel, G., Gottlob, R.: "A New Clinical and Experimental Concept of Fat Embolism". New Eng. J. Med. 276:1192, 1967.

10. Gambino, S. R.: "Critique: Clinical Chemistry "Check Sample" #CC46. Amer. Soc. Clin. Path.
    ALSO:
    Gambino, S. R., Lum, G.: "Correspondence: Alkaline Phosphatase After Pulmonary Infarction". New Eng. J. Med. 287:361, 1972.

11. Gault, M. H., Cohen, M. W., Kahana, L. M., Lelin, F. T., Meakins, J. F., Aronovitch, M.: "Serum Enzymes in Patients with Carcinoma of Lung". Canad. Med. Assoc. 96:87, 1967.

12. Gold, J. A.: "Serum Enzymes in Bronchogenic Carcinoma and other Pulmonary Diseases". Chest 39:62, 1961.

13. Guenter, C. A., Welch, A. H., Hammarsten, I. F.: "Pulmonary Disease in Alpha-1-Antitrypsin Deficiency". J. Lab. Clin. Med. 72:880, 1968.

14. Ishak, K. G., Jenis, E. H., Marshall, M. L., Bolton, B. H., Battistone, G. C.: "Cirrhosis of the Liver Associated with Alpha 1-Antitrypsin Deficiency". Arch. Path. 94:445, 1972.

15. Killip, T., Payne, M. A.: "High Serum Transaminase Activity in Heart Disease, Circulatory Failure, and Hepatic Necrosis". Circulation 21:646, 1960.

16. King, J. S., Jr., Boyce, W. H.: HIGH MOLECULAR WEIGHT SUBSTANCES IN HUMAN URINE C Thomas Co., Springfield, Ill., 1963.

17. Mason, H. H., Wroblewski, F.: "Serum Glutamic Oxalacetic Transaminase Activity in Experimental Disease States". Arch. Intern. Med. 99:245, 1957.

Lung and Heart Disease References (Continued):

18. Miller, L. D., Joyner, C. R., Dudrick, S. J., Eskin, D. J.: "Clinical Use of Ultrasound in the Early Diagnosis of Pulmonary Embolism". Ann. Surg. 166:381, 1967.

19. Nydick, I., Ruegsegger, P., Wroblewski, F., LaDue, J. S.: "Variations in Serum Glutamic Oxalacetic Transaminase Activity in Experimental and Clinical Coronary Insufficiency, Pericarditis, and Pulmonary Infarction". Circulation 15:324, 1957.

20. Oram, S., Davies, J. P. H., Weinbaum, I., Taggert, P., Kitchen, L. D.: "Conversion of Atrial Fibrillation to Sinus Rhythm by Direct Current Shock". Lancet 2:159, 1963.

21. Ostrow, B. H., Polis, G. N., Evans, J. M.: "Serum Glutamic Oxalacetic Transaminase (GOT) in Pulmonary Embolization". Clin. Res. Proc. 4:155, 1956.

22. Perkoff, G. T.: "Demonstration of Creatine Phosphokinase in Human Lung Tissue". Arch. Intern. Med. 122:326, 1968.

23. Polachek, A. A., Zoneraich, S., Zoneraich, O., Sass, M.: "Pulmonary Infarction and Serum Lactic Dehydrogenase". J. Amer. Med. Assoc. 204:811, 1968.

24. Resnick, W. H.: "Preinfarction Angina". Mod. Conc. Cardio-Vasc. Dis. 31:751, 1962.

25. Shubin, H., Weil, M. H.: "Acute Elevation of Serum Transaminase and Lactic Dehydrogenase during Circulatory Shock". Amer. J. Cardiol. 11:327, 1963.

26. Stolbach, L. L., Krant, M. J., Fishman, W. H.: "Ectopic Production of Alkaline Phosphatase Isoenzyme in Patients with Cancer". New Eng. J. Med. 281:757, 1969.

27. Trujillo, N. P., Nutter, D., Evans, J. M.: "The Isoenzymes of Lactic Dehydrogenase. II. Pulmonary Embolus, Liver Disease, the Postoperative State, and other Medical Conditions". Arch. Intern. Med. 119:333, 1967.

28. Wacker, W. E. C., Rosenthal, M., Snodgrass, P. J., Amador, E.: "A Triad for the Diagnosis of Pulmonary Embolism and Infarction". J. Amer. Med. Assoc. 178:8, 1961.

Lung and Heart Disease References (Continued):

29. Wacker, W. E. C., Snodgrass, P. J.: "Serum LDH Activity in Pulmonary Embolism Diagnosis". J. Amer. Med. Assoc. 174:2142, 1960.

30. Walsh, J. R., Humoller, F. L., Gillick, F. G.: "Serum Transaminase in Pulmonary Disease and Multiple Infarctions". Ann. Intern. Med. 46:1105, 1957.

31. Warbasse, J. R., Wesley, J. E., Connolly, V., Galluzzi, N. J.: "Lactic Dehydrogenase Isoenzymes after Electroshock Treatment of Cardiac Arrhythmias". Amer. J. Cardiol. 21:496, 1968.

32. West, M., Gelb, D., Zimmerman, H. J.: "Serum Enzymes in Disease. VII. Significance of Abnormal Serum Levels in Cardiac Failure". Amer. J. Med. Sci. 241:350, 1961.

33. Zimmerman, H. J., West, M.: "Serum Enzyme Levels in the Diagnosis of Hepatic Disease". Amer. J. Gastroenterology 40:387, 1963.

CHAPTER 14

ENZYME ABNORMALITIES IN CENTRAL NERVOUS SYSTEM DISEASE

An effective blood-brain barrier exists which is impenetrable. Abnormalities in serum enzymes may not be reflected in the cerebrospinal fluid; conversely, abnormalities in the cerebrospinal fluid may not be demonstrable in the serum.[4,18] Many studies have demonstrated the independence of the serum and the cerebrospinal fluid relevent to enzyme content; thus, patients with elevated enzymes in the blood arising from, for example, a necrotic liver will frequentl. not manifest a similar enzyme elevation in the cerebrospinal fluid.[2] If the brain or spinal cord is damaged, there will be a release of enzymes from these organs into the spinal fluid,[6,12] and there may be subsequent elevation in the peripheral blood. The enzymes that have been studied in regard to central nervous system disease are GOT, LDH, ICD, and CPK.[14,33]

Necrosis

GOT is present in large amounts in the brain. Necrosis of the brain due to cerebral thrombosis and cerebral hemorrhage, results in an elevation of GOT in the spinal fluid and may at times cause a con comitant rise in the peripheral blood. Unlike the relationship in myocardial infarction, there is no correlation between the degree of elevation of GOT in the spinal fluid and the magnitude of central nervous system necrosis.[9,22] Another difference between GOT elevations in central nervous system damage and myocardial damage is that GOT in the cerebrospinal fluid usually reaches its maximum by the fourth or fifth day and is maintained over a long period of time, sometimes up to a few months.[25,29] GOT elevation in myocardial infarction may peak within 2 or 3 days, and there is a rapid decay with the elevation rarely persisting more than several days. Because the transaminases do not pass freely through the blood-brain barrier, there may be a greater GOT elevation in the cerebrospinal fluid than in the blood. Thus, patients with cerebral infarction following cerebral thrombosis or cerebral hemorrhage, develop an increase of GOT in the cerebrospinal fluid and at times, in the peripheral blood. LDH_2 and LDH_3 may also be present in the spinal fluid in increased amounts with central nervous system necrosis.

Enzyme Abnormalities in Central Nervous System Disease (Continued):

Tay-Sachs Disease

LDH is elevated in the cerebrospinal fluid in Tay-Sachs disease. The enzyme level reaches its highest level in the second year of the disease, then declines to normal. In contrast, cerebrospinal fluid LDH is normal in Niemann-Pick disease.[13]

Vascular Disease

Almost half the patients with cerebral thrombosis and cerebral hemorrhage that were studied had elevated levels of CPK; a large concentration of CPK is present in the central nervous system (Fig. 14-1).[2,10] Differences exist between the time of onset of levels of CPK in the blood in patients with cerebral vascular thrombosis and in those with myocardial infarction. In cerebral infarction, a delay of approximately 2 days occurs before the advent of CPK elevation. In myocardial infarction, there will be an earlier CPK rise within hours. CPK elevation appears to last much longer in the blood in cerebral vascular disease than in those who have sustained a myocardial infarction. Thus, serum CPK plays a major role in confirmation of cerebral vascular disease; it becomes elevated during the acute phase of cerebral vascular disease, rising within 2 days after the cerebral damage has occurred, and usually lasts several days in the serum.[12]

Cerebrospinal fluid CPK has been suggested as a diagnostic aid in cerebral vascular disease and in neoplastic disease of the central nervous system.[7,30] However, this measurement has not been universally adopted.

Enzyme Abnormalities in Central Nervous System Disease (Continued):

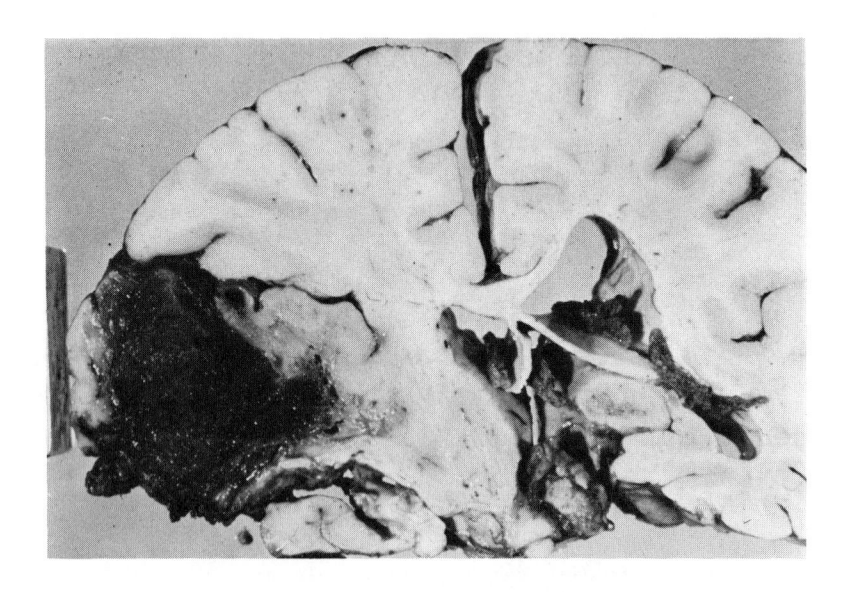

Fig. 14-1 Gross photograph of brain from a patient who died from a hypertensive cerebral hemorrhage. The patient's serum CPK was 284 I.U.

Psychoses

Elevation of CPK has also occurred in patients with schizophrenia. Some studies have indicated that there is leakage of CPK from the cells of the central nervous system in acutely psychotic patients. However, this concept has been challenged, and some workers have suggested that elevation of CPK in these patients results from severe muscular exertion related to the schizophrenia. Some studies have indicated that necrosis of brain tumors also will elevate serum CPK.

Meningitis and Other Inflammations

Patients with acute bacterial meningitis and tuberculous meningitis may have elevated LDH and GOT in the cerebrospinal fluid (Fig. 14-2).[28] CPK elevation may not occur unless abscess formation and necrosis of the brain occurs. The LDH and GOT may be derived from

Enzyme Abnormalities in Central Nervous System Disease (Continued):

inflammatory cells; since exudates that are present in the body cavities as a result of inflammation also have a high LDH activity derived from the inflammatory cell.[32] Furthermore it is also suggested that elevated LDH in the cerebrospinal fluid during the course of an acute bacterial or tuberculous meningitis may be derived from breakdown of the blood-brain barrier and transfer of blood plasma into the cerebrospinal fluid.[33]

Viral meningoencephalitis results in elevated GOT in the cerebrospinal fluid and the peripheral blood. The enzyme elevation in the peripheral blood may arise partially from the central nervous system disease.[17] However, viremia may cause viral myocarditis and viral hepatitis, and elevation of GOT, LDH, and CPK in the peripheral blood may be subsequent to the disease of the heart, liver, or skeletal muscle associated with the central nervous system disease.[1,24,28]

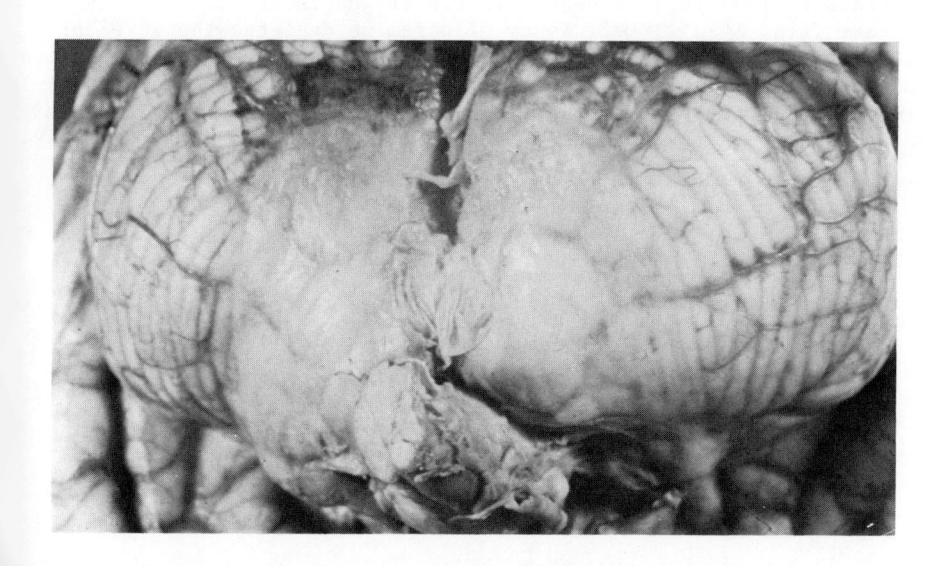

Fig. 14-2 Gross photograph of brain from a patient who died of acute meningococcal meningitis. The LDH of the cerebrospinal fluid was 435 I.U.

Enzyme Abnormalities in Central Nervous System Disease (Continued):

Demyelinating Diseases

LDH, GOT, and CPK are present in the white and gray matter of the central nervous system, but the gray matter contains more of the enzymes. Some investigators have demonstrated a varying pattern of LDH, GOT, and CPK elevation in demyelinating diseases of the central nervous system. Cerebrospinal fluid CPK has been elevated at times in patients with multiple sclerosis. An increase in GOT has been reported in amyotrophic lateral sclerosis.[27] Patients with multiple sclerosis have generally normal GOT and LDH. Ribonuclease may be increased in the cerebrospinal fluid in multiple sclerosis.[15]

Thiamine Deficiency

Korsakoff's psychosis and Wernicke's hemorrhagic encephalopathy may be associated with a decrease in a thiamine-dependent enzyme in the cerebrospinal fluid, transketolase. The assessment of deficiency of transketolase is also useful in the diagnosis of beriberi heart disease (Fig. 14-3). Deficiency of thiamine depletes the level of blood transketolase, and rapid increase in the enzyme occurs in blood and cerebrospinal fluid after thiamine therapy. Chronic renal failure will also cause a transketolase decrease. Uremic serum contains an inhibitor of transketolase. The inhibitor is removed by dialysis.

Enzyme Abnormalities in Central Nervous System Disease (Continued):

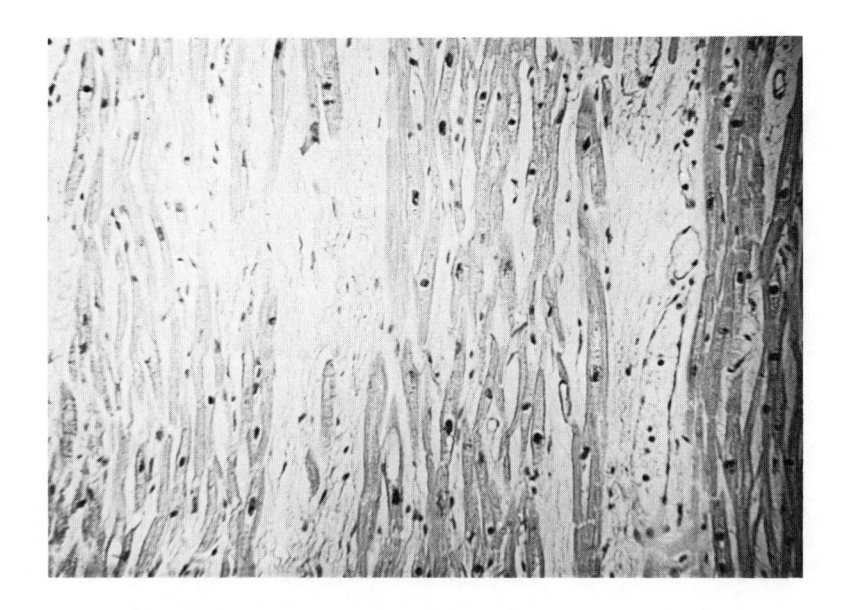

Fig. 14-3 Photomicrograph of section of heart from alcoholic patient who died of acute shoshin beriberi. The transketolase level was markedly depressed.

Neoplasms
───────────

Neoplasms of the central nervous system do not present with a consistent pattern of enzyme. Metastatic neoplasms may or may not elevate LDH, GOT, or CPK in the cerebrospinal fluid or the peripheral blood. The tendency for LDH elevation in the cerebrospinal fluid in patients with metastatic carcinoma or lymphoma has been reported rarely (Fig. 14-4). This elevation is found more in metastatic than in primary neoplasms of the central nervous system.[16] Since leukemic or lymphoma cells tend to produce LDH, especially the fast-moving LDH_2 and LDH_3, these isoenzymes may be measured serially to indicate the efficacy of treatment of the central nervous system leukemia or lymphoma. Beta-glucuronidase may be also increased in the cerebrospinal fluid in neoplasia.[3,20] The cerebrospinal fluid does not contain acid phosphatase. An increase in acid phosphatase occurs in the cerebrospinal fluid in patients with multiple myeloma or metastatic cancer to the vertebrae.

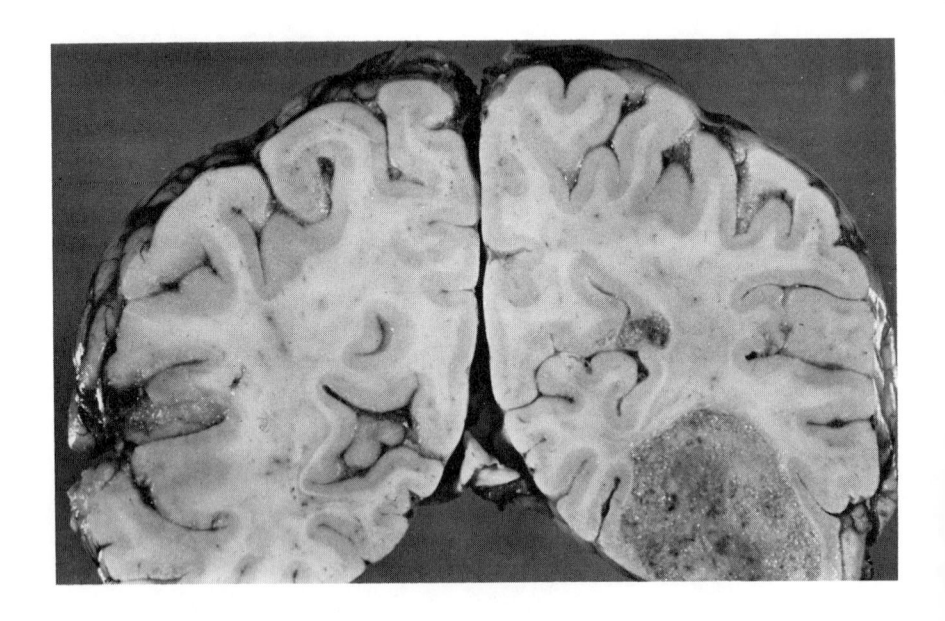

Enzyme Abnormalities in Central Nervous System Disease (Continued):

Fig. 14-4 Gross photograph of brain from a patient with metastases to the brain from a breast cancer. The cerebrospinal fluid LDH level was 355 I.U.

In summary, patients with primary neoplasms of the central nervous system may rarely have elevations of serum or cerebrospinal fluid GOT or LDH. Tumors which are extrinsic to the dura mater usual do not provoke altered enzyme activity unless there is pressure necrosis of the central nervous system. If there is an extrinsic dural neoplasm causing cerebrospinal fluid obstruction, there may be a rise in the cerebrospinal fluid activity of LDH, GOT, or CPK. Necrosis of the central nervous system associated with the neoplasm or necrosis of the neoplasm may cause an elevation of CPK, LDH, or GOT. If hemorrhage occurs in the central nervous system associated with a neoplasm, the GOT and LDH from the red blood cells may cause an elevation of these enzymes in the cerebrospinal fluid or peripheral blood. Convulsions associated with the neoplasm may cause elevation of GOT, LDH, or CPK in the cerebrospinal fluid and peripheral blood associated with leakage of these enzymes from skeletal muscle.[8,21,2]

Enzyme Abnormalities in Central Nervous System Disease (Continued):

Convulsive Disorders

Elevation of CPK, LDH, and GOT have been reported in the cerebrospinal fluid and blood in patients with convulsive disorders. Prolonged seizures such as in status epilepsy causes higher elevations than does a single short-lived attack. The enzyme rise represents enzyme release from prolonged muscle contraction or from central nervous system cells.

REFERENCES

1. Abbassy, A. S., Aboulwafa, M. H.: "Evaluation of Transaminase Activity in the Cerebrospinal Fluid in Paralytic Poliomyelitis". J. Pediat. 59:60, 1961.

2. Acheson, J., James, D. C., Hutchinson, E. C., Westhead, R.: "Serum Creatine Kinase Levels in Cerebral Vascular Disease". Lancet 1:1306, 1965.

3. Anlyan, A. J., Starr, A.: "Beta-Glucuronidase Activity of Spinal and Ventricular Fluids in Humans". Cancer 5:579, 1952.

4. Aronson, S. M.: "Enzyme Determinations in Neurologic and Neuromuscular Diseases of Infancy and Childhood". Pediat. Clin. N. Amer. 7:527, 1960.

5. Aronson, S. M., Saifer, A., Volk, B. W.: "Serial Enzyme Studies of Serum and Cerebrospinal Fluid in Amaurotic Family Idiocy". Amer. Med. Assoc. Diseases of Children 97:684, 1959.

6. Brodell, H. L., Randt, C. T., Morledge, J. H., Goldblatt, D.: "Cerebrospinal Fluid Transaminase Activity in Acute and Chronic Neurologic Diseases". J. Lab. Clin. Med. 53:906, 1959.

7. Connor, R. C. R.: "Creatine Kinase in Intracranial Lesions". Lancet 11:991, 1967.

8. Cunningham, V. R., Phillips, J., Field, E. J.: "Lactic Dehydrogenase Isoenzymes in Normal and Pathological Spinal Fluids". J. Clin. Path. 18:765, 1965.

9. Davson, H.: PHYSIOLOGY OF THE OCULAR AND CEREBROSPINAL FLUIDS Little, Brown & Co., pg. 159, 1956.

Central Nervous System Disease References (Continued):

10. Dubo, H., Park, D. C., Pennington, R. J. T., Kalbag, R. M., Walton, J. N.: "Serum Creatine Kinase in Cases of Stroke, Head Injury and Meningitis". Lancet 11:743, 1967.

11. Embree, L. J., Dreyfus, P. M.: "Blood Transketolase Determinations in Nutritional Disorders of the Nervous System". Trans. Amer. Neurol. Assoc. J. 88:36, 1963.

12. Fleisher, G. A., Wakin, K. G., Goldstein, N. P.: "Glutamic Oxalacetic Transaminase and Lactic Dehydrogenase in Serum and Cerebrospinal Fluid of Patients with Neurologic Disorders". Mayo Clin. Proc. 32:188, 1957.

13. Green, J. B., Oldewurtel, H. A., O'Doherty, D. S., Forester, F. M., Sanchez-Longo, L. P.: "Cerebrospinal Fluid Glutamic Oxalacetic Transaminase Activity in Neurologic Disease". Neurology 7:312, 1957.

14. Herschowitz, N., Cummings, J. N.: "Creatine Kinase in Cerebrospinal Fluid". J. Neurol. Neurosurg. & Psychiat. 27:247, 1964.

15. Houck, J. C.: "Cerebrospinal Fluid Ribonuclease Activity". J. Appl. Physiol. 13:273, 1958.

16. Jakoby, R. K., Jakoby, W. B.: "Lactic Dehydrogenase of Cerebrospinal Fluid in the Differential Diagnosis of Cerebrovascular Disease and Brain Tumor". J. Neurosurg. 15:45, 1958.

17. Junger, G., Junger, I.: "Transaminase Activity in the Serum in Acute Poliomyelitis". Acta. Med. Scand. 166:369, 1960.

18. Katzman, R., Fishman, R. A., Goldensohn, E. S.: "Glutamic Oxalacetic Transaminase Activity in Spinal Fluid". Neurolc 7:853, 1957.

19. Krogsgaard, A. R., Quaade, F.: "Glutamic Acid Oxalacetic Transaminase in the Spinal Fluid in Infectious Diseases of the Central Nervous System". Acta Neurol. Scand. 39:154, 1963.

20. Lehrer, B. M.: "Beta-Glucuronidase in Cerebrospinal Fluid of Patients with Diseases of the Nervous System". Trans. Amer. Neurol. Assoc. 88:244, 1963.

Central Nervous System Disease References (Continued):

21. Lending, M., Slowbody, L. B., Mestern, J.: "Cerebrospinal Fluid Glutamic Oxalacetic Transaminase and Lactic Dehydrogenase Activity in Children with Neurological Disorders". J. Pediat. 65:415, 1964.

22. Lieberman, J., Daiber, O., Dulkin, S. I., Lobstein, O. E., Kaplan, M. R.: "Glutamic Oxalacetic Transaminase in Serum and Cerebrospinal Fluid of Patients with Cerebrovascular Accidents". New Eng. J. Med. 257:1201, 1957.

23. Lonergan, E. T., Semar, M., Lange, K.: "Transketolase Activity in Uremia". Arch. Intern. Med. 126:851, 1970.

24. Marinesco, G., as quoted by Munger, G., and Junger, I.: "Transaminase Activity in the Serum in Acute Poliomyelitis". Acta. Med. Scand. 166:369, 1960.

25. Miyazaki, M.: "Glutamic Oxalacetic Transaminase in Cerebrospinal Fluid". J. Nerv. Ment. Dis. 126:169, 1958.

26. Myerson, R. A., Hurwitz, J. D., Sall, T.: "Serum and Cerebrospinal Fluid Transaminase Concentrations in Various Neurologic Disorders". New Eng. J. Med. 257:273, 1957.

27. Sercl, M., Kovarik, J., Jicha, J., Lichy, J.: "The Problems of Serum Transaminase in Amyotrophic Lateral Sclerosis". Acta. Neurol. Scand. 41:279, 1965.

28. Shuttleworth, E. C., Allen, N.: "Early Differentiation of Chronic Meningitis by Enzyme Assay". Neurology 18:534, 1968.

29. Wakin, K. G., Fleisher, G. A.: "Effect of Experimental Cerebral Infarction on Transaminase Activity in Serum". Mayo Clin. Proc. 31:391, 1956.

30. Wardle, E. N.: "Cerebrospinal Fluid Enzyme Activity in Neurologic Diseases". Lancet 1:1163, 1965.

31. Wroblewski, F.: "The Mechanism of Alteration in Lactic Dehydrogenase Activity of Body Fluids". Ann. N. Y. Acad. Sci. 75:322, 1958.

Central Nervous System Disease References (Continued):

32. Wroblewski, F., Decker, B., Wroblewski, R.: "Activity of Lactic Dehydrogenase in Spinal Fluid". Amer. J. Clin. Path. 28:269, 1957.

33. Wroblewski, F., Decker, B., Wroblewski, R.: "The Clinical Implications of Spinal Fluid Lactic Dehydrogenase Activity". New Eng. J. Med. 258:635, 1958.

CHAPTER 15

ENZYME ABNORMALITIES IN SURGERY
AND
FOLLOWING THERAPEUTIC PROCEDURES

Serum enzymology is extremely pertinent to the practice of surgery. One important enzyme is serum pseudocholinesterase. The succinylcholine compounds inhibit acetylcholinesterase. Prolonged paralysis is occasionally found in patients with low serum cholinesterase activity. This may be extremely important in patients with diseases associated with low cholinesterase activity or in apparently healthy individuals with an inherited deficiency of serum cholinesterase.

Since pseudocholinesterase hydrolyzes the succinylcholine, individuals with low levels or with a genetically altered cholinesterase (atypical cholinesterase variant) may develop prolonged apnea when the drug is administered in the course of anesthesia or during psychiatric therapy. Determination of this enzyme is important in suspected decrease due to poisoning by organic phosphate insecticides. Enzyme levels also are low in hepatic disease. If succinylcholine is utilized before surgery in patients with hepatic disease where there are low enzyme levels, patients may suffer undesirable "overdose" type effect.[22,41,48]

A rapid test to qualitatively determine the plasma level of cholinesterase has been developed. It is a test paper known as Acholest Test Paper and is utilized as a simple screening method for plasma cholinesterase.

The determination of plasma cholinesterase may be extremely useful in the surgical patient. The lab kit consists of a test strip impregnated with a special substrate, and a control strip is a measure of cholinesterase activity which is reported as normal or decreased. The plasma cholinesterase acts on the substrate to liberate choline and acetic acid. The acetic acid changes the color of the test strip containing an acid base indicator. The best determination, however, is a quantative determination of the enzyme.

Surgical Damage to Muscle

Increased GOT activity following surgical procedures occurs regularly.[5,33] LDH and GPT activity may also increase following surgical procedures.[6,11] The activity of all three may be increased

Enzyme Abnormalities in Surgery and Therapeutic Procedures (Continued):

due to the surgical trauma to skeletal muscle. These enzymes will also be elevated during cardiovascular surgery because of cardiac muscle injury. CPK activity is also increased due to the surgical trauma of skeletal and cardiac muscle.

GOT elevation may also indicate liver injury due to anesthesia, prolonged hypotension or surgery. Enzymes such as CPK, GPT, GOT, and LDH will leak out of ischemic muscle subjected to prolonged hypotension associated with surgery.

Blood Transfusions

LDH_1 and LDH_2 may be increased if patients have received multiple blood transfusions at the time of surgery, especially if old blood bank blood has been used. Hemolysis may occur and erythrocyte GOT and LDH will be present in the plasma. The LDH elevation will be greater than GOT since there is more LDH in the red blood cell than GOT. To determine if myocardial necrosis has occurred during the postoperative period, determination of α-hydroxybutyrate dehydrogenase or CPK is more suitable than GOT.[32,45]

Neurosurgery

Neurosurgical procedures may result in elevation of serum CPK and to a lesser extent, LDH or GOT. These enzymes are present in the central nervous system and, with surgery of the brain or spinal cord may be elevated in the blood or cerebrospinal fluid.[1,14,23,40]

Prostate Surgery and Massage

Surgery of the prostate gland will result in elevation of serum acid phosphatase, especially the type inhibitable by L-tartrate.[13] Prostatic massage during a physical examination will cause a transient rise of the prostatic type of serum acid phosphatase; consequently the clinician should not order serum acid phosphatase test immediately after a rectal examination.[12,16,18,26] It is important to send an unhemolyzed specimen to the laboratory for determination of acid phosphatase since this enzyme is present in large amounts in the red cell.[7,37] GOT and LDH are also present in red cells and will be elevated in a hemolyzed specimen. The normal LDH/GOT ratio is 5:1. In hemolysis the ratio is 10:1 or greater.

Enzyme Abnormalities in Surgery and Therapeutic Procedures (Continued):

Bone Surgery

Surgical procedures involving bone will result in elevation of serum alkaline phosphatase.[20] Patients undergoing thoracic surgery may develop an elevated serum alkaline phosphatase following the surgical procedure since ribs will be removed and, with bone healing, serum alkaline phosphatase will rise.[48] Procedures such as the placement of intramedullary nails or repair of fractures will result in alkaline phosphatase elevation because of osteoblastic activity.[8,31] Bone alkaline phosphatase is heat labile.[39,49]

Pancreatic Disease

An acute abdominal condition in which serum enzymology is important is acute pancreatitis. In this condition, serum amylase and lipase may be elevated.[2,42] Acute pancreatitis is a medical condition which is usually treated by various modalities other than surgery. If elevation of the serum amylase or lipase is prolonged, the presence of a pancreatic pseudocyst or abscess requiring surgical therapy should be suspected.[17,35] GOT and LDH activity are significantly elevated in patients with acute pancreatitis.[44] Serum alkaline phosphatase and leucine aminopeptidase may also be elevated because of swelling of the pancreatic gland and obstruction of the common bile duct.[34] GOT, LDH, and serum alkaline phosphatase elevation in patients with acute pancreatitis may also suggest the presence or coexistance of hepatic disease.

Trauma to the abdomen or intra-abdominal surgery near the pancreas may cause pancreatitis with liberation of amylase and lipase in approximately 25 percent of patients with these conditions.[25] In the patient with painless obstructive jaundice and an elevated serum amylase, one should suspect the possibility of carcinoma of the head of the pancreas resulting in pancreatitis.[28] Alkaline phosphatase with increase in the hepatobiliary isoenzyme, as well as 5'-nucleotidase and leucine aminopeptidase may rise in the serum in carcinoma of the head of the pancreas.[21]

Approximately one-fourth of patients with acute choledocholithiasis develop elevations of serum amylase. The elevation in serum amylase may represent pancreatitis secondary to obstruction of the Sphincter of Oddi.[46] In addition to elevated serum amylase, GOT and alkaline phosphatase may rise. The GOT increase is related to

Enzyme Abnormalities in Surgery and Therapeutic Procedures (Continued):

pancreatitis, ascending cholangitis, or increase in intra-biliary pressure with release of liver GOT.

If a duodenal ulcer perforates or penetrates into the pancreas, a pancreatitis may result with subsequent elevation of serum amylase and lipase (Fig. 15-1).

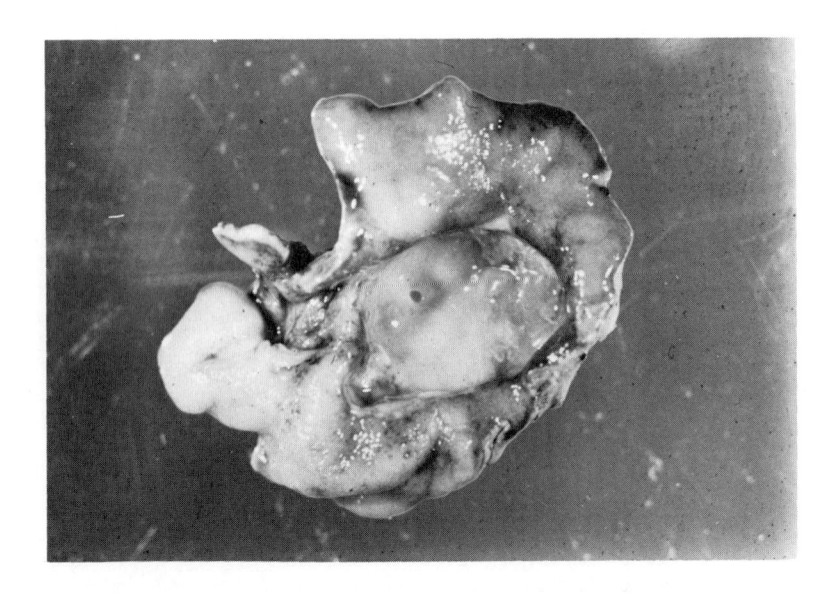

Fig. 15-1   Gross photograph of active duodenal ulcer which penetrated into pancreas and caused an elevation of serum amylase of 410 Somogyi Units and a lipase of 3.6 I.U.

Oviduct Disease

Serum amylase is elevated as a result of perforation of an ectopic pregnancy in the oviduct. Tubal ovarian abscess may also cause a rise in the serum amylase. The oviduct mucosa contains amylase and with inflammation or rupture of the oviduct in which there is an ectopic pregnancy, the serum enzyme level will rise.[19]

Enzyme Abnormalities in Surgery and Therapeutic Procedures (Continued):

Intestinal Obstruction

Intestinal obstruction may result in an elevated serum amylase and lipase. The increase in these enzymes may occur as a result of pancreatitis from perforation of the intestine into the pancreas. Another explanation for the elevated enzyme levels is that there is liberation of lipase and amylase from the intestinal mucosa.[29,30]

Postoperative Chest Pain

CPK, GOT, and LDH assays should be requested in patients who develop chest pain following surgery. It is essential to differentiate between acute myocardial infarction and pulmonary embolism with pulmonary infarction. Since there is very little GOT and CPK in the lung, postoperative patients with chest pain who have developed a pulmonary infarct may have elevated LDH_2 and LDH_3. Patients with acute myocardial infarction may have elevated CPK, GOT, and LDH_1.[9,15] Acute pulmonary infarction may result in elevation of CPK and GOT, if myocardial infarction ensues subsequent to the pulmonary infarction. However, at the onset, CPK and GOT will generally not be elevated; only LDH_2 or LDH_3 will be elevated in pulmonary infarction. GOT may be elevated from red blood cells present in a hemorrhagic pulmonary infarct,[10] and the patient may become jaundiced.

Hemolytic Anemia Related to Cardiac Valves

GOT and LDH may be elevated in patients who have hemolytic anemia subsequent to cardiac valve disease. Furthermore, hemolytic anemia may occur following insertion of prosthetic cardiac valves as a result of the red blood cells striking the valve.[3,38,47]

Gastric Carcinoma and Ulcer

Patients with carcinoma of the stomach may have elevated LDH or beta-glucuronidase in the gastric juice in contrast to benign peptic ulcer in which the LDH activity of the gastric juice is normal. The surgeon should request this determination as part of the diagnostic workup in patients who present with a gastric ulcer.[36,43]

Enzyme Abnormalities in Surgery and Therapeutic Procedures (Continued):

Fat Embolism

The serum lipase will be elevated in patients who sustain extensive fracture of bone and develop fat embolism. The lipase elevation results from lung lipase being liberated subsequent to fat embolism.[4,2]

Intramuscular Injections

Patients who receive intramuscular injections of drugs will have an increased CPK, GOT, and LDH because of skeletal muscle damage.[5] Patients with traumatic hematomas may exhibit increased SGOT and LDH arising from red blood cells in the hematoma.

REFERENCES

1. Acheson, J., James, D. C., Hutchinson, E. C., Westhead, R.: "Serum Creatine-Kinase Levels in Cerebral Vascular Disease". Lancet 1:1306, 1965.

2. Ambrovage, A. M., Howard, J. M., Pairent, F. W.: "The Twenty-four Hour Excretion of Amylase and Lipase in the Urine: Correlation with the Functional State and Operative Injury of the Pancreas". Ann. Surg. 167:539, 1968.

3. Andersen, M. N., Gabieli, E., Zizzi, J. A.: "Chronic Hemolysis in Patients with Ball-valve Prostheses". J. Thorac. Cardiovasc. Surg. 50:501, 1965.

4. Armstrong, H. J., Kuenzig, M. C., Peltier, L. F.: "Lung Lipase Levels in Normal Rats and Rats with Experimentally Produced Fat Embolism". Proc. Soc. Exp. Biol. Med. 124:959, 1967.

5. Ayres, P. R., Willard, T. B.: "Serum Glutamic Oxalacetic Transaminase Levels in Two Hundred and Sixty-Six Surgical Patients". Ann. Intern. Med. 52:1279, 1960.

6. Baer, H., Blount, S. G.: "The Response of the Serum Glutam Oxalacetic Transaminase to Open Heart Operation". Amer. Heart J. 60:867, 1960.

7. Bases, R.: "Elevations of Serum Acid Phosphatase in Certai Myeloproliferative Diseases". New Eng. J. Med. 266:538, 19

Surgery and Therapeutic Procedure References (Continued):

8. Botterell, E. H., King, E. J.: "Phosphatase in Fractures". Lancet 1:1267, 1935.

9. Coodley, E. L.: "Use of Enzymes in Cardiac Diagnosis". Angiology 16:209, 1965.

10. Coodley, E. L.: "Enzyme Profiles in the Evaluation of Pulmonary Infarction". J. Amer. Med. Assoc. 207:1307, 1969.

11. Crawford, D. T.: "Serum Lactate Dehydrogenase in the Immediate Postoperative Period". Amer. Surg. 30:690, 1964.

12. Daniel, O., Van Zyl, J. S.: "Rise of Serum Acid Phosphatase Level following Palpation of Prostate". Lancet 1:998, 1952.

13. Doe, R. P., Seal, U. S.: "Acid Phosphatase in Urology". Surg. Clin. N. Amer. 45:1455, 1965.

14. Dubo, H., Park, D. C., Pennington, R. J. T., Kalbag, R. M., Walton, J. N.: "Serum Creatine Kinase in Cases of Stroke, Head Injury, and Meningitis". Lancet 2:743, 1967.

15. Duma, R. J., Siegel, A. L.: "Serum Creatine Phosphokinase in Acute Myocardial Infarction". Arch. Intern. Med. 115:443, 1965.

16. Dybkaer, R., Jensen, G.: "Acid Phosphatase Levels following Massage of the Prostate". Scand. J. Clin. Lab. Invest. 10:349, 1958.

17. Ebbesen, K. E., Schoenebeck, J.: "Prolonged Amylase Elevation". Acta. Chir. Scand. 133:61, 1967.

18. Glenn, J. F., Spanel, D. L.: "Serum Acid Phosphatase and Effect of Prostatic Massage". J. Urol. 82:240, 1959.

19. Green, C. L.: "Identification of Alpha-Amylase as a Secretion of the Human Fallopian Tube, and 'Tube-like' Epithelium of Mullerian and Mesonephric Duct Origin". Amer. J. Obstet. & Gynec. 73:402, 1957.

20. Gutman, A.: "Serum Alkaline Phosphatase Activity in Diseases of the Skeletal and Hepatobiliary Systems: A Consideration of the Current Status". Amer. J. Med. 27:875, 1959.

Surgery and Therapeutic Procedure References (Continued):

21. Harkness, J., Roper, B. W., Durant, J. A., Miller, H.: "The Serum Leucine Aminopeptidase Test. An Appraisal of its Value in Diagnosis of Carcinoma of Pancreas". Brit. Med. J. 1:1787, 1960.

22. Harris, H., Whittaker, M.: "Differential Inhibition of Human Serum Cholinesterase with Fluoride: Recognition of Two New Phenotypes". Nature 191:496, 1961.

23. Herschowitz, N., Cumings, J. N.: "Creatine Kinase in Cerebrospinal Fluid". Neurol. Neurosurg. Psychiat. 27:247, 1964.

24. Hess, J. W., MacDonald, R. P.: "Serum Creatine Phosphokinase Activity: A New Diagnostic Aid in Myocardial and Skeletal Muscle Disease". J. Mich. Med. Soc. 62:1095, 1963.

25. Jones, R. C., Shires, G. T.: "The Management of Pancreatic Injuries". Arch. Surg. 90:502, 1965.

26. Kendall, A. R.: "Acid Phosphatase Elevation following Prostatic Examination in the Earlier Diagnosis of Prostatic Carcinoma". J. Urol. 86:442, 1961.

27. Korn, E. D.: "Clearing Factor, a Heparin-Activated Lipoprotein Lipase". J. Biol. Chem. 215:1, 1955.

28. Lundh, G.: "Pancreatic Exocrine Function in Neoplastic and Inflammatory Disease. A Simple and Reliable New Test". Gastroenterology 42:275, 1962.

29. MacFate, R. P.: "Amylase in Biological Fluids". In F. W. Sunderman and F. W. Sunderman, Jr. MEASUREMENT OF EXOCRINE AND ENDOCRINE FUNCTIONS OF THE PANCREAS Lippincott, pg. 14, 1961.

30. McGeachin, R. L., Potter, B. A., Lindsey, A. C.: "Puromycin Inhibition of Amylase Synthesis in the Perfused Rat Liver". Arch. Biochem. 104:314, 1964.

31. Mitchell, C. L.: "Serum Phosphatase in Fracture Repair". Ann. Surg. 104:304, 1936.

32. Nutter, D., Trujillo, N. P., Evans, J. M.: "The Isoenzymes of Lactic Dehydrogenase: Myocardial Infarction and Coronary Insufficiency". Amer. Heart J. 72:315, 1966.

Surgery and Therapeutic Procedure References (Continued):

33. Person, D. A., Judge, R. D.: "Effect of Operation on Serum Transaminase Levels". Arch. Surg. 77:892, 1958.

34. Pineda, E. P., Goldbarg, J. A., Banks, B. M., Rutenburg, A. M.: "Serum Leucine Aminopeptidase in Pancreatic and Hepatobiliary Diseases". Gastroenterology 38:698, 1960.

35. Pinkham, R. D.: "Pancreatic Collections (Pseudocysts) following Pancreatitis and Pancreatic Necrosis. Reviews and Analysis of Ten Cases". Surg. Gynec. & Obstet. 3:227, 1945.

36. Piper, D. W., Macoun, M. L., Broderick, F. L., Fenton, B. H., Builder, J. E.: "The Diagnosis of Gastric Carcinoma by the Estimation of Enzyme Activity in Gastric Juice". Gastroenterology 45:614, 1963.

37. Pirofsky, B.: "Hemolysis in Valvular Heart Disease". Ann. Intern. Med. 65:373, 1966.

38. Pirofsky, B., Sutherland, D. W., Starr, A., Griswold, H. E.: "Hemolytic Anemia Complicating Aortic Valve Surgery". New Eng. J. Med. 272:235, 1965.

39. Posen, S., Neale, F. C., Clubb, J. S.: "Heat Interaction in the Study of Human Alkaline Phosphatases". Ann. Intern. Med. 62:1234, 1965.

40. Rosalki, S. B.: "Creatine Kinase and Brain Damage". Lancet 2:722, 1967.

41. Rose, L., Davis, D. A., Lehmann, H.: "Serum Pseudocholinesterase in Depression with Notable Anxiety". Lancet 2:563, 1965.

42. Saxon, E. I., Hinklay, W. C., Vogel, W. C., Zieve, L.: "Comparative Value of Serum and Urinary Amylase in the Diagnosis of Acute Pancreatitis". Arch. Intern. Med. 99:607, 1957.

43. Symrniotis, F., Schenker, S., O'Donnell, J., Schiff, L.: "Lactic Dehydrogenase Activity in Gastric Juice for the Diagnosis of Gastric Cancer". Amer. J. Dig. Dis. 7:712, 1962.

Surgery and Therapeutic Procedure References (Continued):

44. Ticktin, H. E., Trujillo, N. P.: "Serum Enzymes in Diagnosis". Disease-A-Month. June, 1966.

45. Trujillo, N. P., Nutter, D., Evans, J. M.: "The Isoenzymes of Lactic Dehydrogenase. II. Pulmonary Embolism, Liver Disease, the Postoperative State, and other Medical Conditions". Arch. Intern. Med. 119:333, 1967.

46. Webster, P. D., Zieve, L.: "Alterations in Serum Content of Pancreatic Enzymes". New Eng. J. Med. 267:604, 654, 1962.

47. Westring, D. W.: "Aortic Valve Disease and Hemolytic Anemia". Ann. Intern. Med. 65:203, 1966.

48. Wetstone, H. J., Honeyman, M. S., McComb, R. B.: "Kinetic Control of the Quantitative Activity of a Serum Enzyme in Man". J. Amer. Med. Assoc. 192:1007, 1965.

49. Wilkins, W. E., Regan, E. M.: "Course of Phosphatase Activity in Healing of Fractured Bones". Proc. Soc. Exp. Biol. Med. 32:1373, 1935.

CHAPTER 16

ENZYME ABNORMALITIES IN GENITOURINARY TRACT DISEASE

Carcinoma of the Prostate

The main interest in enzymology relevant to disease of the genitourinary tract is acid phosphatase determination in carcinoma of the prostate. This subject has already been extensively reviewed in Chapters 2 and 8. No further reference to acid phosphatase will occur in this Chapter.

Before the subject of carcinoma of the prostate is dismissed, it should be emphasized that other enzymes have been found to be raised in carcinoma of the prostate besides acid phosphatase; LDH and aldolase are also elevated in cancer of the prostate, with the slow migrating LDH_4 and LDH_5 increased.[7,25,45] After successful treatment by orchiectomy and stilbestrol, LDH activity should decrease. A subsequent increase indicates a relapse.[16]

Renal Disease

Renal disease will cause abnormalities in serum enzymes. Alkaline phosphatase, GOT, and LDH will be elevated in certain renal lesions because the renal tubules contain these enzymes. When there is necrosis of renal tubular cells, the enzymes exit from the cells into the serum (Fig. 16-1).[1]

Enzyme Abnormalities in Genitourinary Tract Disease (Continued):

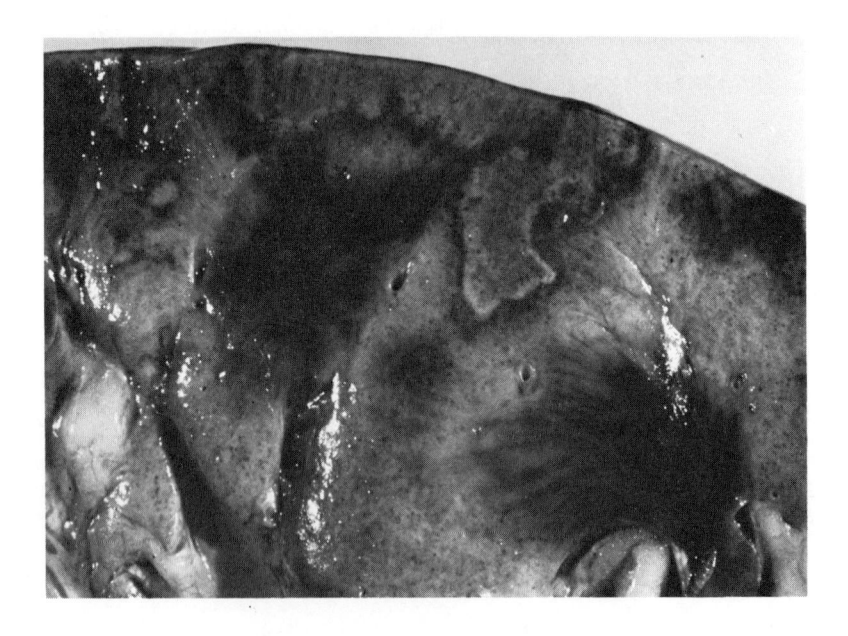

Fig. 16-1 Gross photograph of recent infarct of kidney which was associated with elevated serum enzymes. The LDH was 385 I. U., GOT 280 I.U., and the alkaline phosphatase measured 270 I.U.

Serum alkaline phosphatase may rise within 5 days after renal tubular necrosis,[27,62] and may remain elevated for 2 weeks after necrosis of the tubules has occurred. The enzyme may also increase in the urine,[29] becoming elevated within 48 hours after death of the renal tubules and remaining higher for approximately one week.[18,23] Renal tubular necrosis secondary to obstructive uropathy may also elevate serum alkaline phosphatase.[2]

Serum GOT may rise within 18 to 24 hours after necrosis of renal tubules.[20] Peak levels are reached within 48 hours and become normal 4 to 5 days later.

LDH may rise in the serum following renal tubular necrosis.[33] It may elevate within 24 hours, peak between 4 to 5 days, and remain elevated for approximately 2 weeks. LDH activity in the urine may

Enzyme Abnormalities in Genitourinary Tract Disease (Continued):

also be increased. LDH_1 and LDH_2 are derived from cortical renal tubules whereas LDH_4 and LDH_5 arise from medullary renal tubules. The type of serum isoenzyme elevation will depend on whether renal tubular cells from cortex or medulla become necrotic.

Renal transplant patients will develop slight elevation of serum LDH immediately after the transplant.[66] LDH_1, LDH_2, LDH_4, or LDH_5, but particularly the last two, may be elevated with transplant rejection.[50]

Renal artery stenosis will cause elevation of serum LDH of the fast types, LDH_1 and LDH_2.[6,65] In addition, serum alkaline phosphatase may increase because of the renal cortical damage.

Acute pyelonephritis may cause elevated serum and urinary LDH, GOT, and beta-glucuronidase due to renal tubular damage (Fig. 16-2).[12]

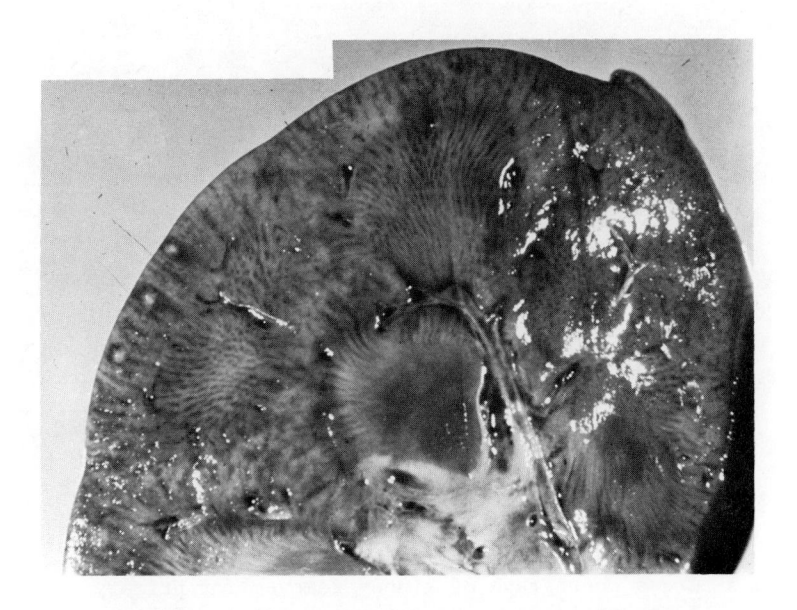

Fig. 16-2 Gross photograph of acute pyelonephritis which resulted in an elevated serum LDH of 310 I.U.

Enzyme Abnormalities in Genitourinary Tract Disease (Continued):

Acute glomerulonephritis is frequently accompanied by elevated serum LDH and alkaline phosphatase.[48] The serum acid phosphatase may also increase. Urinary LDH, GOT, muramidase, and alkaline phosphatase may be elevated.[2,57] Chronic glomerulonephritis causes an increase in serum alkaline phosphatase and LDH. With the alkaline phosphatase increase related to secondary hyperparathyroidism associated with this condition, the cause of LDH elevation is multifactorial. It is not excreted as well and may be derived from multiple pathological sites.[52]

Diabetic glomerulosclerosis and systemic lupus erythematosus may cause elevation of serum and urinary alkaline phosphatase and LDH. In addition, carcinoma of the kidney may cause elevated serum LDH_5 (Fig. 16-3).[35,37]

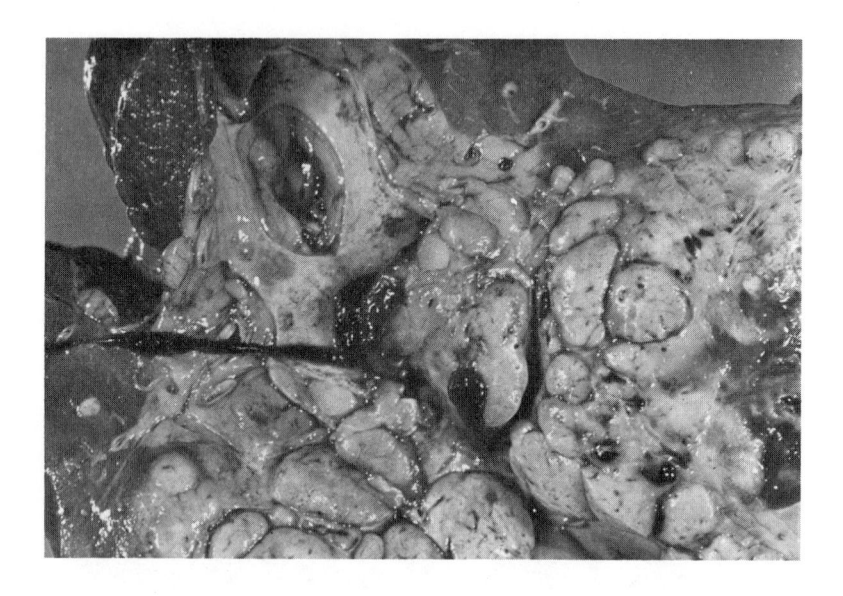

Fig. 16-3 Gross photograph of kidney removed for primary adenocarcinoma. The serum LDH was 655 I.U.

The nephrotic syndrome is associated with a deficiency of lipoprotein lipase. It is thought that this enzyme deficiency contribute

Enzyme Abnormalities in Genitourinary Tract Disease (Continued):

to the milky serum, with an increase in serum cholesterol and triglycerides.

Determination of enzyme activities in urine yields much valuable information. It is more difficult to assay enzymes in the urine because of the presence of urinary enzyme inhibitors, which must be removed. Removal is accomplished by dialysis. The level of urinary enzymes is generally much less than that in serum, but some enzymes, such as acid phosphatase, are found in greater amounts in urine.

Enzymes with a molecular weight of less than 80,000, may be excreted into the urine. Serum enzymes such as LDH and amylase are excreted by the kidneys. Thus, in uremia, both of these enzymes are elevated in the serum.

Renal tubular cells contain a wide variety of enzymes. Necrobiosis of these cells may thus cause their enzymes to be contributed to the urine. In addition, enzymes from the tubular luminal aspects of the cells enter the urine with ease. An insignificant amount of urinary enzyme is derived from epithelial cells of the urinary tract such as the renal pelvis or urinary bladder.

The glands in the genitourinary tract contribute enzymes to the urine. The best example is the prostate gland in the male. The acid phosphatase of urine in the male is twice the amount as in the female as the result of prostatic secretion.[56]

The sources of urinary enzymes due to abnormalities of the genitourinary tract are multifactorial. Enzymes are released into the urine with necrosis of renal tubular cells due to various etiologies. Neoplasms of the genitourinary tract will release enzymes into the urine by excess production with secretion and by necrosis of the neoplastic cells. Other contributing factors are enzymes derived from leukocytes, erythrocytes, and bacteria associated with inflammatory exudates.

Inhibitors and activators of urinary enzymes are present in the urine and are derived from the kidney, glands of the genitourinary tract, and serum. The inhibitors may be removed by dialysis. Drugs and their metabolites may also act as enzyme inhibitors.[28,55]

Many enzymes are present in the urine normally. They belong to the classes of oxidoreductases, transferases, and hydrolases. Only a few are of diagnostic importance. These are amylase, LDH, alkaline

Enzyme Abnormalities in Genitourinary Tract Disease (Continued):

phosphatase, LAP, beta-glucuronidase, and acid phosphatase.[36,49,63]

It is not important to determine urinary enzyme activity per milliliter of urine. The concentration of urinary enzymes in a 24 hour specimen should be determined. A circadian rhythm exists for excretion of urinary enzymes. More enzymes are excreted at night than during the day.

Acute diuresis results in an increase in excretion in urinary enzymes. Hyponatremia and especially hypokalemia causes an increase in urinary enzyme excretion.

Amylase and Uropepsinogen

The primary source of urinary amylase is the pancreas. Determination of urinary amylase is a better diagnostic aid than serum amylase in pancreatic disease. When the serum amylase is elevated, the renal tubule responds with increased excretion. The urinary amylase tends to remain increased longer than serum amylase in acute pancreatitis. Serum amylase elevation is more evanescent. Kidney function must be normal for utilization of the urinary amylase test in pancreatitis. Determination of urinary lipase for pancreatic disease is not useful.[21,30,38]

Uropepsinogen excretion may be of diagnostic importance in patients with peptic ulcer. Good correlation of gastric secretion of uropepsinogen and urine concentration of this enzyme has been recognized. Peptic ulcer patients may have twice the concentration of this enzyme in the urine as normal individuals.[14,15]

Muramidase

Minimal muramidase activity is present in normal urine. Muramidase is also known as lysozyme and is present in lysosomes in renal tubular cells.[42] A marked increase in urinary muramidase indicates acute myelogenous or acute monocytic leukemia. Monoblasts and myeloblasts produce a large amount of this enzyme resulting in elevated levels in the serum and urine. Muramidase may be injurious to renal tubular cells.[26,43]

Renal tubular disease will result in increased urinary muramidase. Nephrosis and Fanconi syndrome are known to cause a prominent increase in urinary muramidase.[44]

Enzyme Abnormalities in Genitourinary Tract Disease (Continued):

A kidney transplant will leak muramidase for a few days and if the graft is not rejected, the urinary muramidase returns to normal. However, if rejection or infarction occurs, urinary muramidase will again increase. Acute pyelonephritis will also result in increased urinary muramidase.[67]

Catalase

Urinary catalase activity is extremely low normally, the minimal enzyme levels being derived from renal tubular cells.

An increase in urinary catalase may develop from renal tubular damage. However, the most important cause for elevated urinary catalase is from bacterial infection.[22] Determination of this enzyme may thus be important to detect urinary tract infection. The enzyme is derived from bacteria, neutrophils, and erythrocytes. The presence of peroxidase in the urine indicates the presence of erythrocytes and neutrophils in the urine associated with various genitourinary diseases.[9,11]

Nitrate Reductase

Patients with pyelonephritis caused by bacteria able to reduce nitrate to nitrite may have increased levels of nitrate reductase in the urine. These bacteria are E. coli, Pseudomonas and Klebsiella. Staphylococci or Streptococci do not reduce nitrates.[59] At present there are no satisfactory tests for nitrate reductase in the urine.[60]

GOT and GPT

Normal urine contains low levels of GOT and GPT.[12] Both of these enzymes will be increased in the urine when serum enzymes are increased in diseases such as acute myocardial infarction or acute viral hepatitis.

Various genitourinary diseases such as infections, immunologic conditions, acute glomerulonephritis, and acute renal infarction will result in increase in GOT.[32] GOT activity in the urine thus, does not persist and may decrease rapidly within hours. The rapid decrease may be related to enzyme inhibitors.

Enzyme Abnormalities in Genitourinary Tract Disease (Continued):

Sulfatase

Renal neoplasms or infections such as tuberculosis will result in increased urinary sulfatase. Sulfatase is normally present to some extent in urine. However, arylsulfatase is absent in urine in metachromatic leukodystrophy.[10,61]

Urokinase

Normal urine contains urokinase which is secreted into the urine from the kidney. Elevated urinary urokinase may result from acute renal damage as is found in acute renal infarction.[19] The determination of Kallikrein in the urine is only of importance in assessing rejection of a renal transplant. Increased Kallikrein activity occurs in the urine with acceptance of the renal transplant but decreases with rejection.

Lactic Dehydrogenase

Several organs contribute to urinary lactic dehydrogenase. It is not essential to determine urinary LDH isoenzymes since it is not diagnostically meaningful.[17,24,34,37,40,51,52]

Various renal diseases will cause elevated urinary lactic dehydrogenase:

1. Acute glomerulonephritis
2. Kimmelstiel-Wilson disease in diabetics
3. Systemic lupus nephritis
4. Acute and chronic pyelonephritis
5. Acute tubular necrosis
6. Acute renal infarction
7. Polycystic renal disease
8. Acute rejection of a kidney transplant[46]
9. Renal tuberculosis
10. Presence of erythrocytes
11. Schistosomiasis[53]
12. Malignant neoplasms of the genitourinary tract

Even though various organs contribute LDH to the urine, a majority of urinary LDH arises from kidney tissue. Dialyzed urine can be stored for 24 hours at refrigerator temperature without loss of LDH

Enzyme Abnormalities in Genitourinary Tract Disease (Continued):

activity. If erythrocytes are present, LDH activity will be increased.[16,45,53]

Many investigators have attempted to utilize urinary LDH as a screening test for detection of malignant neoplasms of the genitourinary tract (Fig. 16-4). An increase in urinary LDH associated with a malignant neoplasm develops from erythrocytes in urine, inflammation within the neoplasm and production of LDH by the tumor. Some silent neoplasms may be detected by the presence of elevated urinary LDH. The size and duration of the neoplasm are not related to elevation of the urinary LDH.[4,5,48,64]

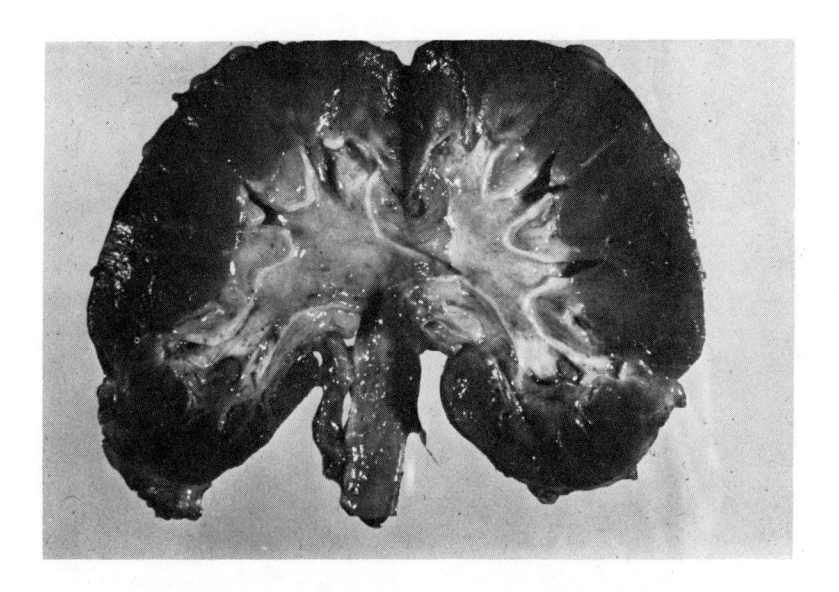

Fig. 16-4 Gross photograph of transitional cell carcinoma of the renal pelvis which caused a marked elevation of urinary LDH. Carcinoma is in upper ureter and renal pelvis.

Enzyme Abnormalities in Genitourinary Tract Disease (Continued):

Aminopeptidases

Four aminopeptidases are present in human urine. The most important is leucine aminopeptidase. The others are cystine aminopeptidase, glycine aminopeptidase, and alanine aminopeptidase.[8,41]

The causes for elevated leucine aminopeptidase in the urine are:

1. Renal tubular damage caused by drugs such as sulfa drugs and antibiotics.

2. Acute tubular necrosis secondary to hypoxia. Proximal tubule contain the most enzyme.

3. Diagnostic procedures such as intravenous pyelography. The radiographic dye induces a stimulation of leucine aminopeptidase secretion into the urine.

4. Pregnancy.

5. Acute glomerulonephritis.

6. Acute pyelonephritis.

7. Malignant neoplasms of the female genital tract.

8. Malignant neoplasms of the kidney.

9. Malignant neoplasms of the head of the pancreas. Renal damage usually is present concommitantly with the pancreatic lesion resulting in increased activity of urinary leucine aminopeptidase from pancreatic and renal sources.

10. Activation of kidney plasminogen which subsequently activate renal peptidases. Thus streptokinase and vasopressin increa urinary leucine aminopeptidase while antifibrinolytic drugs such as salicylic acid and Epsilon Amino Caproic Acid (EACA) decrease urinary leucine aminopeptidase.

Beta-Glucuronidase

Urinary beta-glucuronidase is derived from the lysosomes of renal tubular cells and epithelium of the urinary tract. No correlation exists between serum and urinary beta-glucuronidase. Many

Enzyme Abnormalities in Genitourinary Tract Disease (Continued):

diseases will result in increased urinary beta glucuronidase with the main etiology being carcinoma of the urinary bladder or kidney (Fig. 16-5).[31,54]

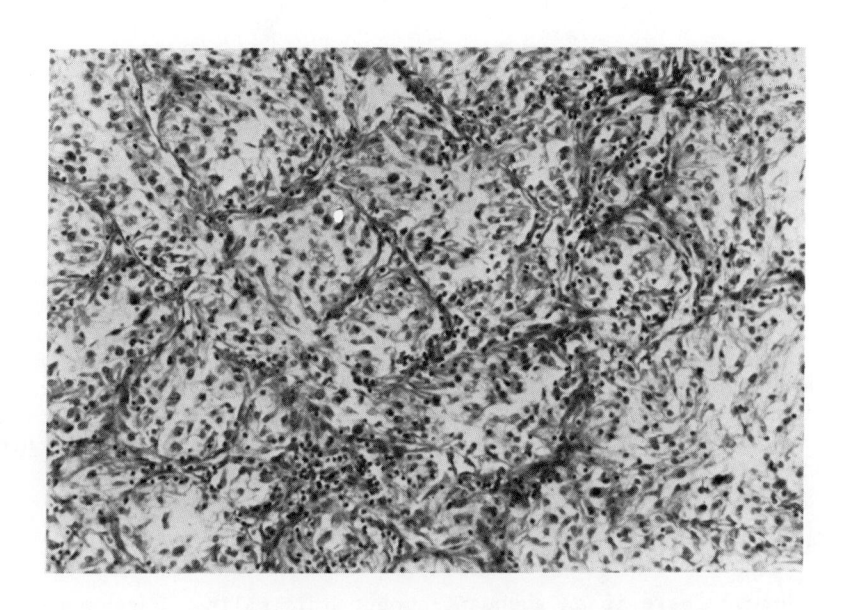

Fig. 16-5 Photomicrograph of section of adenocarcinoma of kidney. The urinary beta-glucuronidase was prominently elevated.

The causes for increased urinary beta-glucuronidase are:

1. Carcinoma, urinary bladder.

2. Carcinoma, kidney.

3. Acute pyelonephritis.

4. Renal tuberculosis.

5. Systemic lupus erythematosus involving the kidney.

Enzyme Abnormalities in Genitourinary Tract Disease (Continued):

6. Acute renal tubular necrosis.

7. Pregnancy.

8. Acute rejection of a renal transplant.

9. Polycystic disease.

10. Schistosomiasis of the urinary bladder.

Bilharziasis of the urinary bladder results in increased urinary beta-glucuronidase from irritation of the bladder epithelium by the ova.

Carcinoma of the kidney, urinary bladder and cervix may result in elevated urinary beta-glucuronidase; the enzyme elevation is secondary to the malignant neoplasm. Some investigators have suggested that the beta-glucuronidase contributes to the development of the neoplasm. Employees of the aniline dye industry have a high incidence of urinary bladder cancer. The carcinogens are excreted in the urine as glucuronides. Urinary beta-glucuronidase hydrolyzes the glucuronide and liberates the carcinogen.[39]

Alkaline Phosphatase

Because there is an abundant amount of alkaline phosphatase in the renal tubules, there is normally a large amount in the urine.[5,13] Inhibitors of alkaline phosphatase are present in the urine. The presence of a large amount of inorganic phosphate also tends to inhibit the enzyme's activity in the urine. Alkaline phosphatase activity of urine is stable at room temperature for approximately 4 hours if the urine is undialyzed. After dialysis it is stable for 24 hours at refrigerator temperature. Urinary alkaline phosphatase can only be determined after the inhibitors have been removed.

Many diseases may be associated with increased urinary alkaline phosphatase. These are:

1. Acute glomerulonephritis.

2. Systemic lupus erythematosus with renal involvement.

3. Acute tubular necrosis.

Enzyme Abnormalities in Genitourinary Tract Disease (Continued):

4. Acute renal infarction.
5. Acute pyelonephritis.
6. Pregnancy.
7. Carcinoma of the kidney.

Acid Phosphatase

The major source of acid phosphatase in the urine of males is the prostate.[3] Approximately 50 times the activity of this enzyme is present in urine over that of serum.[58] In women, acid phosphatase in the urine is approximately half that of men and is derived mainly from the renal tubules.

Acid phosphatase activity of ureteral urine may indicate the kidney which is abnromal. Higher acid phosphatase is present in ureteral urine from the diseased kidney.

Determinations of urinary enzymes are easy to perform. Inhibitors must be removed. Biochemical evidence indicates that these determinations are valuable.

Testicular Disease

Cancer of the testes may cause an increased urinary beta-glucuronidase activity.[10] Marked elevation of the serum and urine LDH may also occur.[68] LDH_1 and LDH_2 or LDH_5 are the elevated isoenzymes.

Acid phosphatase activity of the seminal fluid is decreased in Klinefelter's syndrome,[47] and in patients with bilateral cryptorchidism. The acid phosphatase of seminal fluid is related to the phosphate metabolism of spermatozoa.

Enzyme Abnormalities in Genitourinary Tract Disease (Continued):

REFERENCES

1. Abe, N., Shibuya, M., Nakamura, S.: "Urinary Lactic Dehydrogenase Activity with Special Reference to Urinary Tract Malignancy". Jap. J. Urol. 56:58, 1965.

2. Amador, E., Dorfman, L. E., Wacker, W. E. C.: "Urinary Alkaline Phosphatase and LDH Activities in the Differential Diagnosis of Renal Disease". Ann. Intern. Med. 62:30, 1965.

3. Amador, E., Marshall, G., Price, J. W.: "Serum Acid Alpha-Naphthyl Phosphatase Activity". Amer. J. Clin. Path. 51:202 1965.

4. Amador, E., Wacker, W. E. C., Harrison, J. H.: "Urinary LDH Activity in Neoplastic and Inflammatory Diseases of the Urinary Tract". Amer. J. Clin. Path. 49:271, 1968.

5. Amador, E., Zimmerman, T. S., Wacker, W. E. C.: "Urinary Alkaline Phosphatase Activity. I. Elevated Urinary LDH and Alkaline Phosphatase Activities for the Diagnosis of Renal Adenocarcinomas". J. Amer. Med. Assoc. 185:769, 1963.

6. Aukland, K., Krog, J.: "Influence of Various Factors in Urine Oxygen Tension in the Dog". Acta. Physiol. Scand. 52:350, 1961.

7. Baker, R., Gowan, D.: "The Effect of Hormonal Therapy of Prostatic Cancer on Serum Aldolase". Cancer Res. 13:141, 1953.

8. Bergmann, H., Scheler, F.: "Der Nachweis Tubularer Funktions Storungen der Niere durch Bestimmung der Aminopeptidase-Aktivitatim Harn". Klin. Wschr. 42:275, 1964.

9. Bialestock, D., MacDonald, J. J.: "Screening Test for the Detection of Urinary Tract Disease". Med. J. Aust. 49:704, 1962.

10. Boyland, E., Wallace, D. M., Williams, D. C.: "Activity of Sulphatase and Beta-Glucuronidase in Urine, Serum, and Bladder Tissue". Brit. J. Cancer 9:62, 1955.

Genitourinary Tract Disease References (Continued):

11. Braude, A., Berkowitz, H.: "Detection of Urinary Catalase by Disk Flotation". J. Lab. Clin. Med. 57:490, 1961.

12. Brenner, B. M., Gilbert, V. E.: "Elevated Levels of Lactic Dehydrogenase, Glutamic Oxalacetic Transaminase, and Catalase in Infected Urine". Amer. J. Med. Sci. 245:65, 1963.

13. Butterworth, P. J., Moss, E. W., Pitkanen, E., Pringle, A.: "Some Characteristics of Alkaline Phosphatase in Human Urine". Clin. Chim. Acta. 11:220, 1965.

14. Bolt, R. J., Pollard, H. M., Carballo, A.: "Determination of Gastric Secretory Function by Measurement of Substance Excreted by the Kidneys. I. Uropepsin Excretion in Health and Disease". J. Lab. Clin. Med. 43:335, 1954.

15. Cummins, A. J.: "Uropepsin Excretion in the Differential Diagnosis of Gastrointestinal Bleeding". Ann. Intern. Med. 52:1213, 1960.

16. Denis, L. J., Prout, G. R.: "Lactic Dehydrogenase in Prostatic Carcinoma". Invest. Urol. 1:101, 1963.

17. Dorfman, L. E., Amador, E., Wacker, W. E. C.: "Urinary Lactic Dehydrogenase Activity. III. An Analytical Validation of the Assay Method". J. Amer. Med. Assoc. 184:1, 1963.

18. Duggan, M. L.: "Acute Renal Infarction". J. Urol. 90:669, 1963.

19. Egeblad, K., Astrup, T.: "Fibrinolysis and the Trypsin Inhibitor in Human Urine". Scand. J. Clin. Lab. Invest. 18:181, 1965.

20. Frahm, C. J., Folse, R.: "Serum Oxalacetic Transaminase Levels following Renal Infarction". J. Amer. Med. Assoc. 180:209, 1962.

21. Franzini, C., Moda, S.: "Human Urinary Amylolytic Enzymes in Acute Hepatitis". J. Clin. Path. 18:775, 1965.

22. Gagnon, M., Hunting, W. M., Esselen, W. B.: "New Method for Catalase Determination". Anal. Chem. 31:144, 1959.

Genitourinary Tract Disease References (Continued):

23. Gault, M. H., Steiner, G.: "Serum and Urinary Enzyme Activity after Renal Infarction". Canad. Med. Assoc. J. 93:1101, 1965.

24. Gelderman, A. H., Gelboin, H. V., Peacock, A. C.: "Lactic Dehydrogenase Isoenzymes in Urine from Patients with Malignancies of the Urinary Bladder". J. Lab. Clin. Med. 65:132, 1965.

25. Goldman, R. D., Kaplan, N. O., Hall, T. C.: "Lactic Dehydrogenase in Human Neoplastic Tissues". Cancer Res. 24:389, 1964.

26. Haystell, J. P., Perillie, P. E., Finch, S. C.: "Urinary Muramidase and Renal Disease". New Eng. J. Med. 279:506, 1968.

27. Highman, B., Thompson, E. C., Roche, J., Atland, P. D.: "Serum Alkaline Phosphatase in Dogs with Experimental Splen and Renal Infarcts and with Endocarditis". Soc. Exp. Biol. Med. 95:109, 1957.

28. Hilliard, S. D., O'Donnell, J. F., Schenker, S.: "On the Nature of the Inhibitor of Urinary Alkaline Phosphatase". Clin. Chem. 11:570, 1965.

29. Hoxie, J. H., Coggin, C. B.: "Renal Infarction". Arch. Intern. Med. 65:587, 1940.

30. Hobbs, J. R., Aw, S. E.: "Urinary Isoamylases". In H. Huber CURRENT PROBLEMS IN CLINICAL BIOCHEMISTRY Berne, Vol. 2, pg. 281, 1968.

31. Kallet, H. A., Lapco, L.: "Urine Beta-Glucuronidase Activi in Urinary Tract Disease". J. Urol. 97:352, 1967.

32. Kalmansohn, R. B., Kalmansohn, R. W.: "Acute Myocardial Infarction. Urine Glutamic Oxalacetic Transaminase Activity Calif. Med. 95:165, 1961.

33. Kemp, E., Laursen, T.: "Investigation of the Excretion of Enzymes in Urine". Scand. J. Clin. Lab. Invest. 12:463, 1960.

34. Kiser, W. S., Riggins, R. S.: "Clinical Significance of Increased Activity of Urinary Lactic Dehydrogenase". J. Urol. 96:559, 1966.

Genitourinary Tract Disease References (Continued):

35. Lee, D. A., Cockett, A. T. K., Caplan, B. M., Chiamori, N.: "Urinary Lactic Acid Dehydrogenase Activity in the Diagnosis of Urologic Neoplasms". J. Urol. 95:77, 1966.

36. Levy, A. L., Rottino, A.: "Effect of Disease States on the Ribonuclease Concentrations of Body Fluids". Clin. Chem. 6:43, 1960.

37. Macalalag, E. V., Prout, G. R.: "Confirmation of the Source of Elevated Urinary Lactic Dehydrogenase in Patients with Renal Tumor". J. Urol. 92:416, 1964.

38. McGeachim, R. L., Hargan, L. A.: "The Renal Clearance of Amylase". J. Appl. Physiol. 9:129, 1956.

39. Melicow, M. M., Uson, A. C., Lipton, R.: "Beta-Glucuronidase Activity in the Urine of Patients with Bladder Cancer and other Conditions". J. Urol. 86:89, 1961.

40. Mirabile, C. S., Bowers, G. N. Jr., Berlin, B. B.: "Urinary Lactic Dehydrogenase". J. Urol. 95:79, 1966.

41. Mullan, D. P.: "Urinary Excretion of Leucine Aminopeptidase in Pregnancy". J. Clin. Path. 20:660, 1967.

42. Osserman, E. F., Lawlor, D. P.: "Serum and Urinary Lysozyme (Muramidase) in Monocytic and Monomyelocytic Leukemia". J. Exp. Med. 124:921, 1966.

43. Perillie, P. E., Finch, S. C.: "Muramidase Studies in Philadelphia-Chromosome-Positive and Chromosome-Negative Chronic Granulocytic Leukemia". New Eng. J. Med. 283:457, 1970.

44. Prockop, D. J., Davidson, W.: "A Study of Urinary and Serum Lysozyme in Patients with Renal Disease". New Eng. J. Med. 270:269, 1964.

45. Prout, G. R., Macalalag, E. V., Denis, L. J., Preston, L. W.: "Alterations in Serum Lactate Dehydrogenase and its Fourth and Fifth Isoenzymes in Patients with Prostatic Carcinoma". J. Urol. 94:451, 1965.

46. Prout, G. R., Macalalag, E. V., Hume, D. M.: "Serum and Urinary Lactic Dehydrogenase in Patients with Renal Homotransplants". Surgery 56:283, 1964.

Genitourinary Tract Disease References (Continued):

47. Raboch, J., Homolka, J.: "Acid Phosphatases in the Ejaculate of Men with Disturbances of Somatosexual Development". Fertil. Steril. 12:368, 1961.

48. Ramkissoon, R. A., Chamberlain, N., Boker, E. L., Jennings, E. R.: "Diagnostic Significance of Urinary Lactic Acid Dehydrogenase". J. Urol. 93:603, 1964.

49. Richet, G., Villiers, H., Ardillou, R.: "Activite Tripeptidasique due Plasma au cours de l'Insuffisance Renale". Rev. Franc. Etud. Clin. Biol. 2:808, 1957.

50. Richterich, R., Schafroth, P., Franz, H. E.: "The Isolated Glomerulum of the Rat Kidney. III. Heterogeneity of Lactate Dehydrogenase in the Kidney Cortex, Kidney Medulla, and Glomerulum". Enzymol. Biol. Clin. 1:114, 1961.

51. Riggins, R. S., Kiser, W. S.: "Lactic Dehydrogenase Isozymes in Urine". Invest. Urol. 2:30, 1964.

52. Rosalki, S. D., Wilkinson, J. H.: "Urinary Lactic Dehydrogenase in Renal Disease". Lancet 2:327, 1959.

53. Sayed, W. A., Bassily, S., Mohran, Y., Wassef, S. A., Ghaffar, Y. A.: "Urinary LDH Activity in Urinary Bilharzias and its Complications". Brit. J. Cancer 23:73, 1969.

54. Schapiro, A., Paul, W., Gonick, H.: "Studies of Human Kidney and Urine Beta-Glucuronidase". Enzymol. Biol. Clin. 8:47, 1967.

55. Schoenenberger, G. A., Wacker, W. E. C.: "Peptide Inhibitors of Lactic Dehydrogenase". Biochemistry 5:1375, 1966.

56. Schoenfeld, M. R.: "Acid Phosphatase Activity in Ureteral Urine. A Test for Unilateral Renal Disease". J. Amer. Med. Assoc. 193:618, 1965.

57. Schoenfeld, M. R., Woll, F.: "Serum Acid Phosphatase as an Index of Kidney Disease". J. Urol. 90:373, 1963.

58. Scott, W. W., Huggins, C.: "The Acid Phosphatase Activity of Human Urine, an Index of Prostatic Secretion". Endocrinology 30:107, 1941.

Genitourinary Tract Disease References (Continued):

59. Smith, L. G., Schmidt, J.: "Evaluation of Three Screening Tests for Patients with Significant Bacteriuria". J. Amer. Med. Assoc. 181:159, 1962.

60. Smith, L. G., Thayer, R., Malta, E. M., Utz, J.: "Relationship of the Griess Nitrite Test to Bacterial Culture in the Diagnosis of Urinary Tract Infection". Ann. Intern. Med. 54:66, 1961.

61. Stumpf, D., Austin, J.: "Metachromatic Leukodystrophy (MLD). IX. Qualitative and Quantitative Differences in Urinary Arylsulfatase A in Different Forms of MLD". Arch. Neurol. 24:117, 1971.

62. Verrilli, R. A., Viek, N. F., Uhlman, R. C.: "Studies of Alkaline Phosphatase in Various Pathological Conditions of the Kidney. II. Renal Vascular and Ureteral Obstruction". J. Urol. 86:525, 1961.

63. Vorhaous, L. J., Kark, R. M.: "Serum Cholinesterase in Health and Disease". Amer. J. Med. 14:707, 1953.

64. Wacker, W. E. C., Dorfman, L. E.: "Urinary Lactic Dehydrogenase Activity. I. Screening Method for Detection of Cancer of Kidneys and Bladder". J. Amer. Med. Assoc. 181:972, 1962.

65. Wacker, W. E. C., Dorfman, L. E., Amador, E.: "Urinary Lactic Dehydrogenase Activity. IV. Screening Test for Detection of Renal Disease Dissociated and in Association with Arterial Hypertension". J. Amer. Med. Assoc. 188:671, 1964.

66. Williams, M. A., Tyler, H. M., Morton, M., Nemeth, A., Dempster, W. J.: "Some Biochemical Changes in the Transplanted Kidney". Brit. Med. J. 2:1215, 1962.

67. Wilson, T., Hadley, W. P.: "Urinary Lysozyme". J. Pediat. 36:39, 45, 199, 1950.

68. Zondag, H. A.: "Enzyme Activity in Dysgerminoma and Seminoma. A Study of Lactic Dehydrogenase Isoenzymes in Malignant Diseases". Rhode Island Med. J. 47:273, 1964.

CHAPTER 17

ENZYME ABNORMALITIES IN BONE DISEASE

When an abnormality of bone is discovered, the clinician's interest in serum enzymes usually turns to alkaline phosphatase. Alkaline phosphatase in bone is a reflection of osteoblastic activity, which is usually dependent on minerals such as calcium, vitamin D, growth hormone, protein balance, and mechanical stress. An elevated serum alkaline phosphatase derived from bone suggests increased osteoblastic activity.[23] If disease of the bile ducts and liver can be excluded, an elevation of serum alkaline phosphatase in an adult usually is an indication of bone disease.[13,20] However, other etiologies do exist (see Chapter 2) and should be borne in mind. Serum calcium and phosphorus should also be determined when serum alkaline phosphatase is assayed.

Alkaline phosphatase is also present in blood vessel endothelium, especially in lung and spleen, uterine endometrium, renal tubules,[21] bile ducts, liver cells, gastrointestinal mucosal epithelium, placenta, thyroid epithelium, parenchymal cells of the pancreas, myeloid cells of the peripheral blood, and bone marrow.[47] In the child most of the serum alkaline phosphatase is derived from the bones; while in the adult from the liver, gastrointestinal mucosa,[49] and bone. The alkaline phosphatase activity of human sera may be separated into different isoenzymes.[24,28,34,50]

Several causes exist for elevation of serum alkaline phosphatase. If the possibility of hepatic disease be excluded, the usual etiology is a bone disease; however, it must be kept in mind that a number of other important causes may contribute to elevation of serum alkaline phosphatase.[28] Serum alkaline phosphatase may be elevated because of a neoplastic lesion producing the Regan isoenzyme.[42] Gastrointestinal disease such as ulcerative colitis or peptic ulcer may be present with liberation of enzyme from the diseased mucosa (Fig. 17-1).[49,53] Infarction of kidney, lung, or spleen may cause an elevated serum alkaline phosphatase.[7,9,17,19,36,37] The patient may be a growing child and have a physiologic cause of bone growth for the alkaline phosphatase elevation or pregnancy may be responsible with the alkaline phosphatase originating from the placenta.[4]

Enzyme Abnormalities in Bone Disease (Continued):

Fig. 17-1 Gross photograph of total colectomy specimen removed for chronic active ulcerative colitis. The patient had a prominent elevation of serum alkaline phosphatase of 290 I.U. derived from the necrotic colonic mucosa.

The alkaline phosphatase in bone and cartilage is located at the advancing front of calcification.[43] When the bone matrix is deficient, there is a reduction of alkaline phosphatase in the bone.[22]

The alkaline phosphatase isoenzyme from bone is extremely heat labile.[35] It is not inhibited by 0.005 M L-phenylalanine. Its mobility in an electrophoretic field is much faster than that of gastrointestinal or placental alkaline phosphatase.[29] It is slower but may overlap the electrophoretic mobility of hepatic or biliary alkaline phosphatase.

Serum alkaline phosphatase is higher in growing children than in adults. A marked increase in the serum alkaline phosphatase occurs during the neonatal period, with an elevation to three times the adult level at the end of the first month of life. This level is

Enzyme Abnormalities in Bone Disease (Continued):

maintained until approximately the age of 1 year at which time, it becomes approximately twice the adult level. It then remains at this concentration until the onset of puberty at which time, with the spurt of bone growth, the alkaline phosphatase again may attain concentrations of three times the adult range.[31] Males tend to exhibit somewhat higher levels of alkaline phosphatase than females. This difference may be related to differences of skeletal mass but it may be related also to greater physical activity. Serum alkaline phosphatase is higher in females when the endometrium is in a proliferative phase. This change in the serum alkaline phosphatase may be related to estrogenic stimulation.[39]

The highest serum alkaline phosphatase activity associated with bone disease occurs in Paget's disease.[10,30] The enzyme elevation is directly proportional to the severity of the condition. The pathogenesis of this disease is not completely known; it begins with extensive osteolytic activity and increased vascularity. The bone destruction is rapidly followed by intense osteoblastic activity and an irregular mosaic of bone occurs and a dense osteoblastic lesion results. In patients with primary or metastatic neoplasms of bone, the nature of the lesion determines whether the serum alkaline phosphatase is increased. If the lesions are prominently osteoblastic, which may be found in osteogenic sarcoma and carcinoma of the prostate (Fig. 17-2) the serum alkaline phosphatase will be increased.[18,51,52] If the neoplasm is primarily osteolytic with no repair, such as might be present in patients with multiple myeloma, the serum alkaline phosphatase will remain normal[8] while the acid phosphatase rises. Since extensive bone destruction occurs in multiple myeloma, the serum alkaline phosphatase is normal because there is no osteoblastic activity. If the serum alkaline phosphatase activity is elevated, hepatic disease may be present or bone repair is beginning following successful therapy.

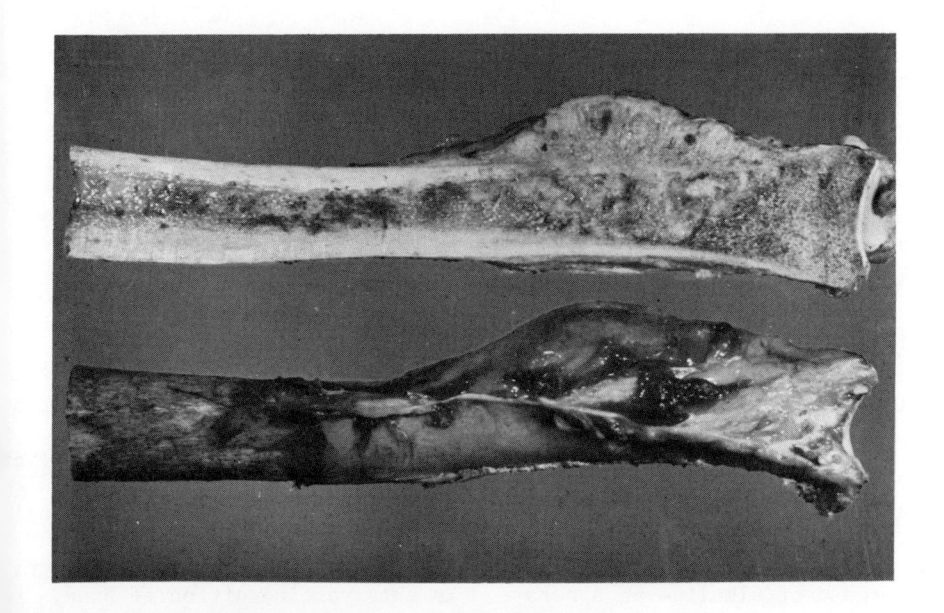

Enzyme Abnormalities in Bone Disease (Continued):

Fig. 17-2 Gross photograph of bone demonstrating the osteoblastic metastases from the prostate. The serum alkaline phosphatase was 410 I.U. and was heat labile.

Metastatic neoplasms arising from breast, kidney, lung, or thyroid may cause an elevation of alkaline phosphatase or rarely no change from normal. Lymphoma and the histiocytoses involving bone will cause a rise in serum alkaline phosphatase, and the increase in alkaline phosphatase is dependent on the secondary osteoblastic activity.

Increased serum alkaline phosphatase activity is the earliest and most sensitive signal of the presence of rickets and may reflect the severity of the disease.[2,40] A child with rickets shows an early rise in the enzyme level. Abnormalities in the serum calcium and phosphorus follow later. The level of elevation of serum alkaline phosphatase parallels the severity of the bone disease.[41]

In osteomalacia there will be an increased serum alkaline phosphatase in contrast to osteoporosis which is not accompanied by

Enzyme Abnormalities in Bone Disease (Continued):

osteoblastic activity, and in which the alkaline phosphatase remains normal.[1] Some causes of osteomalacia may be vitamin D deficiency, thyrotoxicosis, hyperparathyroidism, and malabsorption due to small intestinal disease, or after gastric resection.[5] When treatment is successful, the serum alkaline phosphatase may return to normal quickly. With a bone fracture, the serum alkaline phosphatase is slightly increased.[3,27,32] A slight elevation may be seen at times in postmenopausal senile osteoporosis because some bone repair may occur in this condition.[48]

Congenital hyperphosphatasia causes radiolucent bones with thin cortices. The bones are brittle and fractures are frequent. Bone turnover is prominent and serum alkaline and acid phosphatase are elevated; a synonym for the disease is osteitis deformans of children.

Fracture of a long bone with osteoblastic repair causes elevation of serum alkaline phosphatase as does surgical repair and surgical manipulation of bone; for example, insertion of an intramedullary nai

In osteopetrosis if there is bone repair, there will be a slight increase of alkaline phosphatase. In this rare condition of bone, there usually is no significant abnormality in serum alkaline phospha tase,[11] calcium, or phosphorus; while acid phosphatase will be elevated.[45]

Low serum alkaline phosphatase has been observed in patients ill with chronic anemia, malabsorption with hypomagnesemia, vitamin C deficiency, hypothyroidism, and hypophosphatasia (Fig. 17-3).[38,46] Hypophosphatasia is an autosomal recessive inherited disease.[33] Various clinical types are recognized:

1. Infantile, up to 6 months of age; symptoms of hypercalcemia, persistent vomiting, and soft bones diffusely.

2. Childhood, up to 24 months of age; signs of bow legs, loss of teeth, and milder bone problems.

3. Adult, signs of pathologic fractures, history of vitamin D-resistant rickets, and bone pain.[44]

Enzyme Abnormalities in Bone Disease (Continued):

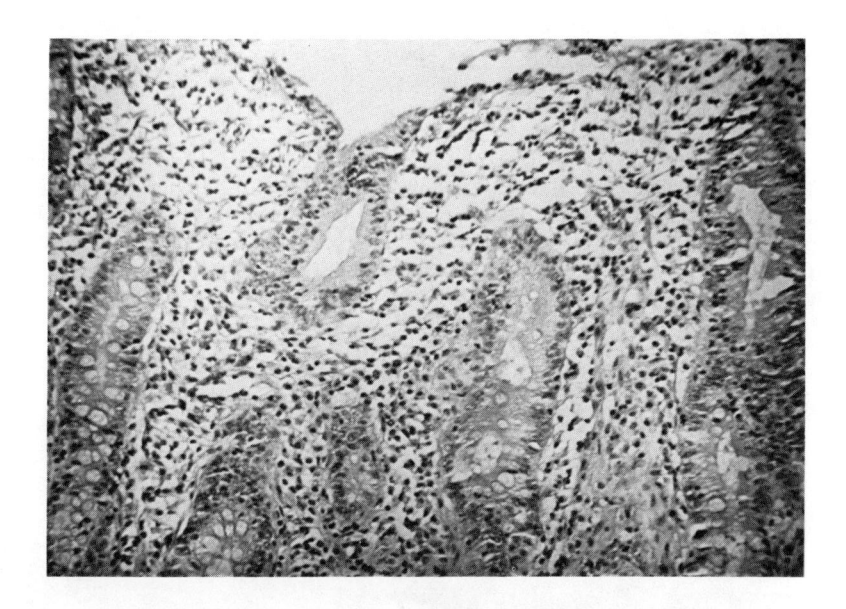

Fig. 17-3 Photomicrograph of section of small intestine from patient with sprue. The serum alkaline phosphatase was increased to 145 I.U.

In hypophosphatasia, there is a marked reduction or absence of the alkaline phosphatase in myeloid leukocytes, serum, and tissues including bone. Clinically, the condition resembles rickets and varies in severity. Patients with a mild condition may survive into adult life and may have bone deformities with increased fragility resulting in frequent fractures. Hypercalcemia with calcinosis of the soft tissues may result. The condition may be extremely severe at birth, and death may result shortly after birth or in early childhood. The diagnostic blood features are low leukocyte alkaline phosphatase and marked depression in serum alkaline phosphatase,[15] which quickly differentiates this lesion from rickets, where the serum alkaline phosphatase is elevated.[25] A characteristic chemical abnormality is the excretion of phosphoethanolamine in the urine.[26] This substance is a natural substrate for alkaline phosphatase; when there is a deficiency of the enzyme, the substrate accumulates and is excreted in the urine. If the disease is present in utero a stillbirth may result. Some spontaneous remissions have occurred, but

Enzyme Abnormalities in Bone Disease (Continued):

the serum alkaline phosphatase is still depressed. Rarely, treatment of the disease with cortisone has caused an increase in the alkaline phosphatase and a decrease in elevated serum calcium. [14,16,44]

There are other minor causes for a lower serum alkaline phosphatase. One is magnesium deficiency, which is associated with malabsorption or with uncontrolled diabetes mellitus or other conditions in which parenteral fluids have been administered without the addition of magnesium. Magnesium is one of the ions necessary for the functioning of alkaline phosphatase. In the presence of a magnesium deficiency the serum alkaline phosphatase may be lowered. A lower serum alkaline phosphatase is also present in malnourished and cachectic individuals and in anemia usually pernicious anemia. It has been postulated that an inadequate amount of vitamin B_{12} impairs osteoblastic activity and this results in the lower alkaline phosphatase (Fig. 17-4)

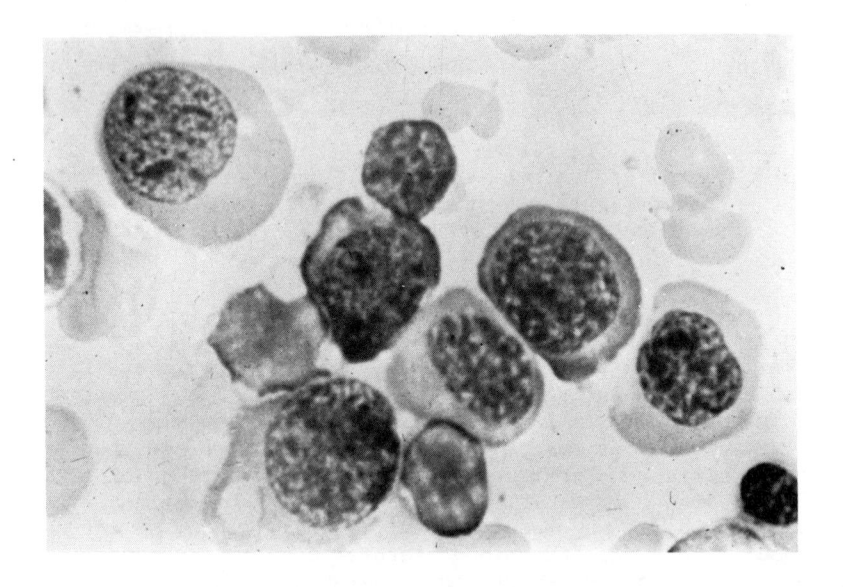

Fig. 17-4 Photomicrograph of smear of bone marrow from a patient with megaloblastic pernicious anemia. The serum alkaline phosphatase was decreased to 15 I.U.

Enzyme Abnormalities in Bone Disease (Continued):

Some investigators have noted a lower serum alkaline phosphatase in hypothyroid patients. This may result from stunted bone growth. The alkaline phosphatase activity of the thyroid gland appears to be related to the functional activity of the gland. Greater alkaline phosphatase activity is present in hyperplastic glands and a lower amount in atrophic ones. In hypothyroidism there is an underlying deficiency in osteoblastic activity resulting in a deficiency of serum alkaline phosphatase level.[5]

Anticoagulants such as oxalate or fluoride inhibit alkaline phosphatase activity. Thus, if these anticoagulants are used and plasma is sent to the clinical laboratory for analysis, there will be a low alkaline phosphatase activity because of the inhibition produced by the anticoagulant.[6]

REFERENCES

1. Bartter F. C.: "Osteoporosis". Amer. J. Med. 22:797, 1957.

2. Bodansky, D., Jaffe, H. L.: "Phosphatase Studies. V. Serum Phosphatase as a Criterion of the Severity and Rate of Healing of Rickets". Amer. J. Dis. Child. 48:1268, 1934.

3. Botterell, E. H., King, E. J.: "Phosphatase in Fractures". Lancet 1:1267, 1935.

4. Boyer, S. H.: "Alkaline Phosphatase in Human Sera and Placentae". Science 134:1002, 1961.

5. Cassar, J., Joseph, S.: "Alkaline Phosphatase Levels in Thyroid Disease". Clin. Chim. Acta. 23:33, 1969.

6. Christian, D. G.: "Drug Interference with Laboratory Blood Chemistry Determinations". Amer. J. Clin. Path. 54:118, 1970.

7. Cordonnier, J. J., Miller, J. A.: "The Relationship between Alkaline Phosphatase in the Kidney and Urinary Calculi". J. Urol. 66:12, 1951.

8. Dillman, C. E., Silverstein, M. N.: "Alkaline Phosphatase in Multiple Myeloma". Amer. J. Med. Sci. 249:445, 1965.

Bone Disease References (Continued):

9. Duggan, M. L.: "Acute Renal Infarction". J. Urol. 90:669, 1963.

10. Eisenberg, E., Gordan, G. S.: "Skeletal Dynamics in Man measured by Nonradioactive Strontium". J. Clin. Invest. 40:1809, 1961.

11. Ensign, D. C.: "Serum Phosphatase in Osteopetrosis". J. Lab. Clin. Med. 32:1541, 1947.

12. Eyring, E. J., Eisenberg, E.: "Congenital Hyperphosphatasia A Clinical, Pathological, and Biochemical Study of Two Cases J. Bone Joint Surg. 50 A:1099, 1968.

13. Fishman, W. H., Ghosh, N. K.: "Isoenzymes of Human Alkaline Phosphatase". ADVANCES IN CLINICAL CHEMISTRY Academic Press Vol. 10, pg. 255, 1967.

14. Fleisch, H., Russell, R. G. G., Straumann, F.: "Effect of Pyrophosphate on Hydroxyapatite and its Implications in Calcium Homeostasis". Nature 212:901, 1966.

15. Fraser, D.: "Hypophosphatasia". Amer. J. Med. 22:730, 195?

16. Fraser, D., Yendt, E. R.: "Metabolic Abnormalities in Hypophosphatasia". Amer. J. Dis. Child. 90:552, 1955.

17. Gault, M. H., Steiner, G.: "Serum and Urinary Enzyme Activity after Renal Infarction". Canad. Med. Assoc. J. 93:1101, 1965.

18. Gutman, A. B.: "Serum Alkaline Phosphatase Activity in Diseases of Skeletal Hepatobiliary Systems: Consideration of Current Status". Amer. J. Med. 27:875, 1959.

19. Highman, B., Thompson, E. C., Roche, J., Atland, P. D.: "Serum Alkaline Phosphatase in Dogs with Experimental Splenic and Renal Infarcts and with Endocarditis". Proc. Soc. Exp. Biol. Med. 95:109, 1957.

20. Hodson, A. W., Latner, A. L., Raine, L.: "Isoenzymes of Alkaline Phosphatase". Clin. Chim. Acta. 7:255, 1962.

21. Hoxie, J. H., Coggin, C. B.: "Renal Infarction". Arch. Intern. Med. 65:587, 1940.

Bone Disease References (Continued):

22. Jeffree, G. M.: "Phosphatase Activity in the Limb Bones of Monkeys (Ilagothrix humboldti) with Hyperparathyroidism". J. Clin. Path. 15:99, 1962.

23. Keroff, J. F.: "A Rapid Serum Screening Test for Increased Osteoblastic Activity". Clin. Chim. Acta. 22:231, 1968.

24. Kowlessar, O. D., Haeffner, L. J., Riley, E. M.: "Localization of Serum Leucine Aminopeptidase, 5-Nucleotidase and Non-specific Alkaline Phosphatase by Starch-Gel Electrophoresis. Clinical and Biochemical Significance in Disease States". Ann. N. Y. Acad. Sci. 94:836, 1961.

25. McCance, R. A., Fairweather, D. V. I., Barrett, A. M., Morrison, A. B.: "Genetic, Clinical, Biochemical and Pathological Features of Hypophosphatasia". Quart. J. Med. 25:523, 1956.

26. McCance, R. A., Morrison, A. B., Dent, C. E.: "The Excretion of Phosphoethanolamine and Hypophosphatasia". Lancet 1:131, 1955.

27. Mitchell, C. L.: "Serum Phosphatase in Fracture Repair". Ann. Surg. 104:304, 1936.

28. Moss, D. W., Campbell, D. M., Anagnostou-Kakaras, E., King, E. J.: "Characterization of Tissue Alkaline Phosphatases and their Partial Purification by Starch-Gel Electrophoresis". Biochem. J. 81:441, 1961.

29. Moss, D. W., King, E. J.: "Properties of Alkaline Phosphatase Fractions separated by Starch-Gel Electrophoresis". Biochem. J. 84:192, 1962.

30. Nagant de Deuxchaisnes, C., Krane, S. M.: "Paget's Disease of Bone: Clinical and Metabolic Observations". Medicine 43:233, 1964.

31. O'Brien, D., Ibbott, F. A., Rodgerson, D. O.: LABORATORY MANUAL OF PEDIATRIC MICROBIOCHEMICAL TECHNIQUES Harper & Row, 4th Edition, pg. 1, 1968.

32. Paterson, C. R., Losowsky, M. S.: "The Bones in Chronic Liver Disease". Scand. J. Gastroenterology 2:293, 1967.

Bone Disease References (Continued):

33. Pimstone, B., Eisenberg, E., Silverman, S.: "Hypophosphatasia: Genetic and Dental Studies". Ann. Intern. Med. 65:722, 1966.

34. Posen, S.: "Alkaline Phosphatase". Ann. Intern. Med. 67:183, 1967.

35. Posen, S., Neale, F. C., Clubb, J. S.: "Heat Inactivation in the Study of Human Alkaline Phosphatases". Ann. Intern. Med. 62:1234, 1965.

36. Pulvertaft, C. N., Luffman, J. E., Robson, E. B., Harris, H., Langman, M. J. S.: "Isoenzymes of Alkaline Phosphatase in Patients operated upon for Peptic Ulcer". Lancet 1:237, 1967.

37. Rosato, F. E., Lazitan, L., Miller, L. D., Tsou, K. C.: "Changes in Intestinal Alkaline Phosphatase in Bowel Ischemia". Amer. J. Surg. 121:289, 1971.

38. Searcy, R.: DIAGNOSTIC BIOCHEMISTRY McGraw-Hill, pg. 41, 1969.

39. Searcy, R.: DIAGNOSTIC BIOCHEMISTRY McGraw-Hill, pg. 44, 1969.

40. Smith, J.: "Plasmaphosphatase in Rickets and Other Disorders of Growth". Arch. Dis. Child. 8:215, 1933.

41. Sterns, G., Warweg, E.: "Studies of Phosphorus of Blood. III. The Phosphorus Partition in Whole Blood and in Serum During Healing of Late Rickets". Amer. J. Dis. Child. 49:79, 1935.

42. Stolbach, L. L., Krant, M. J., Fishman, W. H.: "Ectopic Production of Alkaline Phosphatase Isoenzyme in Patients with Cancer". New Eng. J. Med. 281:757, 1969.

43. Teaford, M. E., White, A. A.: "Alkaline Phosphatase and Osteogenesis in Vitro". Proc. Soc. Exp. Biol. Med. 117:541 1964.

44. Teree, T. M., Klein, L.: "Hypophosphatasia: Clinical and Metabolic Studies". J. Pediat. 72:41, 1968.

Bone Disease References (Continued):

45. Turano, A. F., Fagan, K. A., Cordo, P. A.: "Variations in Clinical Manifestations of Osteopetrosis: Report of Two Cases". J. Pediat. 44:688, 1947.

46. Van Dommelan, C. K. V., Klaassen, C. H. L.: "Cyanocobalamin-dependent Depression of the Serum Alkaline Phosphatase Level in Patients with Pernicious Anemia". New Eng. J. Med. 271:541, 1964.

47. Verrilli, R. A., Viek, N. F., Uhlman, R. C.: "Studies of Alkaline Phosphatase in Various Pathologic Conditions of the Kidney. II. Renal Vascular and Ureteral Obstruction". J. Urol. 86:525, 1961.

48. Wilkins, W. E., Regan, E. M.: "Course of Phosphatase Activity in Healing of Fractured Bones". Proc. Soc. Exp. Biol. Med. 32:1373, 1935.

49. Wilkinson, J.: "Clinical Significance of Enzyme Activity Measurements". Clin. Chem. 16:882, 1970.

50. Wilmer, H. A.: "The Disappearance of Phosphatase from the Hydronephrotic Kidney". J. Exp. Med. Biol. 78:225, 1943.

51. Woodard, H. Q.: "Changes in Blood Chemistry Associated with Carcinoma Metastatic to Bone". Cancer 6:1219, 1953.

52. Woodard, H. Q.: "The Clinical Significance of Serum Acid Phosphatase". Amer. J. Med. 27:902, 1959.

53. Yong, J. M.: "Cause of Raised Serum Alkaline Phosphatase after Partial Gastrectomy and in other Malabsorption States". Lancet 1:1132, 1966.

CHAPTER 18

ENZYME ABNORMALITIES IN SKELETAL MUSCLE DISEASE

Skeletal muscle disease may be associated with alterations in serum enzymes. When there is a suspicion of this type of disease, five enzymes are usually determined; CPK, aldolase, GOT, GPT, and LDH. The serum enzymes are frequently elevated in skeletal muscle disorders associated with dystrophy, necrosis or inflammation.[1] In skeletal muscle disease secondary to atrophy, serum enzymes are usually normal.[2,6,11]

CPK and aldolase are the enzymes that are most often determined in suspected skeletal muscle disease. These two enzymes are usually elevated in patients with muscular dystrophy of the Duchenne type.[10] The serum elevation in related to enzymes leaking from the diseased skeletal muscle cells. Some observers believe that a membrane defect exists in the skeletal muscle cell which permits the enzymes to exit from the cell at a rapid rate.

A correlation exists between the serum enzyme activity of CPK and aldolase and the amount of diseased muscle. Enzyme activity may be elevated early in the course of the disease. There is good evidence that aldolase and CPK are elevated before the clinical diagnosis of muscle disease can be made, especially in family members of patients who have muscular dystrophy.[19,22] Aldolase levels tend to decrease as the disease progresses, especially in young active patients. In contrast, CPK remains elevated as the disease progresses and may remain increased in the late stages of the disease. Elevation of CPK and aldolase is highest in the Duchenne type of dystrophy; it is more moderate in the less rapidly progressive and rare forms of the dystrophic diseases, such as the limb-girdle, facioscapulo-humeral, and myotonic types.[7] GOT, GPT, and LDH are usually elevated in muscular dystrophy.

Serum enzyme activity in patients with skeletal muscle disorders secondary to neurogenic muscular atrophy is normal. The following conditions exhibit normal serum enzymes: Patients with myasthenia gravis, Charcot-Marie-Tooth, Werdnig-Hoffmann disease, demyelinating disorders of the central nervous system such as multiple sclerosis, Parkinson's disease, amyotrophic lateral sclerosis, periodic paralysis congenital amyotonia, and hyperthyroid muscle disease (Table 18-1 and Table 18-2).[18]

Enzyme Abnormalities in Skeletal Muscle Disease (Continued):

TABLE 18-1

CONDITIONS IN WHICH ENZYME ACTIVITY IS NORMAL
1. Muscle atrophy secondary to neurogenic etiology
2. Myasthenia gravis
3. Periodic paralysis
4. Thyrotoxicosis
5. Congenital myotonia
6. Charcot-Marie-Tooth disease
7. Werdnig-Hoffmann disease
8. Multiple sclerosis
9. Parkinson's disease
10. Amyotrophic lateral sclerosis

TABLE 18-2

CONDITIONS IN WHICH MUSCLE ENZYME ACTIVITY IS INCREASED
1. Duchenne's muscular dystrophy
2. Traumatic muscular necrosis
3. Polymyositis
4. Muscle damage associated with surgery
5. Alcoholic myopathy
6. Hypothyroid myopathy
7. Limb-girdle muscular dystrophy
8. Dystrophia myotonica
9. Polymyositis associated with collagen disorders or with other inflammatory etiologies
10. Facioscapulo-humeral muscular dystrophy
11. McArdle's syndrome
12. Injections of drugs into skeletal muscle
13. Severe exercise and strenuous exertion
14. Pregnant myometrium
15. Diffuse muscular necrosis with myoglobinuria

The term myopathy signifies a non-neurogenic muscle lesion and includes polymyositis, endocrine myopathy, and muscular dystrophy. It is essential to differentiate the condition of primary myopathy from muscular atrophy secondary to a neurologic lesion. The term polymyositis refers to a myopathy in which there is a nonspecific inflammatory and degenerative muscle lesion which may be associated with the collagen diseases.[21] Serum CPK, aldolase, GOT, GPT, and LDH are usually elevated in polymyositis (Fig. 18-1). A correlation

Enzyme Abnormalities in Skeletal Muscle Disease (Continued):

exists between the level of enzyme elevation and the severity of the disease. In polymyositis with rapid progression, high enzyme activity occurs; with slow progression there will be a lesser elevation of the serum enzymes. If the polymyositis is effectively treated, the elevated serum enzymes decline.[26]

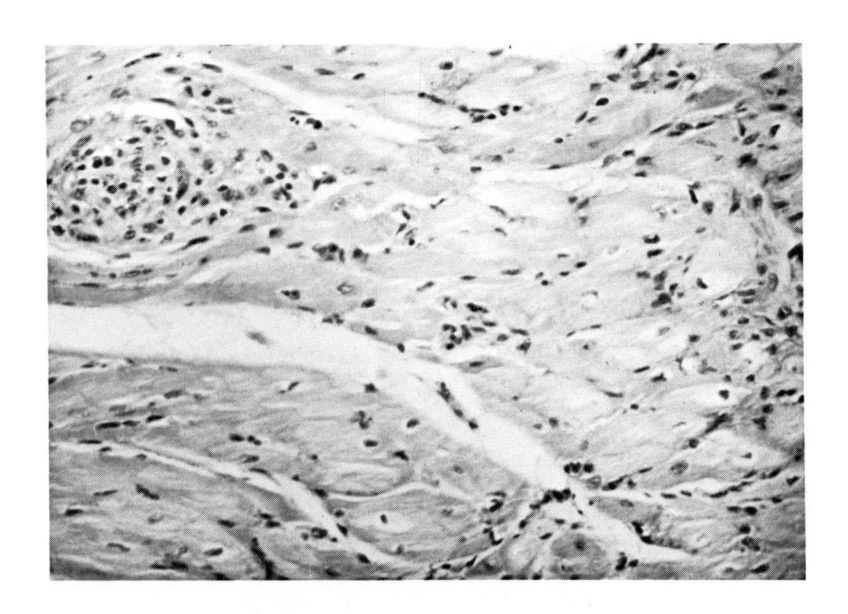

Fig. 18-1 Photomicrograph of section of cardiac muscle from a patient with dermatomyositis. The serum CPK was 340 I.U.

Determination of the isoenzymes of LDH have some value in patients with skeletal muscle disease. Patients with muscular dystrophy usually have a serum elevation of LDH_1 and LDH_2. LDH_5 usually predominates in skeletal muscle of normal individual, but a reversion to LDH_1 occurs in muscular dystrophy and LDH_5 is relatively reduced.[3]

Patients with skeletal muscle necrosis secondary to alcoholism may develop extreme elevations of CPK, LDH, GOT, and GPT.[5] These patients develop a sensitivity to alcohol and have muscle pain tenderness and edema of the affected skeletal muscle.[9] Diffuse necrosis of

Enzyme Abnormalities in Skeletal Muscle Disease (Continued):

skeletal muscle with myoglobinuria and renal failure occur. The clinical symptom picture resembles that of McArdle syndrome, in which there is a phosphorylase deficiency. When alcohol consumption is stopped, the muscle necrosis subsides.[27,31]

A large amount of interest exists to detect the carrier state of muscular dystrophy. CPK is the enzyme that is determined in order to detect carriers; aldolase is of less value,[4,20,23] and GOT and LDH are of no value. Duchenne's muscular dystrophy is a sex-linked recessive genetic disease. Half the males of a carrier mother are at risk in developing the disease, and half the females in affected families may be carriers. CPK and aldolase may be increased in these female carriers. Elevations of CPK have been detected as early as age 6 weeks in infants who may develop the disease (Fig. 18-2). CPK elevation occurs in all females who carry the gene. However, CPK determination is of no value to detect carriers of other rare forms of muscular dystrophy, such as the facioscapulo-humeral or limb-girdle types. It has rarely been possible to detect muscular dystrophy of the Duchenne type during gestation by the finding of large amounts of CPK in aspirated amniotic fluid by amnioncentesis.

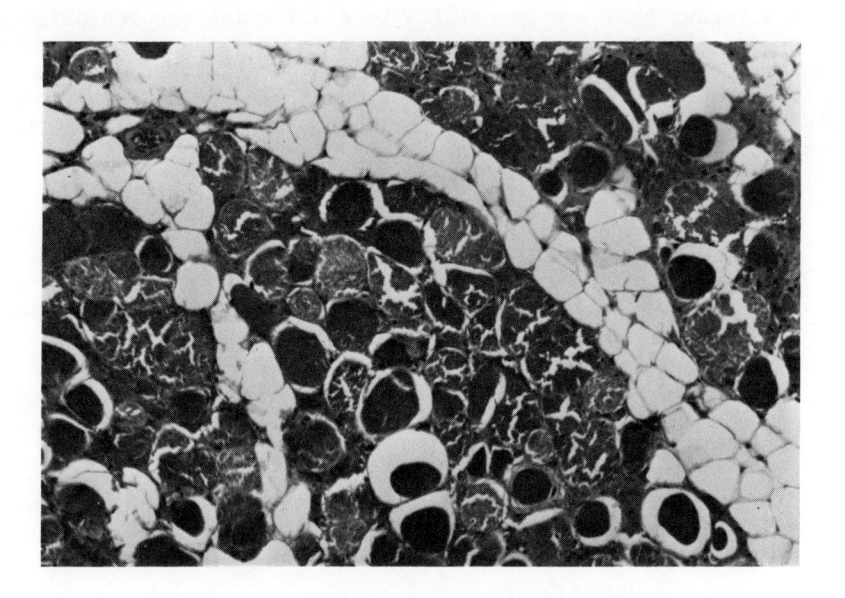

Fig. 18-2 Photomicrograph of section of skeletal muscle of male infant affected with muscular dystrophy. The serum CPK was 490 I.U.

Enzyme Abnormalities in Skeletal Muscle Disease (Continued):

Increased CPK activity may also occur in patients who sustain muscle damage from trauma, surgery involving skeletal muscle, or injections into skeletal muscle. Severe exercise[8], strenuous exertion, or intramuscular injections should be avoided before samples for enzyme assay are obtained in patients suspected of having muscle disease.[17,29]   This is especially true in obtaining CPK samples in suspected muscular dystrophy patients or asymptomatic carriers.

In addition to CPK and aldolase elevation in muscular dystrophy, marked elevation of GPT and GOT are at times present in patients with muscular dystrophy. Out of 500 patients with progressive muscular dystrophy 78 percent had GPT and GOT elevations.[12,28,30]

Extraction studies of the pregnant myometrium have shown that there is a 15 fold rise in CPK activity in the pregnant myometrium in contrast to the nonpregnant myometrium.[13]

Thus, the determination of serum CPK, LDH, GOT and GPT is useful if skeletal muscle disease is suspected. The determination of these enzymes can be useful to confirm the diagnosis often before the clinical signs and symptoms of muscle disease are present such as in muscular dystrophy; they can be helpful to distinguish between primary myopathy and skeletal muscle atrophy. They can be helpful to differentiate atrophy from polymyositis; they can be helpful to detect a carrier of muscular dystrophy. The serum enzymes of skeletal muscle origin may also be used to follow the course of a primary myopathy or dystrophy.[14,25]

REFERENCES

1. Aronson, S. M.: "Enzyme Determinations in Neurologic and Neuromuscular Diseases of Infancy and Childhood". Pediat. Clin. N. Amer. 7:527, 1960.

2. Aronson, S. M., Volk, B. W.: "Studies on Serum Aldolase Activity in Neuromuscular Disorders. I. Clinical Applications". Amer. J. Med. 22:414, 1957.

3. Blanchaer, M. D., Van Wijhe, M.: "Isoenzymes of Lactic Dehydrogenase in Skeletal Muscle". Amer. J. Physiol. 202:827, 1962.

4. Boyer, S. H., Fainer, D. C.: "Genetics and Diseases of Muscle". Amer. J. Med. 35:622, 1963.

Skeletal Muscle Disease References (Continued):

5. Ekbom, K., Hed., R., Kirstein, L., Astrom, K. C.: "Muscular Affections in Chronic Alcoholism". Arch. Neurol. 10:449, 1964.

6. Evans, J. H., Baker, R. W.: "Serum Aldolase and the Diagnosis of Myopathy". Brain 80:557, 1957.

7. Fowler, W. M., Jr., Pearson, C. M.: "Diagnostic and Prognostic Significance of Serum Enzymes. II. Neurologic Diseases other than Muscular Dystrophy". Arch. Phys. Med. 45:125, 1964.

8. Griffiths, P. D.: "Serum Levels of ATP. Creatine Phosphotransferase (Creatine Kinase). The Normal Range and Effect of Muscular Activity". Clin. Chim. Acta. 13:413, 1966.

9. Hed, R., Lundmark, C., Fahlgran, H., Orell, S.: "Acute Muscular Syndrome in Chronic Alcoholism". Acta. Med. Scand. 171:585, 1962.

10. Hess, J. W., MacDonald, R. P., Frederick, R. J., Jones, R. N., Neely, J., Gross, D.: "Serum Creatine Phosphokinase (CPK) Activity in Disorders of Heart and Skeletal Muscle". Ann. Intern. Med. 61:1015, 1964.

11. Hughes, B. P.: "A Method for the Estimation of Serum Creatine Kinase and its Use in Comparing Creatine Kinase and Aldolase Activity in Normal and Pathological Sera". Clin. Chim. Acta. 7:597, 1962.

12. Hughes, B. P.: "Serum Enzymes in Carriers of Muscular Dystrophy". Brit. Med. J. 2:963, 1962.

13. Konttinen, A., Pyorala, T.: "Serum Enzyme Activity in Late Pregnancy, at Delivery, and during Puerperium". Scand. J. Clin. Lab. Invest. 15:429, 1963.

14. Lauryssens, M. G., Lauryssens, M. J., Zondag, H. A.: "Electrophoretic Distribution Pattern of Lactic Dehydrogenase in Mouse and Human Muscular Dystrophy". Clin. Chem. Acta. 9:276, 1964.

15. Markert, C. L., Moller, F.: "Multiple Forms of Enzymes: Tissue, Ontogenetic and Species-specific Patterns". Proc. Nat. Acad. Sci. 45:753, 1959.

Skeletal Muscle Disease References (Continued):

16. Murphy, E. G., Cherniak, M. M.: "Glutamic Oxalacetic Transaminase Activity in the Serum in Muscular Dystrophy and other Neuromuscular Disroders in Childhood". Pediatrics 22:1110, 1958.

17. Nuttall, F. Q., Jones, B.: "Creatine Kinase and Glutamic Oxalacetic Transaminase Activity in Serum: Kinetics of Change with Exercise and Effect of Physical Conditioning". J. Lab. Clin. Med. 71:847, 1968.

18. Okinaka, S., Kumagai, H., Ebashi, S., Sugita, H., Momoi, H., Toyokura, Y., Fujie, H.: "Serum Creatine Phosphokinase Activity in Progressive Muscular Dystrophy and Neuromuscular Diseases". Arch. Neurol. 4:520, 1961.

19. Pearce, J. M. S., Pennington, R. J., Walton, J. N.: "Serum Enzyme Studies in Muscle Disease. II. Serum Creatine Kinase Activity in Muscular Dystrophy and in other Myopathic and Neurologic Disorders". J. Neurol. Neurosurg. Psychiat. 27:96, 1964.

20. Pearce, J. M. S., Pennington, R. J. T., Walton, J. N.: "Serum Enzyme Studies in Muscle Disease. III. Serum Creatine Kinase Activity in Relatives of Patients with the Duchenne Type of Muscular Dystrophy". J. Neurol. Neurosurg. Psychiat. 27:181, 1964.

21. Pearson, C. M.: "Polymyositis and Related Disorders". In J. N. Walton (Eds.) DISORDERS OF VOLUNTARY MUSCLE Little, Brown, & Co. pg. 305, 1964.

22. Pearson, C. M., Chowdury, S. R., Fowler, W.: "Biochemical Detection and Histological Study of Muscular Dystrophy in the Preclinical State". J. Clin. Invest. 40:1070, 1961.

23. Pearson, C. M., Chowdury, S. R., Fowler, W., Jr., Jones, M. H., Griffith, W. H.: "Studies of Enzymes in Serum in Muscular Dystrophy. II. Diagnostic and Prognostic Significance in Relatives of Dystrophic Persons". Pediatrics 28:962, 1961.

24. Pearson, C. M., Kar, N. C.: "Serum Enzymes". In A. S. Cohen (Ed.) LABORATORY DIAGNOSTIC PROCEDURES IN THE RHEUMATIC DISEASES Little, Brown & Co., pg. 252, 1967.

Skeletal Muscle Disease References (Continued):

25. Pearson, C. M., Kar, N. C., Peter, J. B., Munsat, T. L.: "Muscle Lactate Dehydrogenase Patterns in Two Types of X-Linked Muscular Dystrophy". Amer. J. Med. 39:91, 1965.

26. Pennington, R. J.: "Biochemical Aspects of Muscle Disease". In J. N. Walton (Ed.) DISORDERS OF VOLUNTARY MUSCLE Little, Brown & Co., pg. 255, 1964.

27. Perkoff, G. T., Hardy, P., Velez-Garcia, E.: "Reversible Acute Muscular Syndrome in Chronic Alcoholism". New Eng. J. Med. 274:1277, 1966.

28. Richterich, R., Rosen, S., Aebi, U., Rossi, E.: "Progressive Muscular Dystrophy. The Identification of the Carrier State in the Duchenne Type by Serum Creatine Kinase Determination". Amer. J. Hum. Genet. 15:133, 1963.

29. Swaimann, K. F., Awad, E. A.: "Creatine Phosphokinase and other Serum Enzyme Activity after Controlled Exercise". Neurology 14:977, 1964.

30. Walton, J. N.: "Muscular Dystrophy: Some Recent Advances in Knowledge". Brit. Med. J. 1:1271, 1964.

31. Wolf, P. L.: "Alcoholic Skeletal Muscular Necrosis Syndrome". In BIOCHEMICAL AND CLINICAL ASPECTS OF ALCOHOL METABOLISM C. Thomas Co., pg. 283, 1969.

CHAPTER 19

ENZYME PATTERNS AND ABNORMALITIES IN THE PEDIATRIC AGE GROUP

Concentrations of various serum enzymes differ in the pediatric age group and in adults. This Chapter will present a discussion of serum enzyme abnormalities in children due to disease and physiologic differences from adults.

Alkaline Phosphatase

Alkaline phosphatase levels are different in infants and children from those in adults.[28] During the first week of life the levels may be similar to the adult;[6] the next week is characterized by a sharp elevation in alkaline phosphatase activity. By the end of the first month, the level is three times that of the adult.[7] Thus a 1 month-old child may have an alkaline phosphatase of 240 I.U. in contrast to 80 I.U. in an adult. By the age of 1 year the alkaline phosphatase decreases somewhat to twice the adult level. It remains at this concentration until the onset of puberty at which time, with the spurt of bone growth, the alkaline phosphatase again may attain concentrations of three times the adult range. Boys tend to have somewhat higher levels than girls; this difference may be related to differences in skeletal mass[38] or differences in physical activity. The elevated alkaline phosphatase in the serum of infants and children is related to the greater osteoblastic activity of bone. This alkaline phosphatase is heat labile and not inhibited by phenylalanine

Acid Phosphatase

Acid phosphatase activity is higher throughout childhood than in adults.[21] The highest acid phosphatase is found immediately after birth; this may be three times the adult activity. By the age of 2 weeks, the activity is twice the adult level and remains so until the thirteenth to fourteenth year of life. Adult levels are reached by age 16 to 18 years. Increased acid phosphatase activity in the newborn may also be related to physiologic hemolysis. Several other causes exist for elevated serum acid phosphatase in children. Children may have an elevated acid phosphatase due to:

Pediatric Enzyme Patterns and Abnormalities (Continued):

1. Hemolysis
2. Hepatic disease, e.g. hepatitis
3. Bone disease, e.g. osteogenic sarcoma, osteogenesis imperfecta, or juvenile rheumatoid arthritis (Fig. 19-1)
4. Kidney disease, e.g. glomerulonephritis
5. Reticulo-endothelial diseases, e.g. Gauchers disease, Nieman-Pick disease, lymphoma, and Hodgkin's disease
6. Carcinomas, metastatic to bone or liver
7. Thrombocytopenia

Fig. 19-1 Gross photograph of osteogenic sarcoma of bone in a child. The serum acid phosphatase was 5.6 I.U.

Patients with Gaucher's disease have an elevated acid phosphatase which is not inhibited by L-tartrate.[44] Patients with thrombocytopenic purpura may have an elevated serum acid phosphatase, especially if megakaryocytes are present in the bone marrow; with the absence of bone marrow megakaryocytes the acid phosphatase will not be elevated.[23,31]

Pediatric Enzyme Patterns and Abnormalities (Continued):

Aldolase

Serum aldolase activity in infants is twice that found in umbilical cord blood and approximately five times the adult level. The activity during childhood is approximately twice that of the adult. Adult levels are reached at the time of puberty.[10] The increased aldolase in infants and children may be related to hyperplasia of the adrenal gland since there is increased activity of this gland during infancy.[3,14]

Determination of serum aldolase has been found useful in the pediatric age group in differentiating primary myopathy such as Duchenne muscular dystrophy in which the enzyme level is elevated from neurogenic muscle atrophy, in which the aldolase level is normal. Aldolase elevation may also be present in infants and children with polymyositis, hepatitis, and hemolytic anemia.[1,9,12,37,47,48]

Ceruloplasmin

Ceruloplasmin is determined to ascertain if Wilson's disease is present. Ceruloplasmin reaches adult activity levels at 1 year of age; before the age of 1 year the ceruloplasmin activity is extremely low. The normal serum ceruloplasmin is 280 to 570 units in the adult. A decrease in serum enzyme is present in Wilson's disease (Fig. 19-2) Copper is elevated in urine and tissues such as the lentiform nucleus of the brain and the liver leading to cirrhosis and brain damage in Wilson's disease. There is an increased tissue and urine copper but a decreased copper and ceruloplasmin in the blood.[4,5,45] Children with nephrosis may excrete large amounts of α_2-globulin and thus have a lower serum ceruloplasmin (Fig. 19-3).[24] Other causes for elevated ceruloplasmin are usage of estrogens, pregnancy, acute inflammatory diseases, lymphomas such as active Hodgkin's disease, hyperthyroidism and tissue necrosis.[15,52] Thus, determination of of ceruloplasmin is valuable when the clinician wishes to ascertain if a disease is active. It may be used as an acute phase protein and give similar results as the sedimentation rate or C-reactive protein.

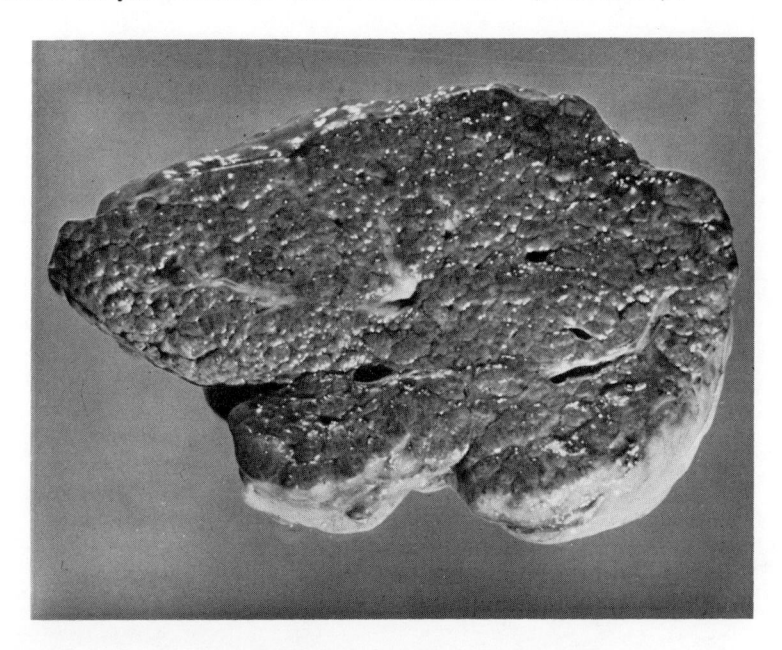

Pediatric Enzyme Patterns and Abnormalities (Continued):

Fig. 19-2 Gross photograph of cirrhotic liver from a patient with Wilson's disease. The serum ceruloplasmin was 180 units.

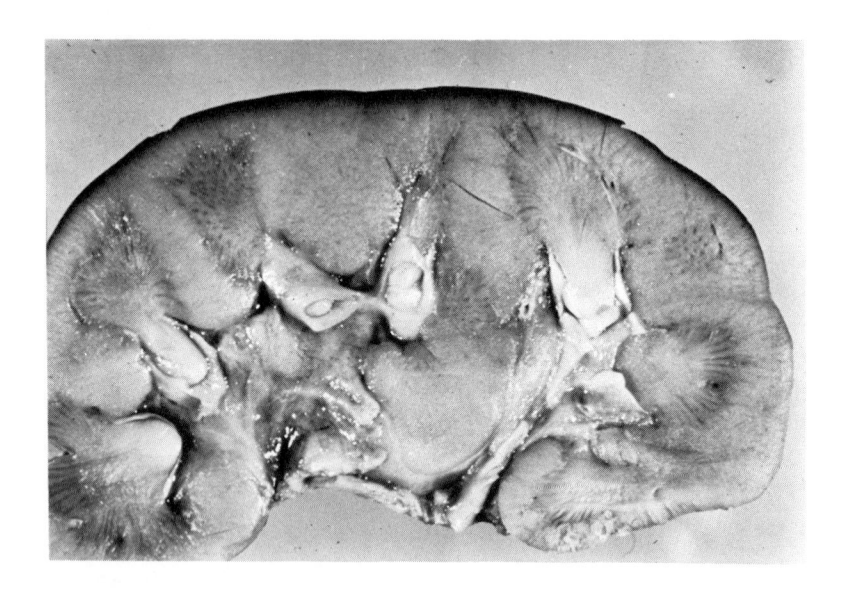

Pediatric Enzyme Patterns and Abnormalities (Continued):

Fig. 19-3 Gross photograph of kidney from a patient who died from nephrosis. There was marked proteinuria and a ceruloplasmin of 90 units.

Creatine Phosphokinase

Determination of creatine phosphokinase in children is important in the differentiation of muscular dystrophy of the Duchenne type from skeletal muscle atrophy.[32] The best laboratory test for this is serum CPK assay. Young women, furthermore, who are carriers of the muscular dystrophy gene will have an elevated CPK.[19,33] Male infants who are in a preclinical state and who are destined to develop muscular dystrophy will also have an elevated CPK. In the Duchenne type of muscular dystrophy, enzyme levels from 100 to 3000 I. U. may occur. Children who are hypothyroid may also have CPK elevation.

Disaccharidases

The determination of lactase deficiency of gastrointestinal tract mucosa is important in pediatrics. The syndrome consists of

Pediatric Enzyme Patterns and Abnormalities (Continued):

malnutrition resulting from chronic diarrhea. The stools are thin and watery and have a pH of 4 to 6. An increased content of lactic and other organic acids occurs in the stools. If lactose is deleted from the diet, a prompt regression of symptoms occurs. The severity of lactase deficiency varies in intensity, and the patient's symptoms are variable relevant to the degree of enzyme deficiency.[8]

Another important disaccharidase deficiency is that of sucrase or isomaltase[2] which clinically resembles lactase deficiency. The diagnosis rests on determining the tissue disaccharidase levels. The normal amount of lactase in the intestinal mucosa is 0.2 to 19 micromoles of substrate split per minute per gram of wet mucosal tissue; the normal amount of maltase is 13 to 54 micromoles; the normal amount of isomaltase is 4 to 13 micromoles; and the normal amount of sucrase is 6 to 17 micromoles.

Lactic Dehydrogenase

In normal infants from birth to 10 days, the normal serum LDH range is 308 to 1780 I.U. with an average of 815 I.U. In the cerebrospinal fluid, the range is 2.3 to 84 I.U. with an average of 22.6 I.U.[27] In adults, the serum range is up to 200 I.U. and in cerebrospinal fluid up to 30 I.U.

LDH in children may be elevated in various conditions. LDH_1 may be elevated in myocarditis, such as rheumatic myocarditis and muscular dystrophy, with LDH_5 decreased in muscular dystrophy. LDH_1 may be elevated in sickle cell anemia or other hemolytic anemias, such as is associated with a G-6-PD deficiency. Children with leukemia have an elevated LDH_2 and LDH_3 level.[51]

Infants with birth injury and intracranial lesions may have increased LDH in the cerebrospinal fluid, with levels ranging from 20 to 600 I.U. (Fig. 12-4)[50] Children with viral hepatitis have elevated LDH_5. Umbilical cord blood LDH activity is higher in jaundiced babies than in normal ones. Following exchange transfusion, the LDH activity declines.[17]

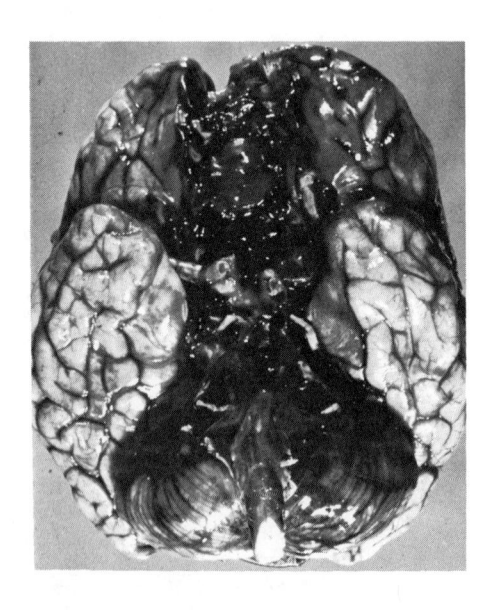

Pediatric Enzyme Patterns and Abnormalities (Continued):

Fig. 19-4   Gross photograph of brain in which a subarachnoid hemorrhage was present and the spinal fluid LDH was 570 I.U.

Acute glomerulonephritis causes an elevated LDH_1 and LDH_2 due to necrosis of the renal cortex. LDH_4 and LDH_5 may be increased in glomerulonephritis if there is necrosis of tubules in the renal medulla.[11] Children with collagen diseases such as dermatomyositis will have elevated LDH_5; and trauma to skeletal muscle will also cause an elevation of LDH_5. Tumors in children may elevate the LDH isoenzymes, either 2, 3, 4, and 5, with the latter being more commonly elevated.[11]

Isocitric Dehydrogenase

Isocitric dehydrogenase activity in umbilical cord blood is between 40 to 250 percent higher than that found in adult serum. This level is maintained for the first 3 months of life and may be elevated throughout the first year.[10] Marked elevations of ICD may be indicative of hepatocellular damage due to various causes.

Pediatric Enzyme Patterns and Abnormalities (Continued):

Leucine Aminopeptidase

Determination of serum leucine aminopeptidase is useful in pediatrics. The normal level is up to 33 I.U. This enzyme is present particularly in the bile ducts. If there is intrahepatic or extra-hepatic biliary obstruction, leucine aminopeptidase will be elevated. Infants with neonatal giant cell hepatitis develop levels up to 500 I.U., and those with biliary atresia have levels above 500 I.U. The leucine aminopeptidase level may be helpful in differentiating these two diseases.[25,26,41]

Lipase

Serum lipase may be elevated in children who have acute pancreatitis secondary to mumps. Lipase levels in these patients tend to rise more slowly and remain elevated longer than serum amylase and may serve as a better test for the diagnosis of this condition. The lipase level in children is up to 1.0 I.U. The iodized oil test for lipase deficiency is valuable in the pediatric age group to determine if pancreatic deficiency disease is present.[16] If the normal child is given iodized oil, the iodine is split off after absorption and excreted into the urine. Children who have lipase deficiency will not be able to absorb the iodized lipiodol; thus, there will be no splitting off of iodine, and no iodine will be excreted in the urine. The iodine is detected when a blue color forms with the addition of starch. The possibility of allergy to iodine should be considered before iodized oil is administered. Patients should not receive iodine-containing medications at the time of this test, or take pancreatic enzymes within 48 hours, and be on a restricted fat intake. The test is performed as follows: 5.0 ml. of Lipiodol is mixed with plain or flavored mild in a Waring blender and administered with breakfast. Urine specimens are examined prior to the test to exclude abnormal iodine excretion and between 8 and 18 hours after the ingestion of the Lipiodol. If a failure of blue color formation occurs when starch is added to the urine, this is an indication of lipase deficiency.[30,42]

Amylase

Amylase may be elevated in children with mumps or acute pancreatitis secondary to mumps. Amylase may be absent normally in the newborn period.[43] Measureable enzyme activity appears about 2 months of

Pediatric Enzyme Patterns and Abnormalities (Continued):

life and reaches adult levels at about 1 year. Renal failure and massive liver necrosis will also cause serum amylase elevations.

Glutamic Oxalacetic Transaminase and Glutamic Pyruvic Transaminase

The normal GOT levels for infants are up to 67 I.U. GPT in one-third of infants is in the range of 27 to 54 I.U. and in two-thirds in the range of 0 to 27 I.U. In the cerebrospinal fluid, infants have a GOT of 1 to 7 I.U. with a mean of 3 I.U. Thus, the GOT in children is somewhat higher than that of the adult.[27] This may reflect the immaturity of the liver or increased muscle activity.[36] Causes for GOT and GPT elevation are numerous. These are as follows: (1). Elevation in the cerebrospinal fluid may signify intracranial damage; (2). GOT and GPT elevation may occur in any type muscle damage, such as Duchenne muscular dystrophy; neurogenic skeletal muscle atrophy does not increase GPT and GOT; (3). Myocarditis may elevate GOT in children (Fig. 19-5); (4). GOT and GPT may be increased in viral or giant cell hepatitis;[13] (5). Pancreatitis due to mumps may elevate the GOT level; (6). Hemolytic anemia results in a high GOT; (7). Children who are burned will have an elevated GOT due to liberation of the enzyme from the skin;[39,54] (8). Those who have collagen diseases such as dermatomyositis may have an elevated GOT level from skeletal muscle; (9). Children undergoing cardiac catheterization or angiograms will have elevated GOT and GPT levels; (10). Children with infectious mononucleosis or leukemia may have elevated GOT and GPT from proliferation of lymphopoietic cells or liver involvement.

Spurious elevation of GOT may occur if the child is taking drugs such as erythromycin which react with the colorimetric dye in the colorimetric chemical assay of the enzyme causing a spurious increase. Diabetic children who develop keto acidosis also show a falsely elevated GOT because of the presence of serum alpha-keto acids which interact with the colorimetric chemicals in the non-enzymatic determination causing spurious elevation.[53]

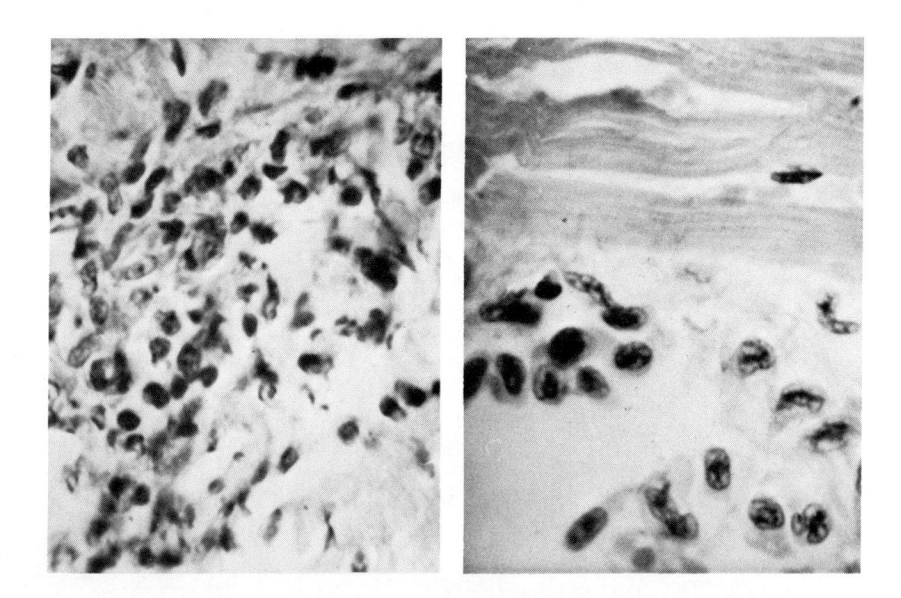

Pediatric Enzyme Patterns and Abnormalities (Continued):

Fig. 19-5 Photomicrographs of section of myocardium from a child who died of active rheumatic myocarditis and heart failure. The SGOT was 330 I.U.

Exocrine Pancreatic Secretion

The exocrine secretion of the pancreas contains enzymes such as lipase, amylase, and trypsinogen. Trypsinogen is transformed to trypsin by enterokinase. The assay for trypsin in duodenal juice was the definitive test for cystic fibrosis until the sweat chloride test was developed.[49] Pancreatic secretions may be impaired in a variety of diseases. Congenital lipase or trypsinogen deficiency may also occur. The normal trypsin content in the duodenal juice is 160 to 180 micrograms per milliliter. The determination of this enzyme may be valuable in the assessment of cystic fibrosis.[34] Under the age of 1 year, failure of the stool to digest gelatin in a dilution of 1:100 is suggestive of cystic fibrosis. Lower values are found, however, in meconium stools and in fecal stools in the first 10 days of life. Above 1 year, many normal children's stools will digest in a dilution of 1:12.5. No digestion at this dilution is again suggestive of mucoviscidosis.[18,29] A majority of children with cystic fibrosis

Pediatric Enzyme Patterns and Abnormalities (Continued):

show no digestion of gelatin with duodenal juice. Occasional titers from 1:50 are recorded. Lower titers of 1:12.5 are not, however, normally seen in normal children, especially over the age of 2 years.

REFERENCES

1. Aronson, S. M., Volk, B. W.: "Studies on Serum Aldolase Activity in Neuromuscular Disorders. I. Clinical Applications". Amer. J. Med. 22:414, 1957.

2. Auricchi, S., Dahlquist, A., Murset, G., Parker, A.: "Isomaltose Intolerance causing Decreased Ability to Utilize Dietary Starch". J. Pediat. 62:165, 1963.

3. Brenner, M. D.: "Studies on the Involution of the Fetal Cortex of the Adrenal Gland". Amer. J. Path. 16:787, 1940.

4. Cantarow, A., Trumper, M., Saunders, W.: CLINICAL BIOCHEMISTRY Saunders Co., pg. 158, 1962.

5. Cartwright, G. E., Markowitz, H., Shields, G. S., Wintrobe, M. M.: "Studies on Copper Metabolism. XXIX. A Critical Analysis of Serum Copper and Ceruloplasmin Concentrations in Normal Subjects, Patients with Wilson's Disease, and Relatives of Patients with Wilson's Disease". Amer. J. Med. 28:555, 1960.

6. Christiansson, G., Berth, J.: "A Study of the Enzyme Pattern in Children and Newborn Infants". Acta. Paediat. Scand. 49:626, 1960.

7. Clark, L. C., Beck, E.: "Plasma 'Alkaline' Phosphatase Activity. I. Normative Data for Growing Children". J. Pediat. 36:335, 1950.

8. Cozzetto, F. J.: "Intestinal Lactase Deficiency in a Patient with Cystic Fibrosis: Report of a Case with Enzyme Assay". Pediatrics 32:228, 1963.

9. Dale, R. A.: "Demonstration of Aldolase in Human Platelets. The Relation to Plasma and Serum Aldolase". Clin. Chim. Acta. 5:652, 1960.

Pediatric Enzyme Pattern References (Continued):

10. Emanuel, B., West, M., Zimmerman, H. J.: "Serum Enzymes in Disease. XII. Transaminases, Glycolytic and Oxidative Enzymes in Normal Infants and Children". Amer. J. Dis. Child. 105:77, 1963.

11. Emery, A. E. H., Sherbourne, D. H., Pusch, A.: "Electrophoretic Pattern of Lactic Dehydrogenase in Various Diseases". Arch. Neurol. 12:251, 1965.

12. Evans, J. H., Baker, R. W.: "Serum Aldolase and the Diagnosis of Myopathy". Brain 80:557, 1957.

13. Fowler, W. M., Jr., Pearson, C. M.: "Diagnostic and Prognostic Significance of Serum Enzymes. II. Neurologic Diseases other than Muscular Dystrophy". Arch. Phys. Med.

14. Friedman, M. M., Lapan, B.: "Serum Aldolase in the Neonatal Period: Including a Colorimetric Determination of Aldolase by Standardization with Dehydroxyacetone". J. Lab. Clin. Med. 51:745, 1958.

15. Gault, M. H., Stein, J., Aronoff, A.: "Serum Ceruloplasmin in Hepatobiliary and Other Disorders: Significance of Abnormal Values". Gastroenterology 50:8, 1966.

16. Henry, R. J., Sobel, C., Berkman, I. S.: "On the Determination of 'Pancreatic Lipase' in Serum". Clin. Chem. 3:77, 1957.

17. Hishikaaw, O.: "Studies of Lactic Dehydrogenase Activities and Transaminase Activities of Serum in Newborn Infants and Children". Acta. Paediat. Jap. 6:36, 1964.

18. Horsfield, A.: "Proteolytic Activity of Stools". J. Med. Lab. Tech. 10:18, 1952.

19. Hughes, B. P.: "Serum Enzymes in Carriers of Muscular Dystrophy". Brit. Med. J. 2:963, 1962.

20. Kove, S., Goldstein, S., Wroblewski, F.: "Activity of Glutamic Oxalacetic Transaminase in the Serum in the Neonatal Period". Pediatrics 20:281, 1960.

21. Laron, Z., Epstein-Halberstadt, B.: "Activity of Acid Phosphatase in the Serum of Normal Infants and Children". Pediatrics 26:281, 1960.

Pediatric Enzyme Pattern References (Continued):

22. Lending, M., Slobody, L. B., Mestern, J.: "Cerebrospinal Fluid Glutamic Oxalacetic Transaminase and Lactic Dehydrogenase Activity in Children with Neurological Disorders". J. Pediat. 65:415, 1964.

23. Lepow, H., Schoenfeld, M. R., Messeloff, C. R., Chu, F.: "Nonprostatic Causes of Acid Hyperphosphatasemia: Report of a Case due to Multiple Myeloma". J. Urol. 87:991, 1962.

24. Markowitz, H., Gubler, C. J., Mahoney, J. P., Cartwright, G. E., Wintrobe, M. N.: XIV. Copper, Ceruloplasmin and Oxidase Activity of Sera of Normal Human Subjects, Pregnant Women and Patients with Infection, Hepatolenticular Degeneration and the Nephrotic Syndrome". J. Clin. Invest. 34:149, 1955.

25. Natoli, G., Natoli, V., Lapi, A. S., Mancini, G., Renzulli, F.: "LAP Isoenzymes in Neonatal Hepatitis and Biliary Atresia". Lancet 2:209, 1969.

26. O'Brien, D., Ibbott, F. A., Rodgerson, D. O.: LABORATORY MANUAL OF PEDIATRIC MICROBIOCHEMICAL TECHNIQUES Harper & Row, 4th Edition, pg. 10, 1968.

27. O'Brien, D., Ibbott, F. A., Rodgerson, D. O.: LABORATORY MANUAL OF PEDIATRIC MICROBIOCHEMICAL TECHNIQUES Harper & Row, 4th Edition, pg. 6, 1968.

28. O'Brien, D., Ibbott, F. A., Rodgerson, D. O.: LABORATORY MANUAL OF PEDIATRIC MICROBIOCHEMICAL TECHNIQUES Harper & Row, 4th Edition, pg. 1, 1968.

29. O'Brien, D., Powell, B. W.: "Tryptic Activity of the Stoo in the Newborn". Ormond Str. J. 3:33, 1952.

30. O'Brien, D., Walker, D. M., Ibbott, F. A.: "Specificity of the Iodized Oil Test Fat Absorption". Pediatrics 23:422, 1959.

31. Oski, F. A., Naiman, J. L., Diamond, L. K.: "Use of Plasm Acid Phosphatase Value in the Differentiation of Thrombocytopenic Stages". New Eng. J. Med. 268:1423, 1963.

32. Pearce, J. M. S., Pennington, R. J., Walton, J. N.: "Seru Enzyme Studies in Muscle Disease. II. Serum Creatine Kir Activity in Muscular Dystrophy and in Other Myopathic and Neurologic Disorders". J. Neurol. Neurosurg. Psychiat. 27:96, 1964.

Pediatric Enzyme Pattern References (Continued):

33. Pearce, J. M. S., Pennington, R. J. T., Walton, J. N.: "Serum Enzyme Studies in Muscle Disease. III. Serum Creatine Kinase Activity in Relatives of Patients with the Duchenne Type of Muscular Dystrophy". J. Neurol. Neurosurg. Psychiat. 27:181, 1964.

34. Richmond, R. C., Schwachmann, H.: "Studies of Fibrocystic Disease of the Pancreas: Mucoviscidosis Chymotrypsin Activity in Duodenal Fluid". Pediatrics 16:207, 1955.

35. Riggins, R. S., Kiser, W. S.: "A Study of Lactic Dehydrogenase in Urine and Serum of Patients with Urinary Tract Disease". J. Urol. 90:594, 1963.

36. Rubin, S. L.: "Serum Glutamic Oxalacetic Transaminase: A Longitudinal Study in Infants". Nebraska Med. J. 44:458, 1959.

37. Schapira, F.: "Fructose-1-Phosphoaldolase of the Serum". Path. Biol. (Paris) 9:63, 1961.

38. Searcy, R.: DIAGNOSTIC BIOCHEMISTRY, McGraw-Hill, pg. 44, 1969.

39. Searcy, R.: DIAGNOSTIC BIOCHEMISTRY, McGraw-Hill, pg. 515, 1969.

40. Sibley, J. A., Fleisher, G. A.: "The Clinical Significance of Serum Aldolase". Mayo Clin. Proc. 29:591, 1954.

41. Siegel, I. A.: "Leucine Aminopeptidase in Pregnancy". Obstet. & Gynec. 14:488, 1959.

42. Silverman, F. N., Shirkey, H. C.: "A Fat Absorption Test Using Iodized Oil, with Particular Application as a Screening Test in the Diagnosis of Fibrocystic Disease of the Pancreas". Pediatrics 15:143, 1955.

43. Somogyi, M.: Diastase Activity in Human Blood". Arch. Intern. Med. 67:678, 1941.

44. Stanbury, J., Wyngaarden, J., Fredrickson, D., Oski, F. A., Norman, J. L., Diamond, L. K.: METABOLIC BASIS OF INHERITED DISEASE McGraw-Hill Co., 2nd Edition, pg. 576, 1960.

Pediatric Enzyme Pattern References (Continued):

45. Sternlieb, I., Scheinberg, I. H.: "The Diagnosis of Wilson's Disease in Asymptomatic Patients". J. Amer. Med. Assoc. 183:747, 1963.

46. Stolbach, M., Krant, M. J., Fishman, W. H.: "Ectopic Production of Alkaline Phosphatase Isoenzyme in Patients with Cancer". New Eng. J. Med. 281:757, 1969.

47. Thomson, R. A., Vignos, P. J.: "Serum Aldolase in Muscle Disease". Arch. Intern. Med. 103:551, 1959.

48. Thomson, W. H. A.: "The Clinical Biochemistry of the Muscular Dystrophies". ADVANCES IN CLINICAL CHEMISTRY Vol. 7, Academic Press, pg. 138, 1964.

49. Townes, P. L.: "Trypsinogen Deficiency Disease". J. Pediat. 66:275, 1965.

50. Van der Helm, H. J., Zondac, H. A., Klein, F.: "On the Source of Lactic Dehydrogenase in Cerebrospinal Fluid". Clin. Chim. Acta. 8:193, 1963.

51. Wieme, R. J.: "Diagnostic Aspects of LDH Isoenzymes". Postgraduate Med. 35:38, 1964.

52. Wolf, P. L., Enlander, D., Dalziel, J., Swanson, J.: "Green Plasma in Blood Donors". New Eng. J. Med. 281:205, 1969.

53. Wolf, P. L., Williams, D., Potolsky, A., Langston, C.: "Ketosis causing Spurious Elevation of SGOT". Clin. Chem. 17:341, 1971.

54. Wroblewski, F., LaDue, J. S.: "Serum Glutamic Oxalacetic Aminopherase (Transaminase) in Hepatitis". J. Amer. Med. 160:1130, 1956.

CHAPTER 20

ENZYME ABNORMALITIES IN BODY FLUIDS

The various body fluids have characteristic levels of enzymes under normal conditions. Determination of the levels is of inestimable value in diagnosis of many disease states and particularly in ascertaining the cause of abnormal fluid collection in the serous cavities.

Transudates and Exudates

The differentiation of a transudate from an exudate is the responsibility of the clinical laboratory. The standard methods include determination of: specific gravity; protein content; glucose content; and lipid content of the fluid.

A transudate is a fluid that arises from leakage of plasma from the circulatory system. Common etiologic mechanisms for production of transudate fluids in peritoneal, pleural, or pericardial space are: Low serum albumin as seen in heart failure and obstruction of a major venous or lymphatic vessel. A transudate is characterized by a low specific gravity below 1.018, a total protein content less than 2.5 gm. per 100 ml., and glucose and lipid contents similar to those of plasma.[5] The cell count is minimal.

In contrast, an exudate is a collection of fluid, usually in the pleural, pericardial, or peritoneal space, with an etiology of necrosis of tissue or inflammatory, immunologic, or neoplastic disease. An exudate usually has a specific gravity greater than 1.018, a total protein content greater than 2.5 gm. per 100 ml., a glucose content less than that of plasma, and a lipid content greater than that of plasma. The cell count tends to be rather prominent. The cells may be neutrophils, lymphocytes, monocytes, or malignant cells involving the mesothelial surface of the cavity.

A chylous effusion denotes that there has been perforation of the thoracic duct and leakage of chyle into the pericardial, thoracic, or peritoneal space. Chylous fluids are milky. Pseudochylous fluids are also milky but do not consist of chyle; they arise as a result of inflammatory processes or neoplastic involvement of mesothelial lined cavities. The large number of cells and protein in an exudate

377

Enzyme Abnormalities in Body Fluids (Continued):

give the fluid an opalescent or milky appearance or the large number of cells bring in lipid and cause a milky appearance to the fluid.

Effusions are not stagnant collections of fluid in a body cavity. There is an active communication with the circulatory system and constant transfer of materials. Albumin is more readily transferred than the globulins, in fact, two to three times that of globulins.

It is frequently difficult to utilize the standard criteria of specific gravity and total protein content to differentiate a transudate from an exudate. Another method of differentiation is based on the LDH content of the fluid.[6] Transudates are characterized by an LDH content similar to that of plasma, 200 I.U. or less; exudates, by a great increase of LDH content above 200 I.U.[28] Exudates usually have a greatly increased LDH content and this amounts to a content over 200 I.U.[4,16,35] One must keep in mind that when LDH content of blood plasma is elevated, the LDH content of a fluid will also be elevated as a reflection of the plasma content. Thus, one may erroneously classify a transudate as an exudate if the patient has a high blood LDH content, with resultant elevated LDH content of a transudate secondary to congestive heart failure. The increased LDH content of an exudate is secondary to the presence of many neutrophils, lymphocytes, monocytes, or malignant cells containing LDH. A small number of malignant effusions may not have a high LDH, and a small number of transudates may have a high LDH because of increased LDH in the blood (Fig. 20-1). When the fluid and serum both contain a high LDH, it is imperative to compare the effusion LDH:serum LDH ratio. A ratio greater than one signifies a malignant effusion; a ratio less than one a benign effusion.

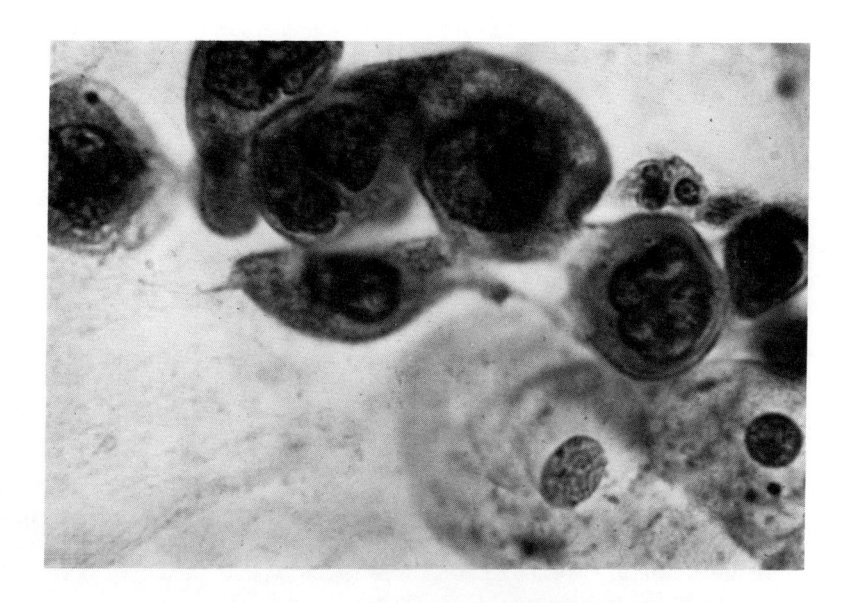

Enzyme Abnormalities in Body Fluids (Continued):

Fig. 20-1 Photomicrograph of Papanicolau-stained smear of pleural fluid containing metastatic undifferentiated malignant bronchogenic cells. The LDH content of the pleural fluid was 780 I.U.

Pleural, Pericardial, and Peritoneal Fluids

Pleural and peritoneal fluids which are secondary to acute pancreatitis contain elevated levels of amylase.[12,20] A diagnostic procedure for acute pancreatitis is aspiration of the abdominal fluid and determination of its amylase content.[2,14,25,30] Thoracic duct fluid in acute pancreatitis will also contain a high amylase and at times study of this fluid will be of diagnostic value. Pleural fluids which result from acute pancreatitis have usually been found on the left side. It has been suggested that a lymphatic pathway brings the amylase into the left pleural cavity.[16] Pleural effusions which result from renal failure may also have a high amylase content. Amylase is excreted by the kidney and with high serum amylase, and renal insufficiency, there will be a lack of amylase excretion.[23] Thus, effusions in the pleural, pericardial, and peritoneal cavity which result from renal insufficiency may have a high amylase content. Some pleural effusions which are exudates produced by malignant

Enzyme Abnormalities in Body Fluids (Continued):

neoplasms metastatic to the pleural cavity have also been shown to contain high amylase content.[8,9] It is thought that the bronchogenic carcinoma obstructs lymphatic channels. Amylase is normally present in the thoracic duct, and with obstruction of the duct, amylase seepage into the pleural cavity with resultant increase in the amylase content of the exudate in the pleural space occurs. Thus, this finding might confuse the clinical picture and lead to an erroneous impression of acute pancreatitis with effusion rather than the true diagnosis of adenocarcinoma of bronchogenic origin with metastases to the pleural space.

Elevation of amylase in a peritoneal fluid is strongly suggestive but not diagnostic of acute pancreatitis. Other conditions such as a ruptured ectopic pregnancy or a perforated intestine will also elevate peritoneal fluid amylase. An abundant amount of the enzyme is present in the oviduct epithelium and, with rupture of an ectopic pregnancy, the amylase is released into the peritoneal cavity.[13] An ample amount of amylase is located in small intestinal epithelium, and intestinal lbstruction with perforation will cause an elevated content of the enzyme in the peritoneal fluid. Increased lipase is also present in peritoneal fluid in patients with acute pancreatitis.

Saliva

Determination of LDH in the saliva has recently been found to be of diagnostic value. The level is elevated in patients who have inflammatory, immunologic, or neoplastic conditions involving the major salivary glands, especially the parotid.[11,22]

Gastric Juice

Determination of LDH or beta-glucuronidase activity in the gastric juice is a reliable index of malignant neoplasm in the stomach. It is important to exclude blood from the specimen because LDH is present in red blood cells and may cause a spurious elevation. The presence of bile in the gastric juice may inhibit enzyme activity The normal level of LDH in the gastric juice is slightly less than 200 I.U., and the primary cause for an increase is carcinoma of the stomach.[32] Benign peptic ulcer does not cause an elevation. LDH activity in the gastric juice is also elevated in patients with pernicious anemia. Beta-glucuronidase is frequently increased in gastric juice in carcinoma of the stomach.[26]

Enzyme Abnormalities in Body Fluids (Continued):

Vaginal Fluid

Determination of enzymes in vaginal fluids may be of diagnostic value in certain conditions. Acid phosphatase in the vaginal fluid suggests sexual intercourse within the preceding 12 hours.[29] The medical examiner employs this test to determine if there has been rape;[10] it is more useful than a search for spermatozoa in the fluid. Spermatozoa degenerate rapidly but acid phosphatase persists for a longer time. Semen has a high content of acid phosphatase of prostatic origin, which is inhibited by L-tartrate. Particularly in suspected rape cases, it is important to use the L-tartrate test to identify acid phosphatase of prostatic origin, since other isoenzymes of acid phosphatase, especially of erythrocyte origin, are not inhibited by L-tartrate.[27]

Determination of 6-phosphogluconic acid dehydrogenase (6-PGD) in the vaginal fluid was originally proposed as a diagnostic test for carcinoma of the uterine cervix.[3] However, it has been shown that inflammatory conditions of the cervix and vagina will cause an elevated 6-PGD which is from leukocytes present in the cervix or vagina associated with the infection.[1,15,21,24] In addition, 6-PGD is present in red blood cells and bleeding lesions will increase the 6-PGD activity in the vaginal fluid.

Beta-glucuronidase is elevated in the vaginal fluid in carcinoma of the cervix, but this is not useful diagnostically because other conditions such as inflammation will also cause an elevation.[18]

Synovial Fluid

The differentiation of rheumatoid arthritis from osteoarthritis by determining enzymes in the synovial fluid is of diagnostic importance.[7] Normal LDH activity is found in osteoarthritis; an elevated LDH is present in synovial fluid from rheumatoid arthritis patients.[33] The rheumatoid arthritis fluid has a greater cell content, and the increased LDH is derived from the large numbers of lymphocytes, plasma cells, and histiocytes in the synovium.[33] Rheumatoid synovial fluid is frequently green because of the high copper content. The protein content and the specific gravity of rheumatoid arthritis synovial fluid are higher than in the osteoarthritis fluid. The glucose content of synovial fluid from rheumatoid arthritis is less than that of synovial fluid in osteoarthritis.[31] Other enzymes usually are markedly elevated in synovial fluid from rheumatoid

Enzyme Abnormalities in Body Fluids (Continued):

arthritis. These are of lysosomal origin. They are: Beta-glucuronidase, acid phosphatase, and beta-galactosidase.[17] In addition, there is increased activity of amino tripeptidase [36] and alkaline phosphatase. The presence of these enzymes contributes markedly to the destruction of the synovia and the articular cartilage, resulting in marked damage to the joint. When the disease is successfully treated, the levels of these enzymes in the synovial fluid fall. The enzymes enter the fluid from the activated lymphocytes, plasma cells, and histiocytes present in the synovia in rheumatoid arthritis. No increase in lysosomal enzymes in synovial fluid in osteoarthritic patients has been reported.

Purulent arthritis secondary to gonococcus, pneumococcus, streptococcus, or staphylococcus organisms causes an increased LDH content of synovial fluid, and LDH is elevated in the synovial fluid in patients with gouty arthritis.

REFERENCES

1. Bell, J. L., Egerton, M. E.: "6-Phosphogluconate Dehydrogenase Estimation in Vaginal Fluid in the Diagnosis of Cervical Cancer". J. Obstet. & Gynec. Brit. Comm. 72:603, 1965.

2. Bolooki, H., Gliedman, M. L.: "Peritoneal Dialysis in the Treatment of Acute Pancreatitis". Surgery 64:466, 1968.

3. Bonham, D. G., Gibbs, D. F.: "A New Enzyme Test for Gynaecological Cancer". Brit. Med. J. 11:823, 1962.

4. Brauer, M. J., West, M., Zimmerman, H. J.: "Serum Enzymes in Disease. XIII. Comparison of Glycolytic and Oxidative Enzymes and Transaminase Values in Benign and Malignant Effusions with those in Serums". Cancer 16:533, 1963.

5. Cantarow, A., Trumper, M.: CLINICAL BIOCHEMISTRY 6th. Edition, Saunders Co., pg. 317, 1962.

6. Chandraaserhar, A. J., Palatoo, A., Dubin A., Levine H.: "Pleural Fluid Lactic Acid Dehydrogenase Activity and Protein Content". Arch. Intern. Med. 128:48, 1969.

7. Cohen, A. S.: "Lactic Dehydrogenase (LDH) and Transaminase (GOT) Activity of Synovial Fluid and Serum in Rheumatic Disease States, with a Note on Synovial Fluid LDH Isoenzyme. Arthritis Rheum. 7:490, 1964.

Body Fluid References (Continued):

8. Ende, N.: "Studies of Amylase Activity in Pleural Effusions and Ascites". Cancer 13:283, 1960.

9. Ende, N.: "Amylase Activity in Body Fluids". Cancer 14:1109, 1961.

10. Fisher, R. S.: "Acid Phosphatase Tests as Evidence of Rape". New Eng. J. Med. 240:738, 1949.

11. Foster, W. S.: "Lactic Dehydrogenase in Mixed Human Saliva". Fed. Proc. 207:2437, 1969.

12. Goldman, M., Goldman, G., Fleischner, F. G.: "Pleural Fluid Amylase in Acute Pancreatitis". New Eng. J. Med. 266:715, 1962.

13. Green, C. L.: "Identification of Alpha-Amylase as a Secretion of the Human Fallopian Tube, and 'Tube-like' Epithelium of Mullerian and Mesonephric Duct Origin". Amer. J. Obstet. & Gynec. 73:402, 1957.

14. Hammarsten, J. F., Honska, W. L., Limes, B. J.: "Pleural Fluid Amylase in Pancreatitis and Other Diseases". Amer. Rev. Tuberc. 79:606, 1959.

15. Hoffman, R. L., Merritt, J. W.: "6-Phosphogluconate Dehydrogenase in Uterine Cancer Detection". Amer. J. Obstet. & Gynec. 92:650, 1965.

16. Horrocks, J. E., King, J., Waind, A. P. B., Ward, J.: "Lactic Dehydrogenase Activity in the Diagnosis of Malignant Effusions". J. Clin. Path. 15:27, 1962.

17. Jaco, R. F., Feldman, A.: "Variations of Beta-Glucuronidase Concentration in Abnormal Human Synovial Fluid". J. Clin. Invest. 34:263, 1955.

18. Kasdon, S. C., Homberger, F., Yorshis, E., Fishman, W. H.: "Beta-Glucuronidase Studies in Women. VI. Premenopausal Vaginal Fluid Values in Relation to Invasive Cervical Cancer". Surg. Gynec. Obstet. 97:579, 1953.

19. Keith, L. M., Jr., Zollinger, R. M., McCleary, R. S.: "Peritoneal Fluid Amylase Determinations as an Aid in the Diagnosis of Acute Pancreatitis". Arch. Surg. 61:930, 1950.

Body Fluid References (Continued):

20. Klaser, M. H., Roth, L. A., Bockus, H. L.: "Relapsing Pancreatitis with Pseudocyst of the Pancreas and Enzyme-containing Pleural Effusion". Gastroenterology 28:842, 1955.

21. Lawson, J. G., Watkins, D. K.: "Vaginal Fluid Enzymes in Relation to Cervical Cancer". J. Obstet. & Gynec. Brit. Comm. 72:1, 1965.

22. Levine I., Augsth, F., Sansur, M., Wroblewski, F.: "Clinical Implications of Lactic Dehydrogenase Activity in Sputum". J. Amer. Med. Assoc. 207:2436, 1969.

23. Hampers, C., Bailey, G., Adrian, K., Merrill, J. P.: "Alterations in Serum Enzymes in Chronic Renal Failure". J. Amer. Med. Assoc. 213, 2263, 1970.

24. Muir, G. G., Canti, G., Williams, D. K.: "Use of 6-Phosphogluconate Dehydrogenase as a Screen Test for Cervical Carcinoma in Normal Women". Brit. Med. J. 11:1563, 1964.

25. Pfeffer, R. B., Mixter, G., Jr., Hinton, W. J.: "Acute Haemorrhagic Pancreatitis. A Safe Effective Technique for Diagnostic Paracentesis". Surgery 43:550, 1958.

26. Piper, D. W., Macoun, M. L., Broderick, F. L., Fenton, B. H. Builder, J. E.: "The Diagnosis of Gastric Carcinoma by the Estimation of Enzyme Activity in Gastric Juice". Gastroenterology 45:614, 1963

27. Reinstein, H., Benotti, N., McBay, A. J.: "The Demonstration of Prostatic Origin of Seminal Stain Acid Phosphatases". Amer. J. Clin. Path. 39:583, 1963.

28. Richterich, R., Zuppinger, K., Rossi, E.: "Diagnostic Significance of Heterogeneous Lactic Dehydrogenases in Malignant Effusions". Nature 191:507, 1961.

29. Riisfeldt, O.: "Acid Phosphatase Employed as New Method of Demonstrating Seminal Spots in Forensic Medicine". Acta. Path. Microbiol. Scand. Suppl. 58:1, 1946.

30. Root, H. D., Hauser, C. W., McKinley, C. R., LaFave, J. W., Mendiola, R. P.: "Diagnostic Peritoneal Lavage". Surgery 57:633, 1965.

Body Fluid References (Continued):

31. Shafer, W.: "Analyses of Extravascular Fluid". J. Amer. Med. Assoc. 210:1087, 1969.

32. Smyrniotis, F., Schenker, S., O'Donnell, J., Schiff, L.: "Lactic Dehydrogenase Activity in Gastric Juice for the Diagnosis of Gastric Cancer". Amer. J. Dig. Dis. 7:712, 1962.

33. Vesell, E. S., Osterland, K. C., Bearn, A. G., Kunkel, H. G.: "Isoenzymes of Lactic Dehydrogenase: Their Alterations in Arthritic Synovial Fluid and Sera". J. Clin. Invest. 41:2012, 1962.

34. West, M., Poske, R. M., Black, A. B., Pilz, C. G., Zimmerman, H. J.: "Enzyme Activity in Synovial Fluid". J. Lab. Clin. Med. 62:175, 1963.

35. Wroblewski, F., Wroblewski, R.: "The Clinical Significance of Lactic Dehydrogenase Activity of Serous Effusions". Ann. Intern. Med. 48:813, 1958.

36. Ziff, M., Simson, J., Scull, E., Smith, A., Shatton, J., Mainland, D.: "Aminotripeptidase Content of Synovial Fluid in Arthritic Disease". J. Clin. Invest. 34:27, 1955.

CHAPTER 21

ENZYME HISTOCHEMISTRY AND ITS APPLICATION TO THE CLINICAL LABORATORY
Elisabeth Von der Muehll, M.T. (ASCP)

As more simplified methods of staining are developed, enzyme histochemistry is becoming increasingly popular as a tool in the chemical laboratory. The importance becomes more obvious as alterations in enzyme patterns in various disease states are clarified. Enzyme histochemistry is utilized primarily to determine enzyme deficiency or accumulation, chiefly in certain congenital and neoplastic diseases.

An example of a congenital enzyme deficiency demonstrated in tissues is Hurler's syndrome, a genetically transmitted disorder of mucopolysaccharide metabolism. It is characterized by excessive visceral storage of chondroiten sulfate B and heparin sulfate and excessive cerebral storage of three gangliosides, GM_3, GM_2, and GM_1, the deficiency of beta-galactosidase being the causative etiology.[16]

The glycogen storage diseases which are inheritable and result in deposition of normal or abnormal glycogen in tissues are:

1. Von Gierke's Cori Type I deficiency glucose-6-phosphatase

2. Pompe's disease Cori Type II deficiency of lysosomal alpha (1, 2, 3, and 4) glucosidase[8,24]

3. Forbe's disease Cori Type III deficiency of amylo-1, 6 glucosidase which is the debranching enzyme

4. Andersen's disease Cori Type IV deficiency of amylo 1, 4-1, 6 transglucosidase which is a branching enzyme

5. McArdles disease Cori Type V deficiency of cardiac and skeletal muscle phosphorylase with normal liver phosphorylase

6. Hers' disease Cori Type VI deficiency of phosphorylase of liver with normal phosphorylase of muscle

In neoplasia one is able to detect and classify malignant cells by the use of histochemistry. By the presence or absence of a high concentration of acid phosphatase in the neoplastic cells, it may be possible to determine whether undifferentiated carcinoma is really of

Enzyme Histochemistry (Continued):

prostatic origin. It is important to inhibit the prostatic acid phosphatase with tartrate to identify the prostatic isoenzyme (Fig. 21-1). Normally gastric mucosa shows no evidence of aminopeptidase activity, but such activity is present in one-third of gastric carcinomas. The enzyme also occurs in bile duct carcinomas.

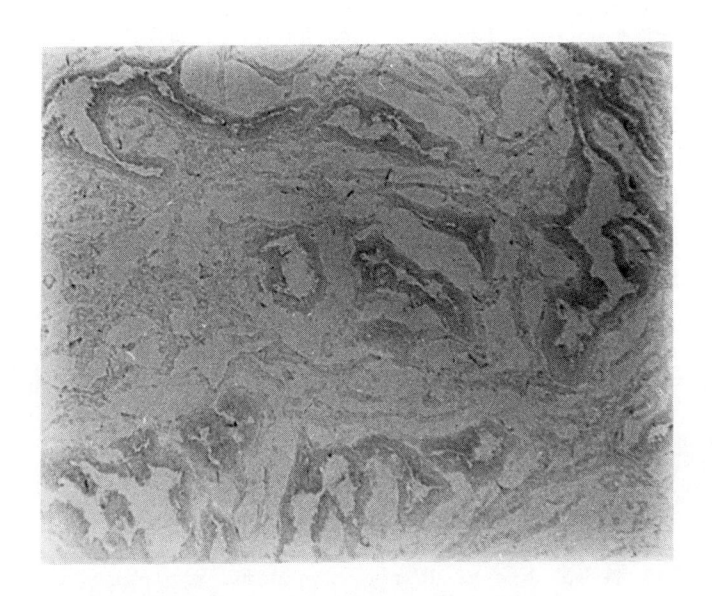

Fig. 21-1 Photomicrograph of section of prostate demonstrating prostatic glands which was stained histochemically with 5-bromo-4-chloro-indolyl-phosphate and was positive for acid phosphatase.

Malignant cells usually show less activity of various enzymes than do non-neoplastic cells. Malignant cells usually contain a higher concentration of beta-glucuronidase than is present in non-neoplastic cells. Increased levels are found in cancer of the breast, colon, lung, cervix, and stomach. This enzyme is present in increased amount in "premalignant" colonic polyps; staining of these tissues for this enzyme is a valuable and practical histochemical procedure.

Enzyme Histochemistry (Continued):

Enzyme histochemistry may be utilized to differentiate a carcinoma from a sarcoma. The neoplasm should be stained for beta-glucuronidase. Sarcomas usually do not have enzymatic activity in contrast to carcinomas which have activity.

Differentiation of a histiocytic lymphoma from an undifferentiated carcinoma at times may be difficult histologically. ATPase activity is pronounced in histiocytic lymphoma as is acid phosphatase and esterase, while anaplastic carcinomas have minimal to absent ATPase, acid phosphatase, and esterase.

Alkaline phosphatase is present in Ewing's sarcoma and absent in neuroblastoma and may be useful to differentiate these neoplasms.

The diagnosis of a non-pigmented malignant melanoma is extremely difficult. Despite the absence of melanin, these tumors may exhibit high tyrosinase activity which is a phenol oxidase-containing copper. Demonstration of tyrosinase thus, is useful and may be determined by autoradiography.

Staining of leukocytes for the detection of alkaline phosphatase is also a very useful procedure.[6] Absence of alkaline phosphatase activity may signify chronic myelogenous leukemia, hypophosphatasia, or paroxysmal nocturnal hemoglobinuria. Increased activity is associated with various leukemoid states, polycythemia vera, and intrauterine and extra-uterine pregnancy, and active Hodgkin's disease.[4,14]

In leukemia, myeloblasts may demonstrate peroxidase or esterase activity while lymphoblasts do not. Leukemic myeloblasts display increased gamma-glutamyl-transpeptidase while non-neoplastic myeloblasts have less activity. Gamma-glutamyl-transpeptidase activity is absent in plasmablasts (myeloma cells) in contrast to marked activity in plasma cells.

Additional uses of enzyme histochemistry include: (1) Identification of prostatic seminal fluid by demonstration of acid phosphatase;[29] (2) Evidence of myocardial infarction when Hematoxylin and Eosin sections are not able to demonstrate early necrosis in tetrazolium-formazan histochemical preparations. Loss of LDH mitochondrial enzyme with coarsening and clumping of the enzyme secondary to anoxia occurs (Fig. 21-2);[20] (3) Evidence of adrenal cortical disease by demonstration of the enzyme 3-beta-hydroxy steroid dehydrogenase. The enzyme is both less active than normal and has shifted intracellularly to the cell border in stress-induced adrenal cortical disease (Fig. 21-3).[27]

Enzyme Histochemistry (Continued):

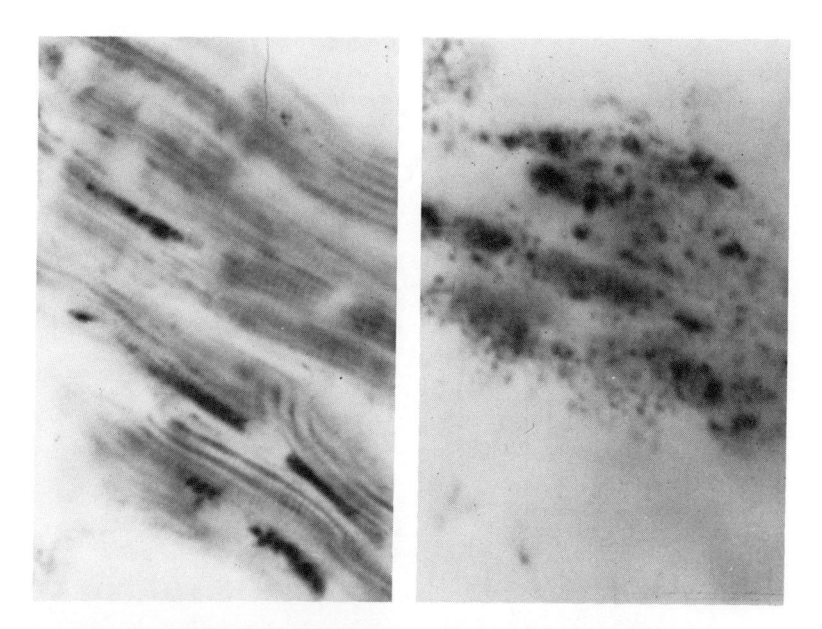

Fig. 21-2 Photomicrograph of tetrazolium stain for LDH of human myocardium demonstrating normal fine granularity of enzyme in mitochondria. The next photomicrograph demonstrates loss of formazan staining in infarcted myocardium with coarsening and clumping of enzyme.

Enzyme Histochemistry (Continued):

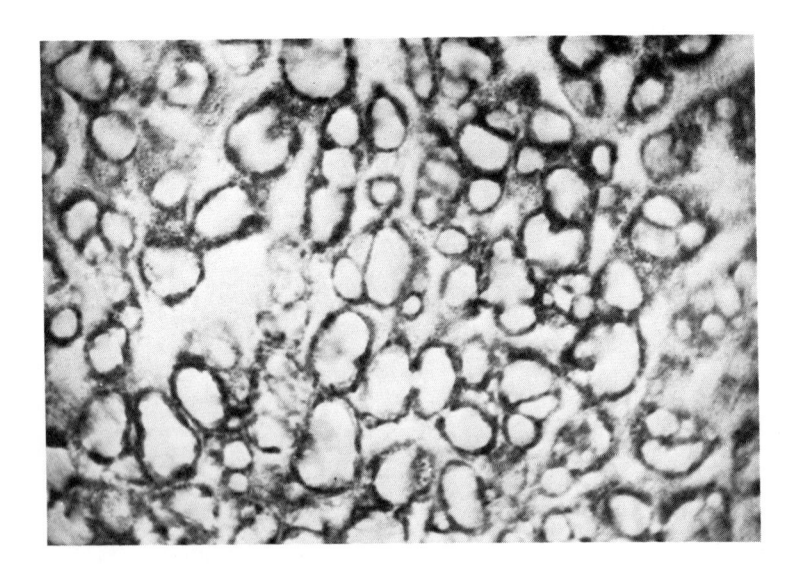

Fig. 21-3   Photomicrograph of diseased human adrenal cortex histochemically stained for 3-beta-hydroxy steroid dehydrogenase. The enzyme has been stained by the tetrazolium formazan technique and shows decreased activity of the enzyme and intracellular peripheral shifting of enzyme due to stress.

Enzyme histochemistry, thus occupies a very important place in the clinical laboratory. In many cases, these studies afford positive evidence of abnormalities and will assist the pathologist in his investigation of a variety of pathologic conditions now inadequately elucidated by means of Hematoxylin and Eosin stained sections.

Enzyme histochemistry assists the pathologist in his investigation of a variety of pathologic conditions now inadequately elucidated by means of H and E stains. It is essential, therefore, to test for a wide spectrum of histochemical procedures and substrates, including indoxyl substrates for glycosidases, esterases, sulfatase, aminopeptidase, and phosphatases. The indoxyl substrates provide the advantages of a simple, rapid means for enzyme histochemistry without the need for a coupling reaction. Hydrolysis of these substrates

Enzyme Histochemistry (Continued):

yields a highly insoluble chromogenic indigo at the site of enzyme deposition.

Enzyme determination of amniotic fluid or amniotic fluid cells in tissue culture is an important advance in the prevention of the birth of infants with mental defects or fatal genetic disease. Cultured fibroblasts from Niemann-Pick patients accumulate more sphingomyelin than control cells.[13] Cultured amniotic fluid fibroblasts have deficiency of a sphingomyelin-cleaving enzyme, sphingomyelinase, and detection of this abnormality now facilitates the diagnosis of this disease in utero. It is not possible to detect heterozygous carriers by this method.[2]

The diagnosis of Gaucher's disease in utero is now possible. The biochemical enzymatic lesion is a deficiency of glucocerebrosidase with the accumulation of lipid glucocerebroside in the reticuloendothelial system. The enzyme has been found to be nearly absent in tissue culture skin fibroblasts. Acid phosphatase activity is increased.[3]

Patients with metachromatic leukodystrophy have a deficiency of arylsulfatase, and sulfatides accumulate in neural tissue, kidney, and the biliary system. A deficiency of the enzyme in tissue culture skin fibroblasts has been demonstrated in this disease.[1,19,28]

Alpha-galactosidase is deficient in the leukocytes of Fabry's disease. Ceramide trihexoside which is a sphingoglycolipid accumulates in the tissues and cultured skin fibroblasts in this disease.[18] Death may occur in the fourth decade from renal failure.[17]

In Tay-Sachs disease, GM_2 ganglioside accumulates in the central nervous system due to the absence of hexosaminidase A. This disease has been diagnosed during the second trimester of pregnancy.[25] An absence of hexosaminidase A activity in the amniotic fluid and the amniotic fluid cells was found. The diagnosis necessitates the determination of both total hexosaminidase and the ratio of hexosaminidase A to B.[26,30,31]

Generalized gangliosidosis is characterized by the accumulation of GM_1 ganglioside in the nervous system. The disease is due to a deficiency of GM_1 beta-galactosidase, which may be detected in cultured skin fibroblasts.[32] One must differentiate GM gangliosidosis from Hurler-Hunter syndrome since in both diseases, gangliosides and mucopolysaccharides accumulate and beta-galactosidase is

Enzyme Histochemistry (Continued):

deficient.[1,12] The enzyme deficiency in the Hurler-Hunter syndrome is not as great as in gangliosidosis.[9] The prenatal diagnosis of GM_1 gangliosidosis is possible if the amniotic fluid or the amniotic cells are deficient in beta-galactosidase.[10,11]

A large number of enzymes have been shown to be active in cultured amniotic fluid cells and are potentially useful for the prenatal diagnosis of specific amino acid disorders. Deficient branched-chain keto-acid decarboxylase deficiency was recently demonstrated in cultured amniotic fluid cells; at birth the infant had maple syrup urine disease.[7]

Ornithine carbamyl transferase deficiency has been detected in non-cultured amniotic fluid cells, making possible the antenatal diagnosis of hyperammonemia Type II.[22]

Cultured fetal cells from the amniotic fluid have been found deficient in galactose-1-pyridyl transferase, making the prenatal diagnosis of galactosemia possible.[15]

Prenatal diagnosis of glycogen storage diseases is now possible. Pompe's disease is characterized by alpha-1, 4-glucosidase deficiency. The deficiency has been demonstrated in cultured amniotic fluid from known maternal heterozygotes.[5,7,23]

Acid phosphatase activity deficiency has recently been described in a 13-week gestation. Cultured amniotic fluid cells were found to be deficient in this lysosomal enzyme in a familial metabolic disorder.[21]

REFERENCES

1. Austin, J., Armstrong, D., Shearer, L.: "Metachromatic Form of Diffuse Cerebral Sclerosis. VI. A Rapid Test for Sulfatase A Deficiency in Metachromatic Leukodystrophy Urine". Arch. Neurol. (Chicago) 14:259, 1966.

2. Brady, R. O.: "Genetics and the Sphingolipidosis". Med. Clin. N. Amer. 53:827, 1969.

3. Brady, R. O., Kanfer, J. N., Shapiro, D.: "Demonstration of a Deficiency of Glucocerebroside-cleaving Enzyme in Gaucher's Disease". J. Clin. Invest. 45:1112, 1966.

Enzyme Histochemistry References (Continued):

4. Clime, A. R. W.: "Neutrophilic Alkaline Phosphatase Test: A Review with Emphasis on Findings in Pregnancy". Amer. J. Clin. Path. 38:95, 1962.

5. Cox, R. P., Douglas, G., Hutzler, J.: "In-utero Detection of Pompe's Disease". Lancet 1:893, 1970.

6. Chang, I. W.: "Clinical Application of the Neutrophil Alkaline Phosphatase Test". Amer. J. Obstet. & Gynec. 86:903, 1963.

7. Dancis, J., Hutzler, J., Cox, R. P.: "Enzyme Defect in Skin Fibroblasts in Intermittent Branched-Chain Ketonuria and in Maple Syrup Urine Disease". Biochem. Med. 2:407, 1969.

8. Dancis, J., Hutzler, J., Lynfield, J.: "Absence of Acid Maltase in Glycogenesis Type 2 (Pompe's Disease) in Tissue Culture". Amer. J. Dis. Child 117:108, 1969.

9. Danes, B. S., Bearn, A. G.: "Hurler's Syndrome: A Genetic Study in Cell Culture". J. Exp. Med. 123:1, 1966.

10. Fratantoni, J. C., Hall, C. W., Neufeld, E. F.: "The Defect in Hurler and Hunter Syndromes. II. Deficiency of Specific Factors involved in Mucopolysaccharide Degradation". Proc. Nat. Acad. Sci. USA 64:360, 1969.

11. Fratantoni, J. C., Neufeld, E. F., Uhlendorf, B. W.: "Intrauterine Diagnosis of the Hurler and Hunter Syndromes". New Eng. J. Med. 280:686, 1969.

12. Ho, M. W., O'Brien, J. S.: "Hurler's Syndrome: Deficiency of a Specific Beta-Galactosidase Isoenzyme". Science 165:611, 1969.

13. Holtz, A. L., Uhlendorf, B. W., Fredrickson, D. S.: "Persistence of a Lipid Defect in Tissue Cultures derived from Patients with Niemann-Pick Disease". Fed. Proc. 23:128, 1964.

14. Kaplow, L. S.: "Histochemical Procedure for Localizing and Evaluating Leukocyte Alkaline Phosphatase Activity in Smears of Blood and Marrow". Blood 10:1023, 1955.

Enzyme Histochemistry References (Continued):

15. Krooth, R. S., Weinberg, A. N.: "Studies on Cell Lines developed from the Tissues of Patients with Galactosemia". J. Exp. Med. 113:1155, 1961.

16. MacBrinn, M., Okada, S., Woollacott, M., Patel, V., Ho, M. W., Tappel, A. L., O'Brien, J. S.: "Beta-Galactosidase Deficiency in the Hurler Syndrome". New Eng. J. Med. 281:338 1969.

17. Mapes, C. A., Anderson, R. L., Sweeley, C. C.: "Galactosylgalactosylglucosylceramide: Galactosyl Hydrolase in Normal Human Plasma and its Absence in Patients with Fabry's Disease". FEBS Letters 7:180, 1970.

18. Matalon, R., Dorfman, A., Dawson, G., et al: Glycolipid and Mucopolysaccharide Abnormality in Fibroblasts of Fabry's Disease". Science 164:1522, 1969.

19. Mehl, E., Jatzkewitz, H.: "Evidence for a Genetic Block in Metachromatic Leukodystrophy". Biochem. Biophys. Res. Commune 19:407, 1965.

20. Morales, A. R., Fine, G.: "Early Human Myocardial Infarctio Arch. Path. 82:9, 1966.

21. Nadler, H. L., Egan, T. J.: "Deficiency of Lysosomal Acid Phosphatase: A New Familial Metabolic Disorder". New Eng. J. Med. 282:302, 1970.

22. Nadler, H. L., Gerbie, A. B.: "Enzymes in Noncultured Amniotic Fluid Cells". Amer. J. Obstet. & Gynec. 103:710, 1969.

23. Nadler, H. L., Messina, A. M.: "In-utero Detection of Type II Glucogenesis (Pompe's Disease)". Lancet 2:1277, 1969.

24. Nitowsky, H. M., Grunefeld, A.: "Lysosomal α-Glucosidase in Type II Glycogenesis: Activity in Leukocytes and Cell Cultures in Relation to Genotype". J. Lab. Clin. Med. 69:472, 1967.

25. O'Brien, J. S.: "Tay-Sachs Disease: Prenatal Diagnosis". Science 172:61, 1971.

Enzyme Histochemistry References (Continued):

26. O'Brien, J. S., Okada, S., Chen, A.: "Tay-Sachs Disease: Detection of Heterozygotes and Homozygotes by Serum Hexosaminidase Assay". New Eng. J. Med. 282:15, 1970.

27. Pearson, B., Wolf, P. L., Grose, F., Andrews, M.: "The Histochemistry of 3β-Hydroxysteroid Dehydrogenase". Amer. J. Clin. Path. 41:256, 1964.

28. Percy, A. K., Brady, R. O.: "Metachromatic Leukodystrophy: Diagnosis with Samples of Venous Blood". Science 161:594, 1968.

29. Reinstein, H., Benotti, N., McBay, A. J.: "The Demonstration of Prostatic Origin of Seminal Stain Acid Phosphatases". Amer. J. Clin. Path. 39:583, 1963.

30. Schneck, L., Friedland, J., Valenti, C.: "Prenatal Diagnosis of Tay-Sachs Disease". Lancet 1:582, 1970.

31. Schneck, L., Volk, B. W.: "Clinical Manifestations of Tay-Sachs Disease and Niemann-Pick Disease". In S. M. Aronson and B. W. Volk (Eds.) INBORN DISORDERS OF SPHINGOLIPID METABOLISM: PROCEEDINGS OF THE THIRD INTERNATIONAL SYMPOSIUM ON THE CEREBRAL SPHINGOLIPIDOSES Pergamon Press, pg. 403, 1967.

32. Sloan, H. R., Uhlendorf, B. W., Jacobson, C. B.: "Beta-Galactosidase in Tissue Culture derived from Human Skin and Bone Marrow: Enzyme Defect in GM_1 Gangliosidosis". Pediat. Res. 3:532, 1969.

33. Wolf, P. L., Silberberg, B., Albert, S., Horwitz, J., Von der Muehll, E.: "Histochemical Similarities between Newborn and Leukaemic Myeloid Cells and their possible Significance". Clin. Path. 22:458, 1969.

CHAPTER 22

MULTIPHASIC TESTING
Elisabeth Von der Muehll, M.T. (ASCP) and Paul L. Wolf, M.D.

Before the introduction of instruments permitting simultaneous multiple biochemical determinations, physicians customarily ordered tests singly or in small groups, and frequently only to confirm a prior clinical impression. Advancing technology, plus a greater awareness of the value of early and presymptomatic diagnosis, however, has led physicians to incorporate various combinations of clinical and chemical determinations into the routine medical work-up. The resulting increase in information has provided further challenge to the physician who must interpret aberrant values.

In this study, we sought to determine what the physician, in fact, did when confronted with the unexpectedly "abnormal" laboratory result and what significance it had for the patient.

Information was obtained from 547 patients seen in the General Medical Clinic of the Stanford University Medical Center. They all had undergone new-patient work-ups by senior medical students under the supervision of faculty internists and were selected only by the criteria that their physicians had chosen to include a routine chemical screening battery in their medical evaluation. Thus, the study population differs from that of the total General Medical Clinic, as well as from the so-called normal population, in that it represents a group of symptomatic patients whose cases physicians sought aid in evaluating through the use of a screening battery of biochemical test

Medical record information was obtained on all patients selected through the computer because their values of a particular chemical determination exceeded two standard deviations from the mean, i.e., lay outside the normal range as indicated by the gray area in the laboratory printout of the computerized screening (Technicon SMA 12/60).

Of the test results available, we chose to investigate the following values: Alkaline phosphatase greater than 80 I.U./ml.; bilirubin above 1.2 mg./100 ml.; SGOT above 50 International Units (I.U.) per ml.; BUN above 20 mg./100 ml. and BUN below 10 mg./100 ml. Medical records of all the patients manifesting these "abnormalities" were carefully reviewed to ascertain whether the physician: 1). acknowledged the presence of the laboratory value in any written for

Multiphasic Testing (Continued):

2). attached any interpretation to it; 3). repeated the test; and, if so, 4). whether the repeated test fell within or without the normal range; and 5). whether this screening "abnormality" led to a previously unrecognized diagnosis.

RESULTS

In the majority of instances, deviations from the normal range led to no specific diagnoses within the 7 to 12 month period of observation following their new-patient work-up. Nor did the extent of deviation from the normal range correlate with the likelihood of obtaining a specific diagnosis, in that both borderline and extreme deviations were seen in the association of, as well as in the absence of, a specific disease entity. Physicians confronted with a screening laboratory value which deviated from the normal range tended to pay little attention to it, and either considered it to be clinically insignificant or made no interpretation. The latter tended to be more likely when the value was slightly elevated. Diagnoses which were made either exclusively or in large part as a consequence of the laboratory result were established promptly and no occult diseases emerged later over the period of time under study.

COMMENT

Although the value of automated multi-test biochemical screening is still controversial, its use, for a variety of reasons -- some medical, some economic -- is undoubtedly spreading. Thus, the contemporary physician will face the ever enlarging problem of interpreting unexpected laboratory "abnormalities". For this reason, he should have sure sense of what is meant by "normal", since the word is employed by both statisticians and clinicians, with meanings which are not at all synonymous. In statistical terms a normal curve is one in which there is a symmetrical distribution of continuous quantitative values, with a peak in the middle, tails extending in opposite directions, and a shape specified by the independent factors of mean and standard deviation. Under the influence of statistics, the notion of "normal" has come to mean any values lying within an arbitrary distance of two standard deviations from the mean (hence, the standard printout of laboratory results).

Multiphasic Testing (Continued):

Sensitivity, specificity, and reliability of laboratory tests must be considered in the interpretation of deviant values. Artifactual variations can be provoked by such factors as drugs, diet, time and method of collection, pregnancy, and errors in the laboratory.

A wide variety of factors might account for the unexpectedly "abnormal" chemical laboratory result. In this study there was little evidence that physicians pursued an explanation for such results in any systematic manner. A similar observation was made by other groups, who noted "an apparently arbitrary acceptance or rejection of abnormal laboratory reports" by physicians. On the basis of our data, we cannot determine how often the observed laboratory variants were artifactual. Although we can look forward to future improvements in sensitivity, specificity, and reliability of laboratory procedures, we must also anticipate additional sources of artifact as our modern pharmacopeia enlarges. Another important consideration in interpretation of laboratory results is that the composition of screening panels has not always taken into account disease prevalence. Thus, a laboratory test might be quite sensitive, specific, and reliable, yet still yield a high proportion of false positives if the disease itself is very rare. Finally, even after a condition is detected, its importance must be weighed in terms of outcome to the patient (i.e., is it really a "disease"?) and usefulness of therapy (i.e., can the patient be helped?). Considerations such as these have led quite naturally to controversies regarding the cost-benefit of routine laboratory screening, and these uncertainties may well be influencing physician response.

Evaluation of the role of routine screening in providing health care will have to begin with the assessment of the clinical significance of screening "abnormalities" by long-term studies on comparison populations looking for factors which correlate with morbidity and mortality. In the meantime, physicians will have to familiarize themselves with potential artifacts affecting the chemical screening battery in order to make reasonable appraisal, exercise discerning judgement, and maintain alertness to possible diagnostic clues.

We have recently produced two new charts at the Stanford University Medical Center stipulating a majority of the causes for high and low values on all of the 12/60 and 6/60 charts. Thus, this differential diagnosis may aid the physician in his practice utilizing these charts. The interspersed charts clarify the detailed lettering lost on the 6/60 and 12/60 charts as reproduced on the following pages. Please do not be concerned about the readability of the charts.

Multiphasic Testing (Continued):

From the physician or student point of view, the important aspects of the chart is the graph lines.

We have produced one hundred-twenty-eight typical patterns of disease and plan to introduce this data into our new computer in the laboratory. In the near future, the laboratory computer will report a differential diagnosis with the SMA 12/60 and 6/60 charts.

STANFORD UNIVERSITY MEDICAL CENTER
Stanford University Hospital
Clinical Laboratory

6/60 Chart

Cl^-	CO_2	K^+	Na^+	BUN	Glu.
	HIGH		HIGH		
	Resp. Acidosis		Dehydration		
	Metab. Alkalosis		Cushing's Syndrome		
	Diuretics		Conn's Syndrome		
	Cushing's Syndrome		Cerebral Damage		HIGH
	Hyperaldosterone		Excess Saline		Diabetes Mellitus
HIGH					Hepatic Disease
Resp. Alkalosis		HIGH			Brain Damage
Metab. Acidosis					Hypothalmic Lesion
Ureterosigmoidostomy		Renal Failure			Pheochromocytoma
Excess Saline		Excess Administration		HIGH	Hyperthyroidism
Renal Failure		Addison's Disease			Cushing's Syndrome
		Trauma		Renal Disease	Acromegaly
		Acidosis		Dehydration	
		Thrombocytosis		Hypotension	
	LOW	Hemolysis		G.I. Hemorrhage	
	Resp. Alkalosis		LOW	Heart Failure	
	Metab. Acidosis	Excess Water			LOW
	Diabetic Ketosis	Inappropriate ADH			
	Renal Failure	Lung Cancer			Excess Insulin
	Diarrhea	Addison's Disease			Insulinoma
	Starvation	Renal Failure			Addison's Disease
		Diarrhea			Hepatic Insuffici.
LOW		Hyperlipemia			Malabsorption
Vomiting					Bulky Neoplasms
Excess Water	LOW				
Resp. Acidosis	Diarrhea				
Metab. Alkalosis	Diuretics				
Renal Disease	Cortisone			LOW	
Inappropriate ADH	Cushing's Syndrome				
Addison's Disease	Conn's Syndrome			Excess in Fluids	
	Periodic Paralysis			Hepatic Insufficiency	
	Glucose Insulin			Pregnancy	

A magnified view of the causes for high and low values of each determination on the Stanford Hospital Clinical Laboratory SMA 6/60 Chart.

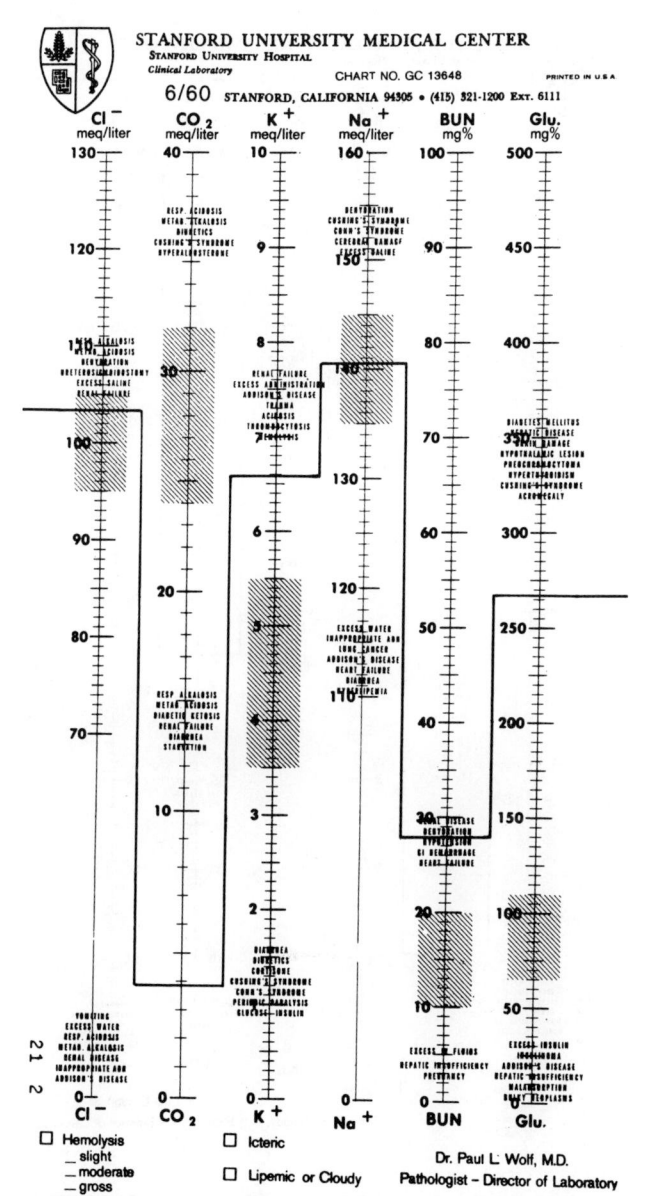

Fig. 22 - 1 BIOCHEMICAL PATTERN OF METABOLIC ACIDOSIS - ANION GAP

1). Extreme decrease in HCO_3^- without reciprocal increase in Cl^-. 2). Lack of equivalence of cations Na^+ and K^+ and anions HCO_3^- and Cl^- with anion gap. The anion gap consists of organic acids such as lactic acid or ketoacids associated with diabetic ketoacidosis, renal failure or ingestion of methyl alcohol or ethylene glycol. 3). Hyperkalemia due to acidosis.

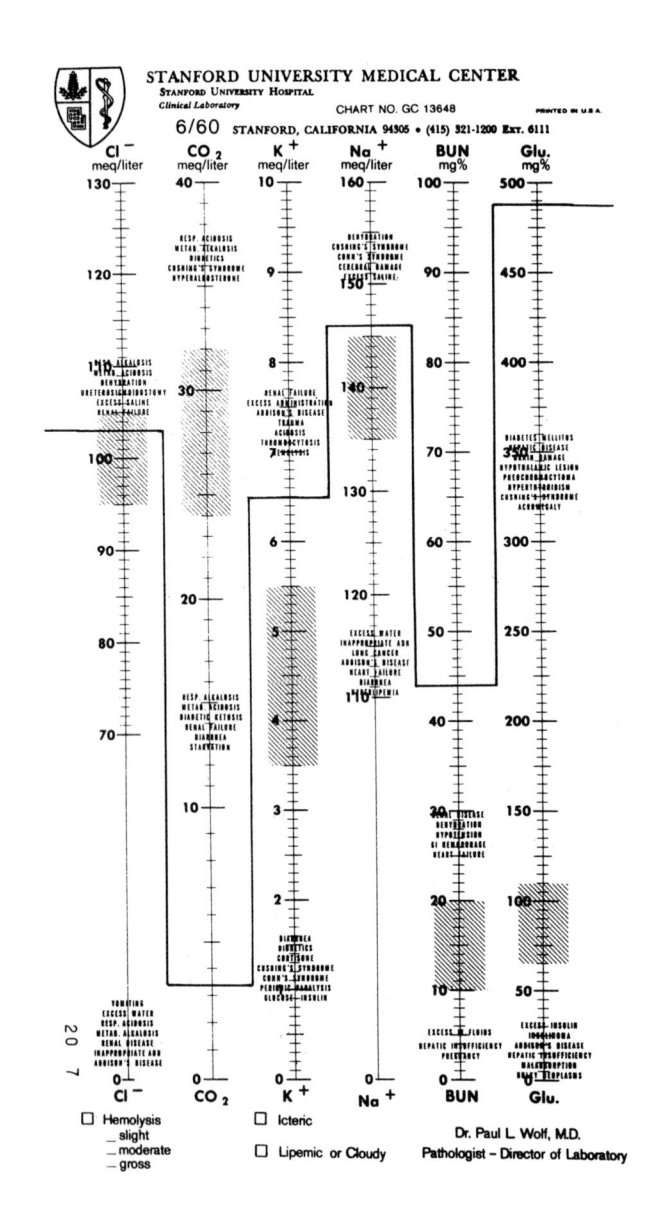

Fig. 22 - 2 BIOCHEMICAL PATTERN OF DIABETIC KETOACIDOSIS

1). Extreme decrease in HCO_3^- without reciprocal increase in Cl^-. 2). Lack of equivalence of cations Na^+ and K^+ anions HCO_3^- and Cl^- with anion gap. The anion gap consists of ketoacids and lactic acid in diabetic ketoacidosis 3). Hyperkalemia is found early in the course of diabetic ketoacidosis as is elevated BUN due to dehydration and prerenal azotemia.

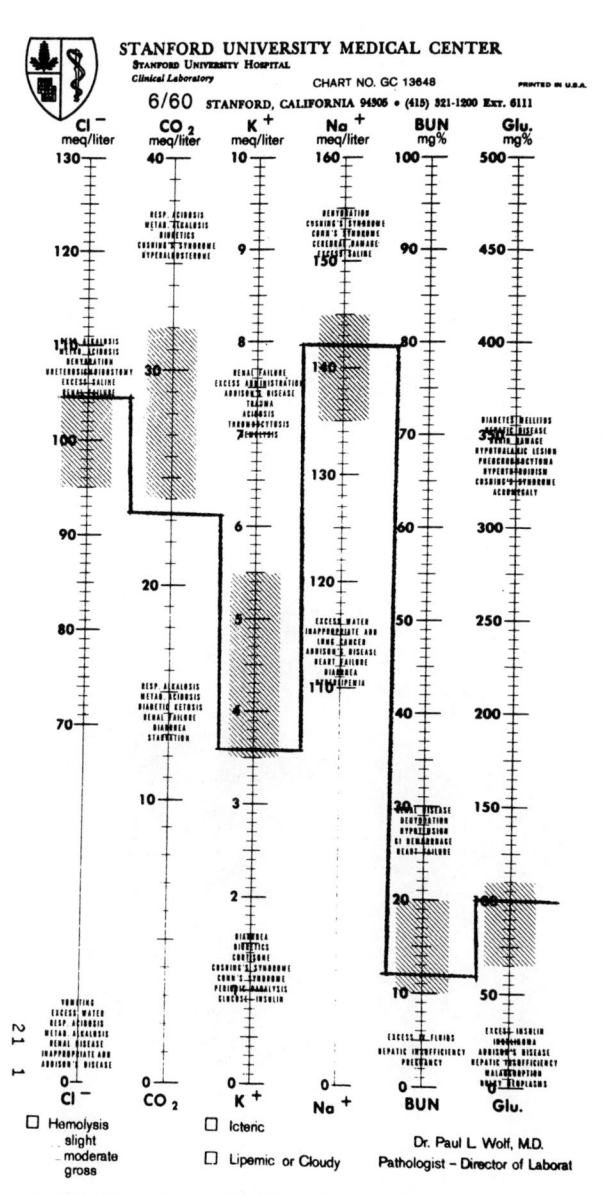

Fig. 22 - 3   BIOCHEMICAL PATTERN OF RESPIRATORY ALKALOSIS

1). Hyperventilation secondary to utilization of a respirator, meningoencephalitis, excessive use of salicylates, pleural pain in pneumonia or pulmonary infarction, psychoneurosis or severe liver disease. 2). Early excessive decrease of CO_2.

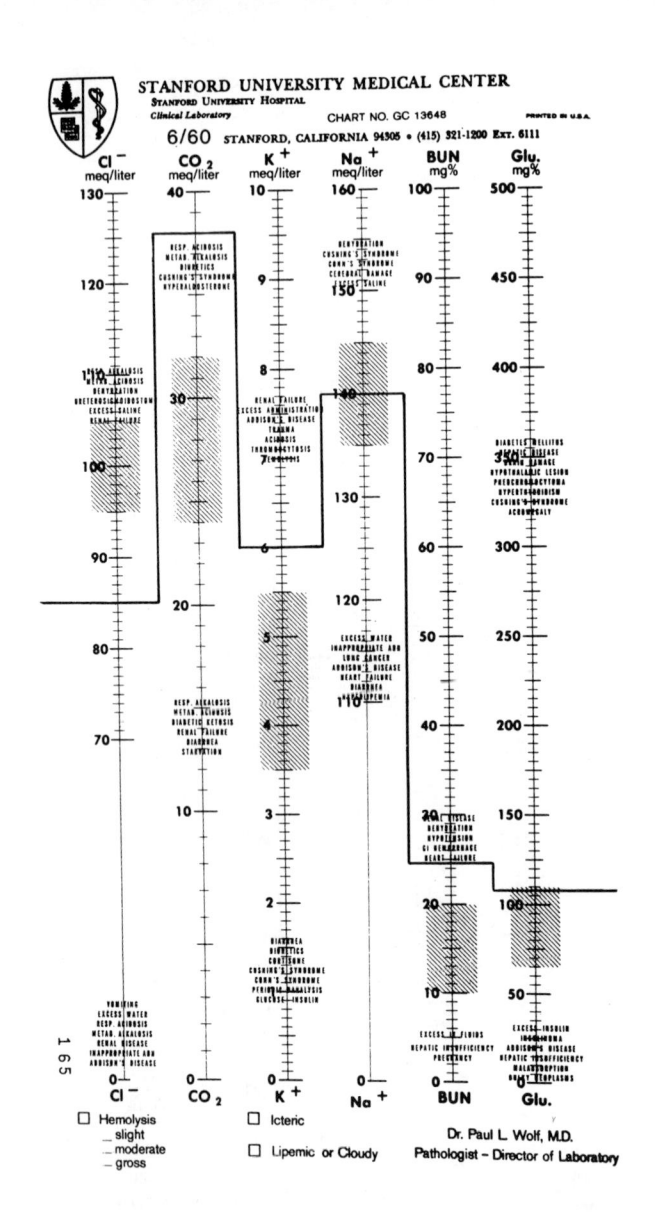

Fig. 22 - 4   BIOCHEMICAL PATTERN OF RESPIRATORY ACIDOSIS

1). Excessive retention of CO_2 with compensatory increase in HCO_3^-. 2). Reciprocal decrease in Cl^-. 3). Elevation in K^+ in acidosis. 4). Pattern consistent with chronic obstructive pulmonary disease, asthma, respiratory depression with narcotics, and Pickwickian syndrome.

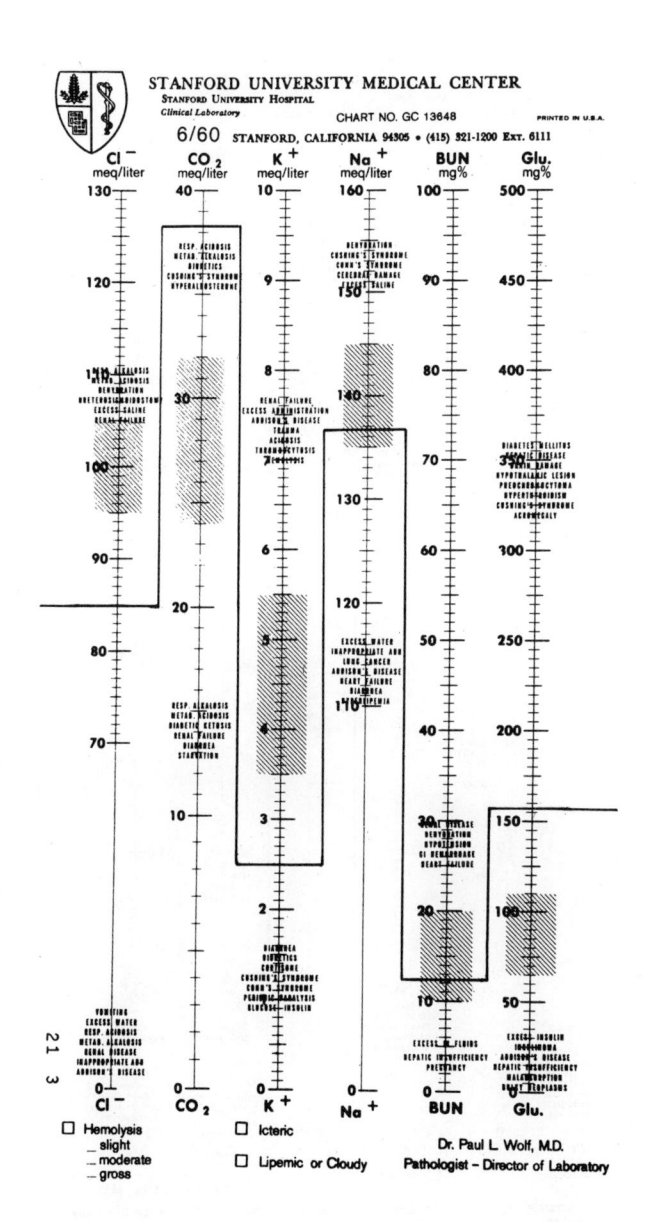

Fig. 22 - 5 BIOCHEMICAL PATTERN OF METABOLIC ALKALOSIS

1). Elevation of HCO_3^- secondary to hypokalemia.
2). Alkalosis is compensatory related to renal tubular attempt to conserve Na^+ and K^+ by hydrogen ion exchange.
3). Hypokalemia results from excessive potassium loss in diuretic therapy, Cushing's syndrome or Conn's syndrome.
4). Decreased Cl^- is reciprocal of HCO_3^- elevation and Cl^- loss as KCl.

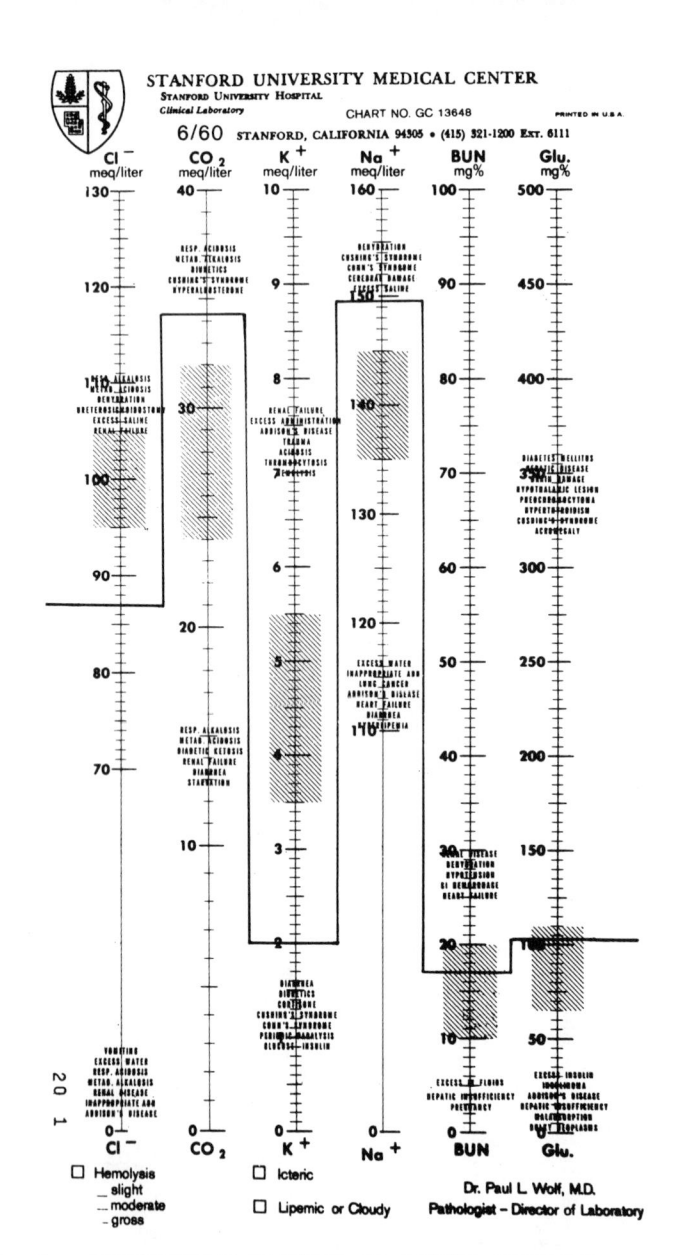

Fig. 22 - 6   BIOCHEMICAL PATTERN OF CONN'S SYNDROME

1). Elevation of HCO_3^- secondary to hypokalemia.
2). Alkalosis is compensatory related to renal tubula attempt to conserve Na^+ and K^+ by hydrogen ion exchange
3). Hypokalemia results from excessive potassium loss due to increase in serum aldosterone. 4). Hypernatre is due to increase in serum aldosterone. 5). Decreas Cl^- is reciprocal of HCO_3^- elevation and Cl^- loss as K

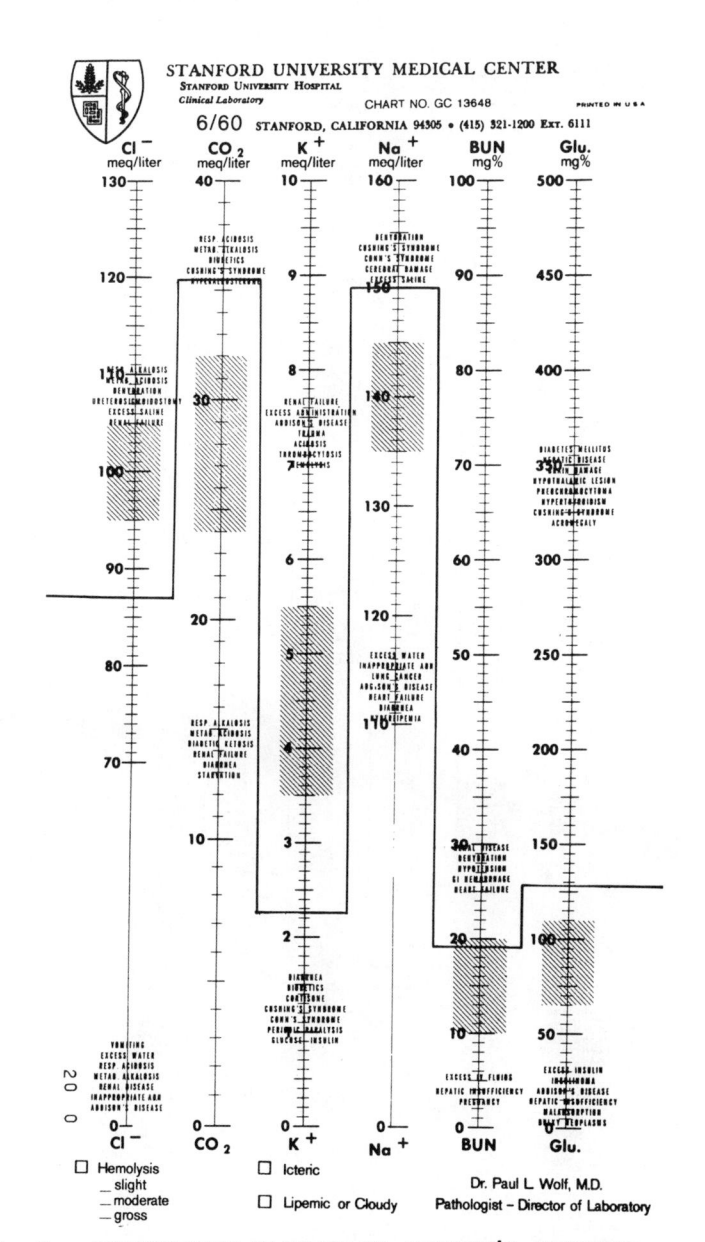

Fig. 22 - 7   BIOCHEMICAL PATTERN OF CUSHING'S SYNDROME

1). Elevation of HCO_3^- secondary to hypokalemia.
2). Alkalosis is compensatory related to renal tubular attempt to conserve Na^+ and K^+ by hydrogen ion exchange.
3). Hypokalemia, hypernatremia, and hyperglycemia is due to increase in serum cortisol. 4). Decreased Cl^- is reciprocal of HCO_3^- elevation and Cl^- loss as KCl.

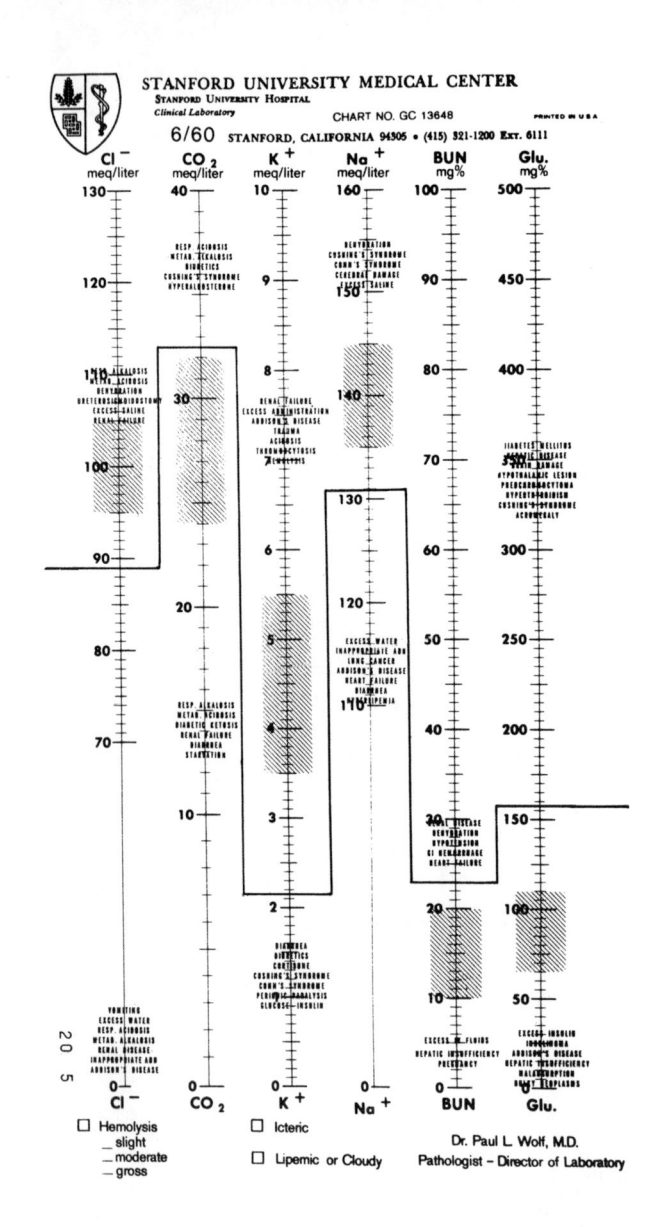

Fig. 22 - 8   BIOCHEMICAL PATTERN OF EXCESSIVE DIURETIC UTILIZATION

1). Elevation of HCO_3^- secondary to hypokalemia.
2). Alkalosis is compensatory related to renal tubular attempt to conserve Na^+ and K^+ by hydrogen ion exchange
3). Hypokalemia, hyponatremia, elevated BUN and hyperglycemia result from excessive use of diuretics.
4). Decreased Cl^- is reciprocal of HCO_3^- elevation and Cl^- loss as KCl.

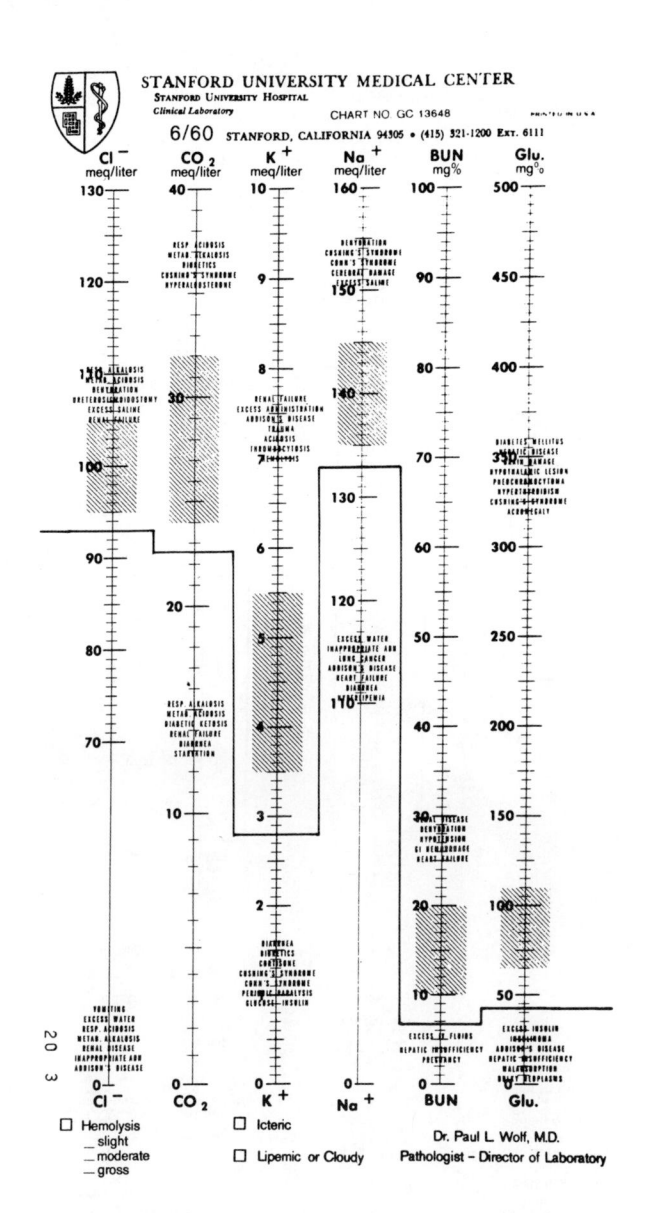

Fig. 22 - 9   BIOCHEMICAL PATTERN OF MALABSORPTION - CHRONIC DIARRHEA

1). Decreased Cl^-, HCO_3^-, K^+, Na^+, due to persistent chronic diarrhea. 2). Loss of base causes metabolic acidosis. 3). Decreased BUN relates to protein malabsorption. 4). Hypoglycemia results from carbohydrate malabsorption.

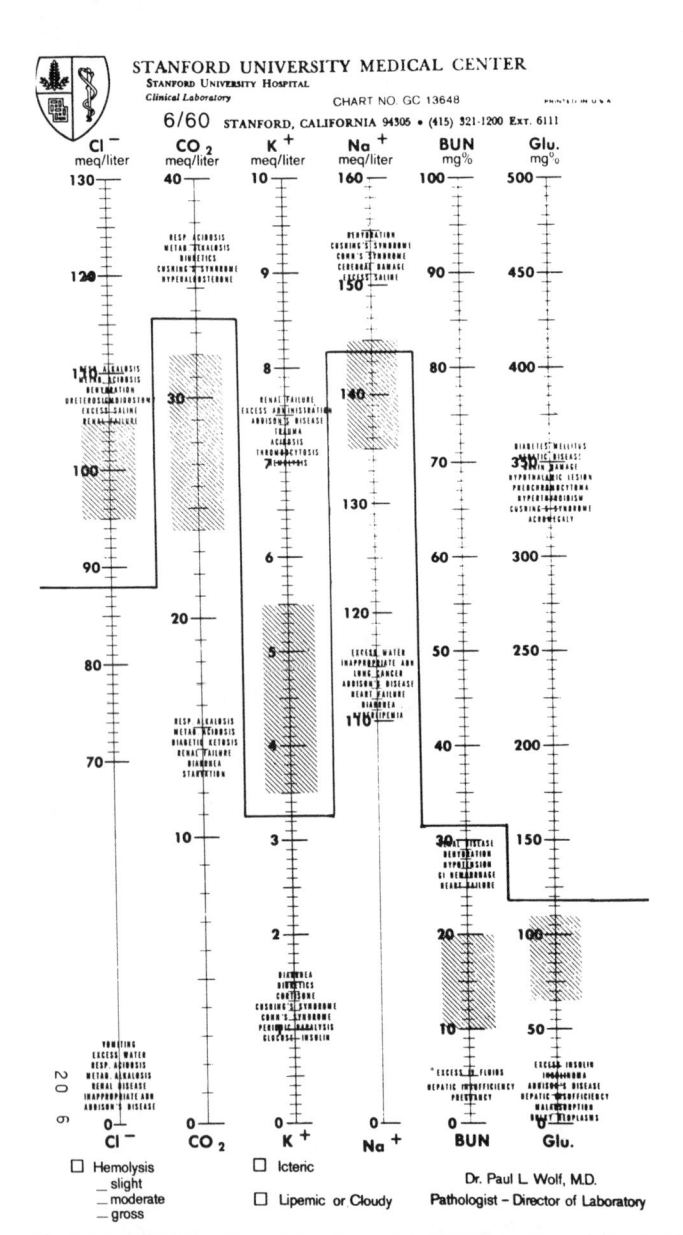

Fig. 22 - 10 BIOCHEMICAL PATTERN OF EXCESSIVE UTILIZATION OF LICORIC:

1). Excessive utilization of licorice from licorice ca⁻ or used as a cathartic induces a pseudo-Conn's syndrome
2). Elevation of HCO_3^- secondary to hypokalemia.
3). Alkalosis is compensatory related to renal tubular attempt to conserve Na^+ and K^+ by hydrogen exchange.
4). Hypokalemia results from excessive potassium loss.
5). Decreased Cl^- is reciprocal of HCO_3^- elevation and Cl^- loss as KCl.

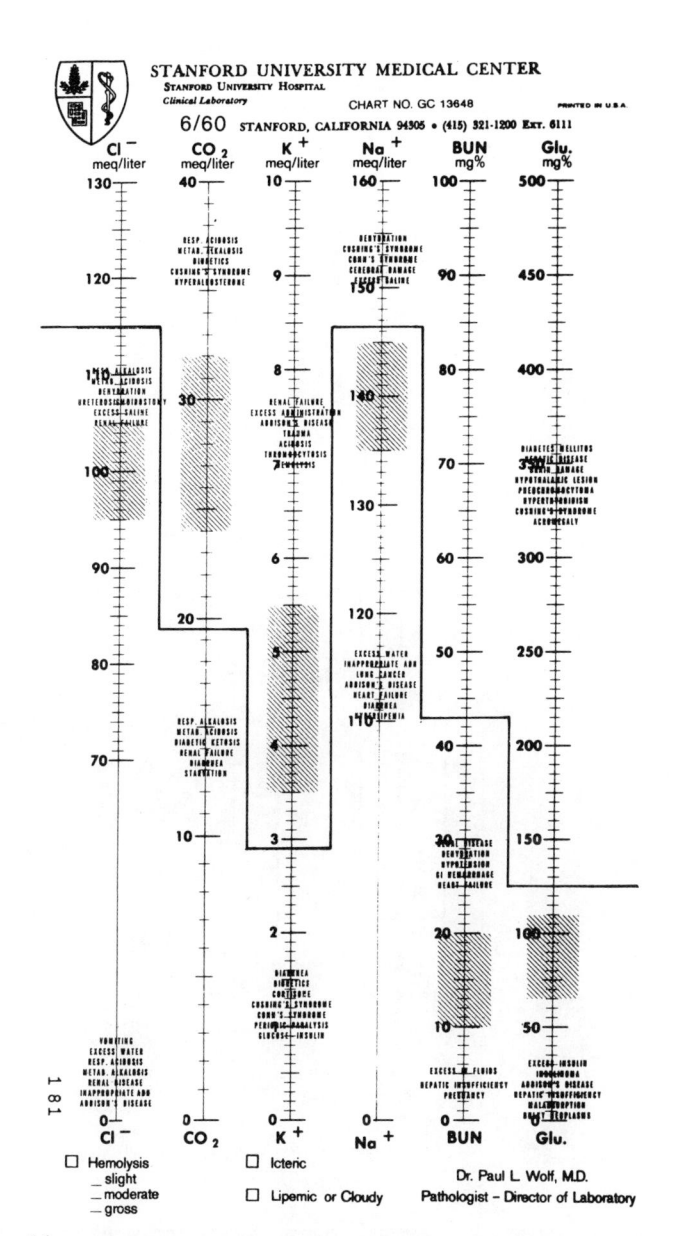

Fig. 22 - 11 BIOCHEMICAL PATTERN OF RENAL TUBULAR ACIDOSIS

1). Hyperchloremic metabolic acidosis with decreased HCO_3^- associated with renal insufficiency. The syndrome may be hereditary or acquired in conditions such as multiple myeloma, amphotericin fungal therapy, or hypokalemic nephropathy.

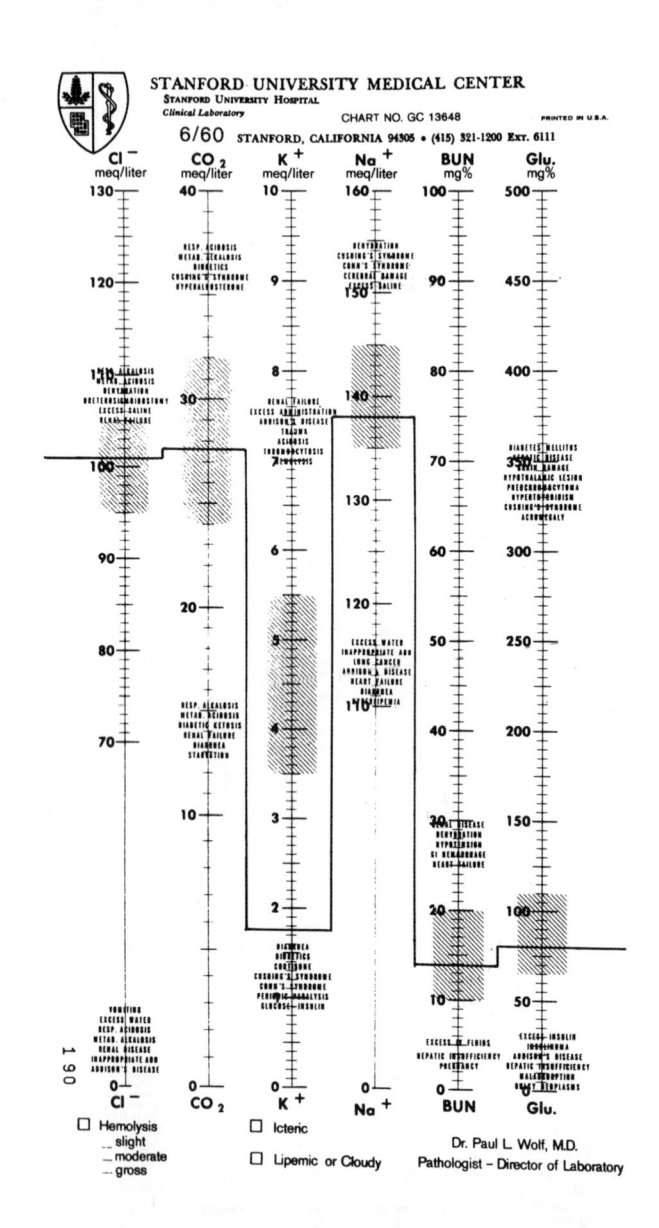

Fig. 22 - 12 BIOCHEMICAL PATTERN OF HYPOKALEMIC PERIODIC PARALYSIS

The hypokalemia results in severe muscle weakness which may result in death from intercostal muscle paralysis. A hyperkalemic periodic paralysis is more unusual.

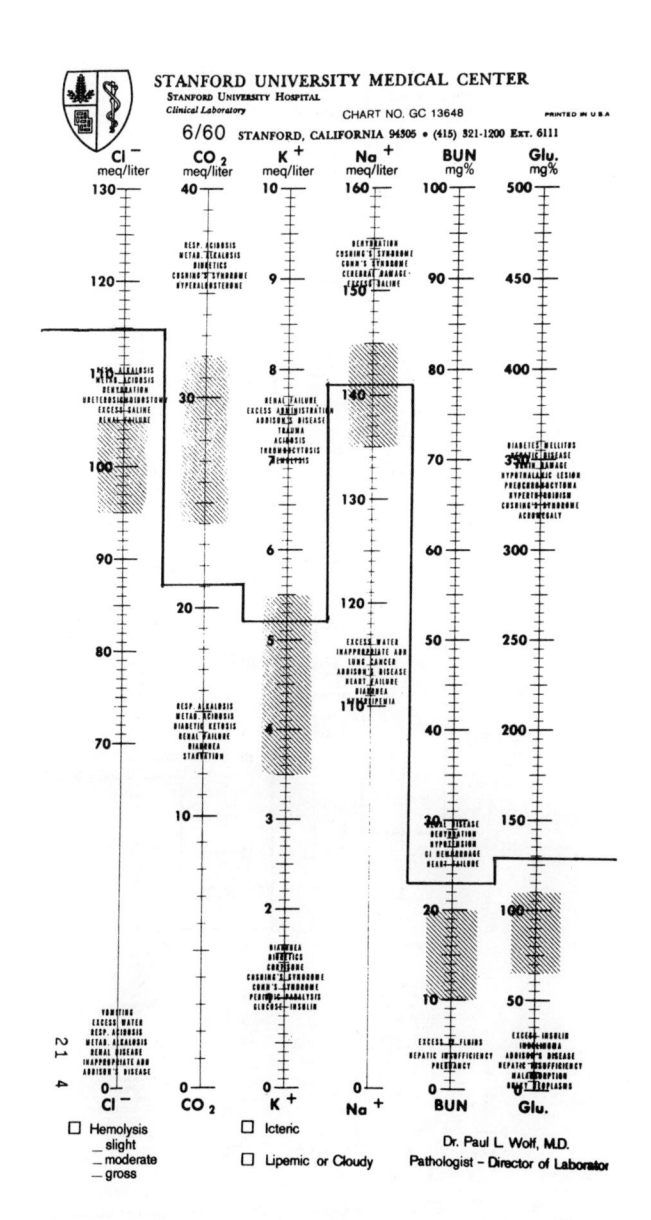

Fig. 22 - 13 BIOCHEMICAL PATTERN OF METABOLIC ACIDOSIS - HYPERCALCEMIA
PARATHYROID ADENOMA OR HYPERPLASIA

This pattern associated with hypercalcemia suggests the
diagnosis of hyperparathyroidism. Excess parathyroid
hormone causes marked HCO_3^- wastage due to action of
PTH on the renal tubule. Hyperchloremic metabolic
acidosis results.

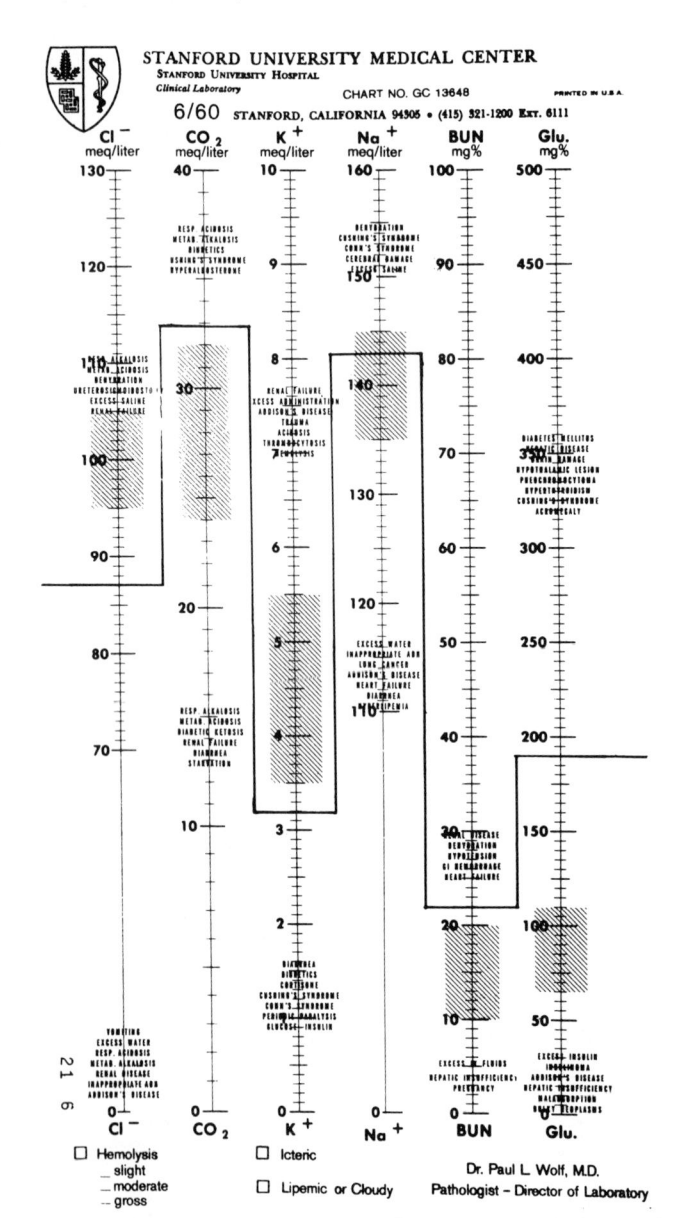

Fig. 22 - 14 BIOCHEMICAL PATTERN OF METABOLIC ALKALOSIS - HYPERCALCE
METASTATIC CANCER TO BONE

In contrast to Fig. 22 - 13, hypercalcemia associated with metastatic bone cancer is associated with metaboli alkalosis unlike the metabolic acidosis caused by exces PTH in hyperparathyroidism.

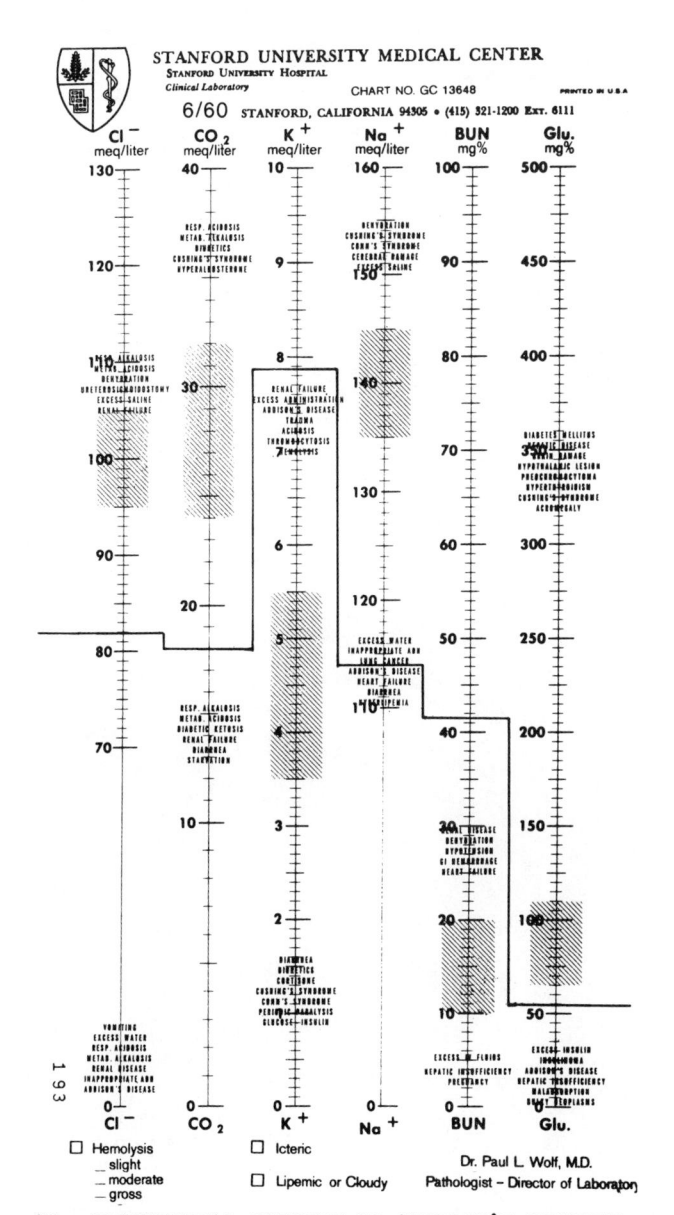

Fig. 22 - 15   BIOCHEMICAL PATTERN OF ADDISON'S DISEASE

Decreased serum cortisol results in marked renal tubular sodium wastage with hyponatremia, metabolic acidosis, and hyperkalemia. Prerenal azotemia occurs with elevated BUN. Hypoglycemia is associated with adrenal insufficiency.

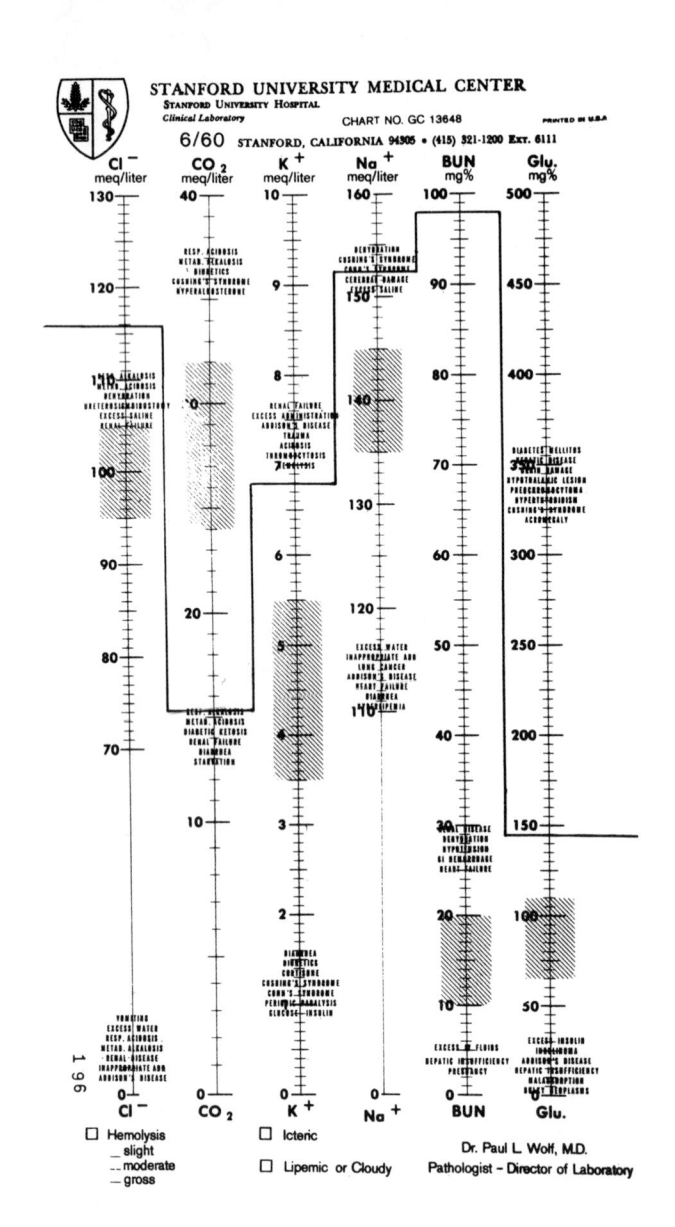

Fig. 22 - 16   BIOCHEMICAL PATTERN OF RENAL INSUFFICIENCY

Renal failure results in hypernatremic, hyperchloremic metabolic acidosis. Hyperkalemia is caused by the acid and potassium retention. Thorn has described a rare syndrome in renal failure which resembles Addison's syndrome and presents with hyponatremia and hyperkalemia. Hyperglycemia results from anti-insulin effects of uremic toxins.

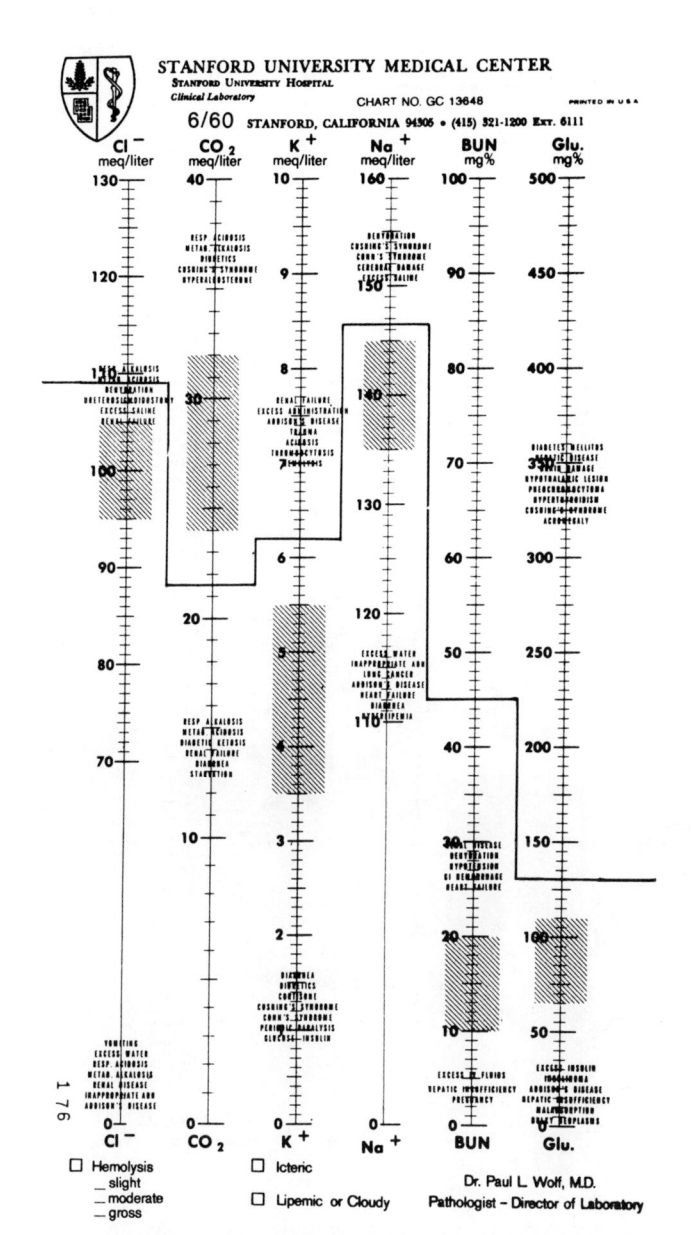

Fig. 22 - 17 BIOCHEMICAL PATTERN OF PRERENAL INSUFFICIENCY

The biochemical pattern of prerenal azotemia, secondary to hypovolemia or cardiac failure, etc., may simulate that of Fig. 22 - 16, renal failure due to intrarenal disease. It is important to compare the BUN/Creatinine ratio. In prerenal azotemia, the ratio is greater than 10:1 while in intrarenal disease, the ratio is preserved at 10:1.

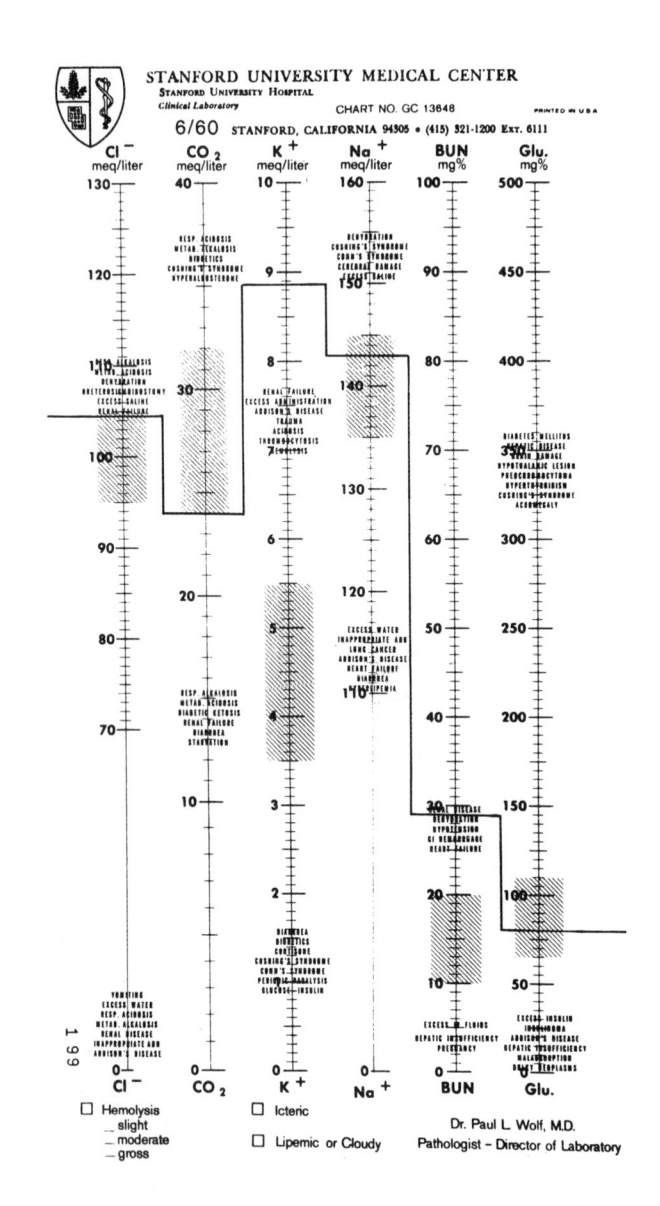

Fig. 22 - 18   BIOCHEMICAL PATTERN OF HEMOLYSIS

Hemolysis will liberate potassium from the erythrocyte and cause hyperkalemia. In addition, renal tubular disease may develop from the acute hemolysis due to hypotension or effect of hemoglobin on the renal tubule

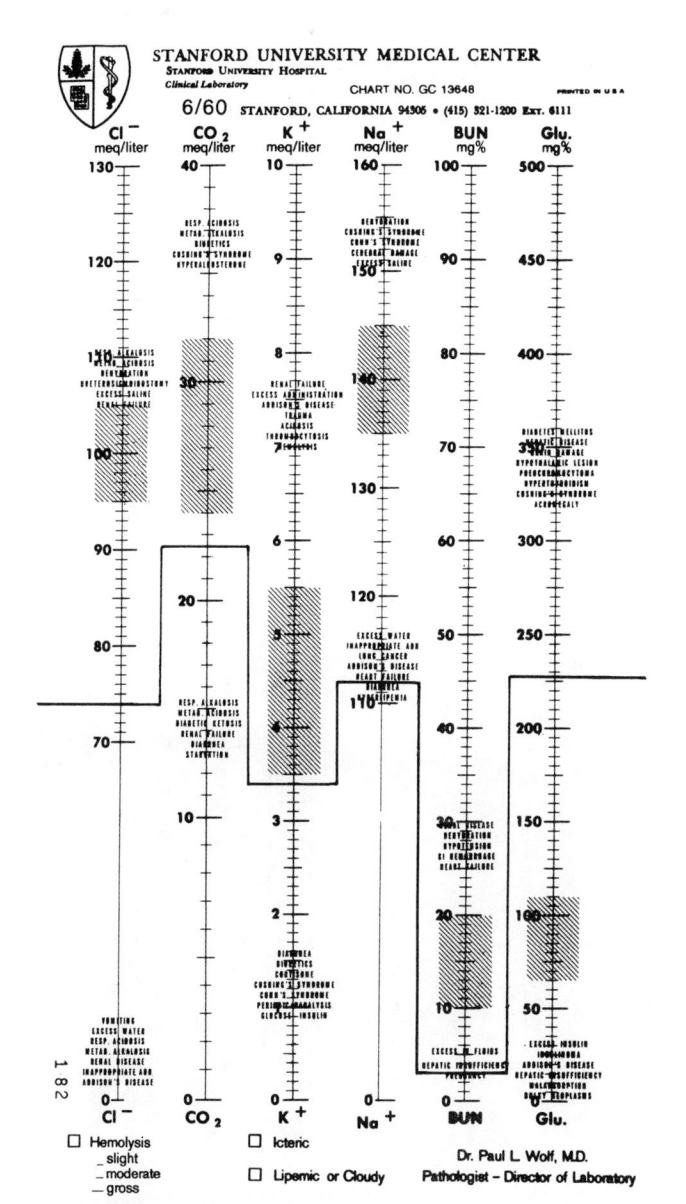

Fig. 22 - 19 BIOCHEMICAL PATTERN OF EXCESSIVE UTILIZATION OF
5 PERCENT GLUCOSE IN WATER

Excessive utilization of 5 percent glucose in water may cause water intoxication with hyponatremia and decreased BUN. If insulin is utilized, a shift of potassium from the serum into the cell occurs. Cerebral function is disturbed by the hyponatremia.

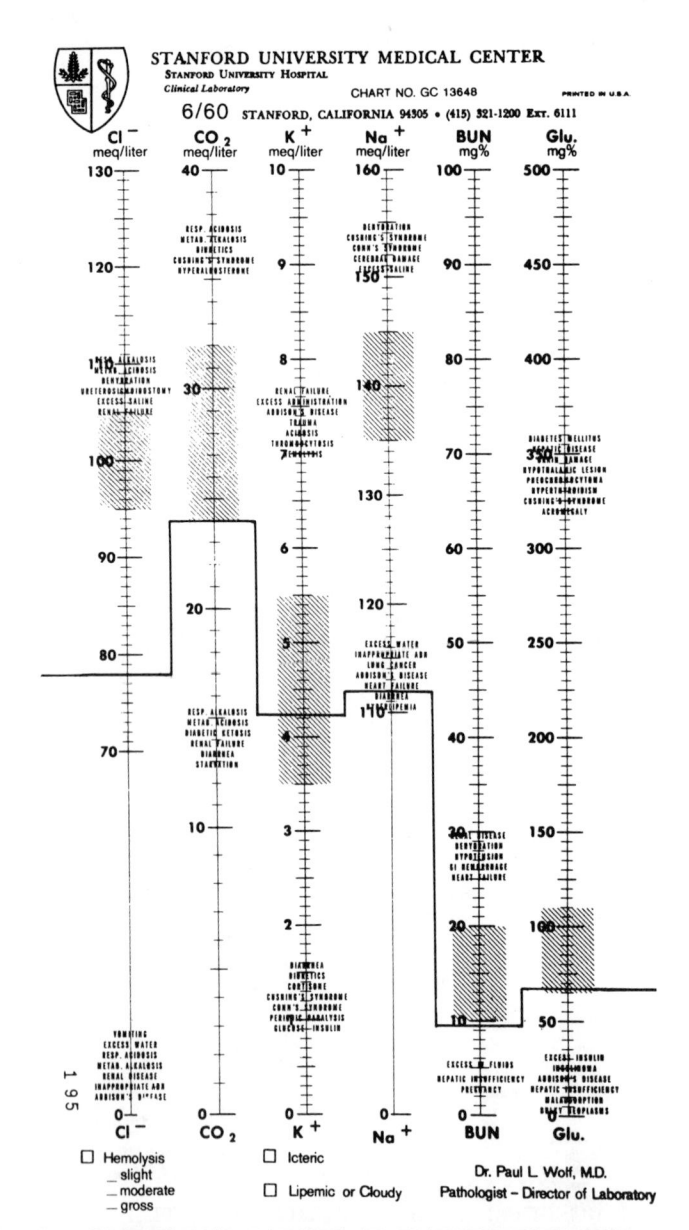

Fig. 22 - 20 BIOCHEMICAL PATTERN OF INAPPROPRIATE ADH

Inappropriate ADH associated with undifferentiated bronchogenic carcinoma, meningitis, hypothalamic lesion, myxedema, acute intermittent porphyria, pulmonary tuberculosis or severe liver disease will cause a marked hyponatremia with stupor, coma, or convulsions.

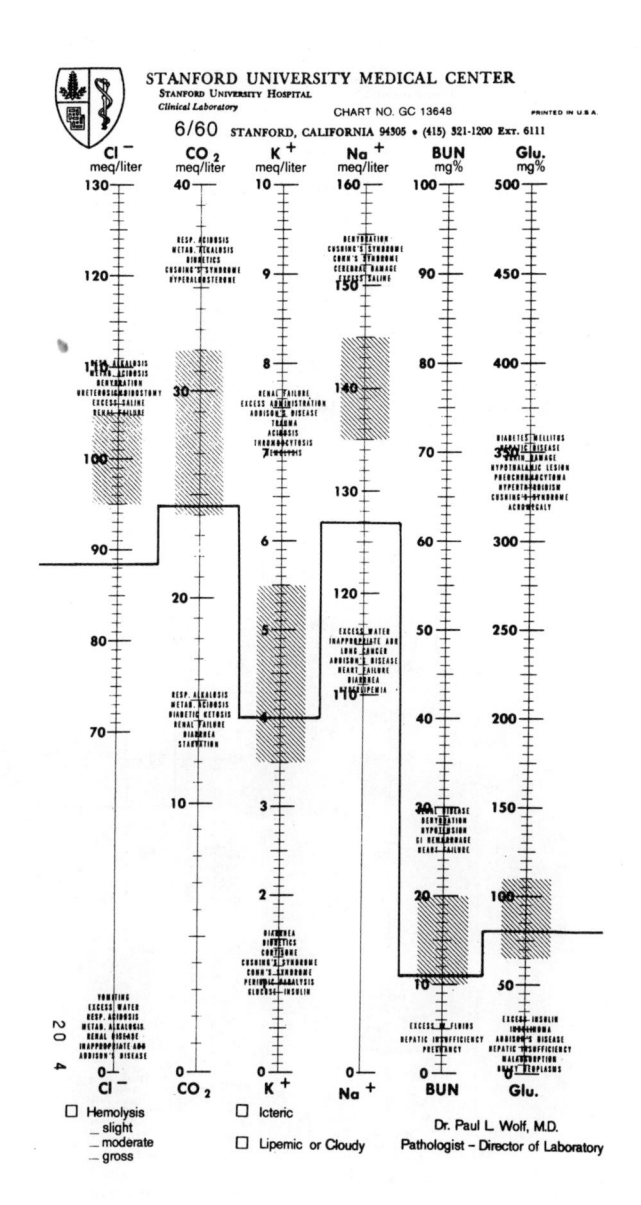

Fig. 22 - 21 BIOCHEMICAL PATTERN OF HYPERLIPEMIA

Hyperlipemia will result in pseudohyponatremia. Lipi
have a hydrophobic action and cause an artifactual
hyponatremia.

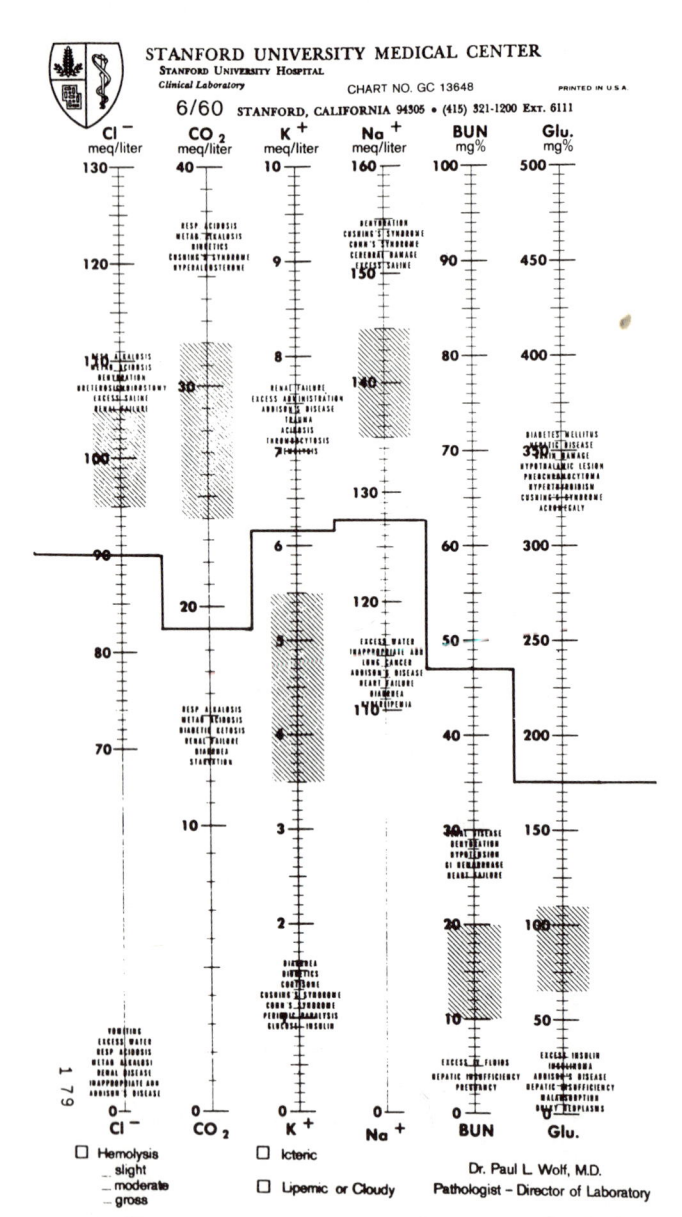

Fig. 22 - 22   BIOCHEMICAL PATTERN OF NEPHROTIC SYNDROME

Hyponatremia is the pseudohyponatremia associated with hyperlipemia of nephrotic syndrome. Lipids exert a hydrophobic action and cause the hyponatremic pattern. The renal insufficiency causes hyperkalemia and metabolic acidosis. Hyperglycemia is due to non-specific reduction of neocuproine reagent by elevated creatinine, uric acid and other uremic toxins.

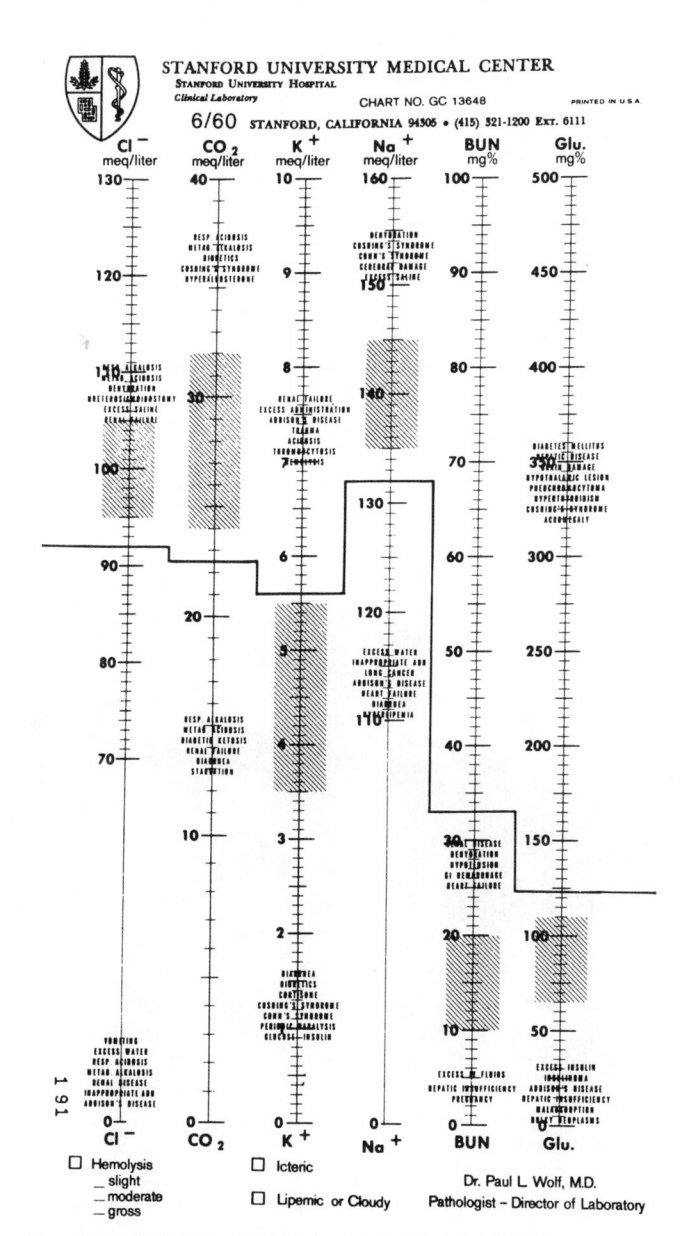

Fig. 22 - 23   BIOCHEMICAL PATTERN OF HEART FAILURE

1). Hyponatremia is found in heart failure. Total body sodium is elevated but a paradoxical hyponatremia occurs due to increased intravascular water. 2). Prerenal azotemia and hyperkalemia is present. 3). Hyperglycemia is caused by decreased serum insulin from the anoxic pancreas in congestive heart failure.

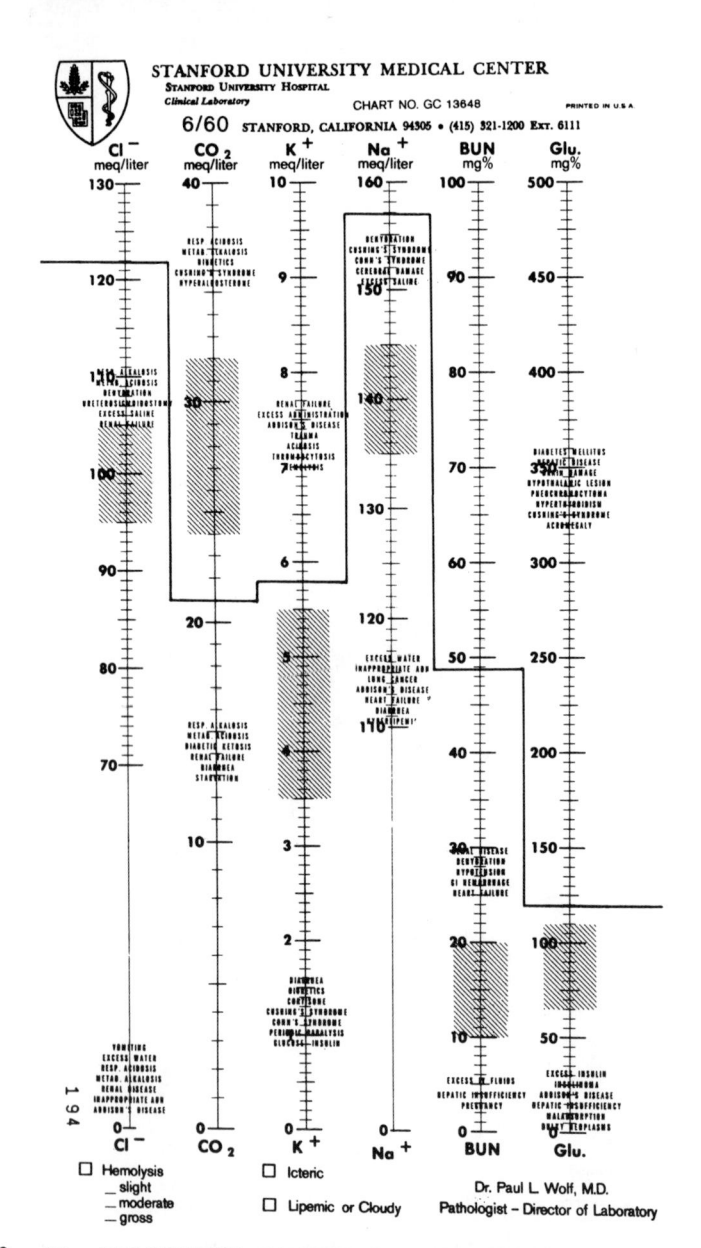

Fig. 22 - 24   BIOCHEMICAL PATTERN OF DEHYDRATION

Hypernatremia and hyperchloremia are associated with severe water dehydration. Prerenal azotemia causes the increased BUN and hyperkalemia.

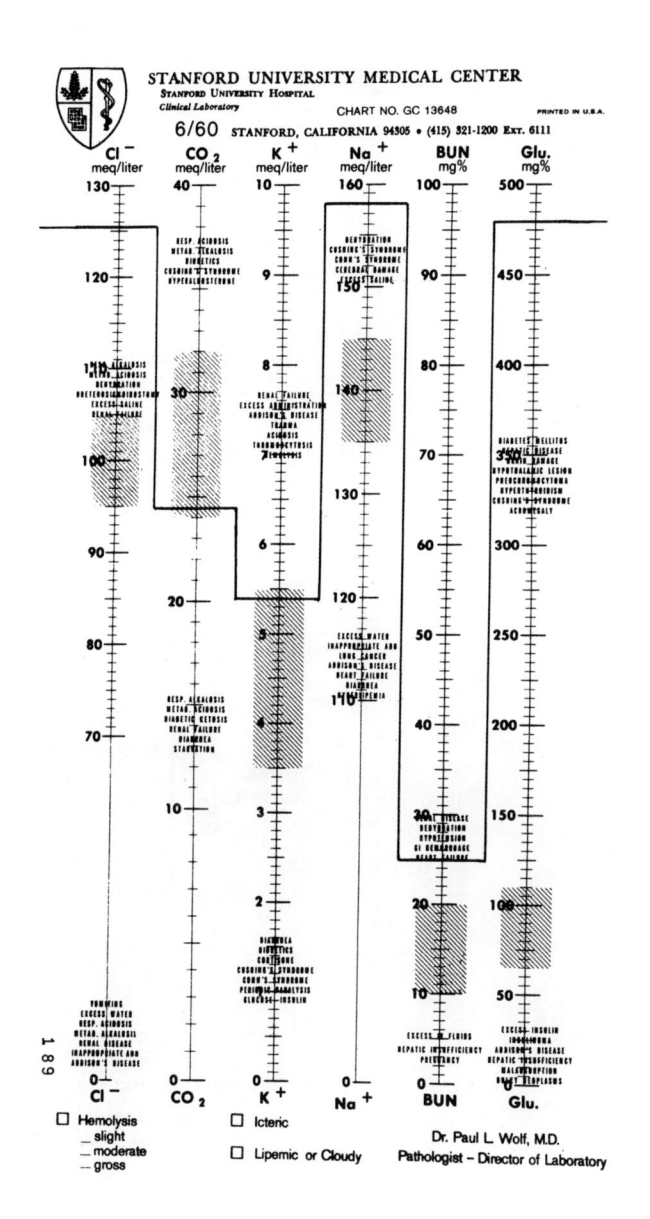

Fig. 22 - 25 BIOCHEMICAL PATTERN OF CEREBRAL DAMAGE

The hypernatremia and hyperchloremia result from hypothalamic damage. Hyperglycemia results from the diabetic pique. Stimulation of the hypothalamic-adrenal medullary axis results in liver glycogenolysis and hyperglycemia.

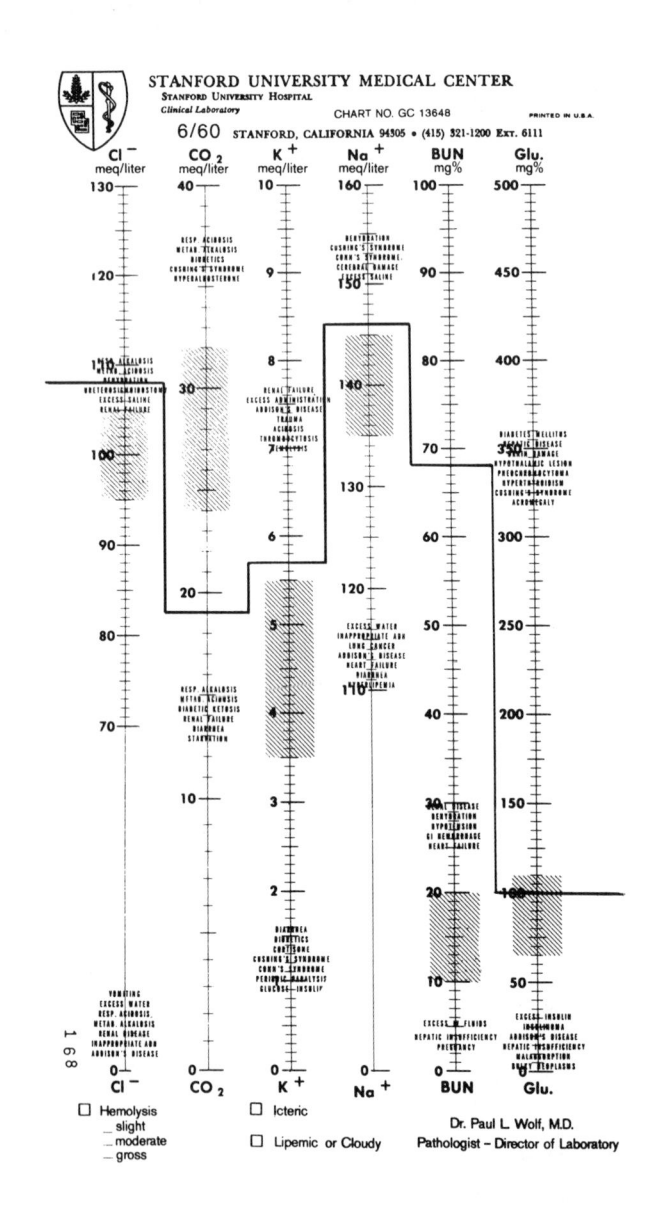

Fig. 22 - 26   BIOCHEMICAL PATTERN OF GASTROINTESTINAL HEMORRHAGE

Blood in the gastrointestinal tract results in elevated BUN. A direct conversion of blood protein to BUN occurs. Hypotension from gastrointestinal hemorrhage causes prerenal azotemia with metabolic acidosis, hyperkalemia and hypernatremia.

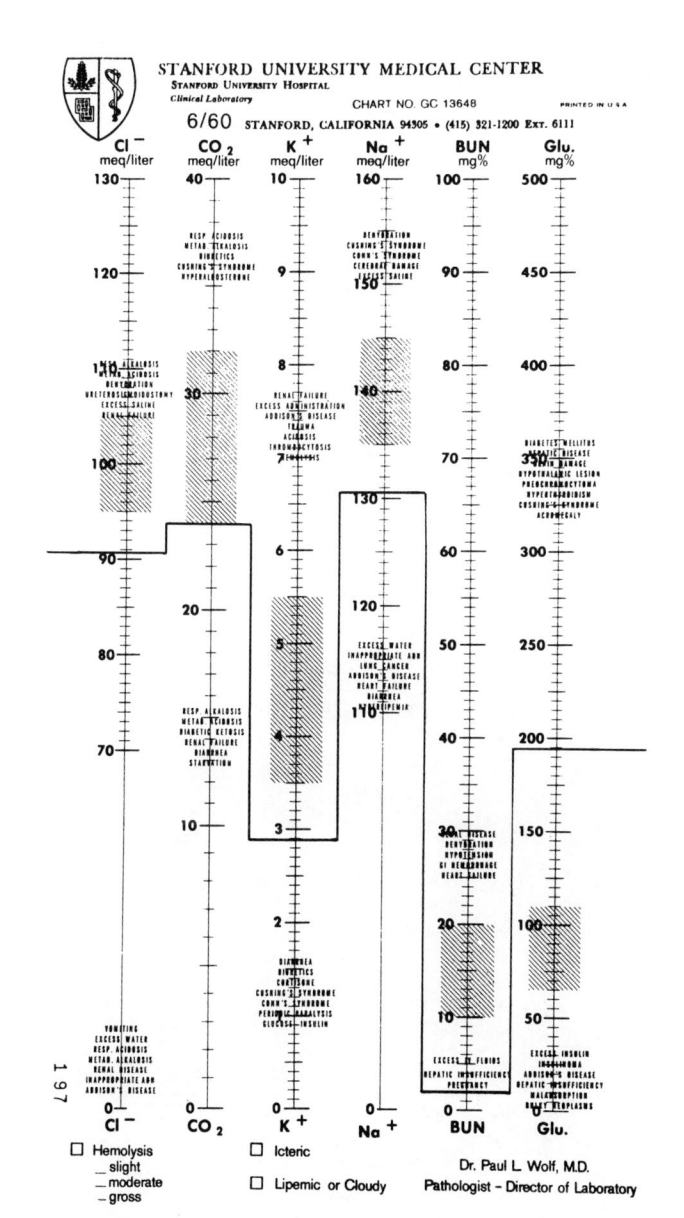

Fig. 22 - 27 BIOCHEMICAL PATTERN OF EXCESSIVE INTRAVENOUS GLUCOSE AND INSULIN

Excessive utilization of intravenous 5 percent glucose in water causes overhyration with hyponatremia, hypochloremia, and low BUN. A shift of intravascular potassium into a cellular location occurs.

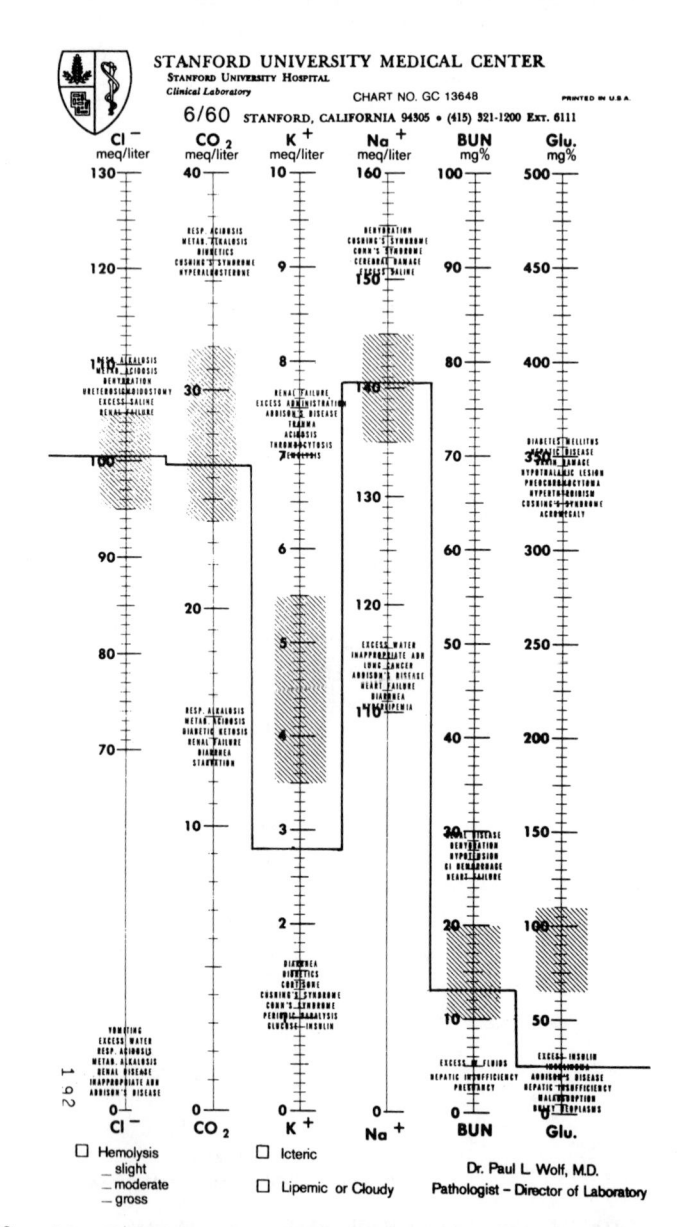

Fig. 22 - 28 BIOCHEMICAL PATTERN OF EXCESSIVE INSULIN

Excessive utilization of insulin in a diabetic causes hypoglycemia and a shift of potassium from the serum to an intracellular location occurs.

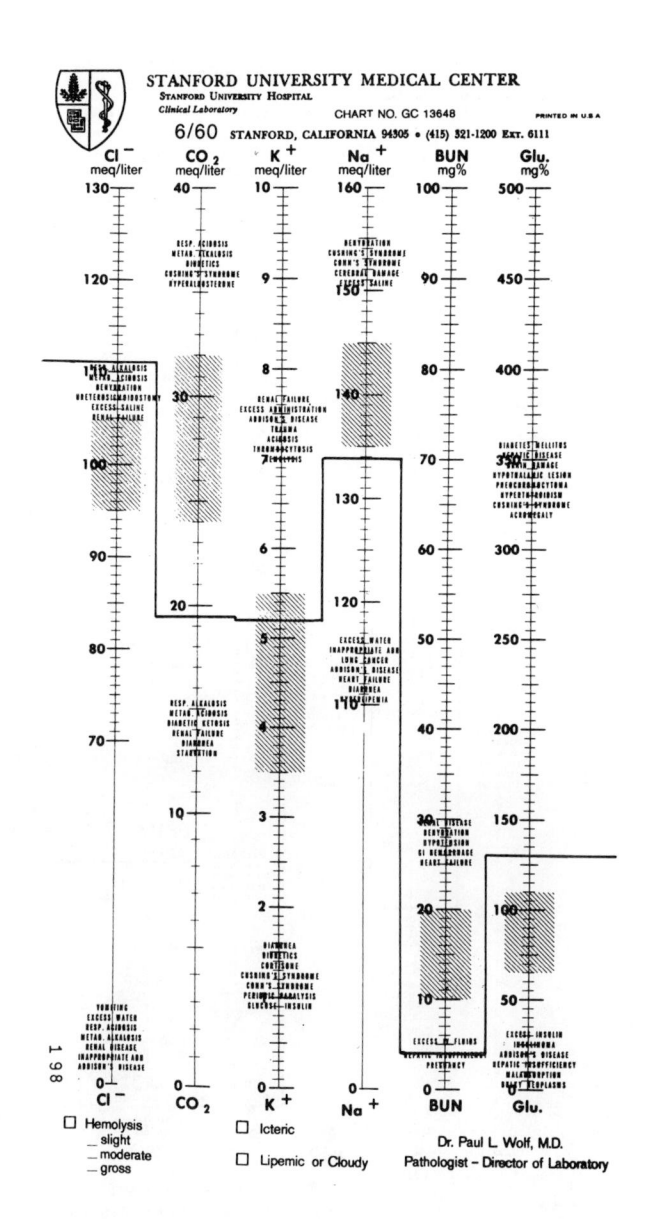

Fig. 22 - 29   BIOCHEMICAL PATTERN OF PREGNANCY - THIRD TRIMESTER

1). Hyperventilation in the third trimester of pregnancy occurs especially during labor. Respiratory alkalosis with hyperchloremia is present. 2). Hyponatremia and low BUN result from physiologic hydremia. 3). Hyperglycemia is present as a result of increased estrogen and progesterone of pregnancy.

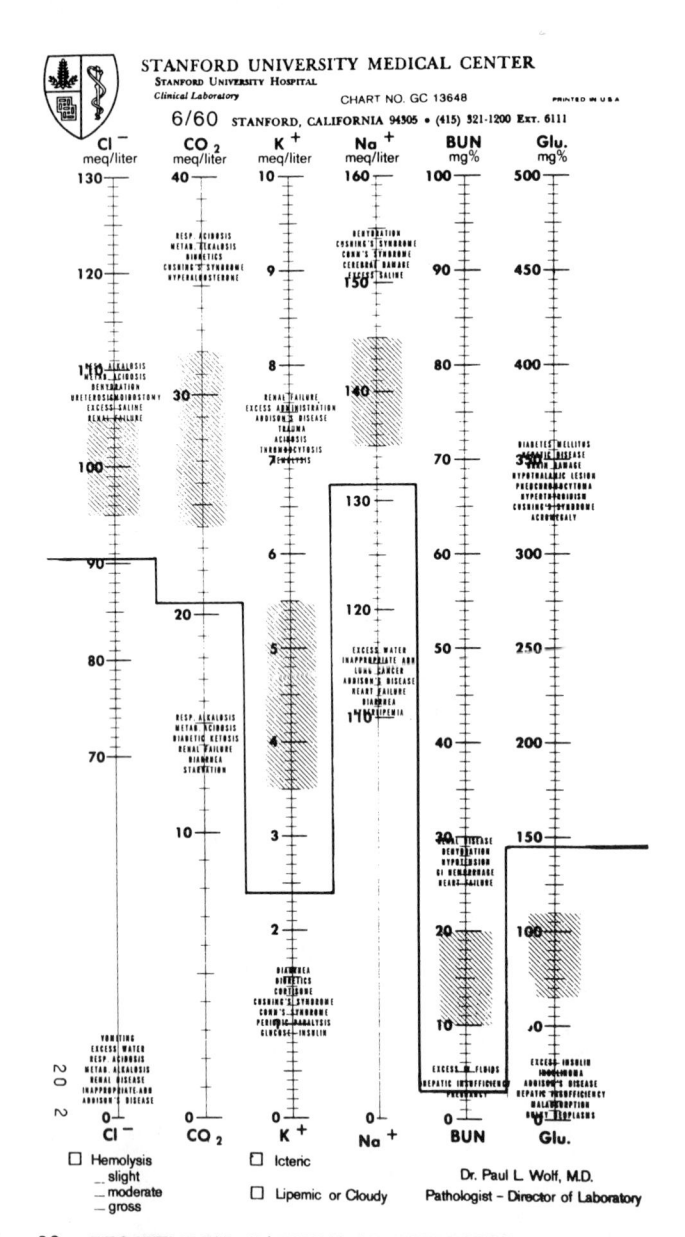

Fig. 22 - 30   BIOCHEMICAL PATTERN OF CIRRHOSIS

Hyponatremia and hypochloremia are caused by excessive ADH function due to lack of inactivation by hepatic cellular dysfunction. Respiratory alkalosis is present which is compensatory due to excess blood lactic acid in hepatic disease. Hypokalemia is due to secondary aldosteronism. Low BUN is due to decreased production in hepatic disease. Hyperglycemia is secondary to glycogenolysis.

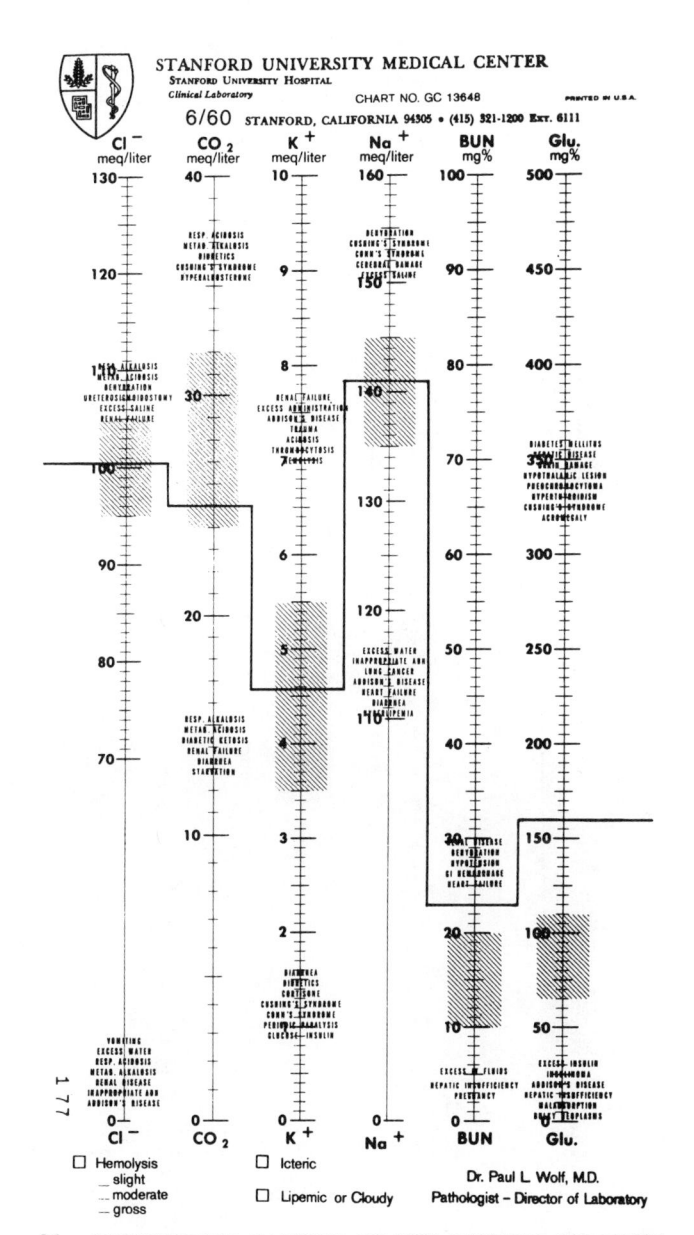

Fig. 22 - 31   BIOCHEMICAL PATTERN OF NON-FASTING SPECIMEN

Hyperglycemia and elevated BUN are caused by obtaining the serum specimen shortly after a high protein and carbohydrate meal.

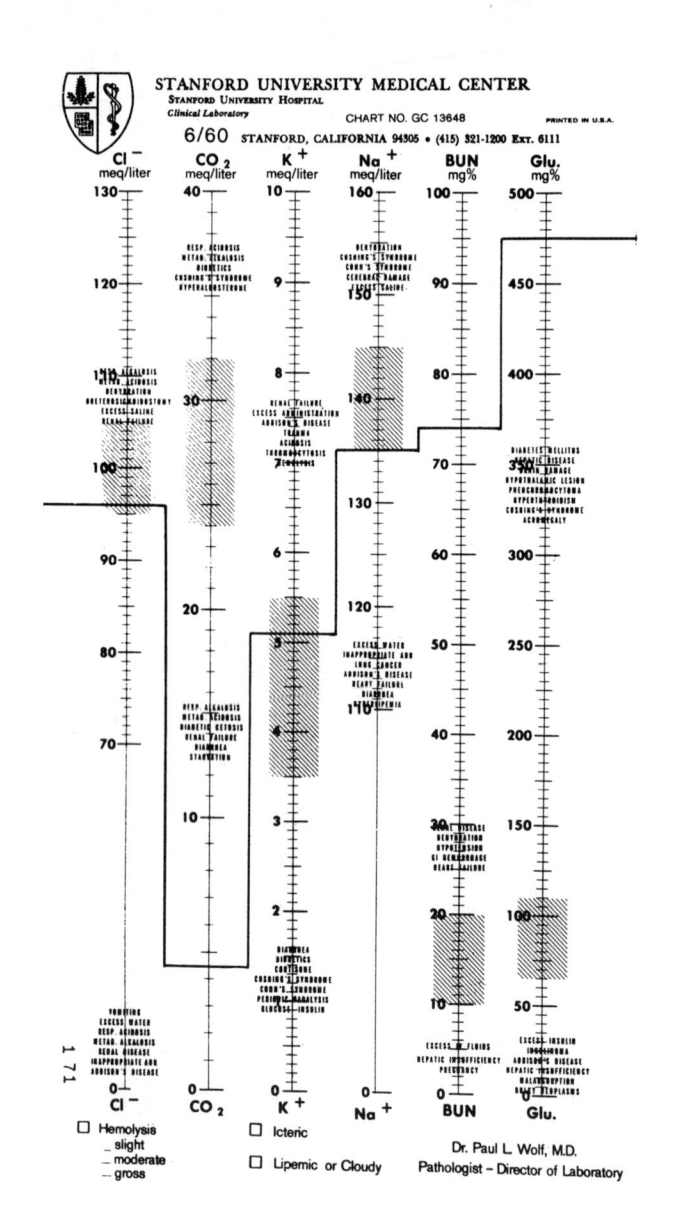

Fig. 22 - 32 BIOCHEMICAL PATTERN OF POST-MORTEM VITREOUS HUMOR
- JUVENILE DIABETES

The vitreous humor reflects the hyperglycemia, elevated BUN, and metabolic acidosis of the serum of a patient who died of diabetic ketoacidosis.

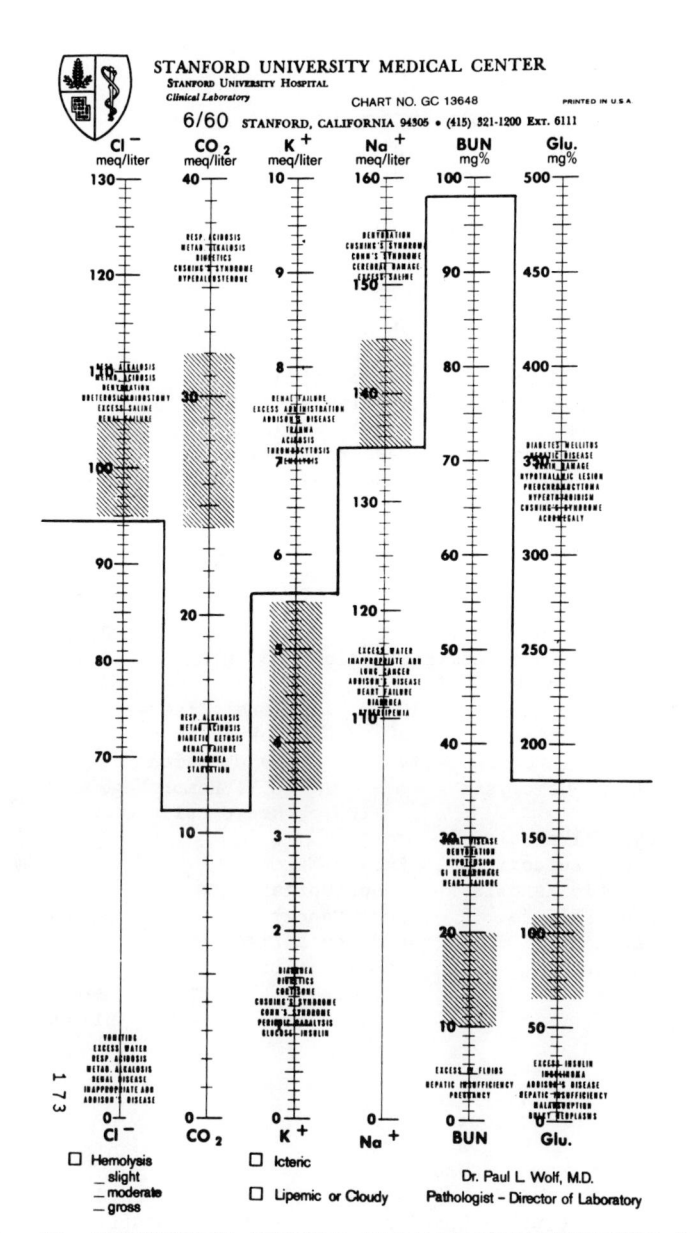

Fig. 22 - 33 BIOCHEMICAL PATTERN OF POST-MORTEM VITREOUS HUMOR
              - ACUTE GLOMERULONEPHRITIS

The vitreous humor reflects the elevated BUN, hyperkalemia, metabolic acidosis, and hyperglycemia of a patient who died of renal failure due to acute glomerulonephritis.

STANFORD UNIVERSITY MEDICAL CENTER
Stanford University Hospital
Clinical Laboratory

6/60 Chart

Cl^-	CO_2	K^+	Na^+	BUN	Glu.
	HIGH		**HIGH**		
	Resp. Acidosis		Dehydration		
	Metab. Alkalosis		Cushing's Syndrome		
	Diuretics		Conn's Syndrome		
	Cushing's Syndrome		Cerebral Damage		**HIGH**
	Hyperaldosterone		Excess Saline		Diabetes Mellitus
HIGH					Hepatic Disease
					Brain Damage
Resp. Alkalosis		**HIGH**			Hypothalmic Lesion
Metab. Acidosis					Pheochromocytoma
Ureterosigmoidostomy		Renal Failure			Hyperthyroidism
Excess Saline		Excess Administration		**HIGH**	Cushing's Syndrome
Renal Failure		Addison's Disease			Acromegaly
		Trauma		Renal Disease	
		Acidosis		Dehydration	
		Thrombocytosis		Hypotension	
	LOW	Hemolysis		G.I. Hemorrhage	
	Resp. Alkalosis		**LOW**	Heart Failure	
	Metab. Acidosis		Excess Water		**LOW**
	Diabetic Ketosis		Inappropriate ADH		
	Renal Failure		Lung Cancer		Excess Insulin
	Diarrhea		Addison's Disease		Insulinoma
	Starvation		Renal Failure		Addison's Disease
			Diarrhea		Hepatic Insufficie
LOW			Hyperlipemia		Malabsorption
					Bulky Neoplasms
Vomiting		Diarrhea			
Excess Water		Diuretics			
Resp. Acidosis		Cortisone			
Metab. Alkalosis		Cushing's Syndrome		**LOW**	
Renal Disease		Conn's Syndrome			
Inappropriate ADH		Periodic Paralysis		Excess in Fluids	
Addison's Disease		Glucose Insulin		Hepatic Insufficiency	
				Pregnancy	

A magnified view of the causes for high and low values of each determination on the Stanford Hospital Clinical Laboratory SMA 6/60 Chart.

STANFORD UNIVERSITY MEDICAL CENTER
Stanford University Hospital
Clinical Laboratory

12/60 Chart

Ca^{++}	I.P.	Glu.	BUN	U.A.	Chol.
		HIGH			HIGH
HIGH		Diabetes Mellitus			Hypothyroidism
		Hepatic Disease			Obstructive Jaundice
Hyperparathyroid		Brain Damage			Nephrosis
Bone Metastases		Hypothalmic Lesion			Diabetes Mellitus
Lymphoma Bone		Pheochromocytoma			Familial
Multiple Myeloma		Hyperthyroidism			Hereditary
Hyperthyroidism		Cushing's Syndrome			
Sarcoid	HIGH	Acromegaly	HIGH		LOW
Excess Vit-D					
Thiazides	Renal Insufficiency		Renal Disease		Hyperthyroidism
	Hypoparathyroidism		Dehydration		Hepatic Failure
	Pseudohypoparathyroid		Hypotension		Anemia
	Hemolysis		G.I. Hemorrhage		Infection
	Diabetes Mellitus		Heart Failure		Inanition
	Bone Growth			HIGH	
LOW	Fracture	LOW			
				Renal Failure	
Hypoparathyroid		Excess Insulin		Gout	
Pseudohypoparathyroid		Insulinoma		Diuretics	
Malabsorption		Addison's Disease		Leukemia	
Pancreatitis		Hepatic Insufficiency		Polycythemia	
Renal Failure		Malabsorption		Acidosis	
Excess I.V. Fluids		Bulky Neoplasms		Psoriasis	
				Hypothyroidism	
	LOW			Tissue Necrosis	
			LOW	Inflammation	
	Hyperparathyroid				
	Excess I.V. Glucose		Excess I.V. Fluids		
	Hypokalemia		Hepatic Insufficiency		
	Fanconi Syndrome		Pregnancy		
	Vit-D Resistant-			LOW	
	Rickets				
	Cirrhosis			Cortisone Usage	
				Allopurinol	
				Wilson's Disease	

A magnified view of the causes for high and low values of each determination on the Stanford Hospital Clinical Laboratory SMA 12/60 Chart

STANFORD UNIVERSITY MEDICAL CENTER
Stanford University Hospital
Clinical Laboratory

12/60 Chart (Continued):

T.P.	Alb.	T. Bili.	A.P.	LDH	SGOT

HIGH — **HIGH** — — **HIGH** —

Dehydration / Dehydration — — Cerebral Damage —
Hyperglobulinemia — — — Myocardial Infarct —
Waldenstrom — — — Lung Infarct —
Multiple Myeloma — — — Muscle Necrosis —
Malignancy — **HIGH** — Hemolysis —
Collagen Disease — — — Kidney Infarct —
Hepatic Disease — Hemolysis — Pernicious Anemia —
Infection — Hepatic Disease — Neoplastic Disease **HIGH**
— — Obstructive Jaundice — Liver Disease —
— — Pulmonary Infarction — Sprue Cerebral Damage
— — Large Hematoma — — Myocardial Infarct
— — Gilbert's Disease — — Viral Hepatitis
— — Dubin-Johnson **HIGH** — Hepatic Disease
— — — — — Muscle Necrosis
— — **LOW** Bone Growth — Hemolysis
— — — Hepatic Disease — Kidney Infarct
— — Overhydration Obstructive Jaundice — Neoplastic Disease
LOW — Hepatic- Osteoblastic Lesions — Burns
— — Insufficiency Paget's Disease — Trauma
Overhydration — Malnutrition Hyperparathyroidism — —
Immunoglobulin- — Malabsorption Pulmonary Infarction — —
 Deficiency — Burns Pregnancy — **LOW**
Hepatic- — Generalized- Peptic Ulcer — —
 Insufficiency — Dermatitis Colitis — Pyridoxine Deficiency
Malnutrition — Nephrosis Renal Infarction — Chronic Dialysis
Malabsorption — — Regan Cancer — Pregnancy
Burns — — — — Beri Beri
Generalized- — — **LOW** — —
 Dermatitis — — — — —
Nephrosis — — Hypophosphatasia — —
— — — Hypothyroidism — —
— — — Anemia — —
— — — Malnutrition — —
— — — Oxalate Anticoagulant — —

— — — — **LOW** —

— — — — Clofibrate Therapy —
— — — — Oxalate Anticoagulant —

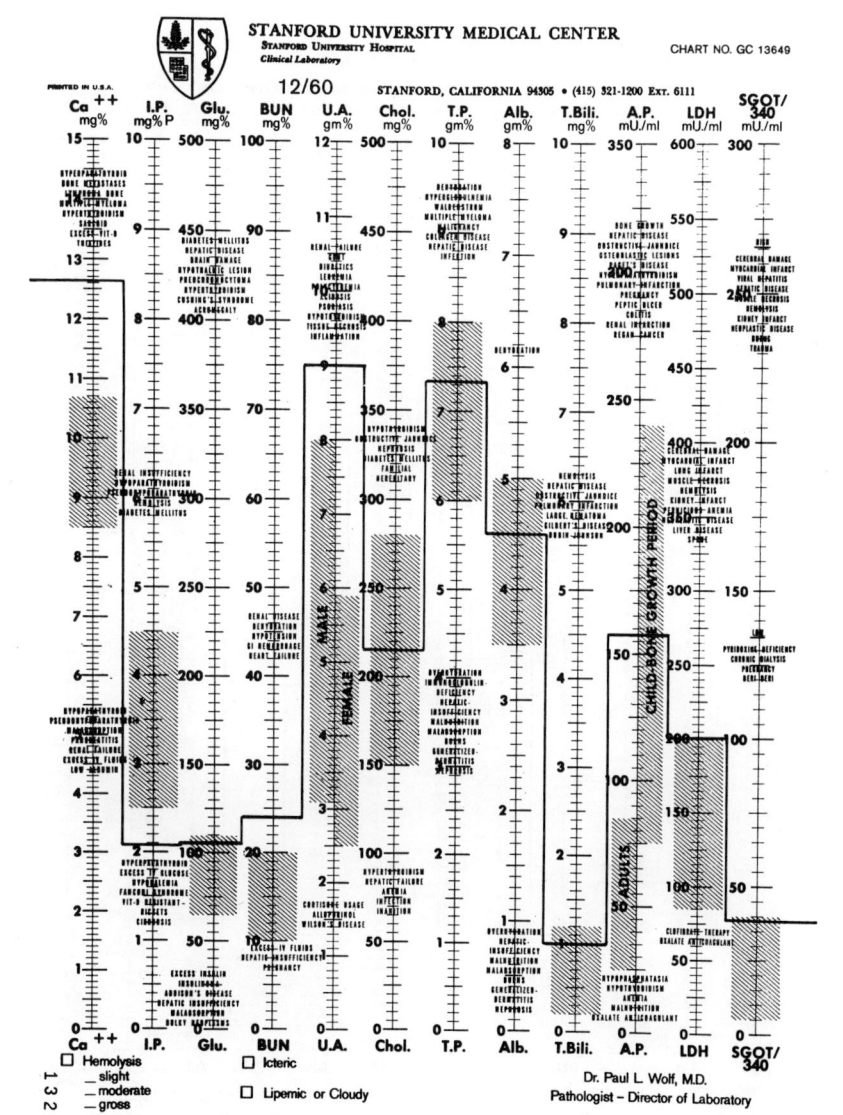

Fig. 22 - 34 BIOCHEMICAL PATTERN OF HYPERPARATHYROIDISM

1). Hypercalcemia due to excessive gastrointestinal calcium absorption and liberation of calcium from bone. 2). Hypophosphatemia due to action of excessive parathormone on renal tubule resulting in pronounced phosphate diuresis. 3). Elevated BUN due to renal calcinosis. 4). Elevated uric acid due to renal disease. 5). Elevated alkaline phosphatase due to osteomalacia.

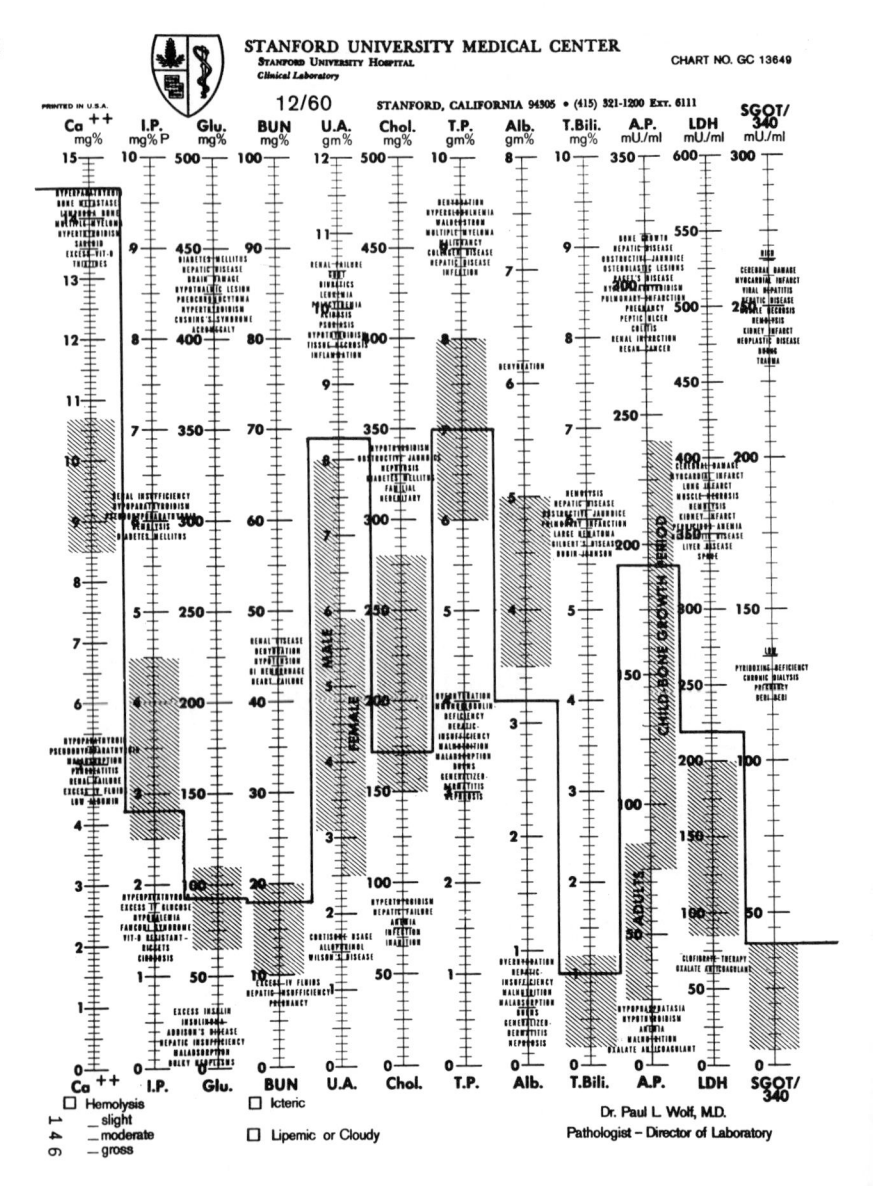

Fig. 22 - 35 BIOCHEMICAL PATTERN OF CANCER METASTATIC TO BONE

1). Hypercalcemia due to metastatic bone cancer. Usu
primary site is lung, breast, prostate, kidney, thyroi
or lymphoma. Malignancy causes bone destruction and h
calcemia. Malignancy may produce hypercalcemia by ect
production of PTH. 2). Phosphate may be within norma
limits. 3). Uric acid is elevated due to proliferati
and necrosis of malignant cells. 4). Hypoalbuminemia
due to malignancy. 5). Elevation of alkaline phospha
is due to osteomalacia. 6). Elevated LDH is derived
proliferating malignant cells.

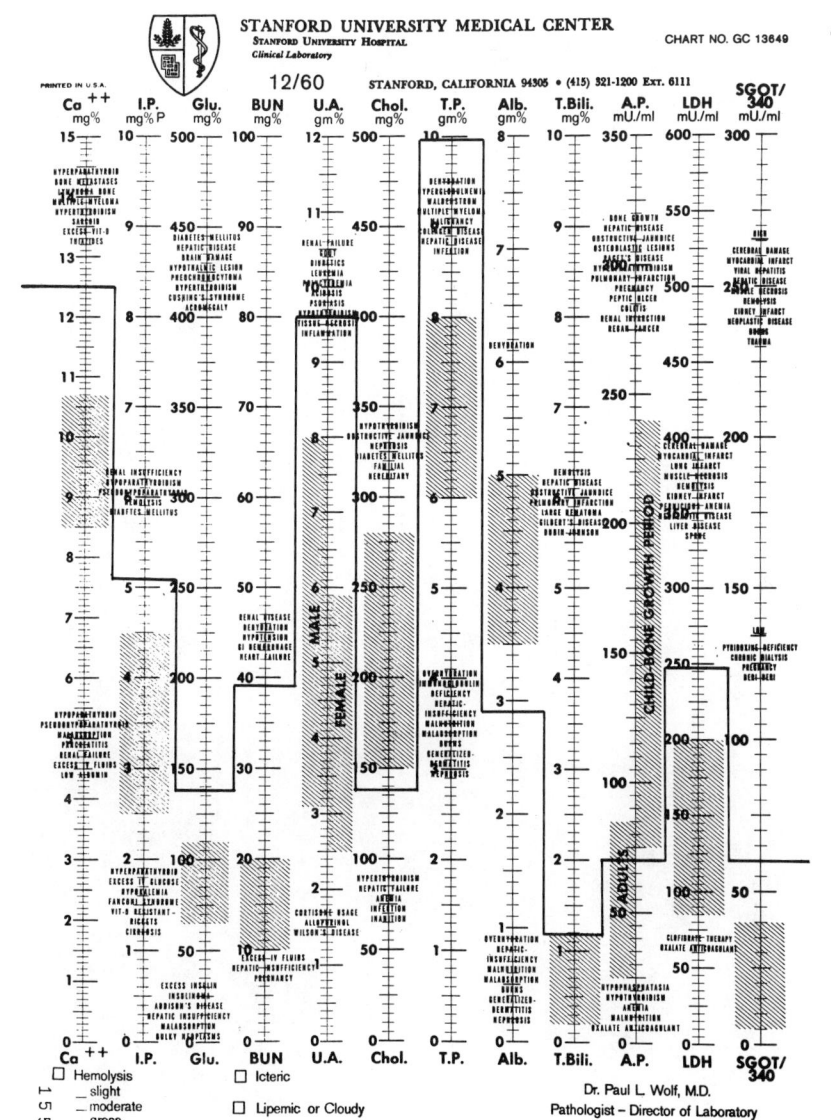

Fig. 22 - 36 BIOCHEMICAL PATTERN OF MULTIPLE MYELOMA

1). Hypercalcemia from extensive osteolysis with liberation of calcium. 2). Elevated BUN, phosphate, uric acid, glucose, and low albumin due to renal insufficiency which results from nephrocalcinosis and renal tubule obstruction by Bence-Jones protein. 3). Low cholesterol due to severe anemia. 4). Elevated uric acid, LDH, and GOT from proliferation and necrosis of plasmablasts. 5). Alkaline phosphatase not elevated because of lack of osteoblastic process. 6). Elevated protein because of monoclonal gammopathy.

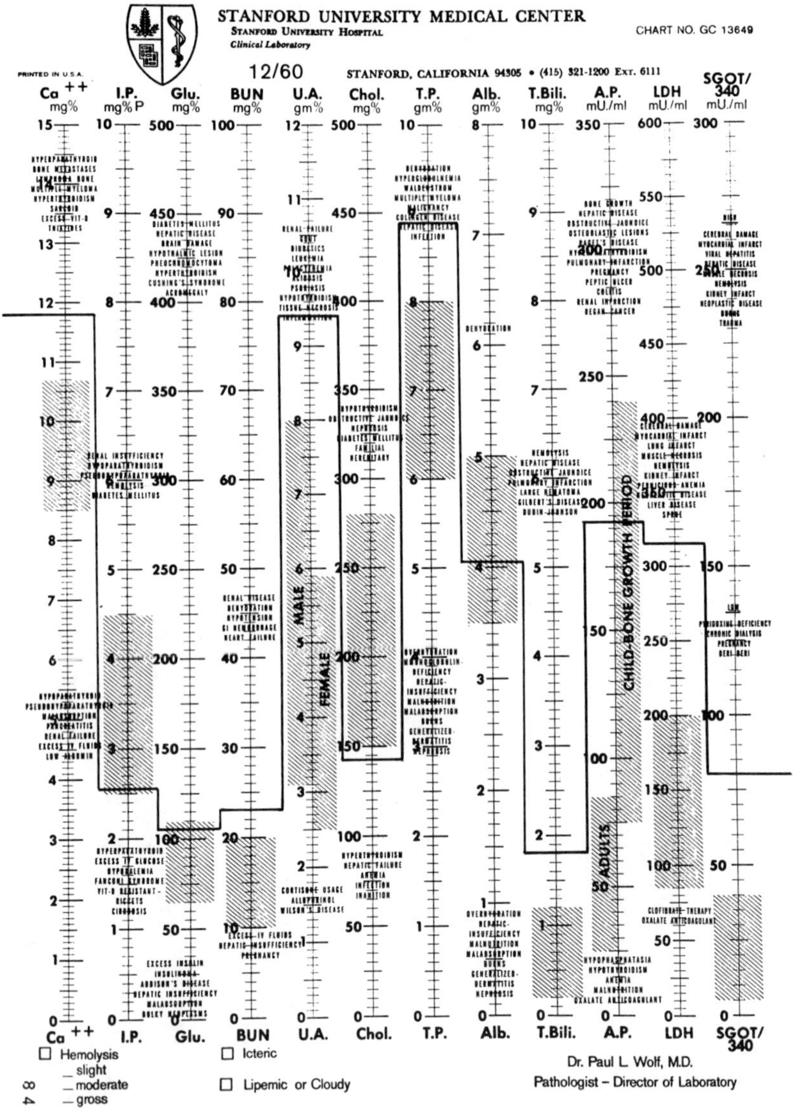

Fig. 22 - 37 BIOCHEMICAL PATTERN OF HODGKIN'S DISEASE INVOLVING BONE

1). Hypercalcemia due to metastatic bone cancer. Usu primary site is lung, breast, prostate, kidney, thyroi or lymphoma. Malignancy causes bone destruction and h calcemia. Malignancy may produce hypercalcemia by ect production of PTH. 2). Phosphate may be within norma limits. 3). Uric acid is elevated due to proliferati and necrosis of malignant cells. 4). Hypoalbuminemi due to malignancy. 5). Elevation of alkaline phospha

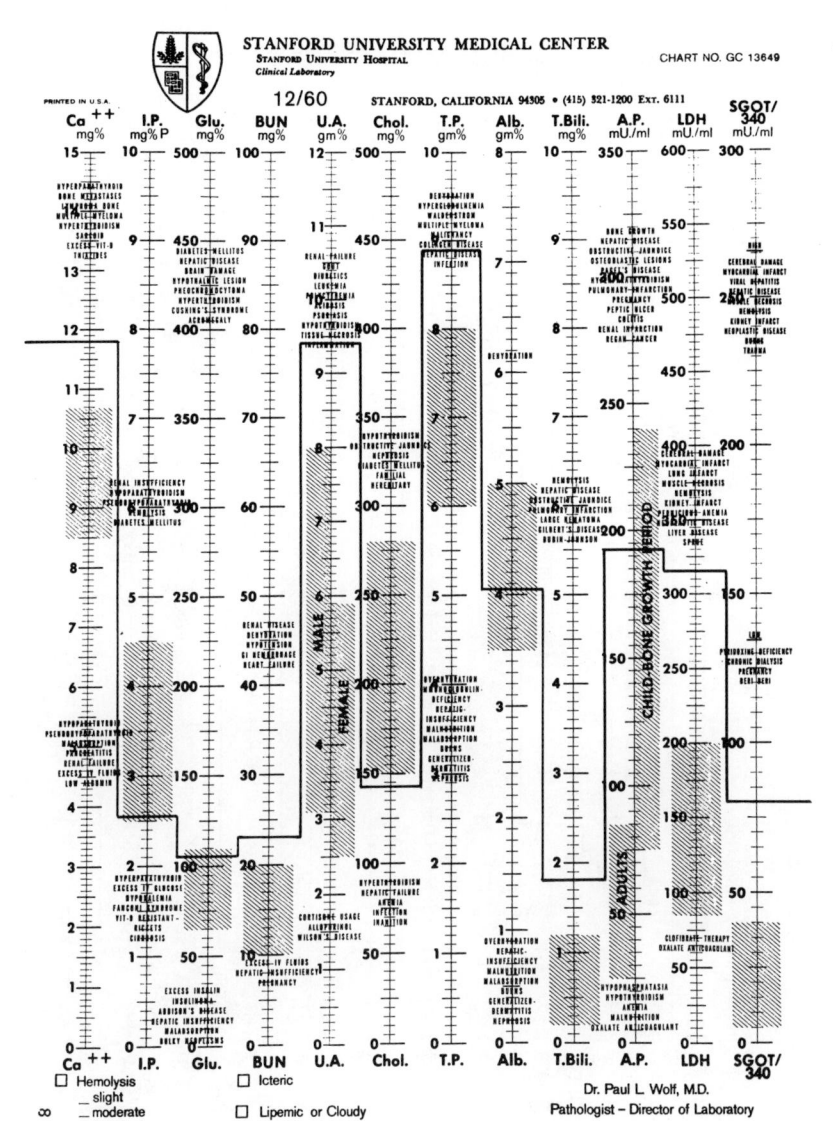

Fig. 22 - 37 (Continued):

is due to osteomalacia or involvement of liver, spleen, kidney, lung, gastrointestinal tract or ectopic production of alkaline phosphatase by neoplastic tissue. 6). Elevated LDH is derived from proliferating malignant cells.

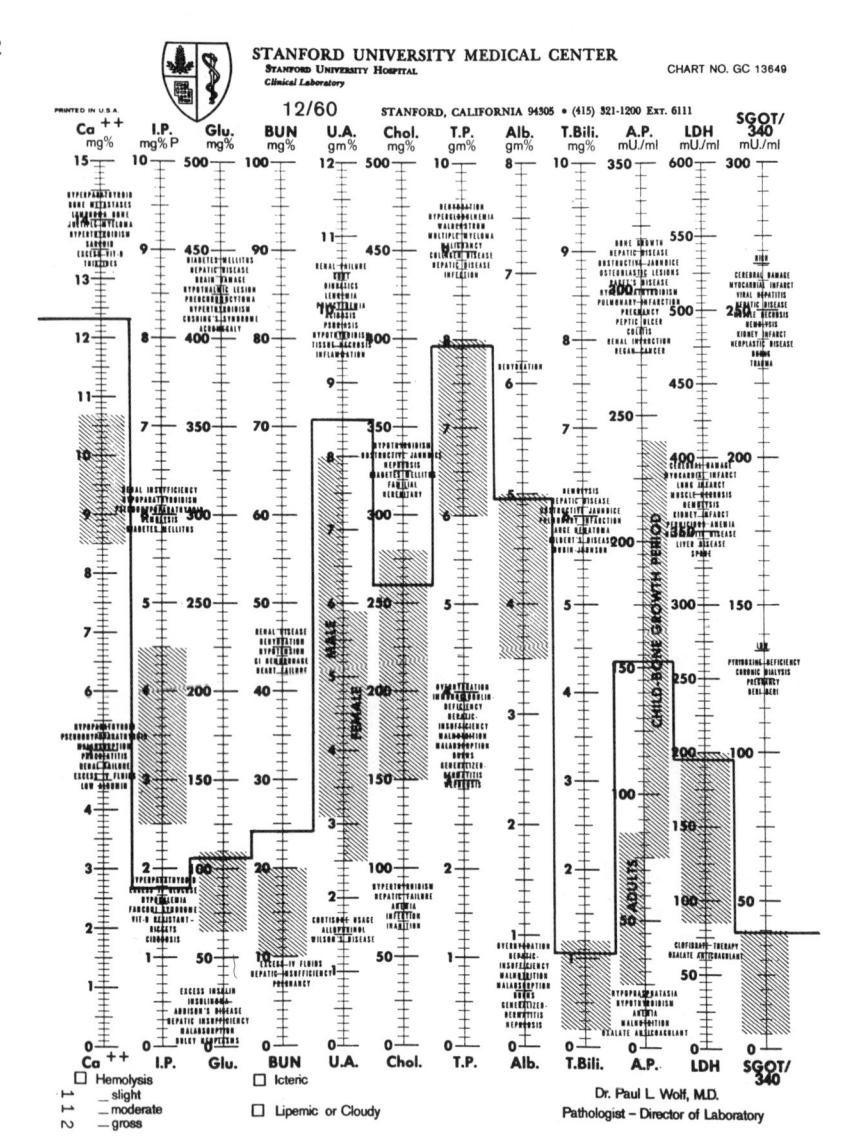

Fig. 22 - 38 BIOCHEMICAL PATTERN OF OSTEOMALACIA INCLUDING RICKETS

1). Hypercalcemia due to liberation of calcium from b
secondary to hypovitaminosis D. 2). Low phosphate du
hypovitaminosis D. 3). Elevated alkaline phosphatase
to osteomalacia. In osteoporosis, the alkaline phosph
is normal. 4). Elevated BUN is due to nephrocalcinos

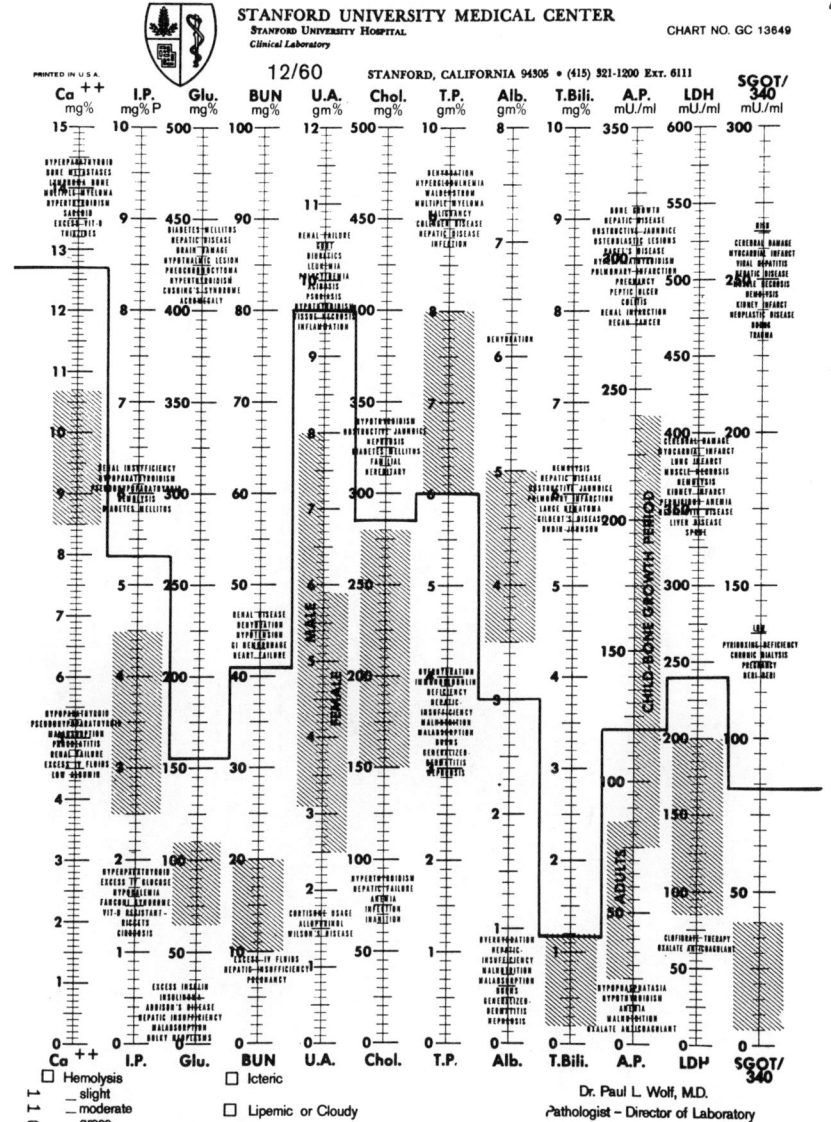

Fig. 22 - 39 BIOCHEMICAL PATTERN OF MILK ALKALI SYNDROME

1). Hypercalcemia due to excessive ingestion of calcium as calcium carbonate and milk in treatment of peptic ulcer - Burnett's syndrome. 2). Elevated BUN, phosphate, uric acid, glucose, and cholesterol due to renal nephrocalcinosis and insufficiency. 3). Decreased albumin due to proteinuria. 4). Elevated alkaline phosphatase, LDH, and GOT due to renal damage, with lack of excretion due to renal failure or liberation from a variety of tissues in uremia.

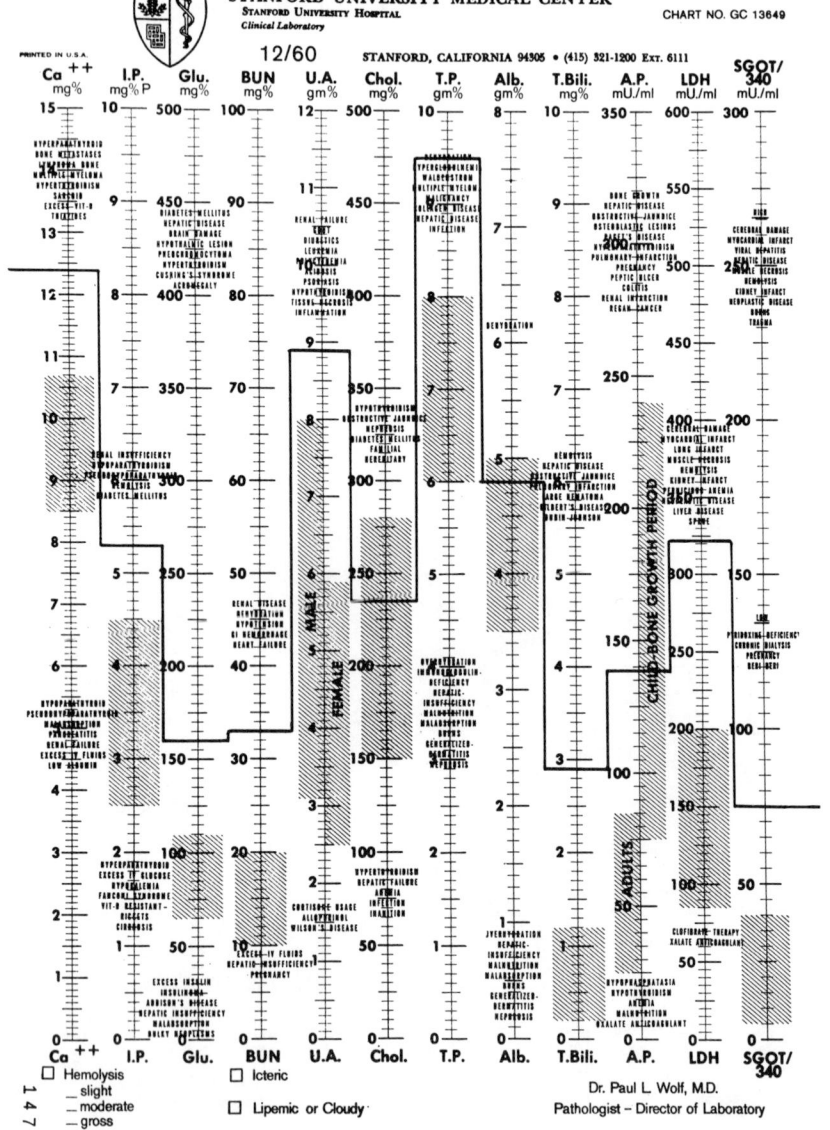

Fig. 22 - 40 BIOCHEMICAL PATTERN OF SARCOIDOSIS

1). Hypercalcemia due to increased sensitivity to PTH and absorption of calcium from gastrointestinal tract. Hypercalcemia also due to elevated serum protein.
2). Elevated BUN, phosphate, uric acid, glucose, and cholesterol due to renal nephrocalcinosis and insuffici
3). Renal disease is also due to polyclonal gammopathy and immunoglobulin deposition in glomeruli.
4). Polyclonal gammopathy causes increase in serum protein.

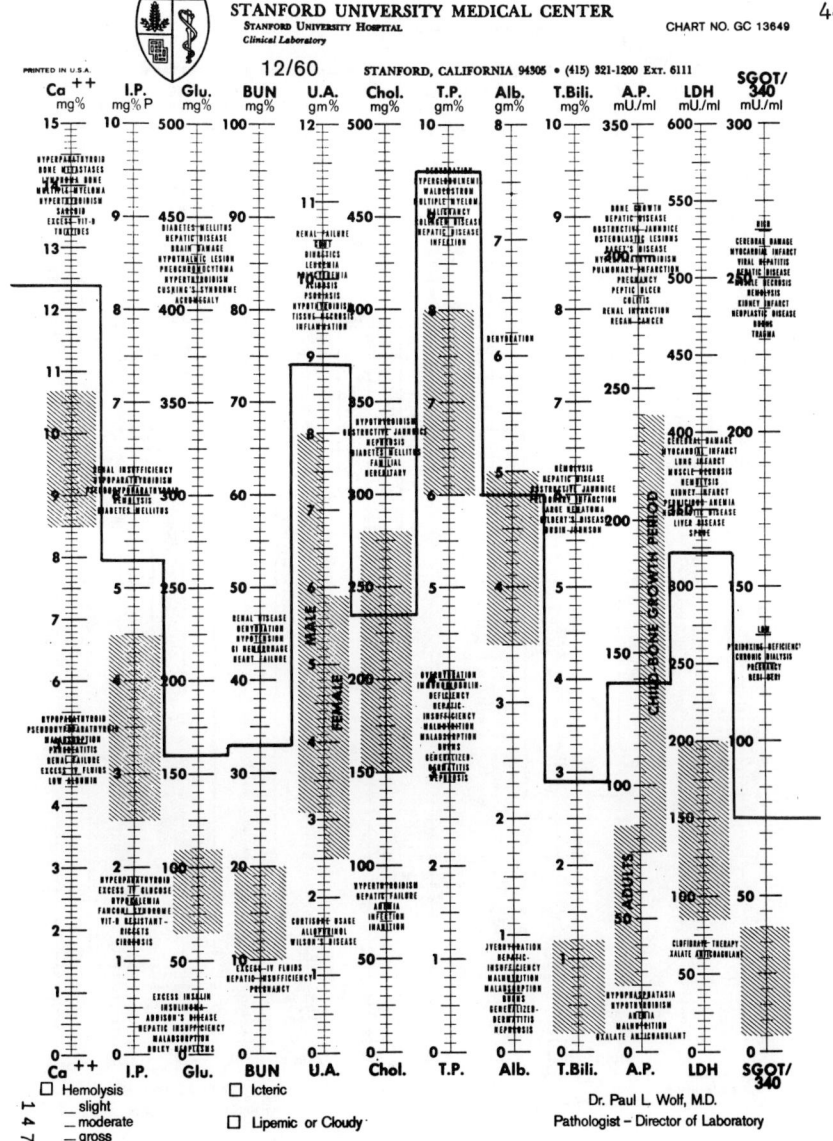

Fig. 22 - 40 (Continued):

5). Sarcoid involvement of liver causes elevated bilirubin, alkaline phosphatase, LDH, and GOT.
6). Elevated alkaline phosphatase, LDH, and GOT due to renal damage, with lack of excretion due to renal failure or liberation from a variety of tissues in uremia.

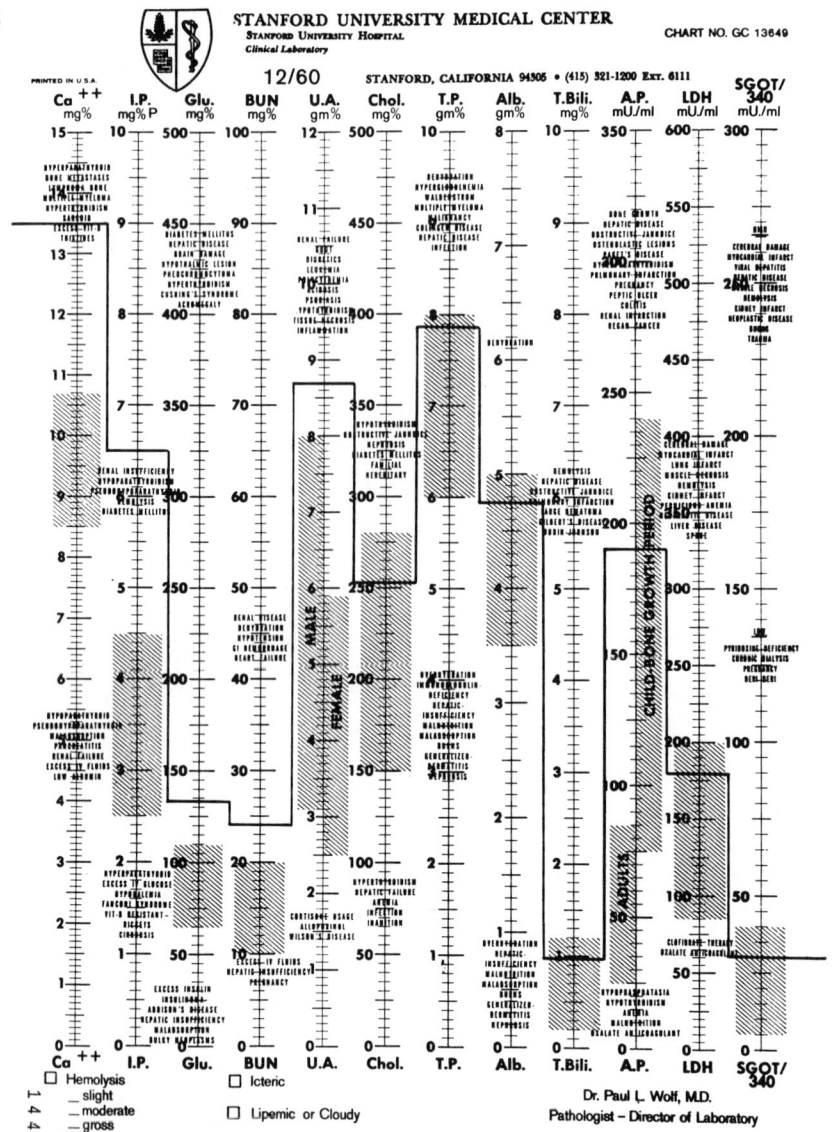

Fig. 22 - 41 BIOCHEMICAL PATTERN OF HYPERVITAMINOSIS D

1). Hypercalcemia and hyperphosphatemia due to hypervitaminosis D. 2). Elevated BUN, phosphate, uric acid glucose, and cholesterol due to renal nephrocalcinosis and insufficiency. 3). Elevated alkaline phosphatase due to osteomalacia secondary to hypervitaminosis D.

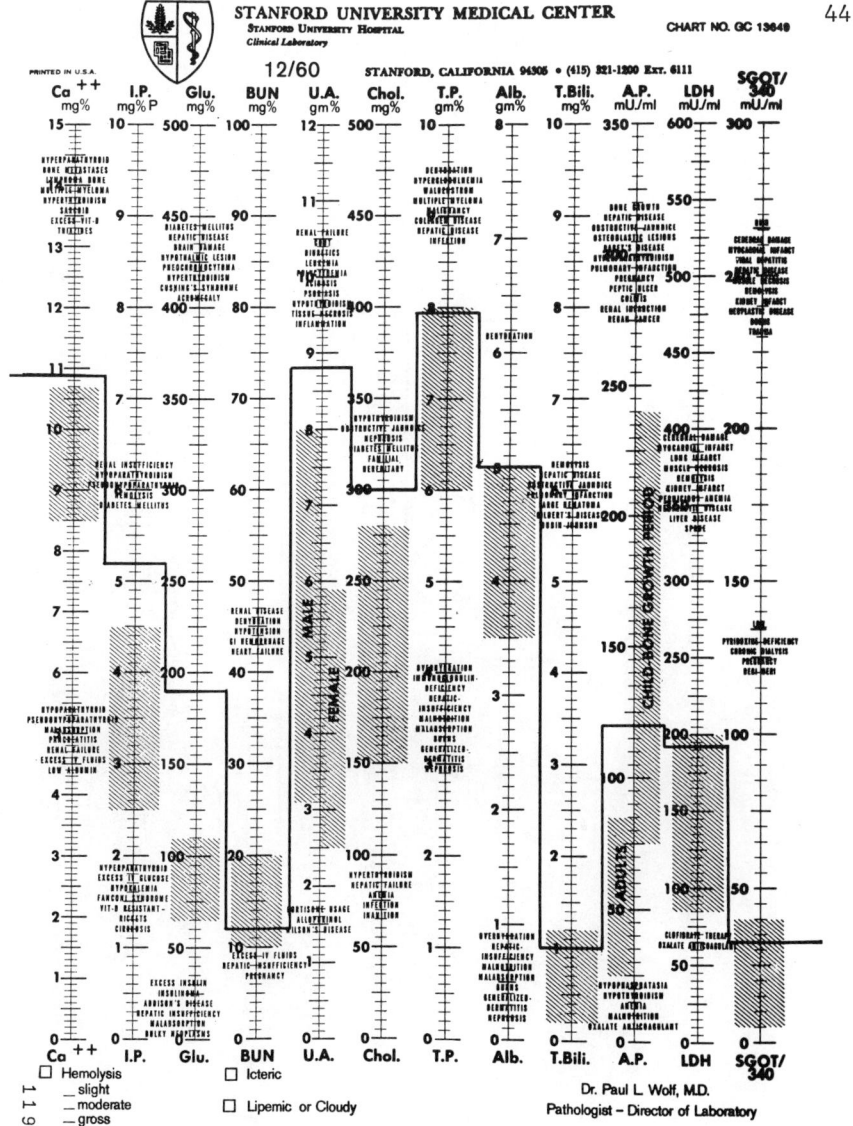

Fig. 22 - 42 BIOCHEMICAL PATTERN OF CUSHING'S SYNDROME

Hypercalcemia and elevated alkaline phosphate due to bone disease in Cushing's syndrome. Elevated glucose and phosphate due to diabetes mellitus which is cortisol induced. Phosphate elevation indicates lack of phosphorylation of glucose. Elevated cholesterol and uric acid due to disturbed carbohydrate and lipid metabolism in Cushing's syndrome.

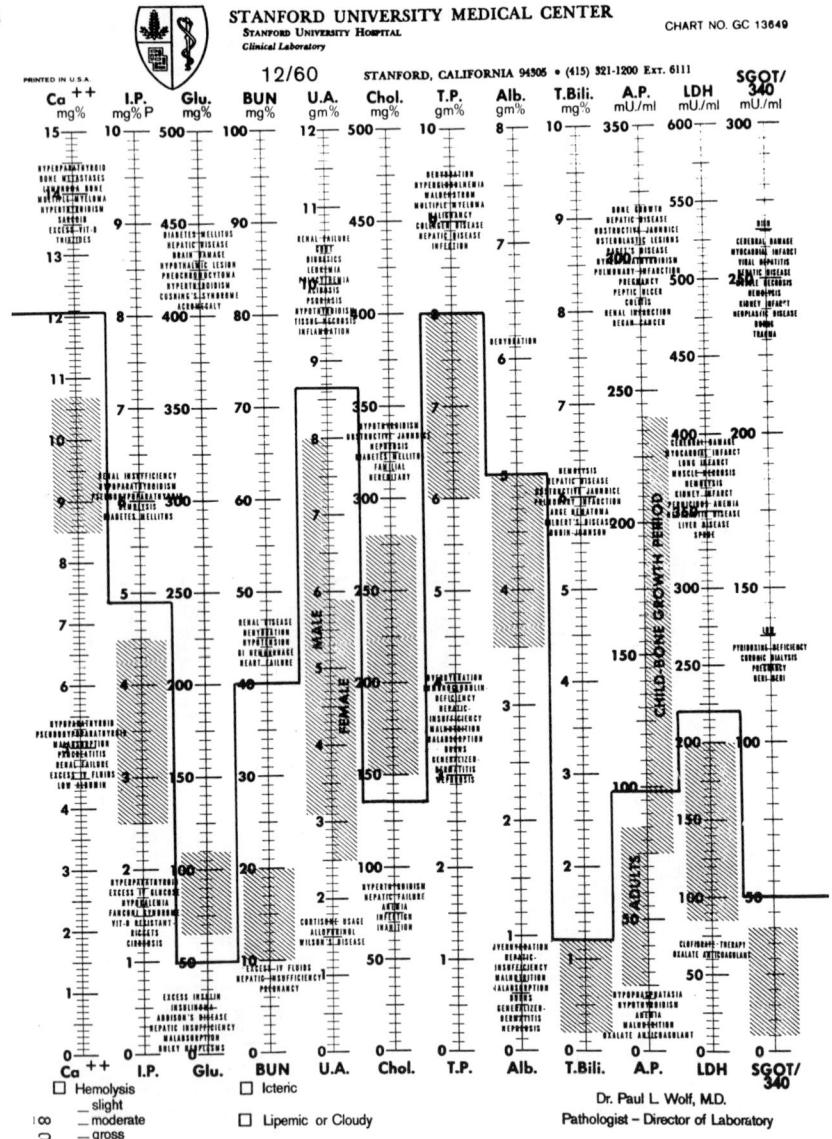

Fig. 22 - 43 BIOCHEMICAL PATTERN OF ADDISON'S DISEASE

1). Hypercalcemia occurs in Addison's disease and is n fully understood. It may be related to increase in ser protein found in Addison's disease due to amyloidosis o tuberculosis of adrenals. 2). Elevated BUN, phosphate uric acid, alkaline phosphatase, GOT, and LDH due to pr renal azotemia in Addison's disease secondary to dehydr tion and hypovolemia. 3). Decreased glucose and chole terol due to low cortisol and innanition.

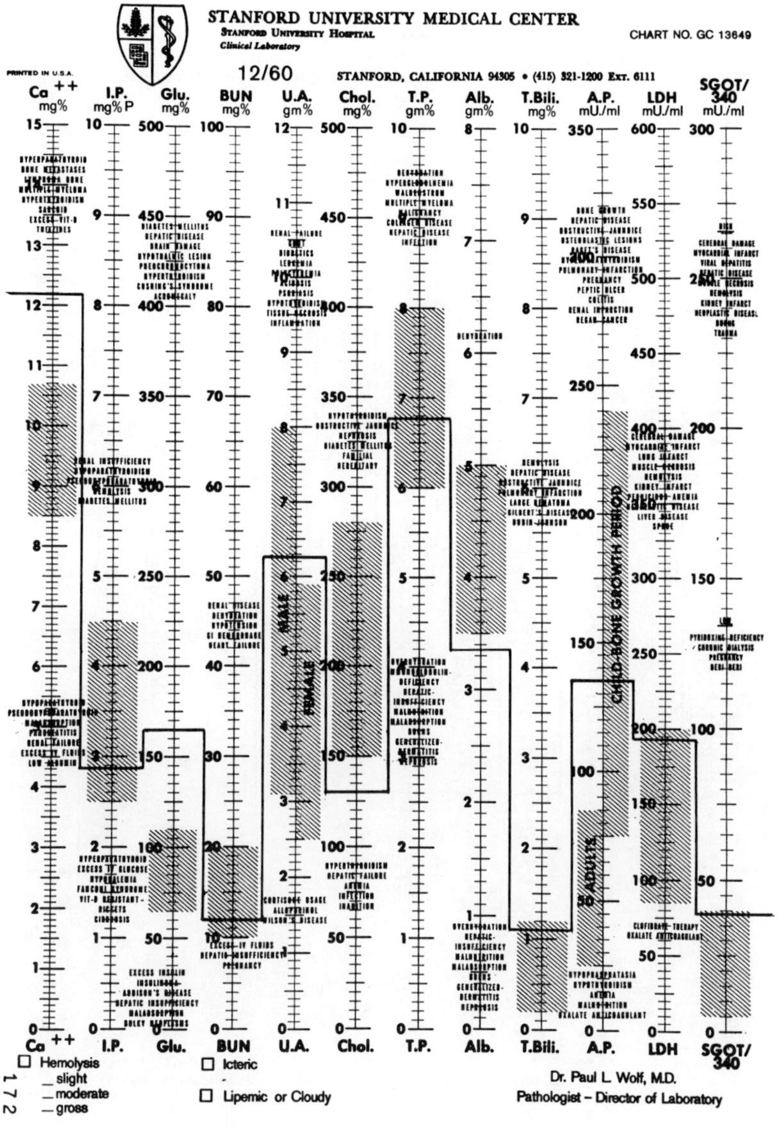

Fig. 22 - 44   BIOCHEMICAL PATTERN OF THYROTOXICOSIS

1). Hypercalcemia and elevated alkaline phosphatase due to bone disease in hyperthyroidism. 2). Hyperglycemia due to excessive glycogenolysis. 3). Low cholesterol and low albumin due to hypermetabolism in thyrotoxicosis.

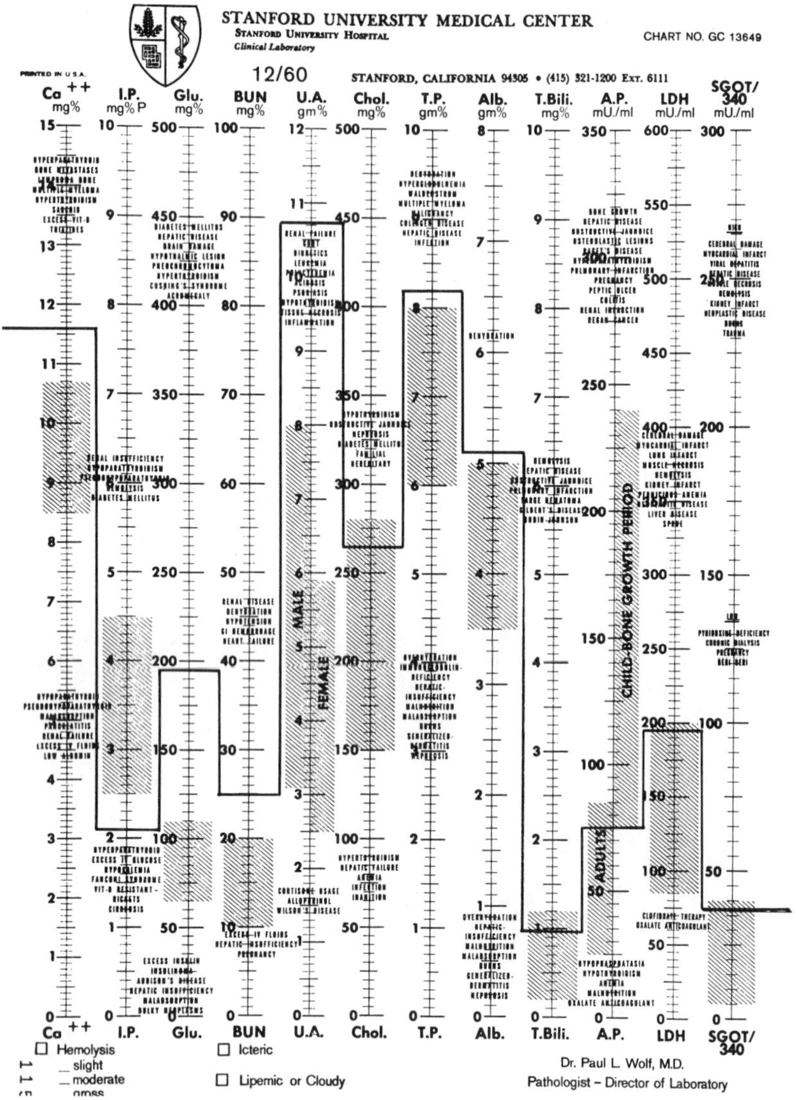

Fig. 22 - 45 BIOCHEMICAL PATTERN OF EXCESSIVE THIAZIDE DIURETIC UTILIZATION

1). Hypercalcemia due to thiazide diuretic usage is not fully understood. It may result from inhibition of excretion or excessive water diuresis. 2). Hyperglycemia is due to glucose intolerance induced by thiazides. 3). phosphate is due to hypokalemia and hyperglycemia. 4). Elevated BUN and protein is due to prerenal dehydration induced by thiazides. 5). Elevated uric acid is to thiazide inhibition of excretion.

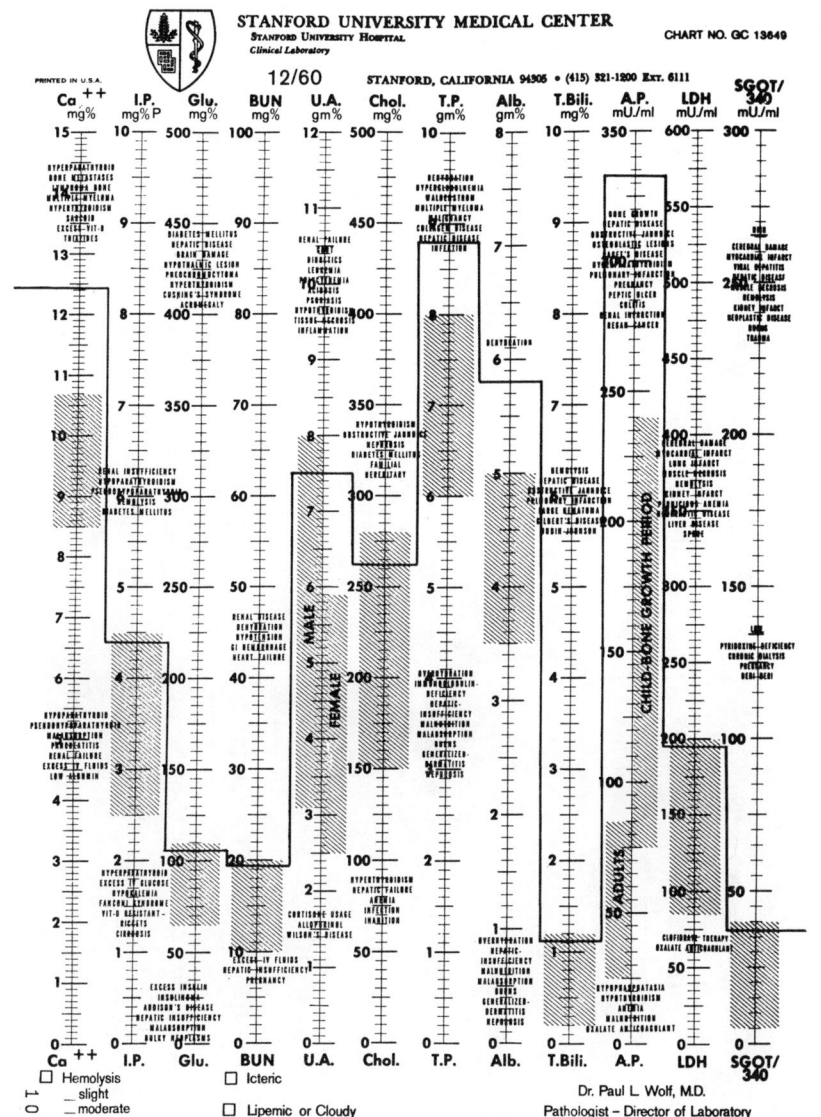

Fig. 22 - 46   BIOCHEMICAL PATTERN OF INTRAVENOUS ALBUMIN FROM PLACENTA

1). Elevated serum albumin is due to excessive intravenous infusion of albumin resulting in elevated total protein and hypercalcemia. 2). Elevated alkaline phosphatase is due to the presence of human placental alkaline phosphatase in the placental blood from which serum albumin is produced by some pharmaceutical companies. 3). Elevated calcium due to increased albumin.

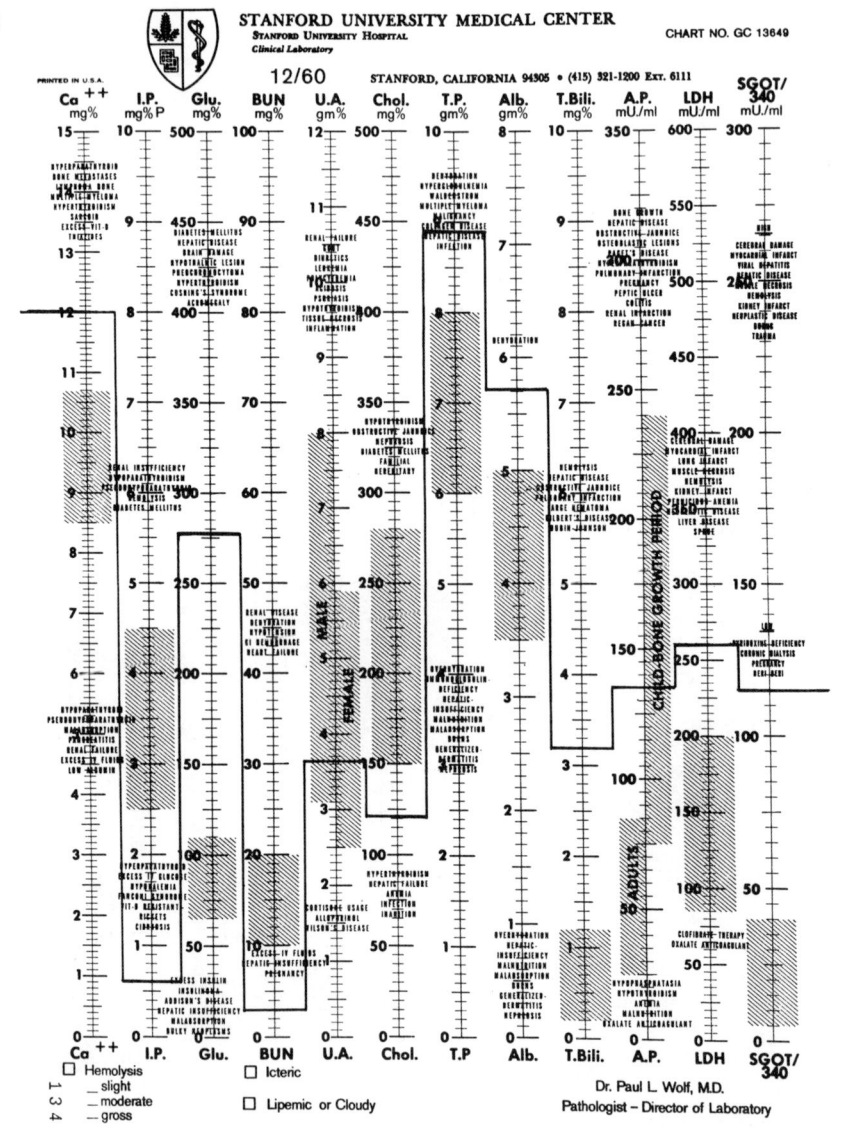

Fig. 22 - 47   BIOCHEMICAL PATTERN OF HYPERALIMENTATION

1). Hypercalcemia caused by increased albumin which was given in large amount intravenously to correct low serum albumin. Hypercalcemia also caused by excessive utilization of vitamin D. 2). Low phosphate due to hyperglyce from continuous utilization of 5 percent glucose in wate 3). Low BUN and cholesterol due to utilization of 5 per glucose in water. 4). Elevated bilirubin, alkaline pho phatase, LDH and GOT due to liver disease caused by

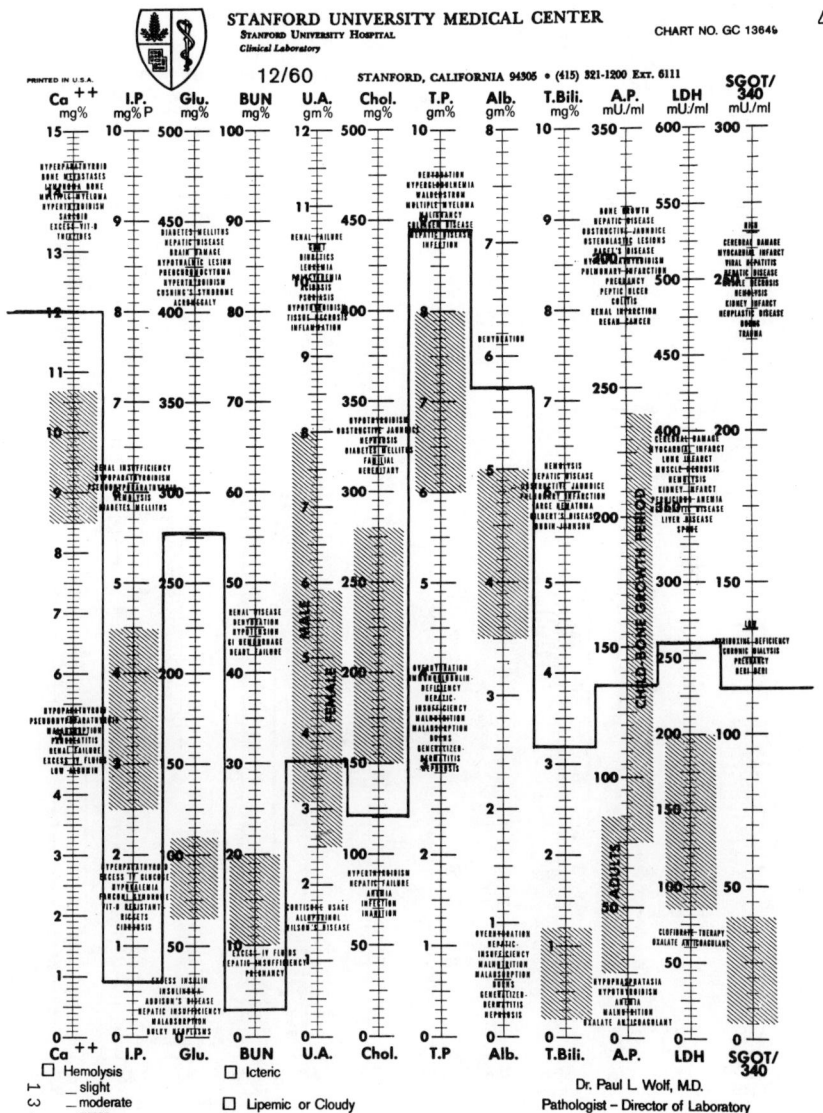

Fig. 22 - 47 (Continued):

hyperalimentation. The utilization of large amounts of protein and carbohydrate causes a "work" hypertrophy of the liver.

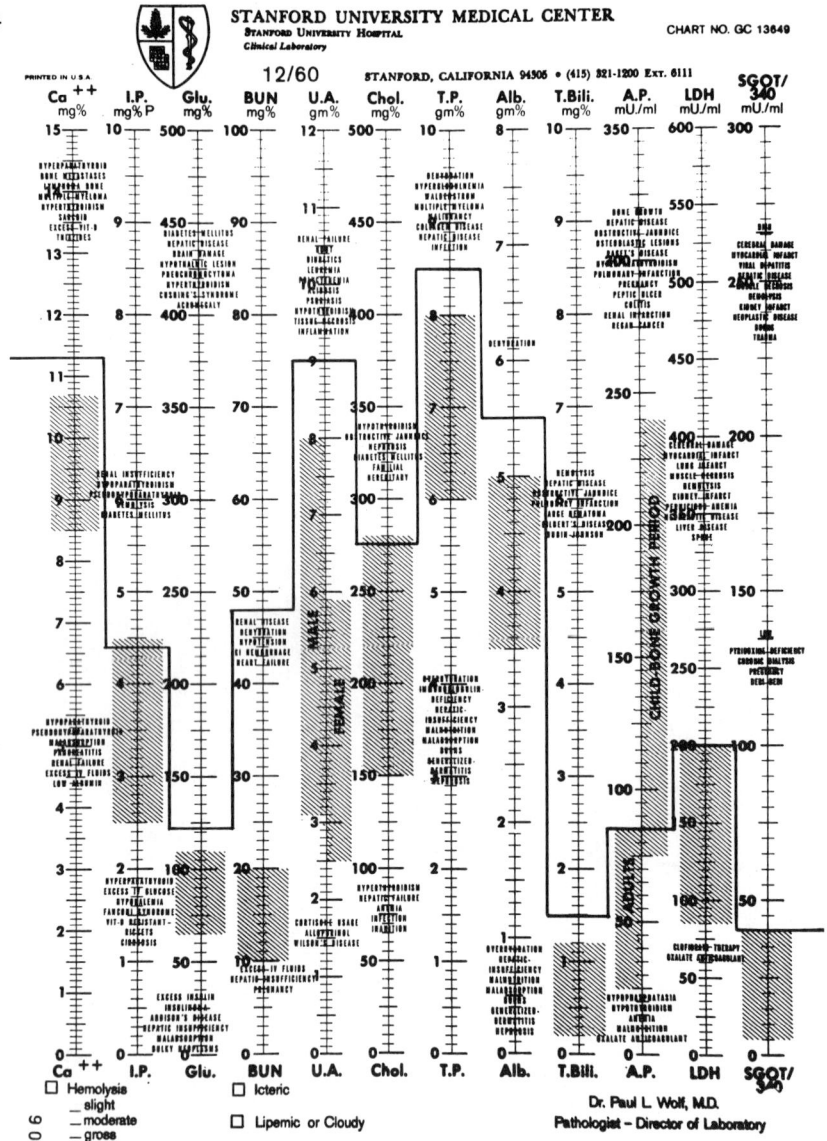

Fig. 22 - 48 BIOCHEMICAL PATTERN OF DEHYDRATION

Dehydration causes an increased serum albumin and protein with resultant hypercalcemia. In addition a prerenal azotemia results with elevated BUN, Uric Acid, and Glucose.

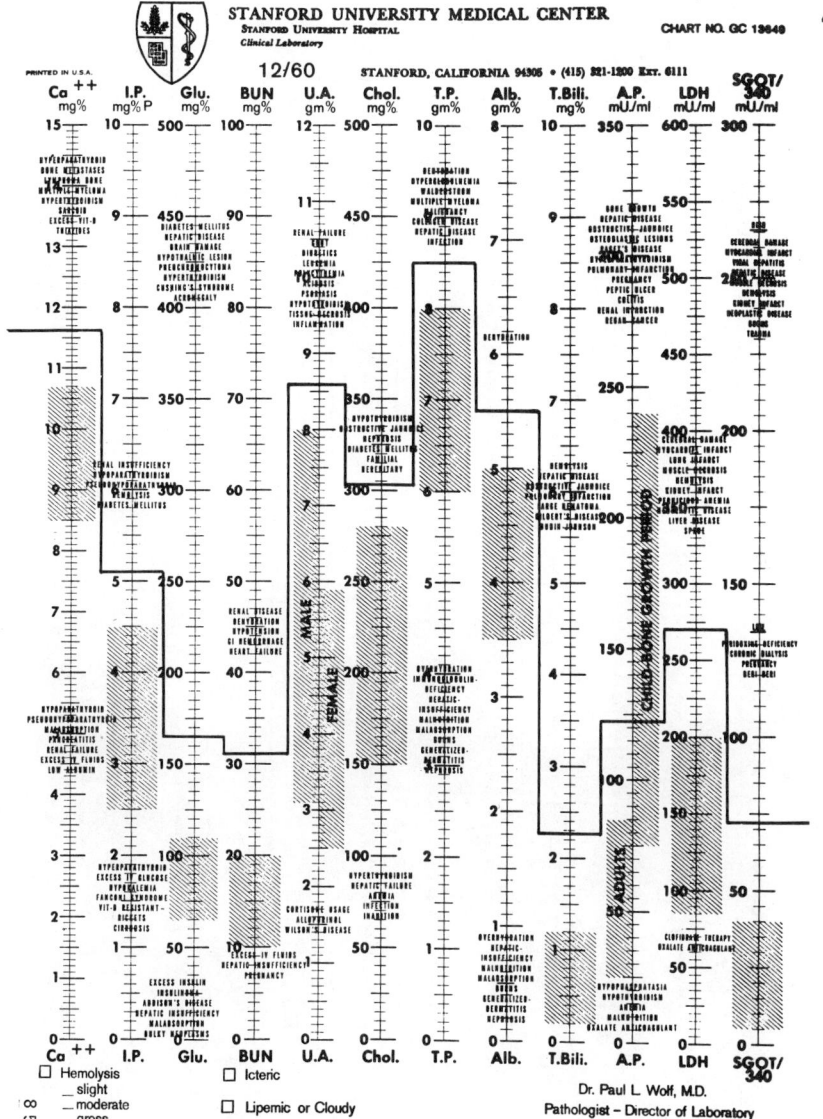

Fig. 22 - 49 BIOCHEMICAL PATTERN OF EVAPORATION OF SPECIMEN

A water evaporation of a serum sample results in artifactual elevation of all of the determinations on the 12/60 or 6/60 Charts.

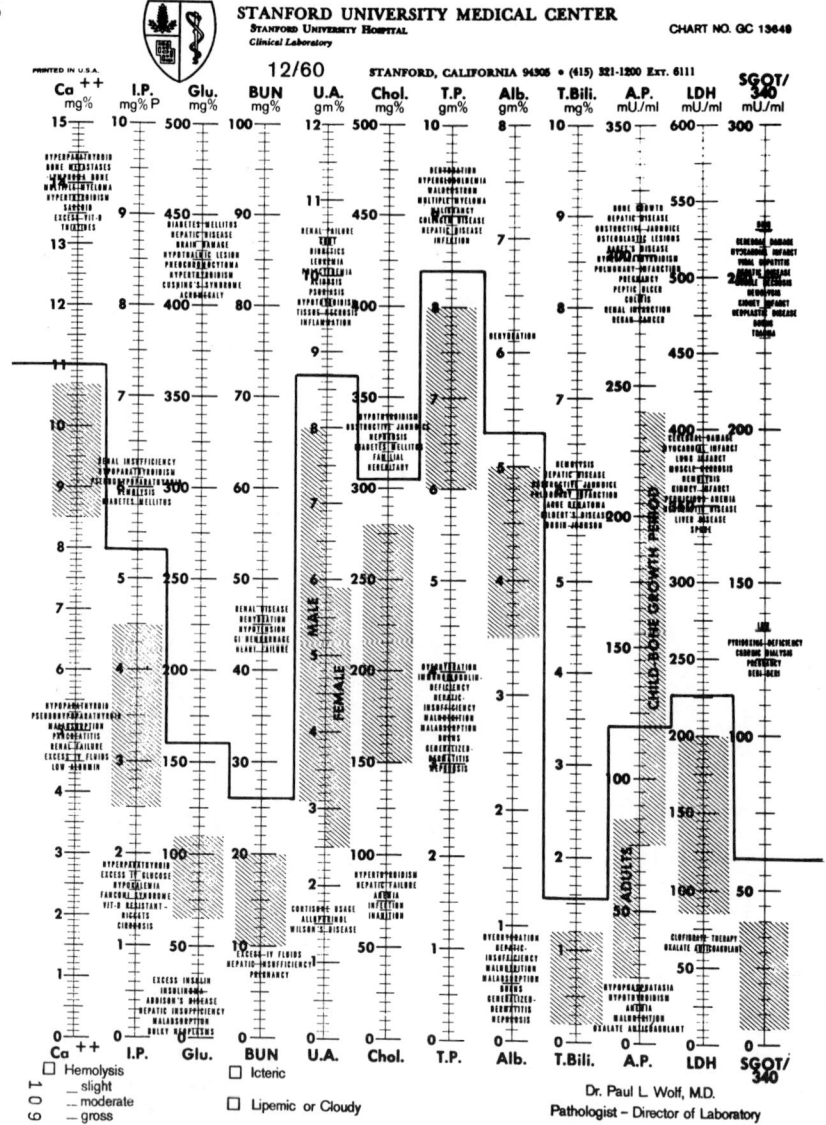

Fig. 22 - 50 BIOCHEMICAL PATTERN OF NON-FASTING LIPEMIC SERUM

1). All serum determinations should be performed on fasting specimens. 2). Non-fasting introduces artifactual elevation of glucose, protein and lipids. 3). The hyperlipemia causes artifactual elevation of calcium, phosphate, BUN, uric acid, bilirubin and enzymes, alkaline phosphatase, LDH and GOT.

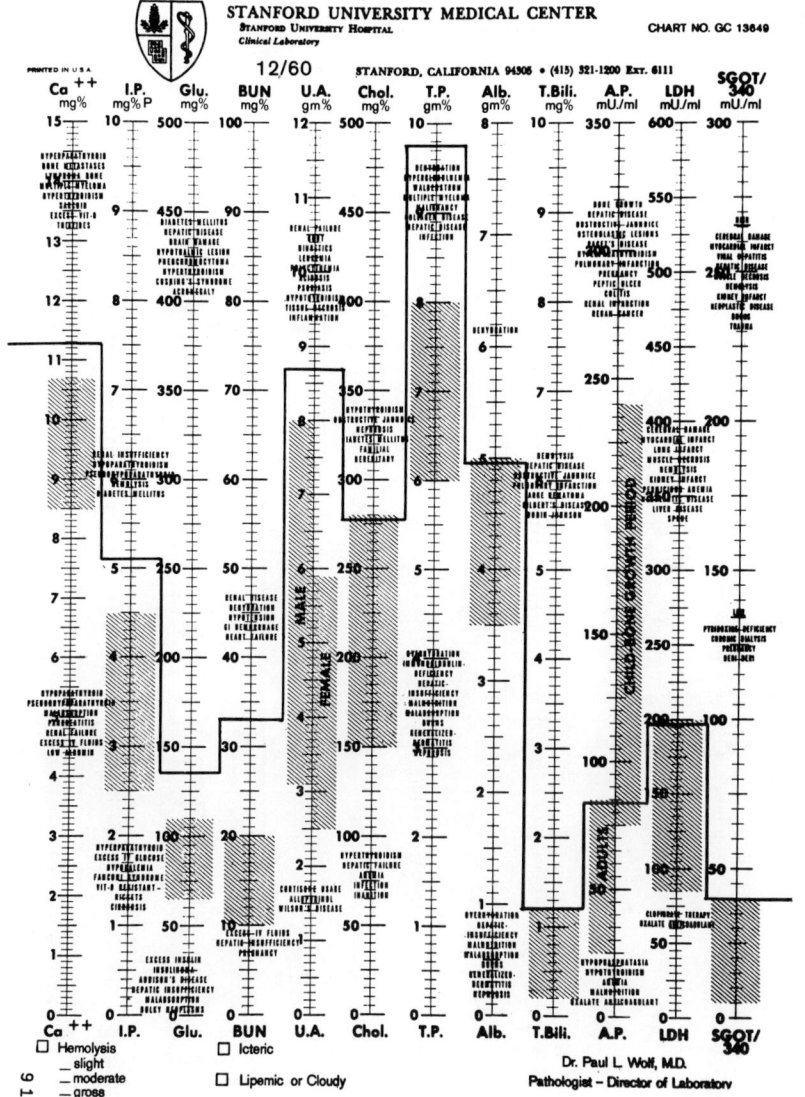

Fig. 22 - 51 BIOCHEMICAL PATTERN OF INCREASED CALCIUM DUE TO
INCREASED SERUM PROTEIN

The elevated calcium is due to increase in serum protein.
The presence of increased serum protein causes increase
in calcium due to increased calcium bound intravascularly.
Increase in gamma globulin may be due to monoclonal or
polyclonal gammapathy. Common causes for polyclonal
states are liver disease, chronic infections, neoplasms
such as lymphoma or sarcoidosis. The monoclonal states

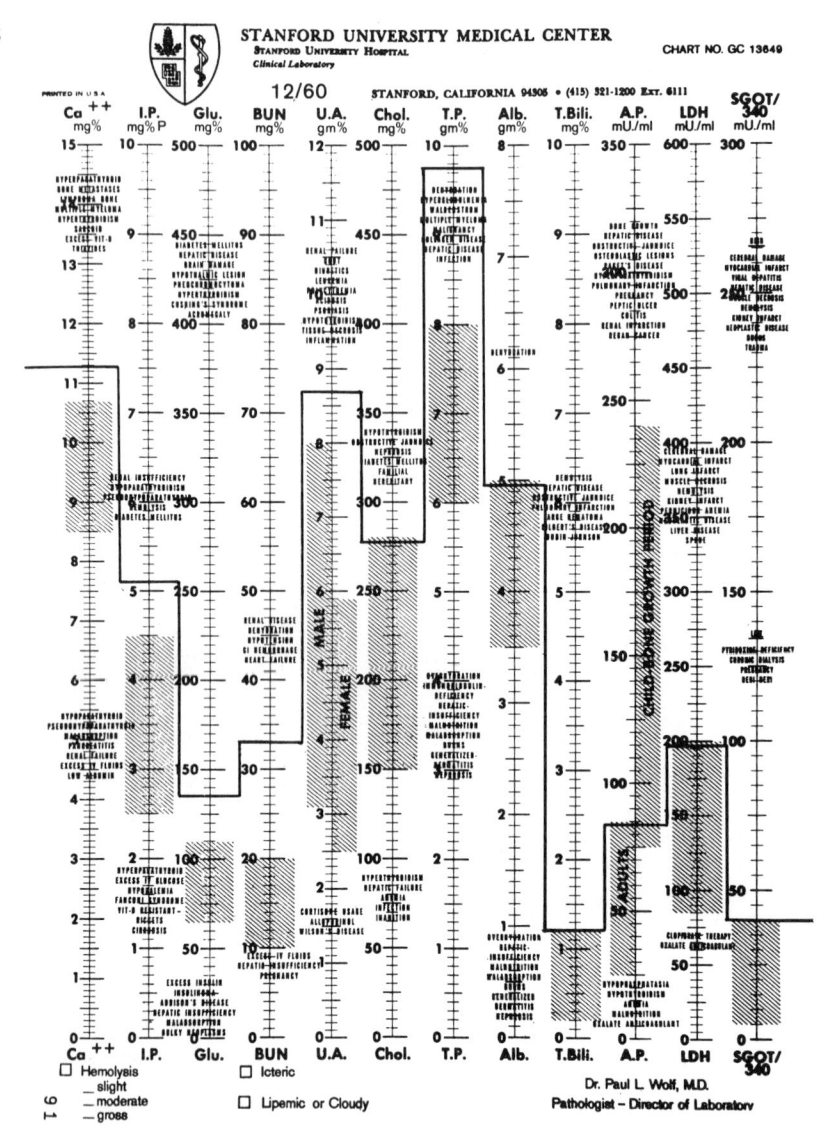

Fig. 22 - 51 (Continued):

are multiple myeloma, Waldenstrom's Macroglobulinemia, or Gaucher's disease. Increase in calcium is due to increased binding of calcium by increase in protein. Renal disease results from deposition of gamma globulin in glomeruli and renal tubules depending on the etiology of the monoclonal or polyclonal gammapathy, with increase in BUN, phosphate, glucose and uric acid.

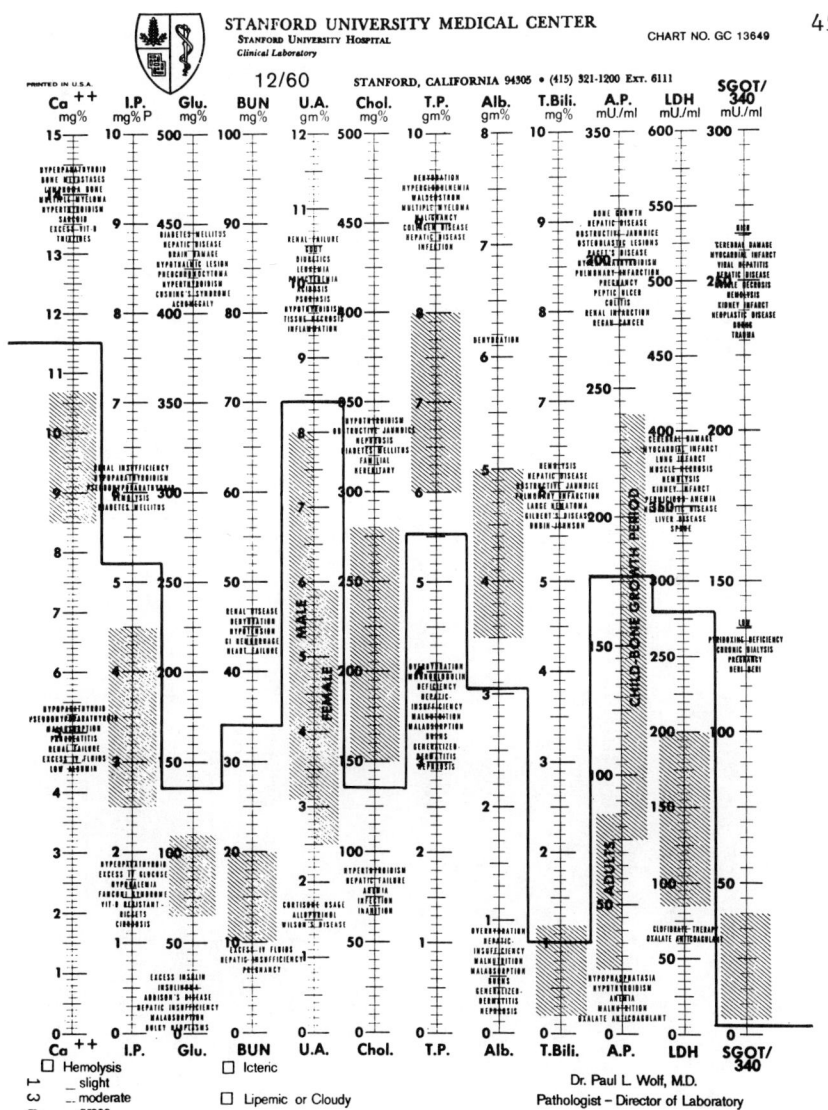

Fig. 22 - 52 BIOCHEMICAL PATTERN OF FREQUENT RENAL DIALYSIS

1). Chronic renal insufficiency causes an elevated BUN, phosphate, glucose, uric acid, and LDH due to retention. 2). The alkaline phosphatase is increased due to secondary hyperparathyroidism. 3). Elevated calcium occurs due to repeated dialysis or after a renal transplant. Usually in chronic renal disease, hypocalcemia is present. However, with renal dialysis, the serum calcium rises. 4). Low proteins occur due to proteinuria. 5). Anemia

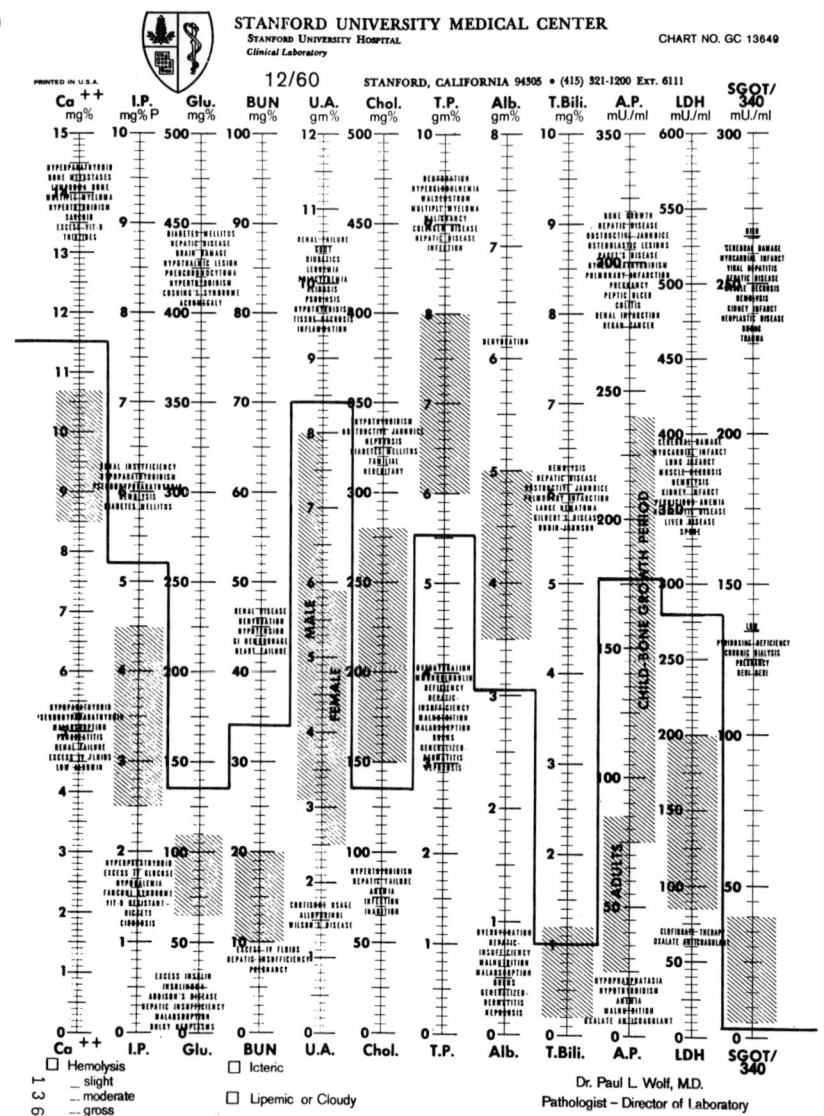

Fig. 22 - 52 (Continued):

causes low cholesterol. 6). An interesting abnormality is low or absent GOT due to decreased pyridoxine which is a cofactor for GOT and is decreased in serum due to the repeated dialysis.

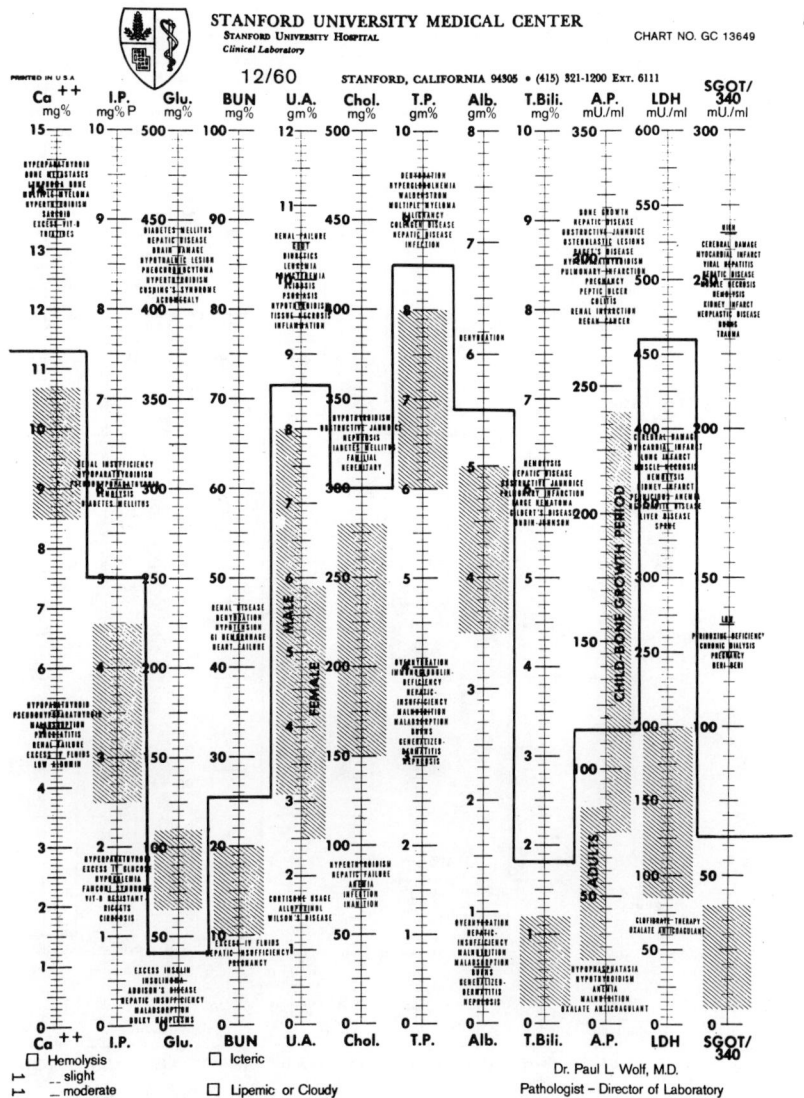

Fig. 22 - 53 BIOCHEMICAL PATTERN OF OLD SERUM ARTIFACT

Utilization of old serum introduces multiple artifacts with elevation of all the values in the SMA 12/60 Chart, except glucose which is low because of glycolysis.

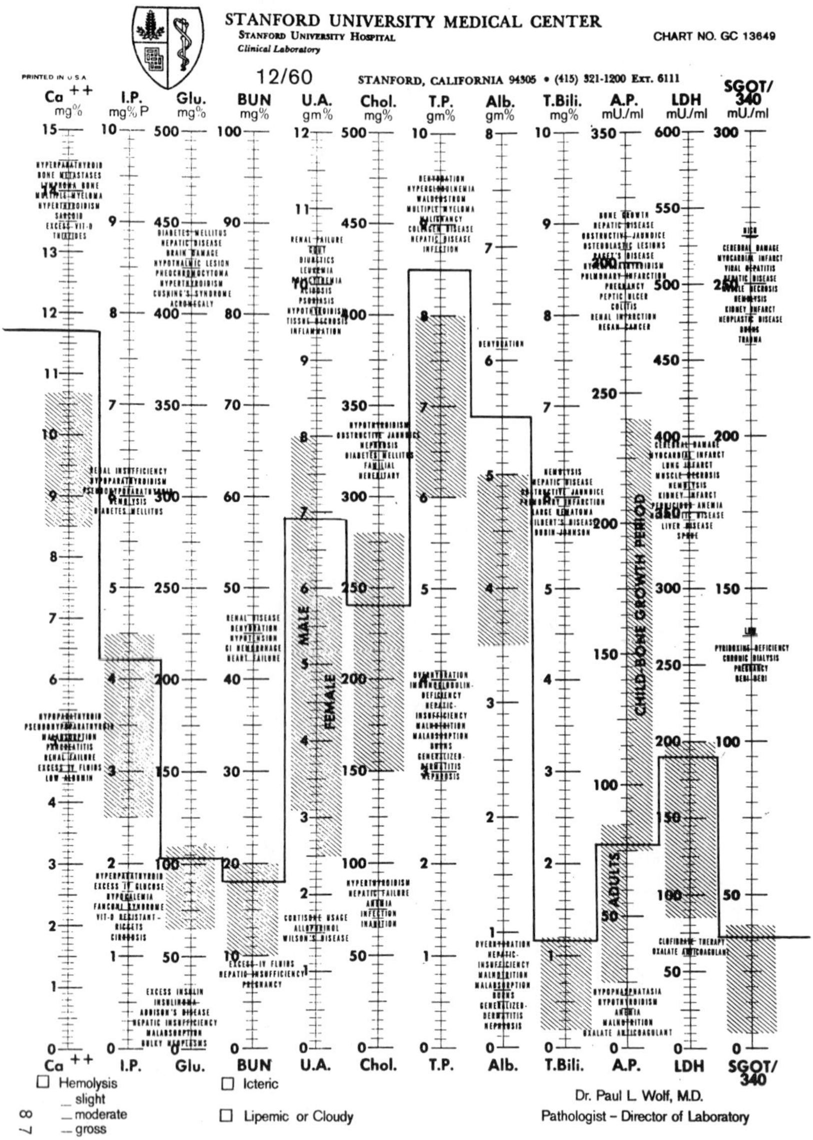

Fig. 22 - 54 BIOCHEMICAL PATTERN OF CONTAMINATION OF SPECIMEN
BY DETERGENT

Test tube detergent contamination of a serum specimen will result in artifactual elevation of albumin, total protein and calcium.

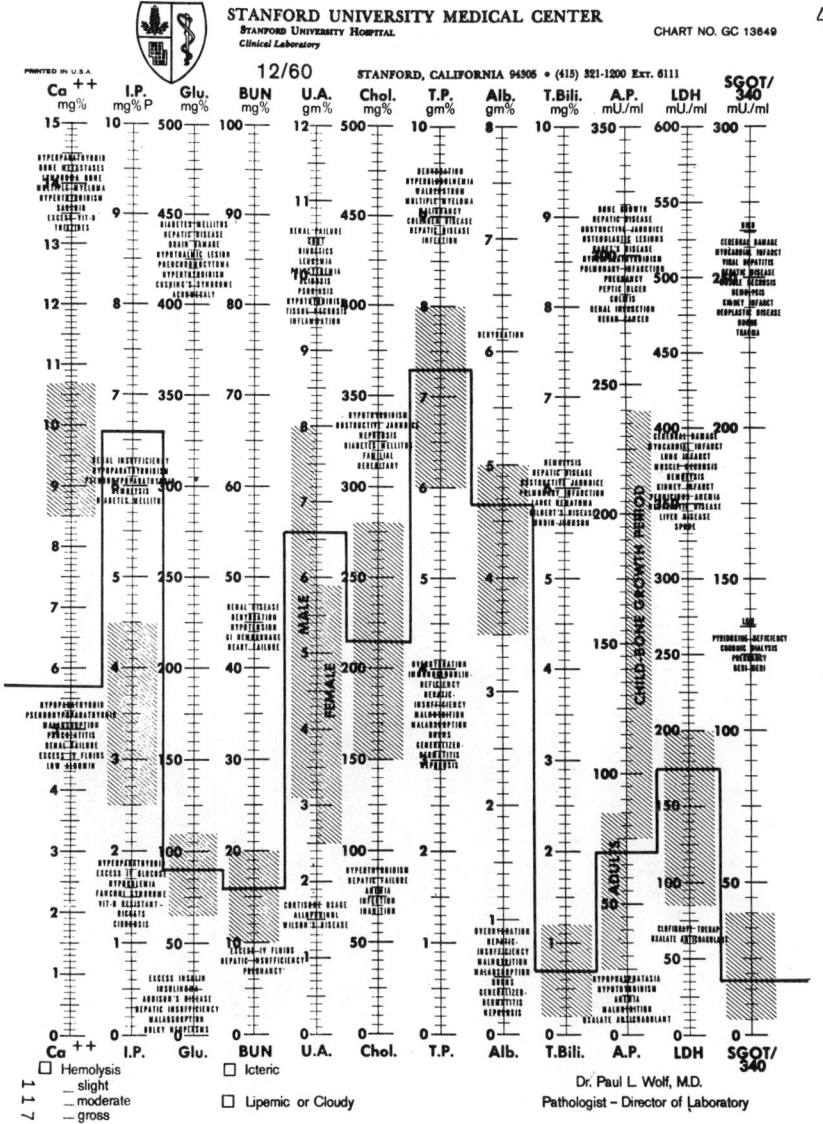

Fig. 22 - 55 BIOCHEMICAL PATTERN OF HYPOPARATHYROIDISM

1). Accidental removal of the parathyroids during thyroid surgery will result in decreased serum parathyroid hormone. 2). Rapid elevation of serum phosphate and decreased calcium will occur. 3). This biochemical pattern is also seen in pseudo-hypoparathyroidism.

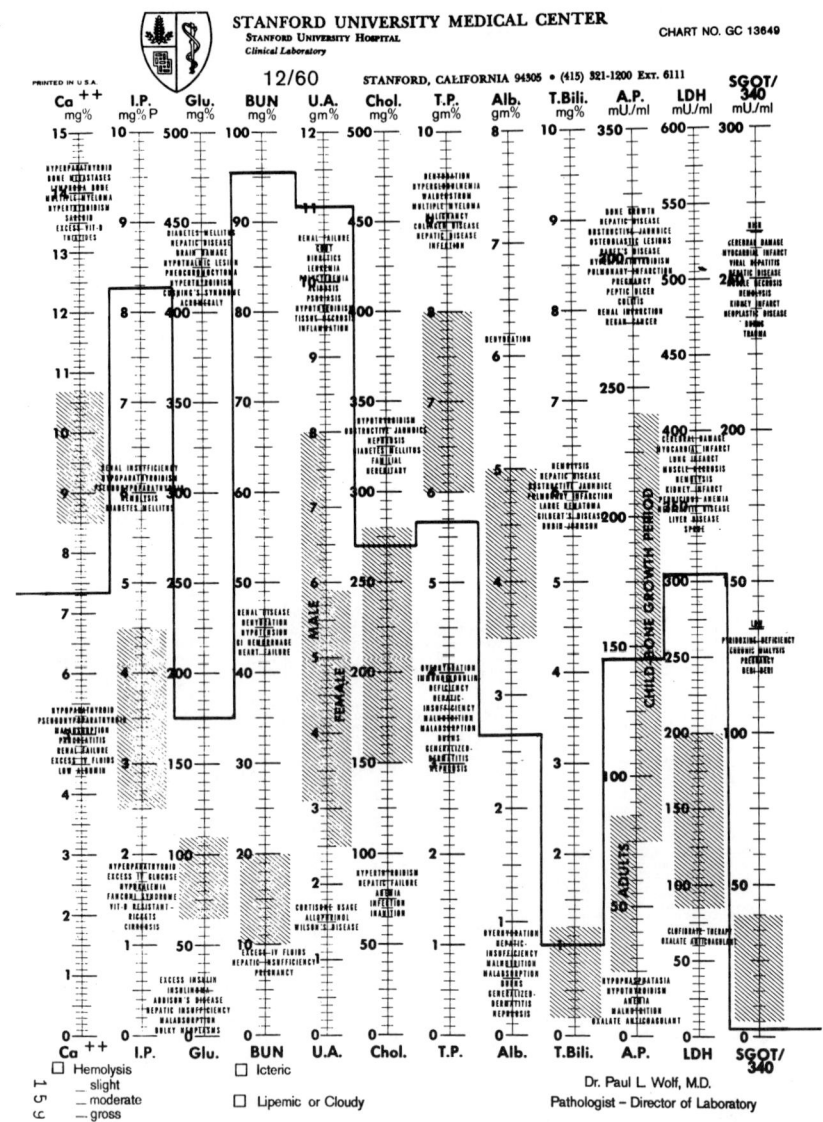

Fig. 22 - 56 BIOCHEMICAL PATTERN OF CHRONIC RENAL INSUFFICIENCY
AND LONG TERM DIALYSIS

1). Chronic renal insufficiency causes elevation of BUN due to non-excretion. 2). In addition, uric acid and phosphate are not excreted and a lower calcium results.
3). The hyperglycemia is artifactual due to nonspecific reduction of neocuproine reagent by retained uremic tox
4). Low albumin is due to proteinuria of renal disease
5). Elevated alkaline phosphatase is due to secondary

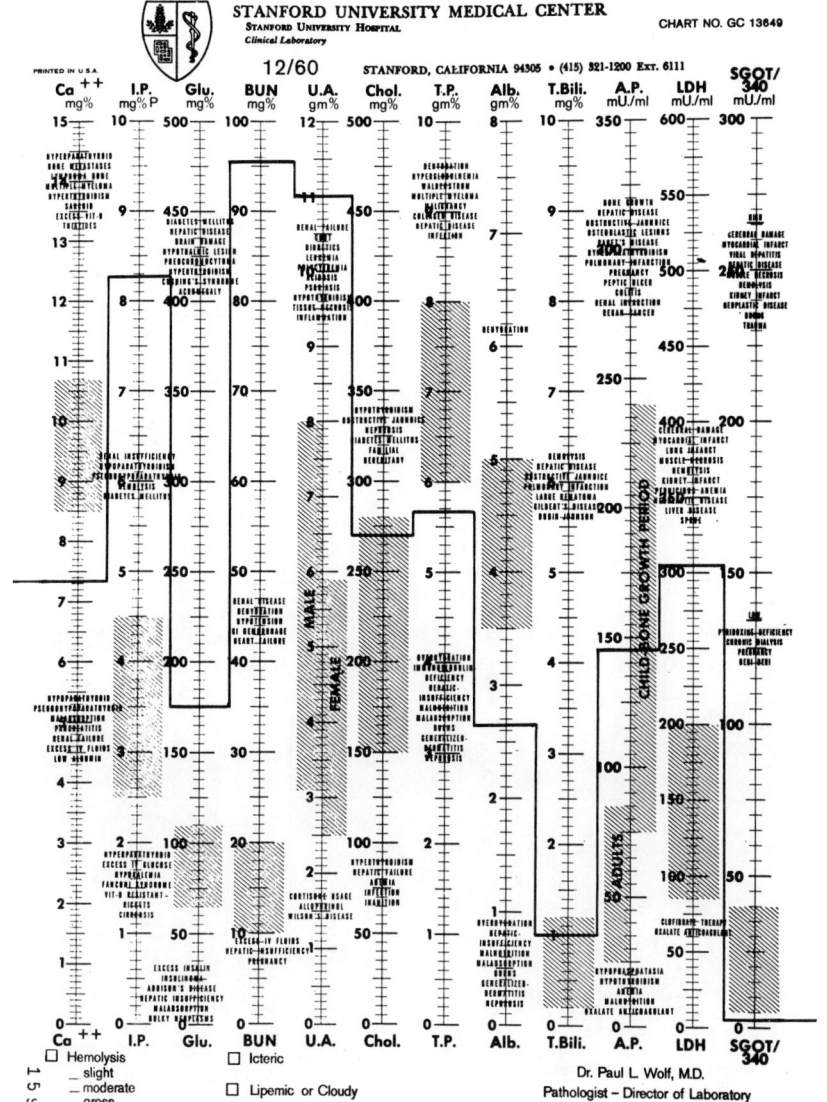

Fig. 22 - 56 (Continued):

hyperparathyroidism. 6). Elevated LDH is due to lack of excretion and renal tubular disease. 7). Low GOT is from long term hemodialysis and low pyridoxine. Pyridoxine is necessary for function of GOT.

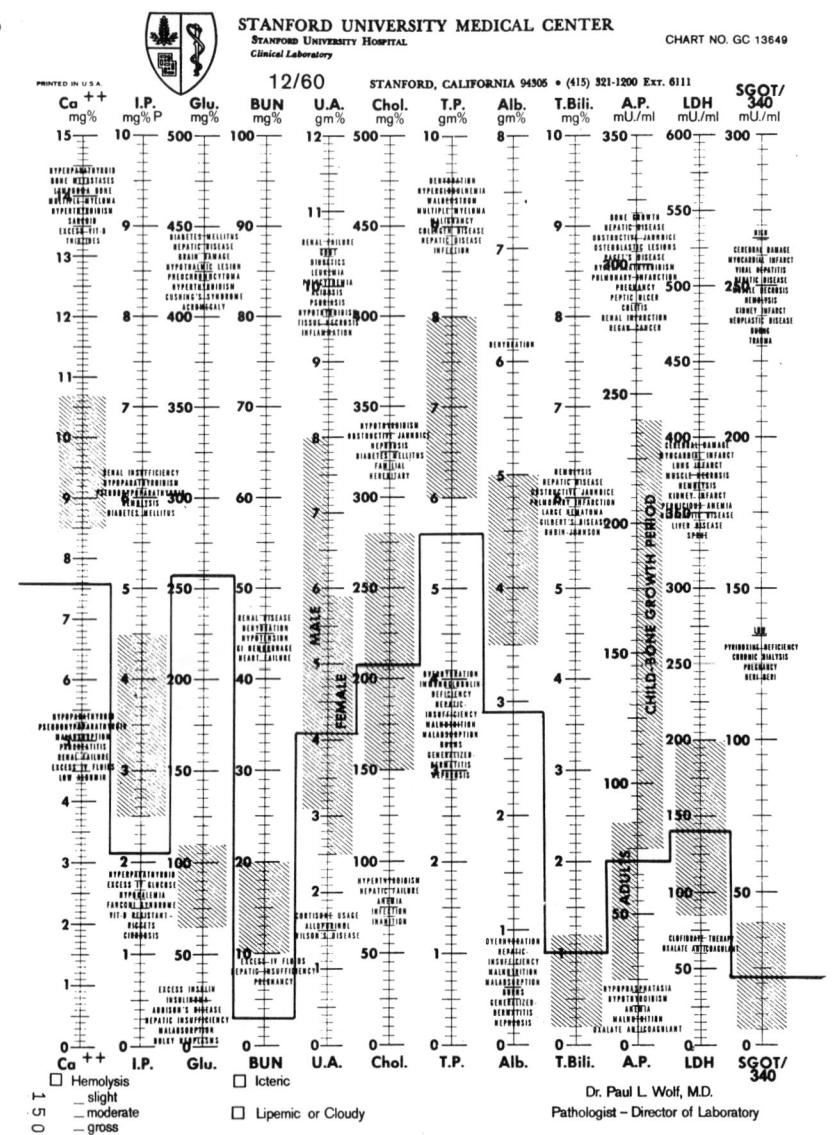

Fig. 22 - 57 BIOCHEMICAL PATTERN OF INTRAVENOUS GLUCOSE ADMINISTRATION OF SEVERAL DAYS DURATION

1). Prolonged administration of intravenous glucose results in hemodilution with lower serum protein and albumin. 2). Consequently a decrease in serum calcium occurs. 3). Hypophosphatemia is present due to phosphorylation of the glucose. 4). The BUN is low due to the hemodilution.

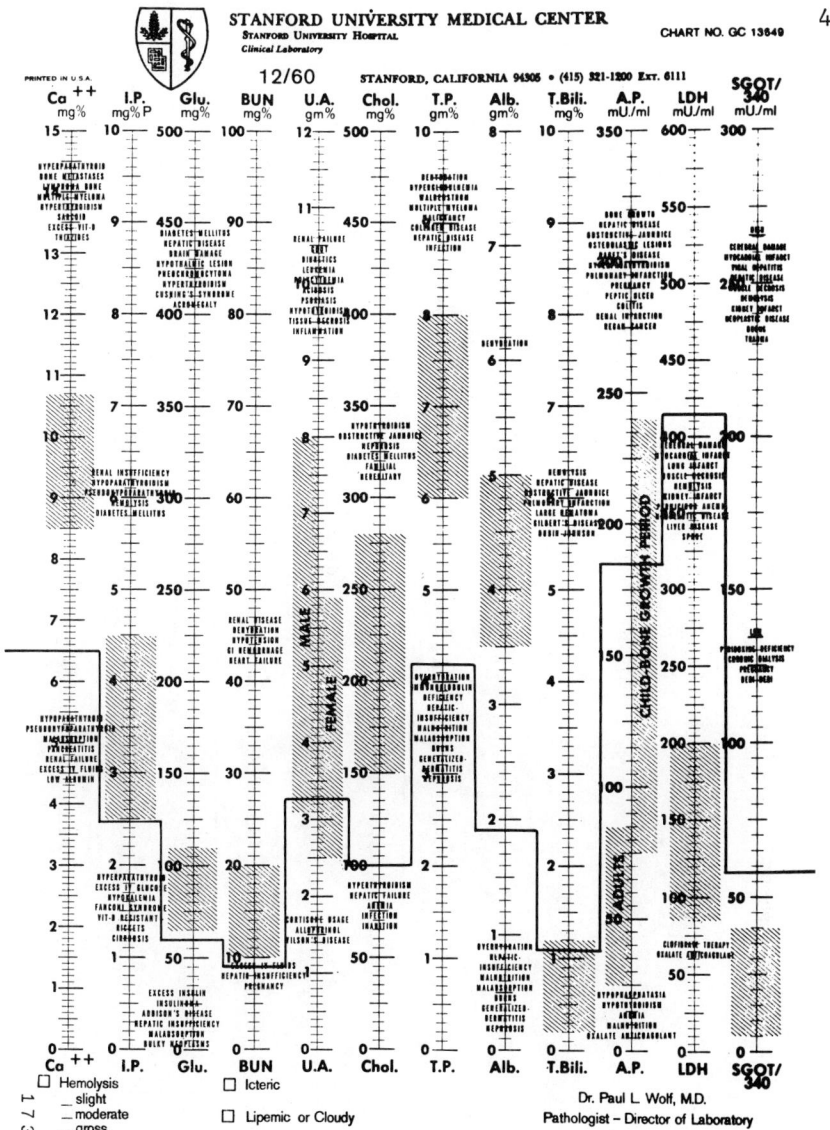

Fig. 22 - 58   BIOCHEMICAL PATTERN OF MALABSORPTION - SPRUE

1). Malabsorption will result in low serum albumin and globulins. 2). A resultant hypocalcemia occurs due to decrease in serum binding proteins and from malabsorption of calcium due to gastrointestinal disease. 3). In addition, sprue will cause low phosphate, glucose, BUN, and cholesterol due to malabsorption. 4). The elevated alkaline phosphatase is from two sources: (a). Diseased mucosa of small intestine, (b). and osteomalacia from

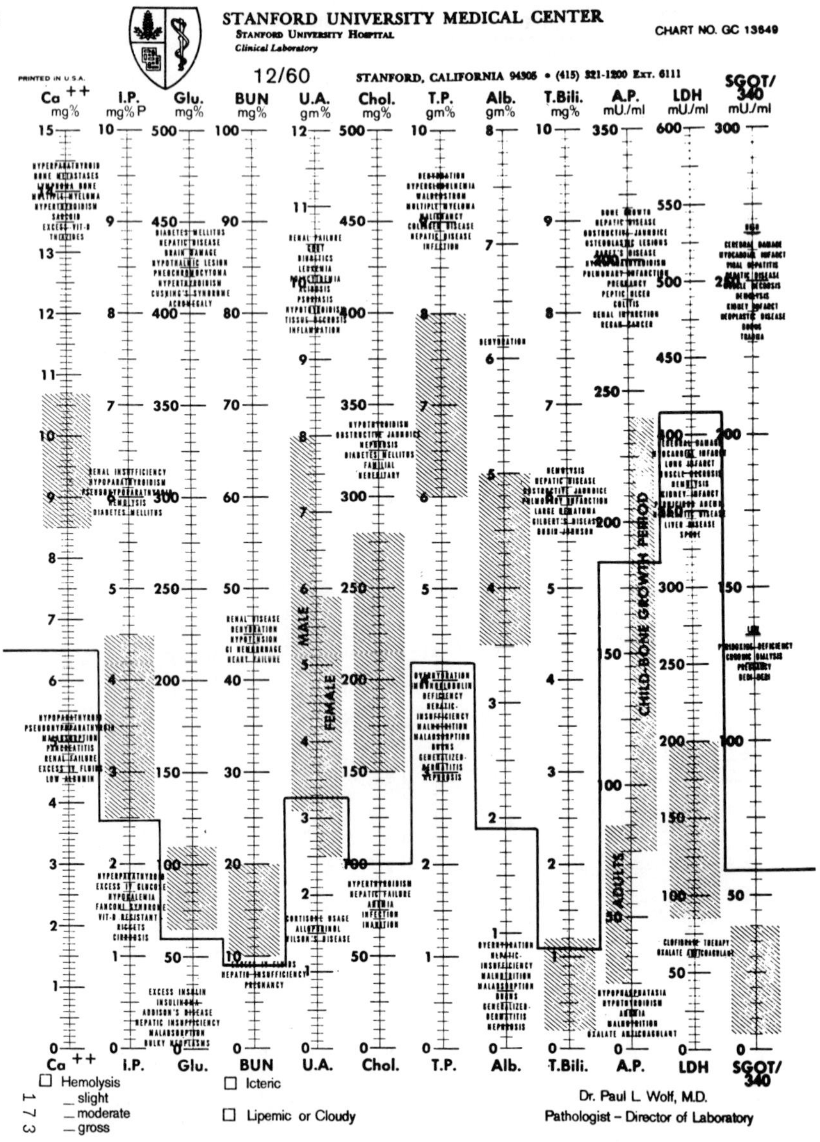

468

Fig. 22 - 58 (Continued): Vitamin D malabsorption. 5). LDH and GOT elevation result from the megaloblastic anemia found in sprue.

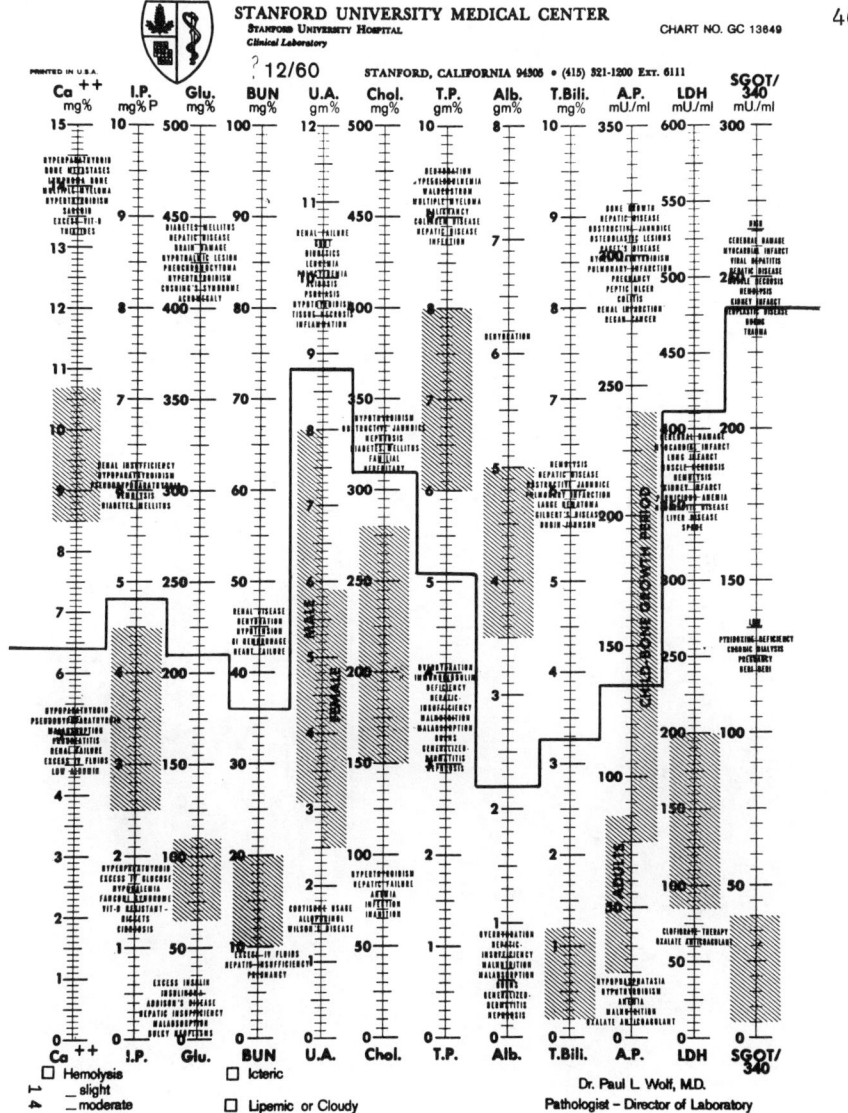

Fig. 22 - 59 BIOCHEMICAL PATTERN OF ACUTE PANCREATITIS

Acute pancreatitis results in liberation of lipase into the abdomen and hydrolysis of fat. Fatty acids are liberated from the triglycerides and the fatty acids and combine with serum calcium causing hypocalcemia. Loss of protein into the abdomen causes low protein and albumin. Destruction of the islets of Langerhan's causes hyperglycemia and elevated phosphate. Prerenal azotemia causes high BUN and uric acid. Hemorrhage and edema of head of pancreas causes

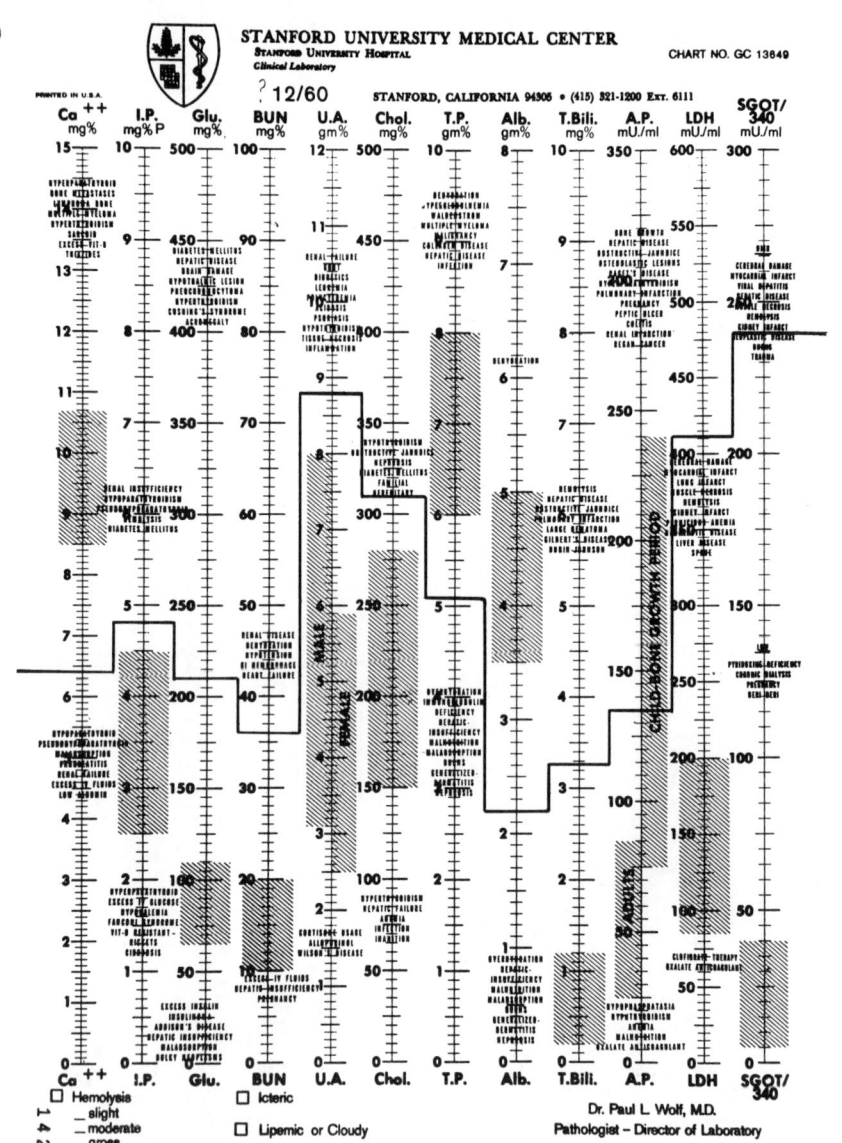

Fig. 22 - 59 (Continued):

obstructive jaundice with elevated alkaline phosphatase bilirubin, and cholesterol. The pancreas contains much LDH and GOT which is liberated with pancreatic necrosis

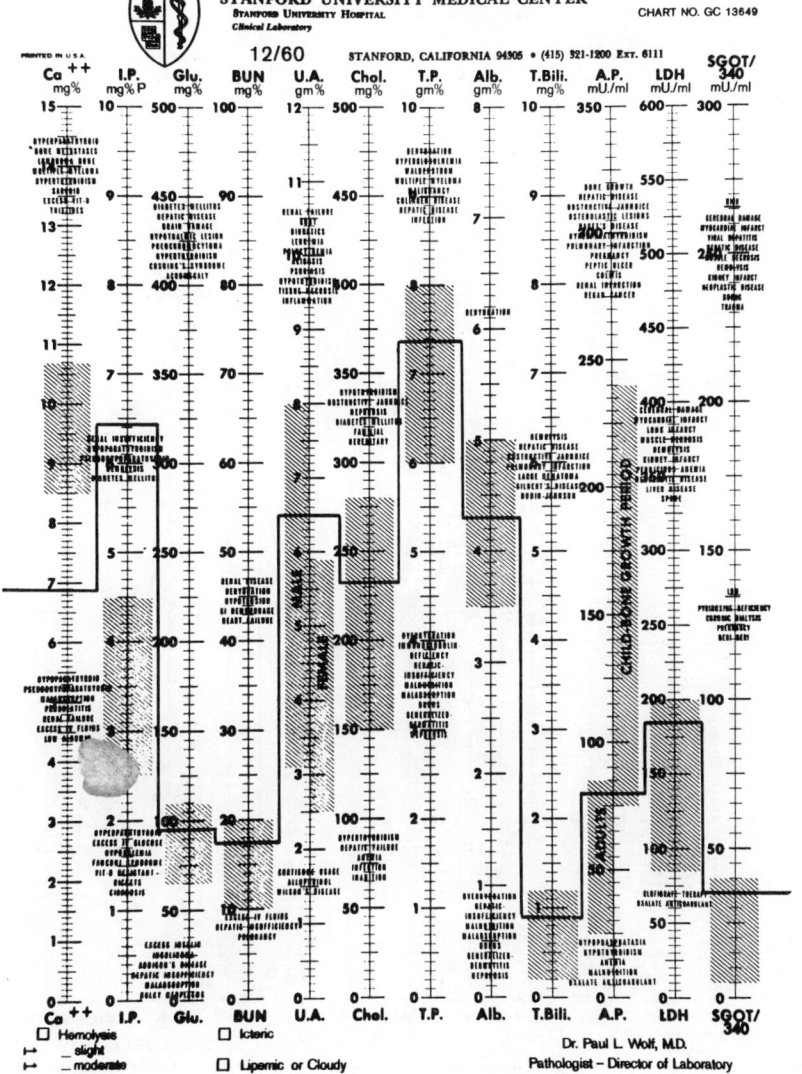

Fig. 22 - 60 BIOCHEMICAL PATTERN OF PSEUDOHYPOPARATHYROID

Pseudohypoparathyroid syndrome is characterized by a
patient with small stature, tetany, short 4th and 5th
metacarpal and metatarsal bones and low serum calcium
and elevated phosphate. Biochemically, the patient is
similar to hypoparathyroidism. However, pseudohypoparathyroidism is associated with normal serum parathyroid
hormone but defective (unresponsive to parathyroid hormone)
renal tubules. The syndrome is hereditary and has

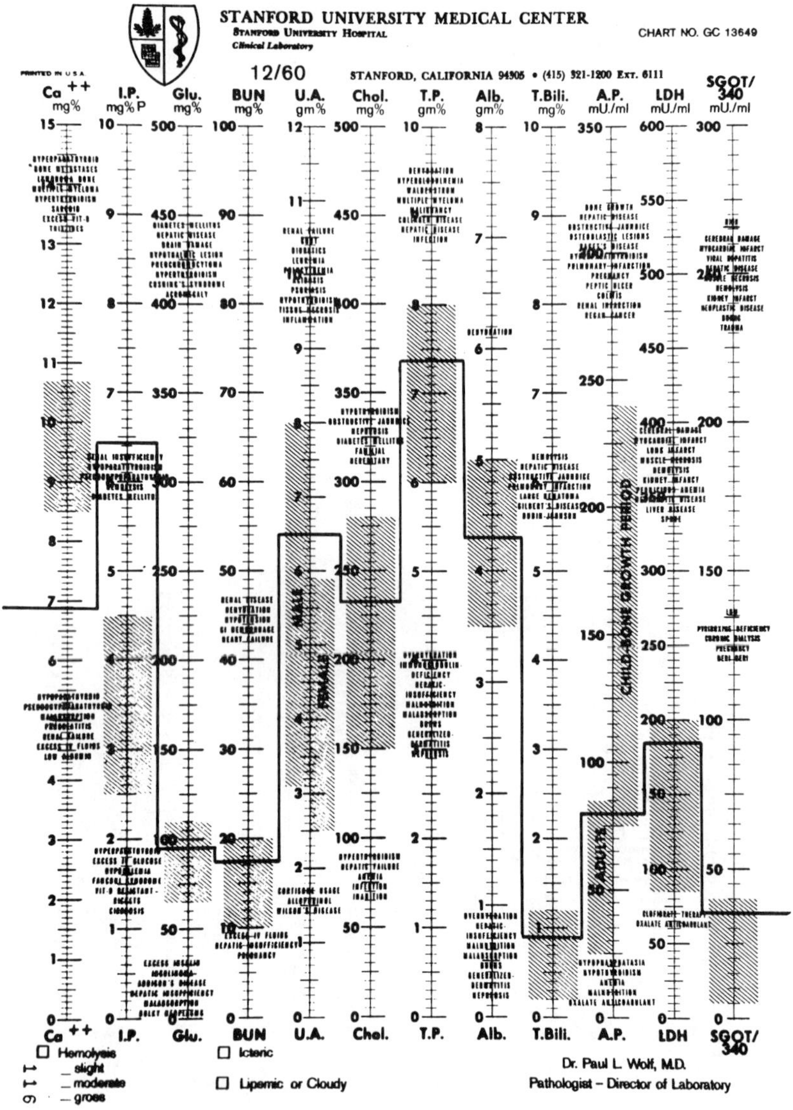

472

Fig. 22 - 60 (Continued):

distinctive morphologic stigmata. It is also known as the Seabright - Bantam syndrome.

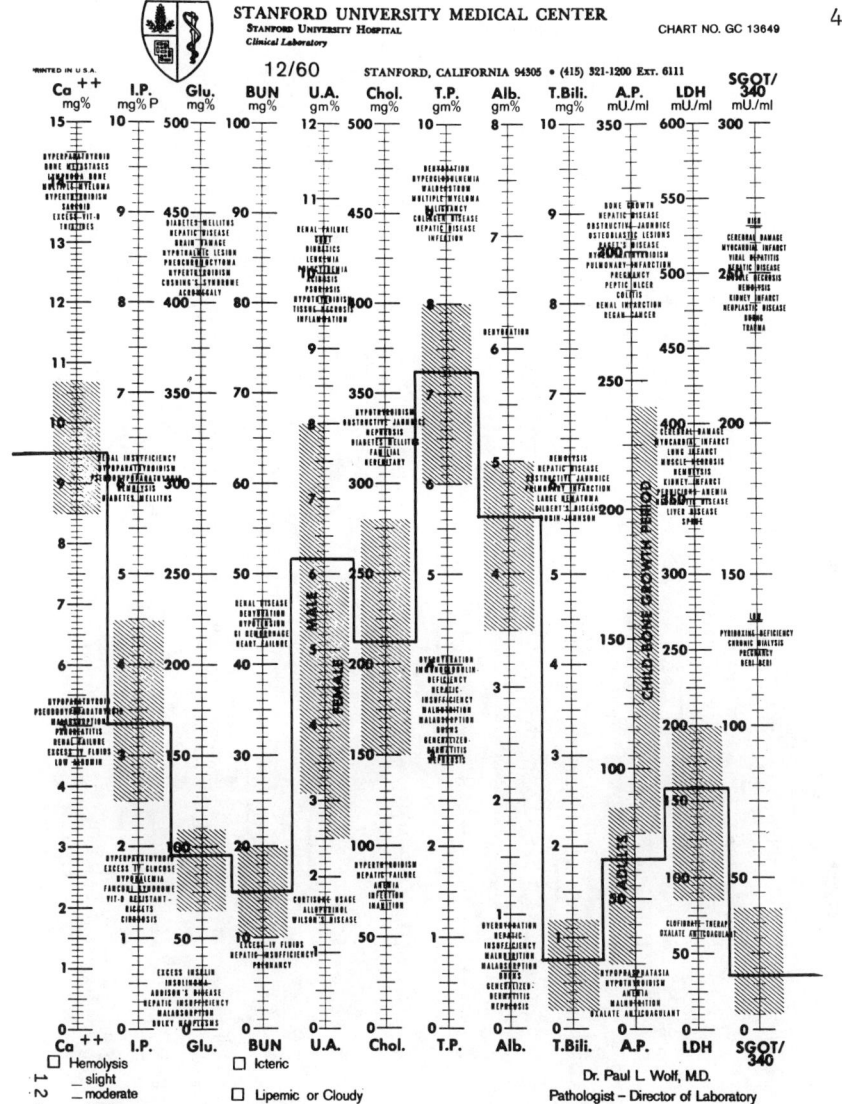

Fig. 22 - 61   BIOCHEMICAL PATTERN OF PSEUDO - PSEUDOHYPOPARATHYROID

See Fig. 22 - 60. The patient with pseudo - pseudohypoparathyroid syndrome has similar morphologic stigmata as pseudohypoparathyroid syndrome, but biochemically is normal, and thus, there is no tetany nor other symptoms of hypocalcemia. For unexplained reasons, the renal tubule has reverted to normal responsiveness to parathyroid hormone.

474

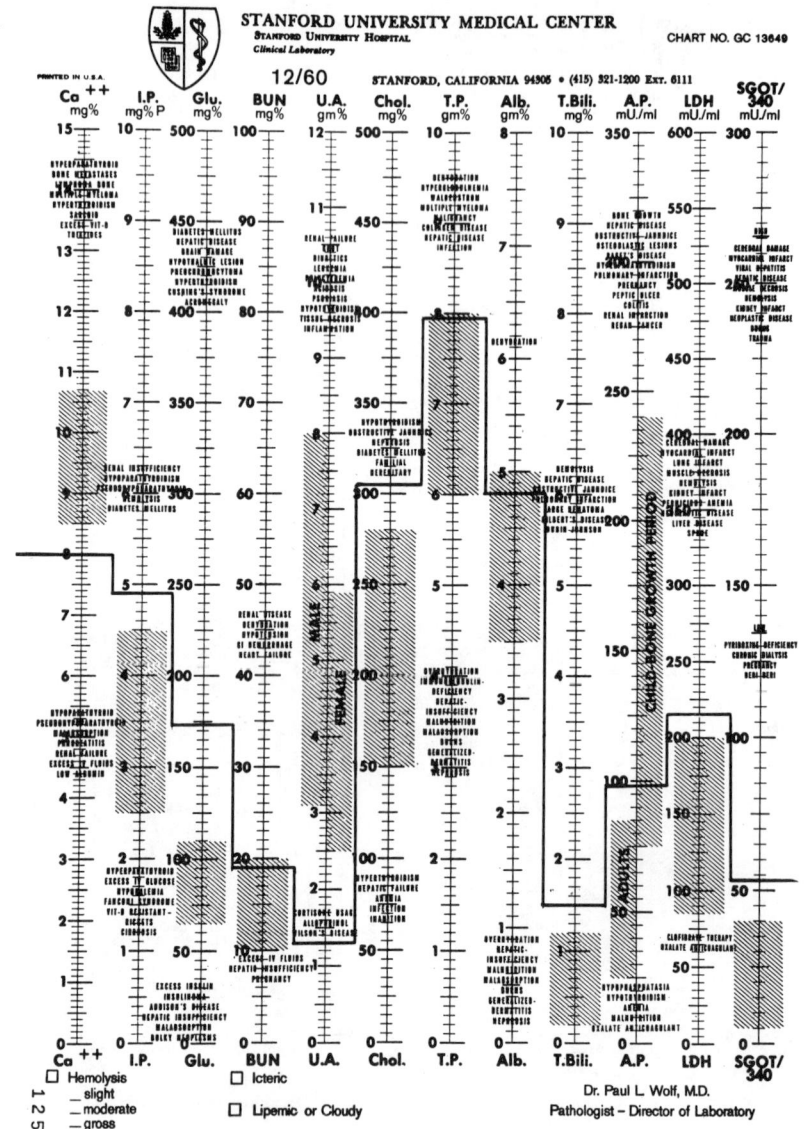

Fig. 22 - 62   BIOCHEMICAL PATTERN OF EXCESSIVE CORTISONE

Excessive utilization of cortisone will lower serum
calcium and uric acid. Steroid induced diabetes mellitu
occurs with hyperglycemia and elevated phosphate. Lipid
metabolism is distorted and elevated cholesterol and tri
glycerides occur. Marked fatty metamorphosis of the li
occurs with elevated bilirubin, alkaline phosphatase, LI
and GOT. Osteomalacia may occur with elevated alkaline
phosphatase.

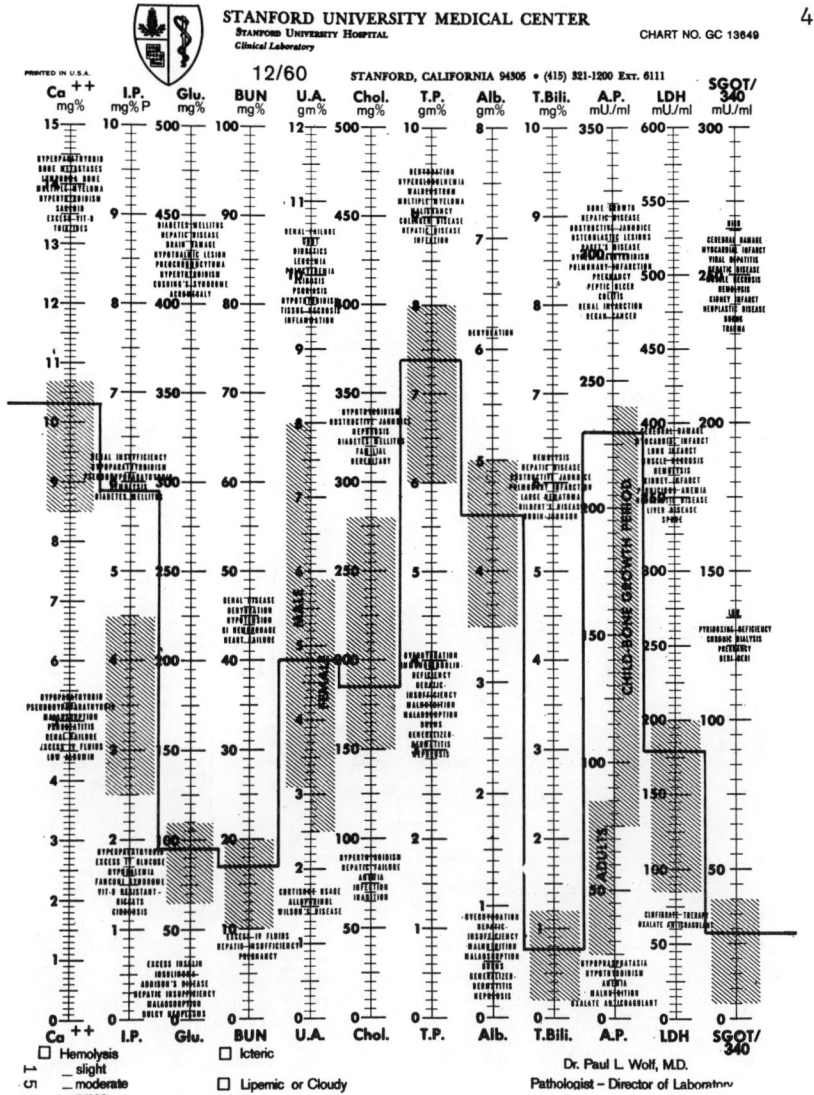

Fig. 22 - 63   BIOCHEMICAL PATTERN OF PHYSIOLOGICAL BONE GROWTH

Physiological bone growth up to the age of approximately 16 years results in elevated serum phosphate and alkaline phosphatase which is the heat labile type.

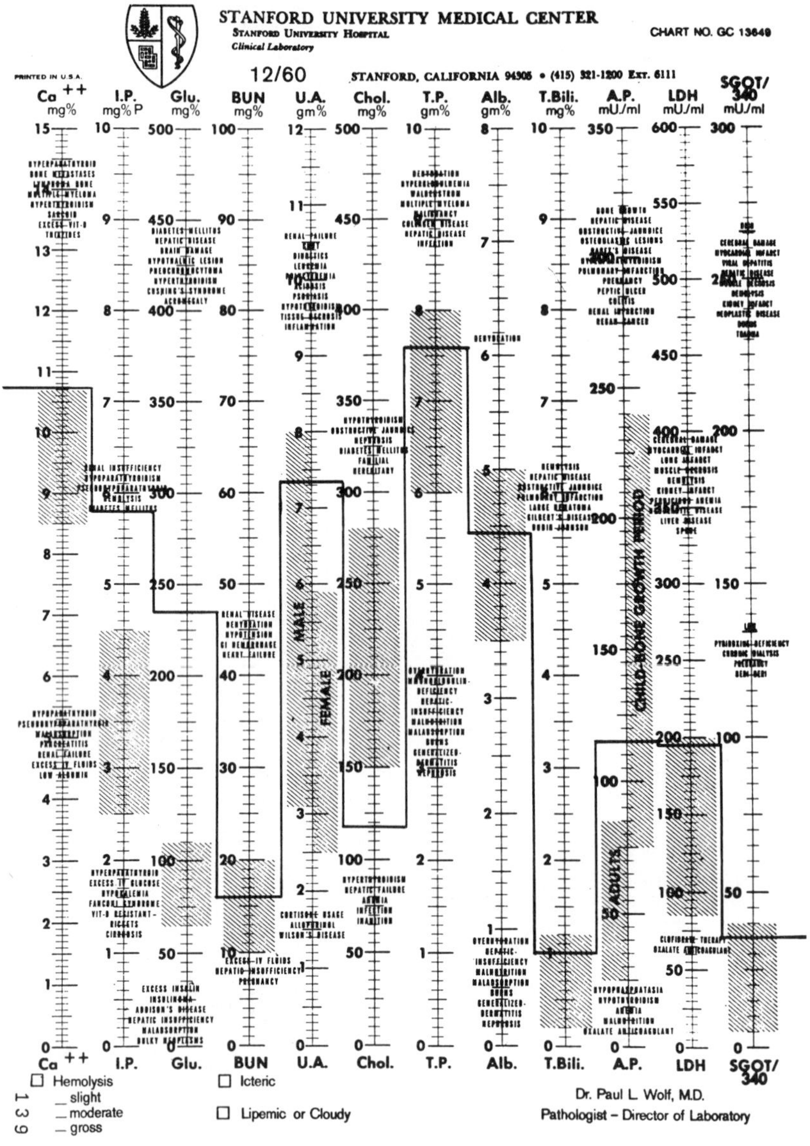

476

Fig. 22 - 64   BIOCHEMICAL PATTERN OF ACROMEGALY

Acromegaly is associated with excessive bone growth with resultant elevated phosphate and alkaline phosphatas The increased serum growth hormone causes glucose intolerance with hyperglycemia. A high incidence of hyperthyroidism is associated which causes low cholesterol.

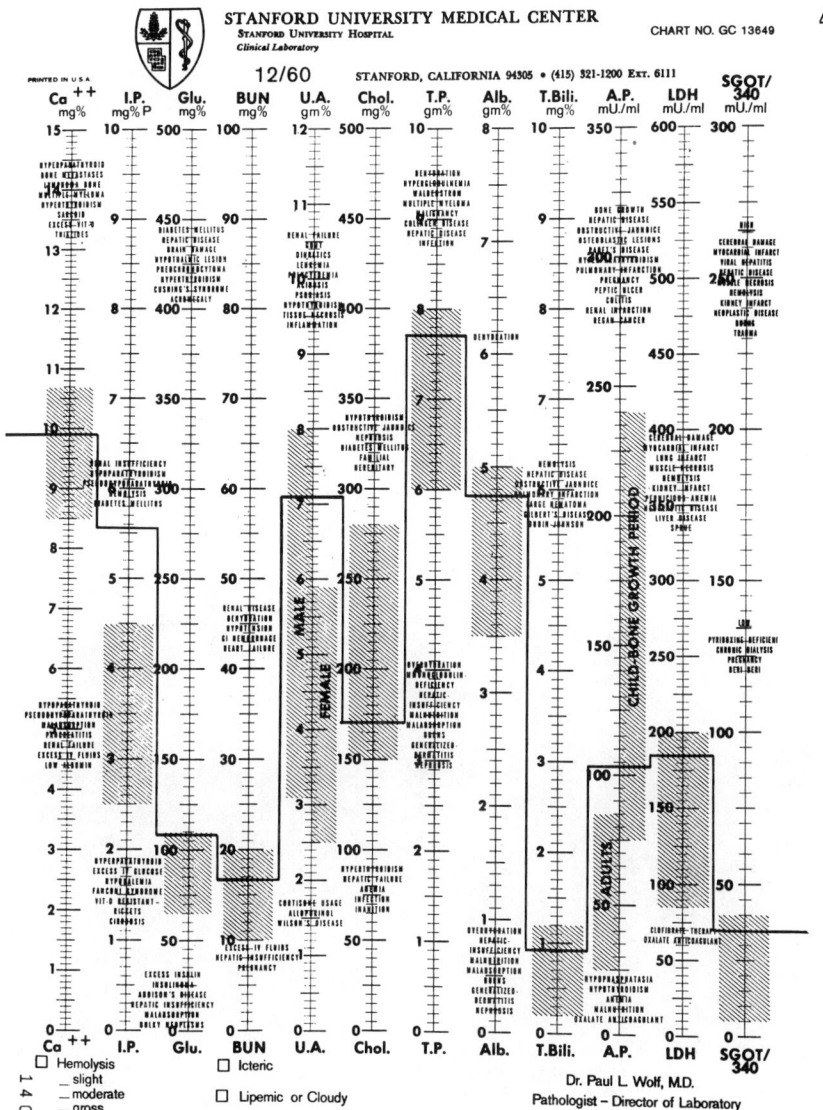

Fig. 22 - 65   BIOCHEMICAL PATTERN OF RECENT BONE FRACTURE

A recent bone fracture usually results in elevated serum phosphate and a minimal increase in a heat labile type of alkaline phosphatase due to increased osteoblastic activity at the fracture site.

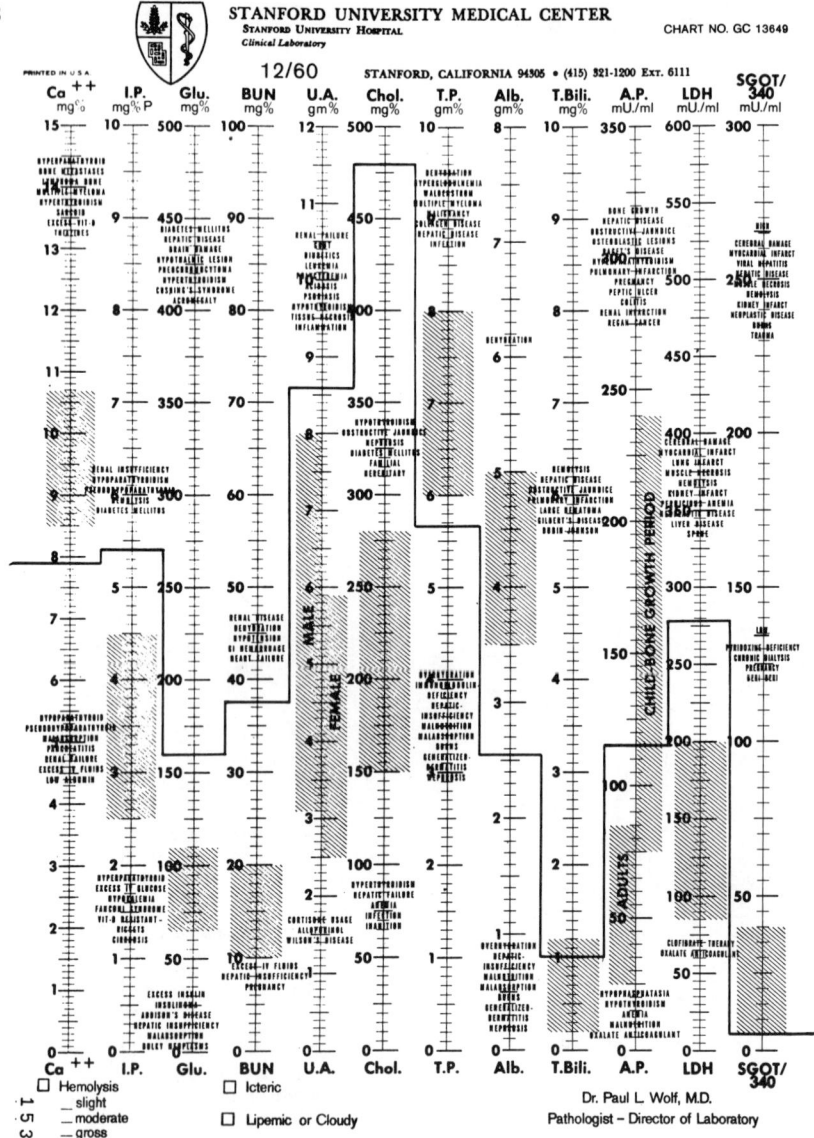

Fig. 22 - 66   BIOCHEMICAL PATTERN OF NEPHROTIC SYNDROME

1). Nephrotic syndrome causes elevation of BUN due to non-excretion. 2). Uric acid and phosphate are not excreted and a lower calcium results. 3). The hyperglycemia is artifactual due to nonspecific reduction of neocuproine reagent by retained uremic toxins. 4). Low albumin is due to proteinuria of renal disease. 5). Elevated alkaline phosphatase is due to secondary hyperparathyroidism. 6). Elevated LDH is due to lack of excretion and

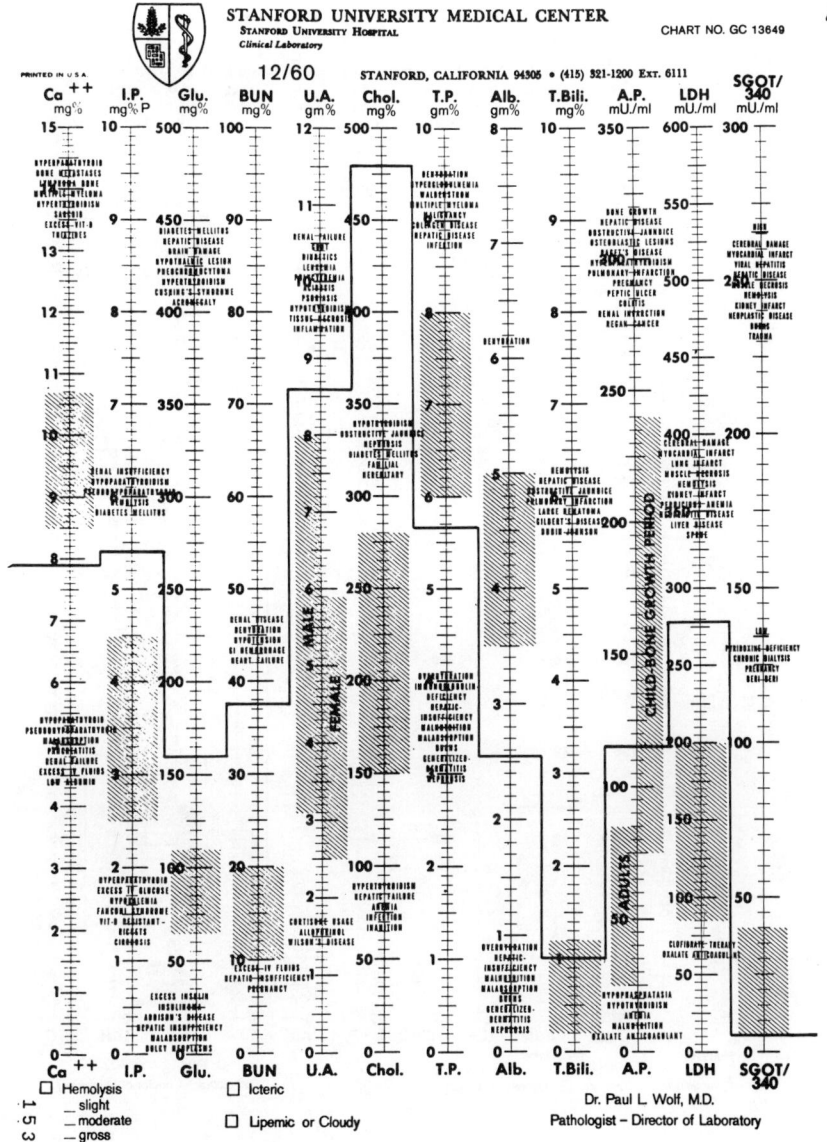

Fig. 22 - 66 (Continued):

renal tubular disease. 7). Low GOT is from long term hemodialysis and low pyridoxine. Pyridoxine is necessary for function of GOT. 8). High cholesterol may be due to (a). Decrease in serum lipoprotein lipase activity. (b). Excessive hepatic synthesis of cholesterol esters to compensate for low albumin.

480

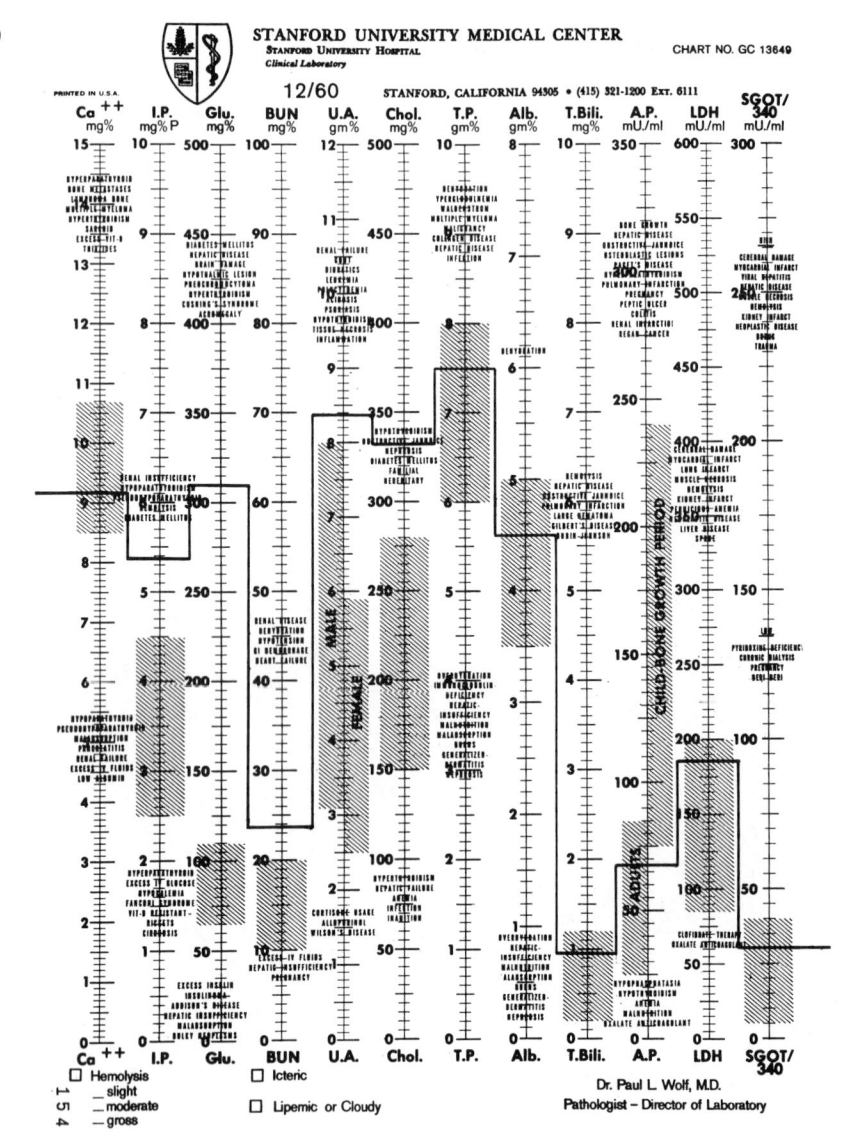

Fig. 22 - 67 BIOCHEMICAL PATTERN OF UNCONTROLLED DIABETES MELLITUS

Uncontrolled diabetes mellitus causes hyperglycemia and elevated phosphate. Dehydration results in elevated BUN on a prerenal basis. Ketoacidosis interferes with renal tubular excretion of uric acid with resultant hyperuricemia. Distorted lipid metabolism results in elevated cholesterol and triglycerides.

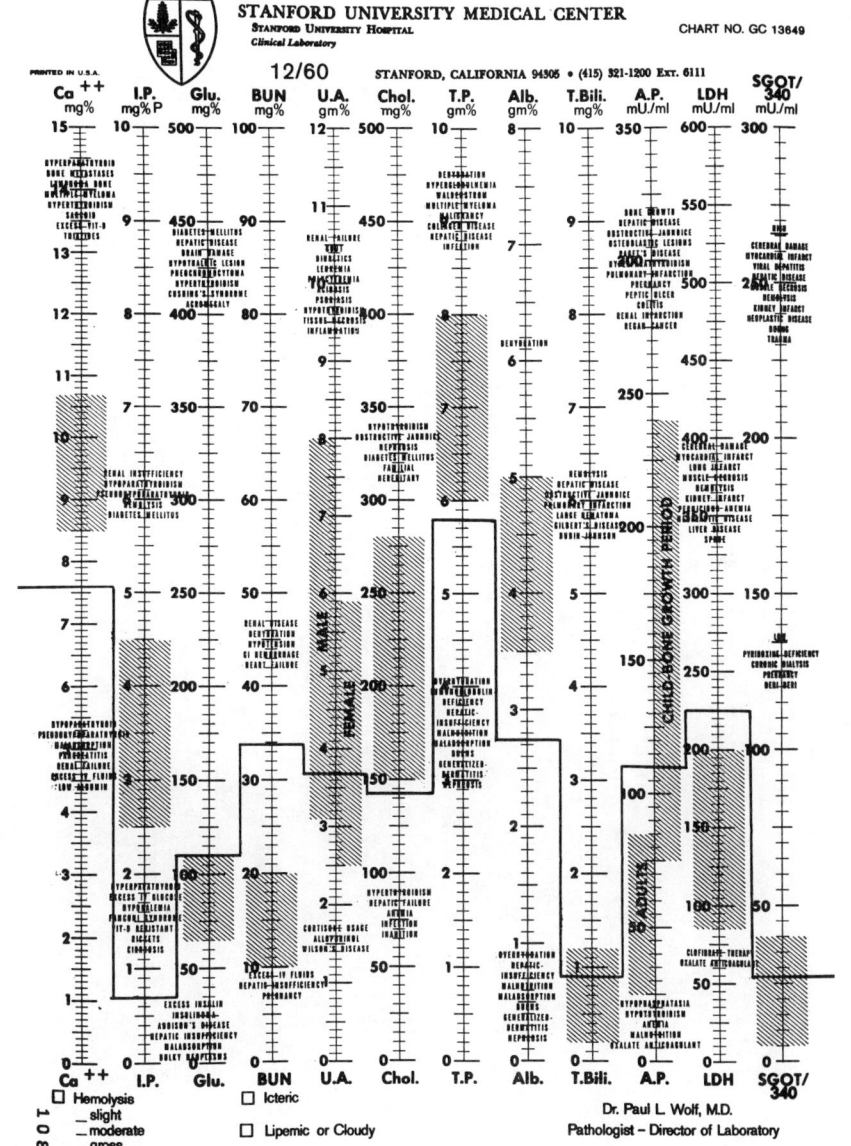

Fig. 22 - 68 BIOCHEMICAL PATTERN OF RENAL TUBULAR ACIDOSIS

1). Renal tubular acidosis either due to a congenital or acquired etiology is associated with an osteomalacia and an elevated alkaline phosphatase. 2). A renal tubular wasting of phosphate occurs. 3). The renal tubular disease results in an elevated BUN, decreased protein due to proteinuria, and elevated LDH due to lack of excretion. 4). The hypocalcemia is secondary to the low serum protein.

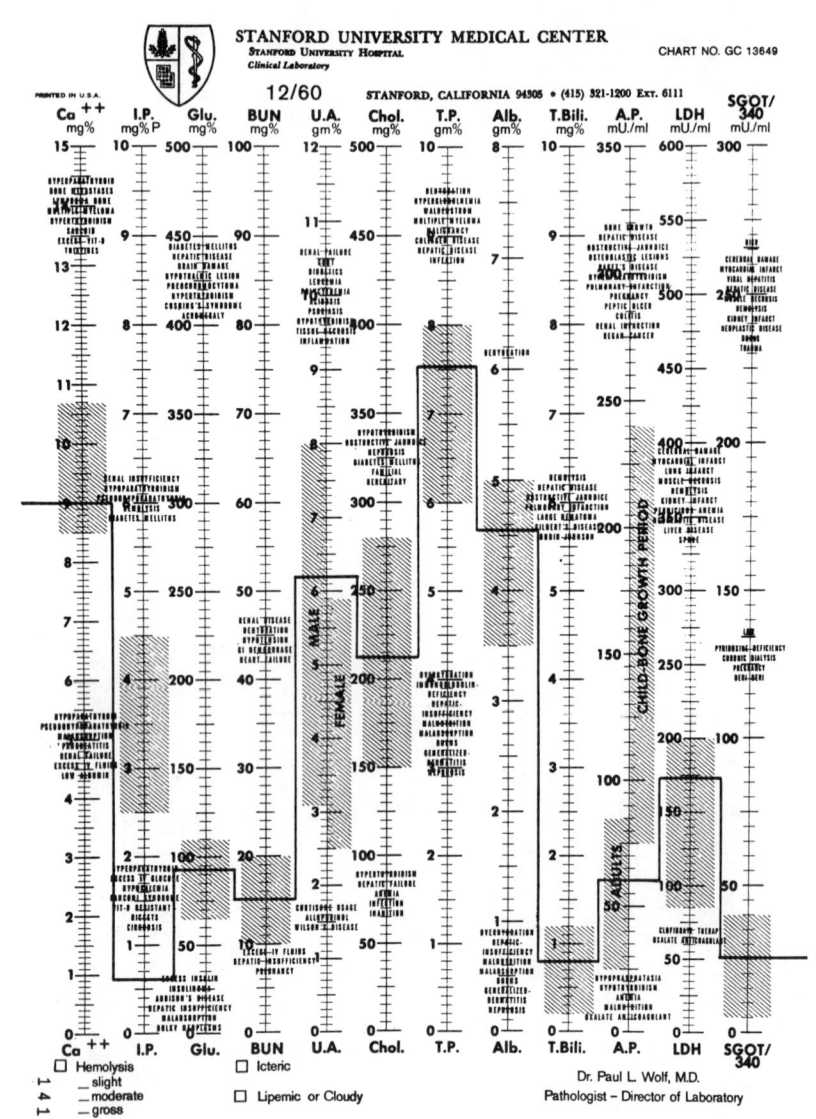

Fig. 22 - 69 BIOCHEMICAL PATTERN OF EXCESSIVE UTILIZATION OF ANTACID

Low serum phosphate occurs in patients with peptic ulcer disease who ingest excessive amounts of antacids, such as aluminum or magnesium hydroxide. These antacids bind phosphate in the gastrointestinal tract and prevent absorption. The 2 - 3 DPG enzyme of red blood cells may suffer from low serum phosphate and interfere with oxygen delivery to tissues. Hypophosphatemia may cause weakness and confusion.

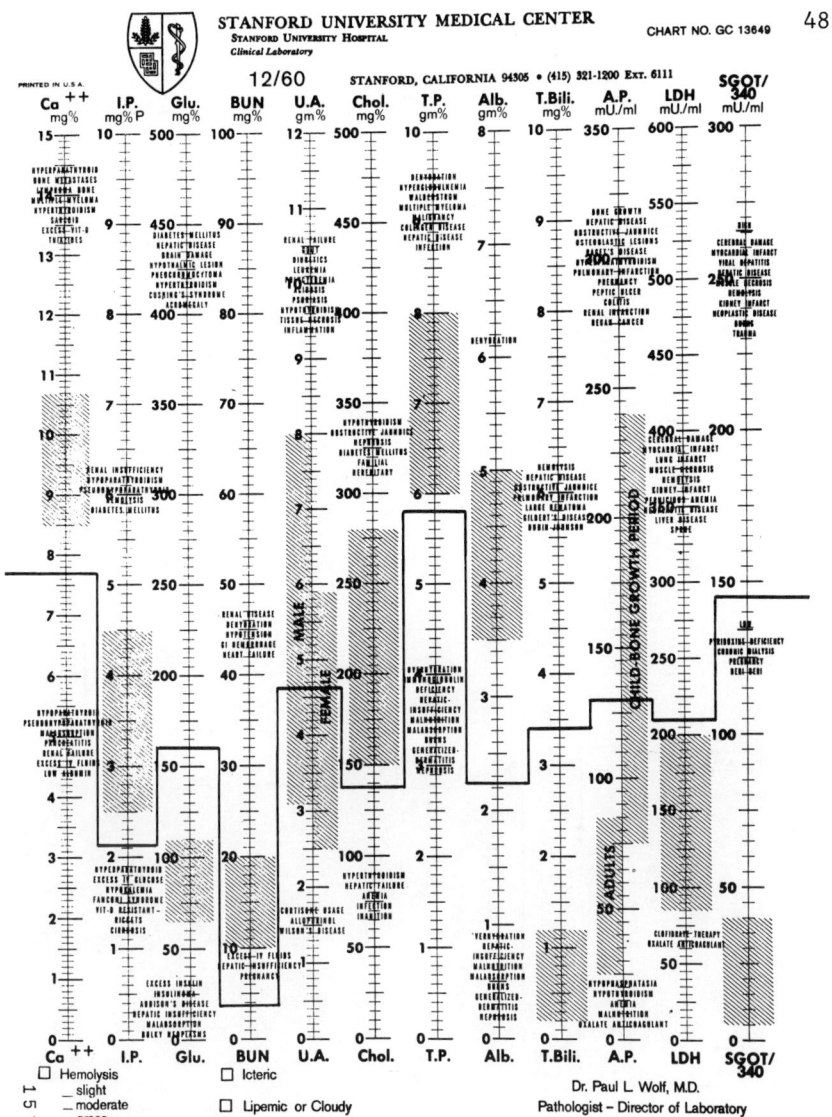

Fig. 22 - 70 BIOCHEMICAL PATTERN OF CIRRHOSIS

1). Cirrhosis is characterized by minimal elevation of alkaline phosphatase, GOT, and LDH. 2). Bilirubin elevation is of the mixed type with elevation of unconjugated and conjugated fractions. 3). Serum albumin is low due to decreased synthesis resulting in decreased calcium due to decreased binding capacity. 4). Decreased cholesterol results from decreased synthesis of cholesterol esters. 5). Decreased synthesis of BUN causes a low BUN.

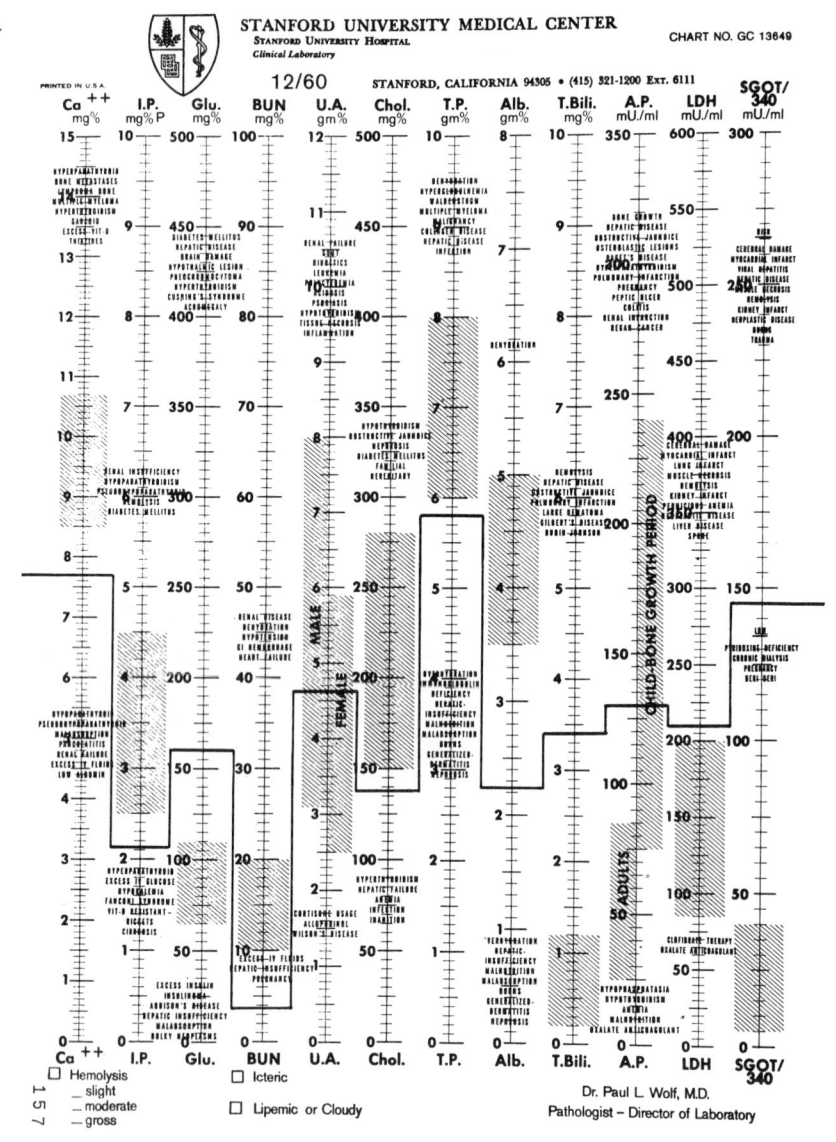

Fig. 22 - 70 (Continued):

6). Increased glycogenolysis causes hyperglycemia with low phosphate due to increased phosphorylation. Low phosphate also results from low potassium due to a renal tubule defect.

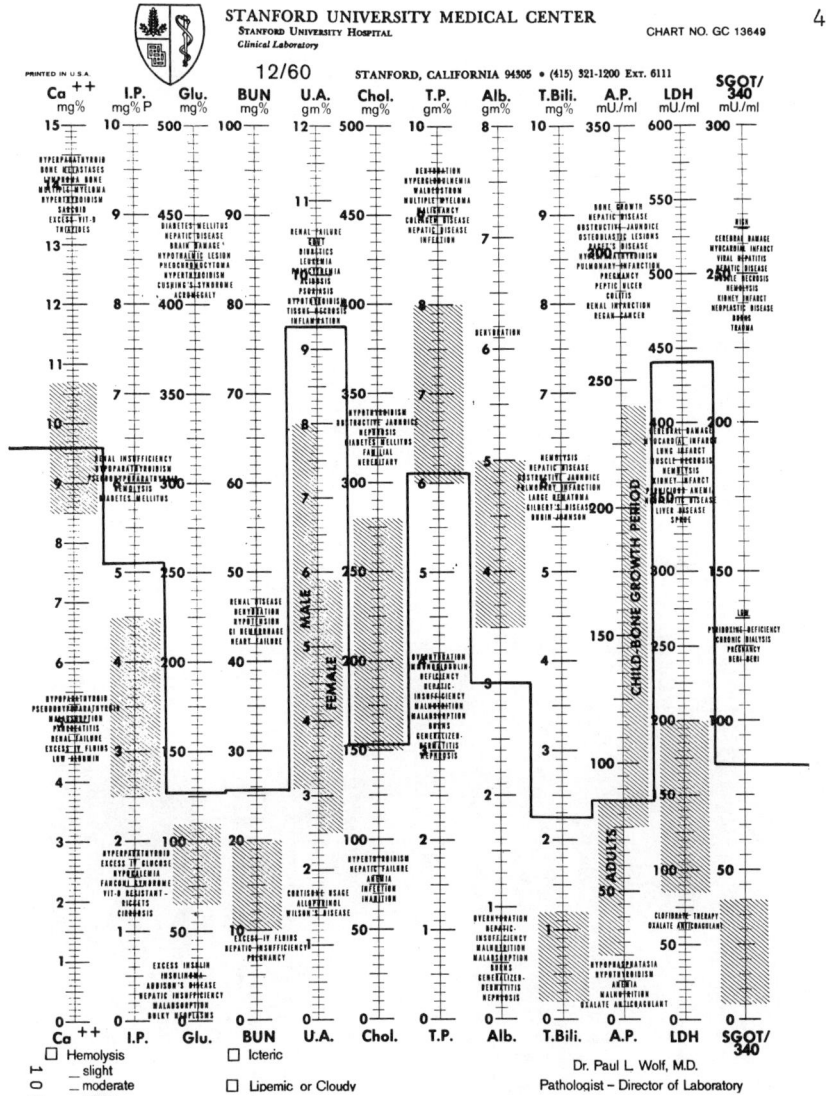

Fig. 22 - 71 BIOCHEMICAL PATTERN OF EXTENSIVE TISSUE NECROSIS AND ACUTE HEMOLYSIS

1). Acute hemolysis or extensive tissue necrosis will cause prominent elevation of LDH and moderate GOT elevation. 2). The erythrocyte or tissue damage will result in elevated uric acid. 3). Increased bilirubin results from the erythrocyte destruction or due to fatty metamorphosis of the liver. 4). Renal damage occurs from hemolysis or from the extensive tissue necrosis.

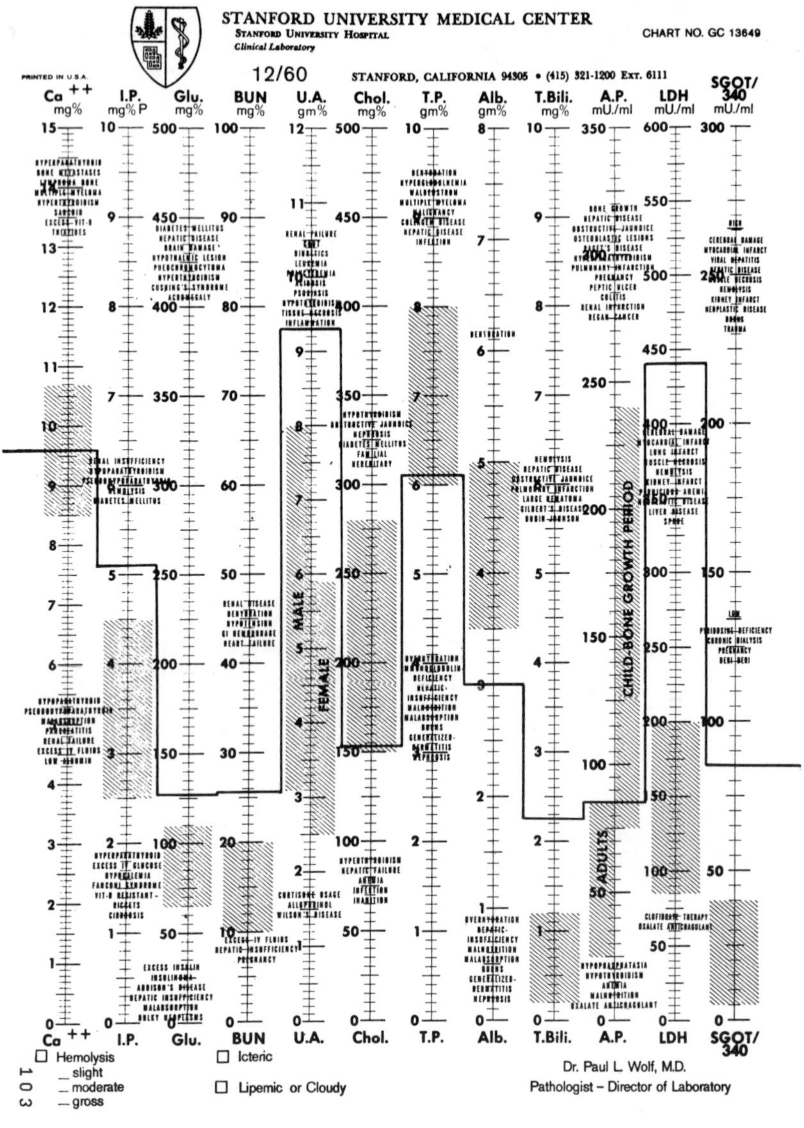

486

Fig. 22 - 71 (Continued):

5). The renal insufficiency results in elevated BUN, phosphate, glucose, and uric acid. 6). Proteinuria causes low serum protein.

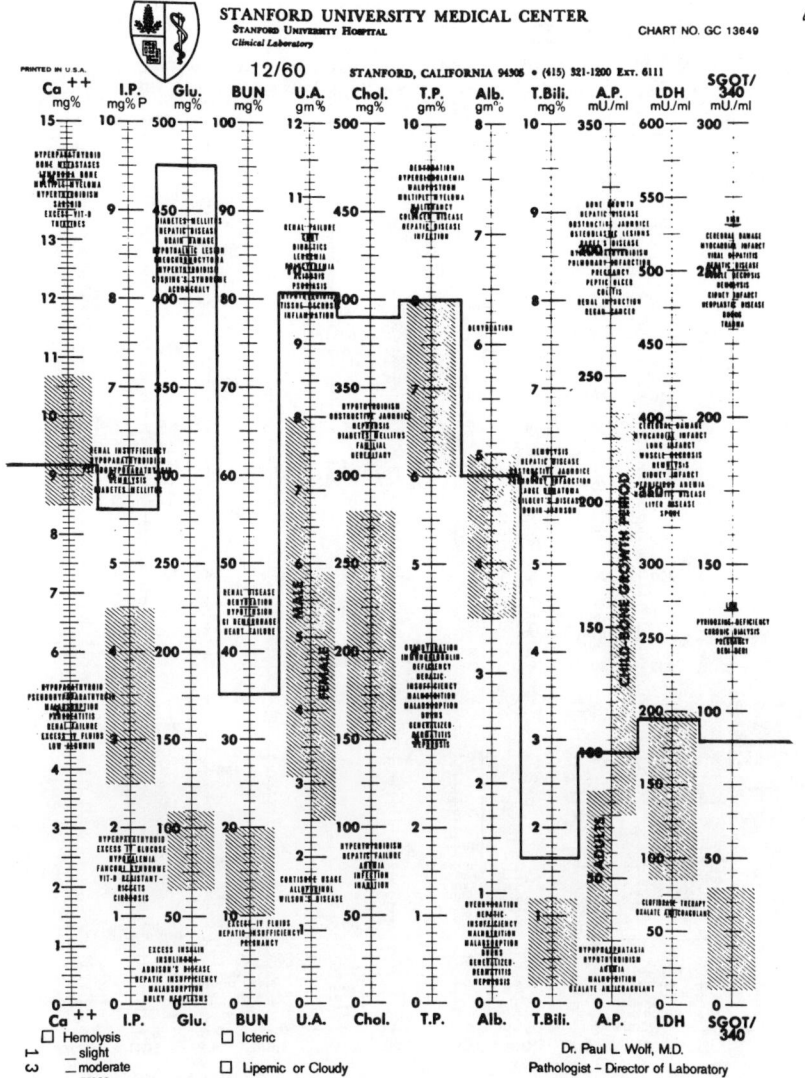

Fig. 22 - 72   BIOCHEMICAL PATTERN OF DIABETES MELLITUS WITH KETOACIDOSIS

1). Uncontrolled diabetes mellitus with ketoacidosis is characterized by marked hyperglycemia with elevated phosphate since phosphorylation of glucose does not occur. 2). Dehydration is frequent and prerenal azotemia is present with elevated BUN. 3). Ketoacidosis inhibits the excretion of uric acid causing hyperuricemia. 4). Uncontrolled diabetes mellitus causes a disturbance in lipid metabolism and elevated cholesterol and triglycerides. 5). A fatty liver is present and bilirubin, alkaline phosphatase, and GOT are elevated.

488

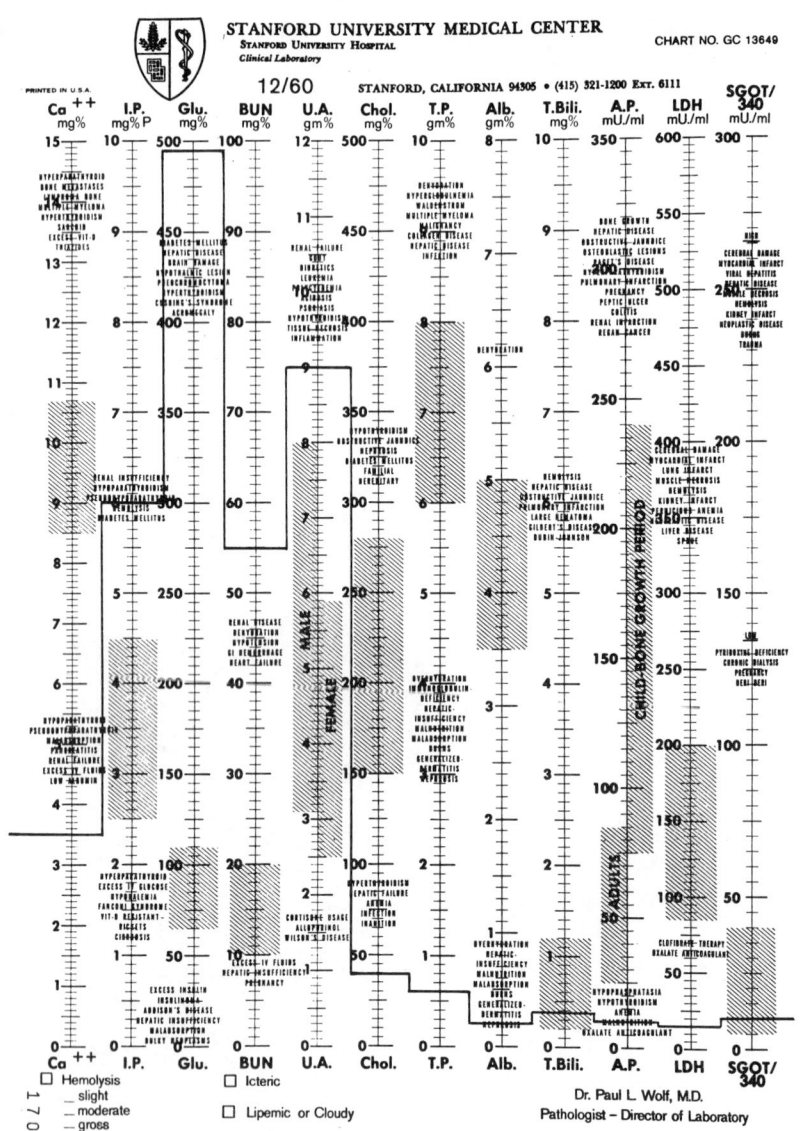

Fig. 22 - 73 BIOCHEMICAL PATTERN OF POST-MORTEM VITREOUS HUMOR - DIABETES KETOACIDOSIS

The Medical Examiner or Hospital Pathologist may determine that uncontrolled diabetes mellitus with ketoacidosis may have caused death by performing a biochemical screening profile on the vitreous humor of the eye. The profile indicates extreme elevation of glucose with prominent elevation of BUN, uric acid, and phosphate due to accompanying renal insufficiency.

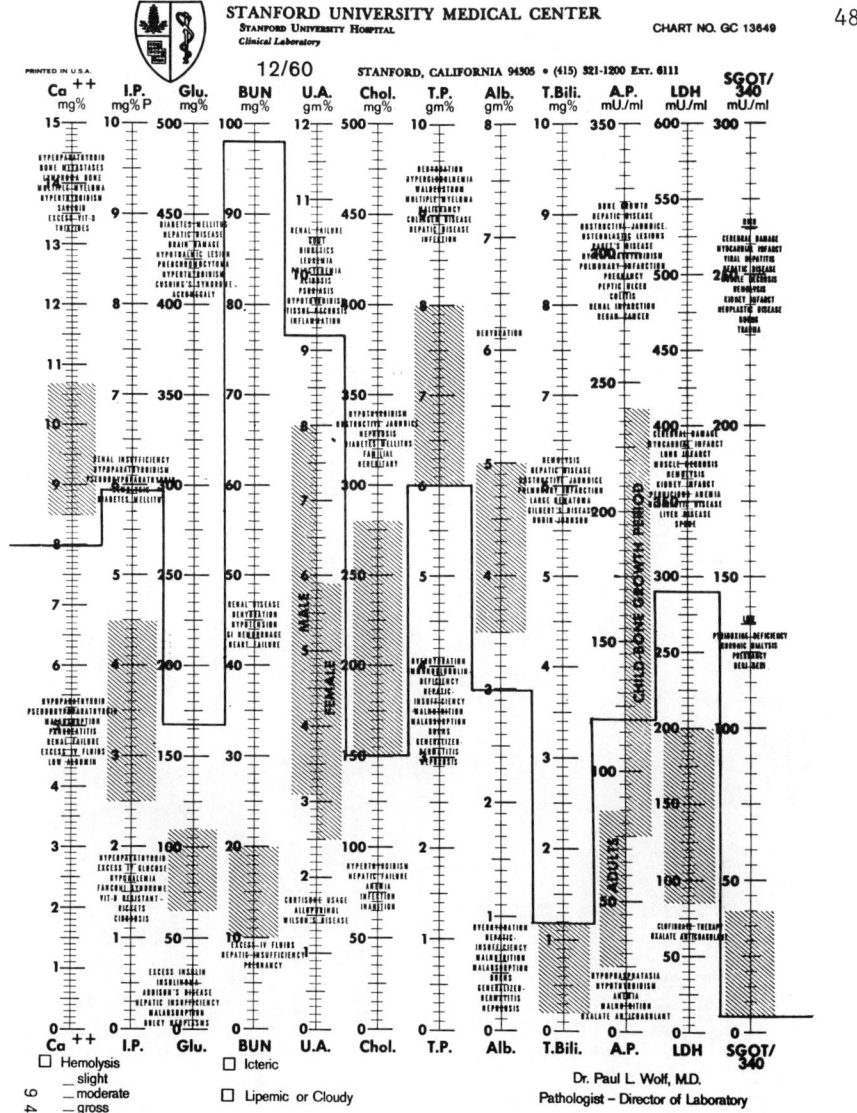

Fig. 22 - 74 BIOCHEMICAL PATTERN OF INFLUENCE OF UREMIA ON GLUCOSE

Uremia is characterized by elevation of certain substances in plasma which will non-specifically reduce neocuproine chemical reagent and cause a spurious elevation of glucose. These reducing substances are uric acid and creatinine. This artifact may be obviated by performing the glucose determination by a glucose oxidase reaction. The uremic toxins are anti-insulin and thus may cause an elevated glucose. BUN is elevated due to retention. Albumin is

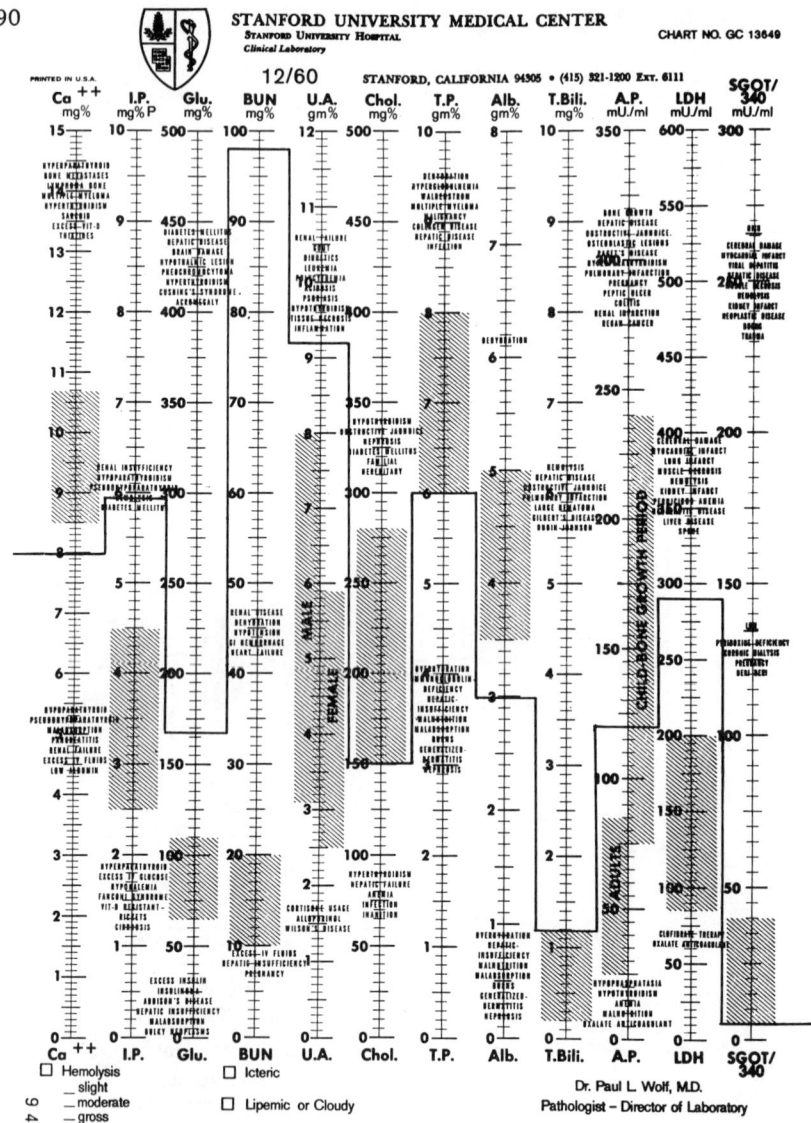

Fig. 22 - 74 (Continued):

decreased due to marked proteinuria. Elevated phosphate, uric acid, and LDH are caused by retention in renal insufficiency. Low calcium results from low albumin while elevated alkaline phosphatase is due to secondary hyperparathyroidism in renal failure.

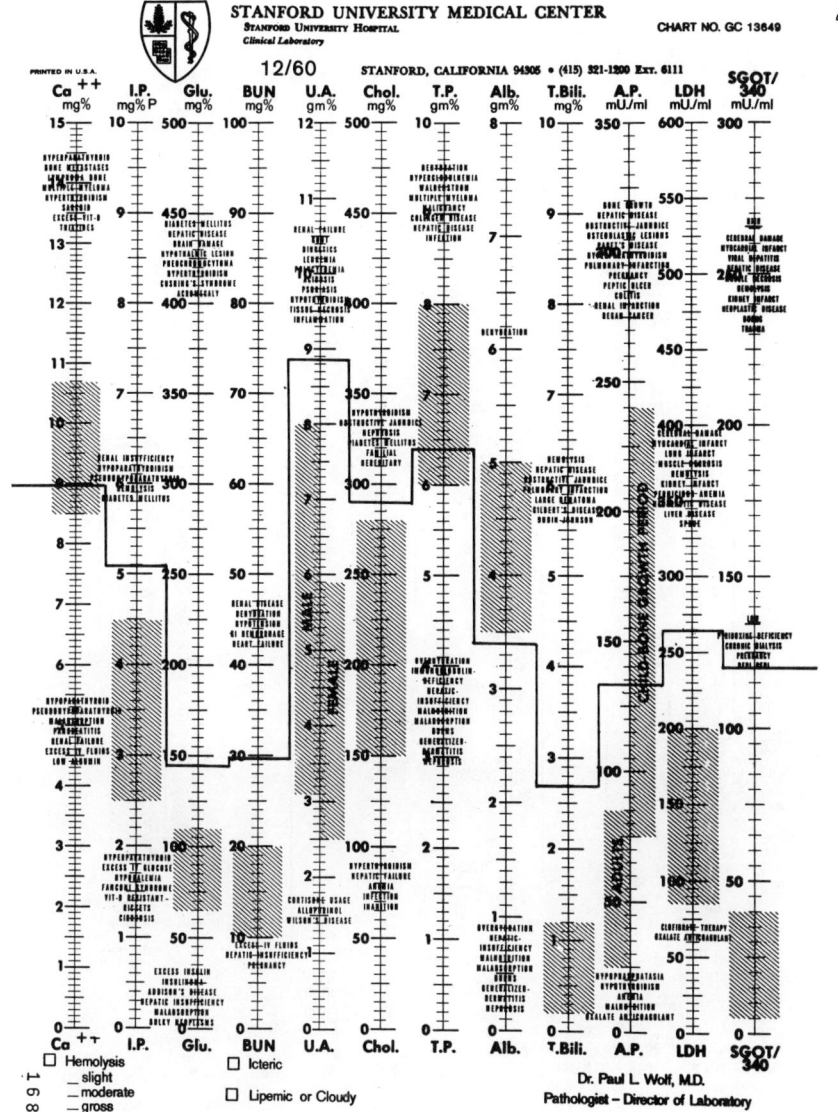

Fig. 22 - 75   BIOCHEMICAL PATTERN OF CONGESTIVE HEART FAILURE

Congestive heart failure will cause passive congestion of the liver resulting in elevated bilirubin, alkaline phosphatase, LDH and GOT. Prerenal azotemia which is frequently present in heart failure will be manifested biochemically by elevation of BUN, phosphate, and uric acid. Hypoxia of the pancreas in heart failure causes hyperglycemia. The elevated cholesterol and uric acid are biochemical abnormalities associated with arteriosclerotic heart disease.

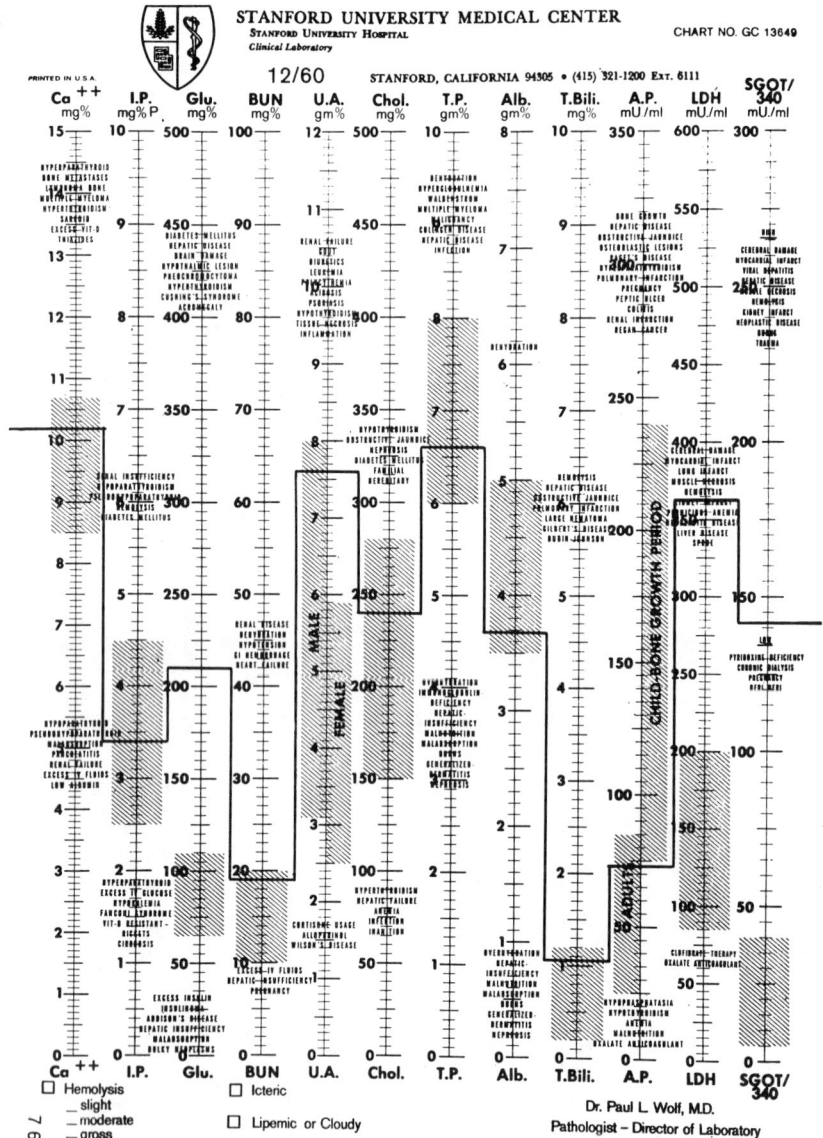

Fig. 22 - 76 BIOCHEMICAL PATTERN OF RECENT CEREBRAL DAMAGE

Recent cerebral damage secondary to trauma results in hyperglycemia related to the diabetic pique response. Cerebral damage may stimulate a hypothalamic - adrenal medullary physiologic response causing catecholamine release and liver glycogenolysis causing hyperglycemia. The elevated LDH and GOT is derived from damaged cerebral tissue.

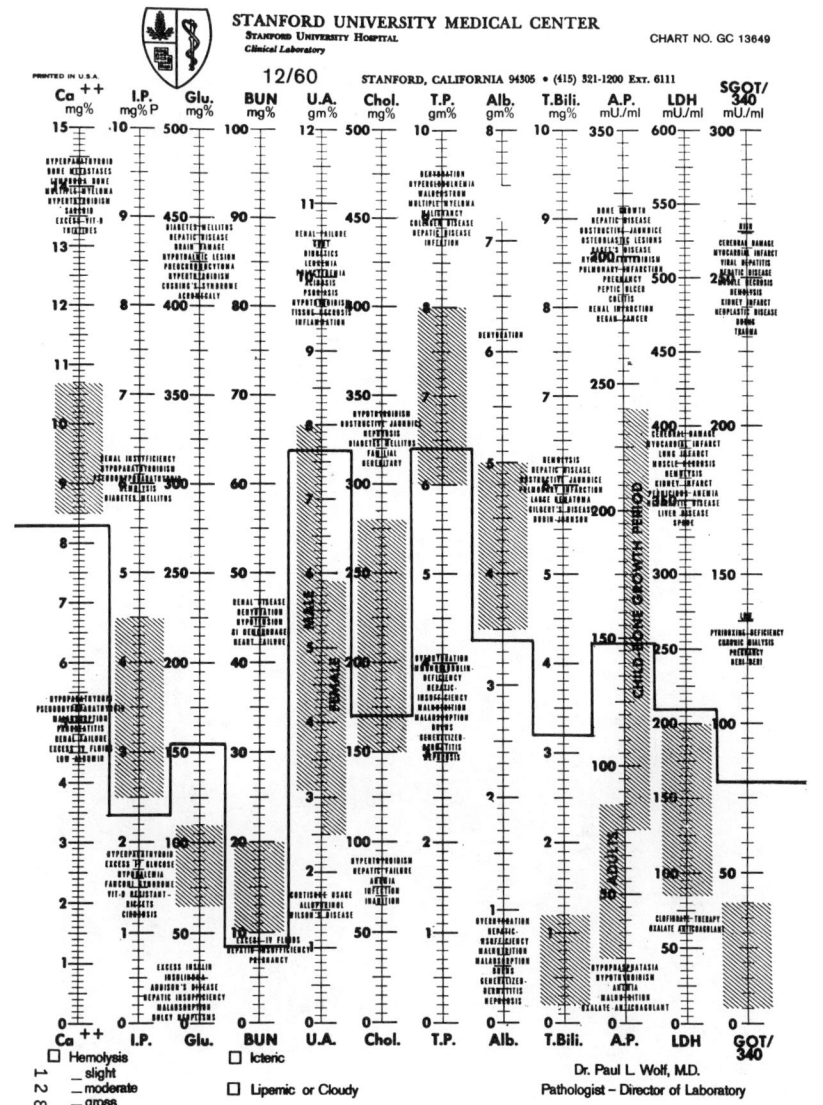

Fig. 22 - 77   BIOCHEMICAL PATTERN OF DRUG INDUCED LIVER DISEASE

Drug induced liver disease may be of three types: 1). Cholestatic, 2). Cytotoxic, or 3). Combination of cholestatic and cytotoxic. The liver disease is associated with elevated bilirubin, alkaline phosphatase, LDH, and GOT. 4). Albumin and BUN production are impaired and both are low. 5). Two hour post prandial hyperglycemia is present due to the liver disease. Hyperglycemia causes increased utilization of phosphate due to increased glucose phosphorylation.

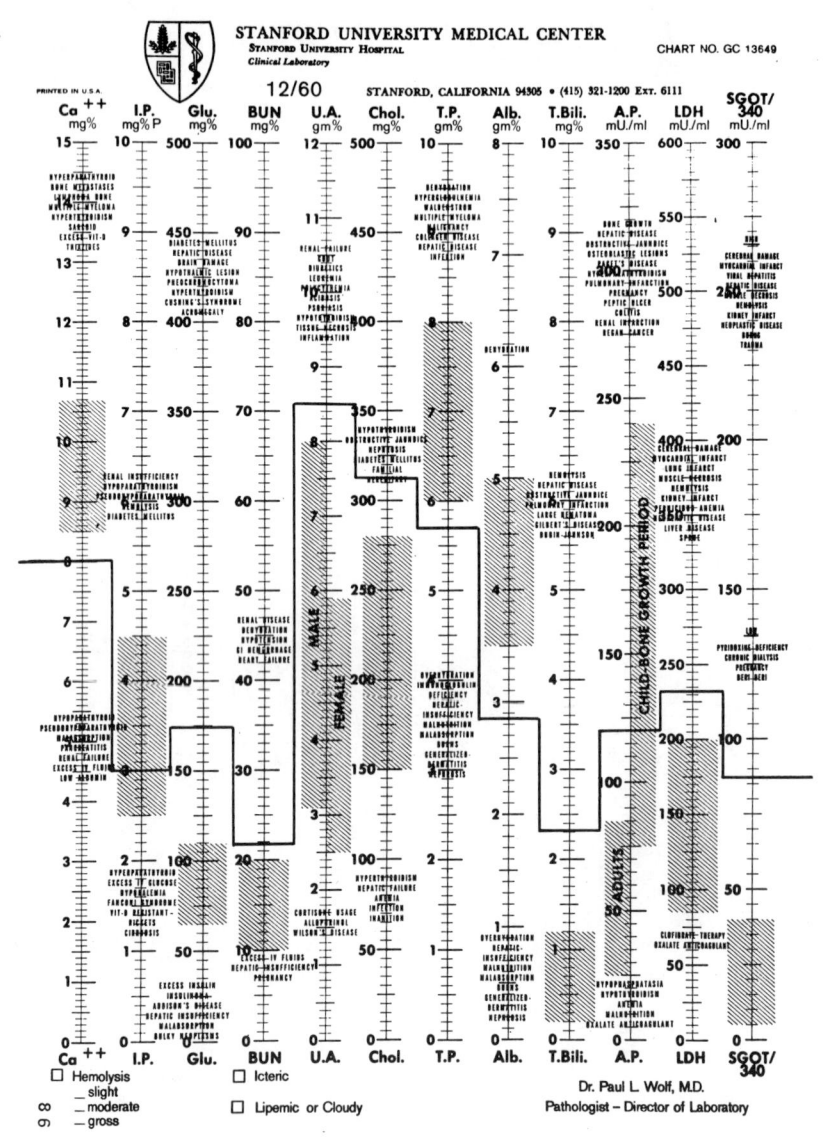

Fig. 22 - 78   BIOCHEMICAL PATTERN OF ORAL CONTRACEPTIVE USAGE

1). The "Pill" causes several abnormalities in serum biochemistry. Ceruloplasmin is elevated due to an increase in alpha$_2$ globulin. 2). Beta-globulins increase and thus transferrin is elevated. 3). Glucose metabolism is disturbed with resultant hyperglycemia. 4). Cholesterol and triglycerides are increased due to induced abnormalities in lipid metabolism. 5). A moderate incidence of drug induced hepatic disease is present

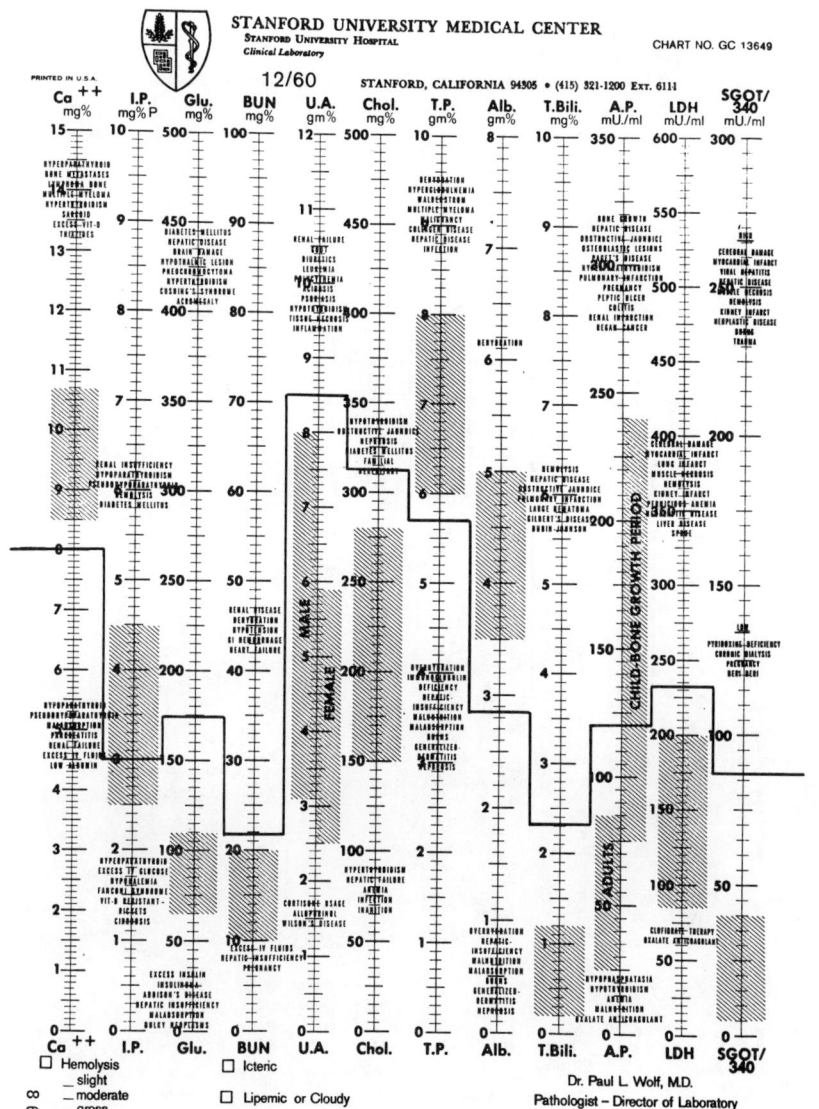

Fig. 22 - 78 (Continued):

especially during the first two cycles of the "Pill" with elevated bilirubin, alkaline phosphatase, LDH, and GOT. 6). Serum proteins and calcium are low.

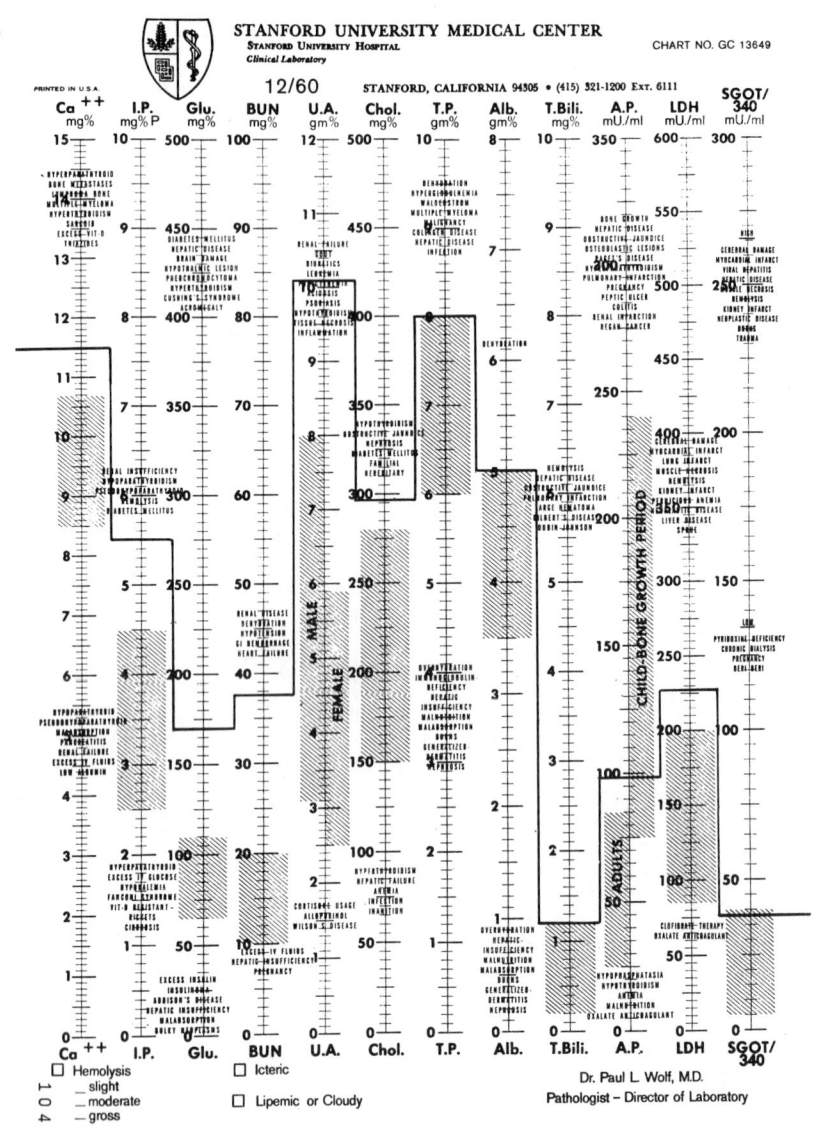

Fig. 22 - 79 BIOCHEMICAL PATTERN OF THIAZIDE DIURETIC USAGE

Excessive utilization of thiazide diuretics causes hyperglycemia due to excessive water dehydration or inhibition of calcium diuresis at the level of the renal tubule. Prerenal azotemia may be induced by hypovolemia and dehydration with elevated BUN, phosphate, uric acid, alkaline phosphatase and LDH. Thiazide diuretics induce glucose intolerance through an unknown mechanism causing hyperglycemia and elevated cholesterol.

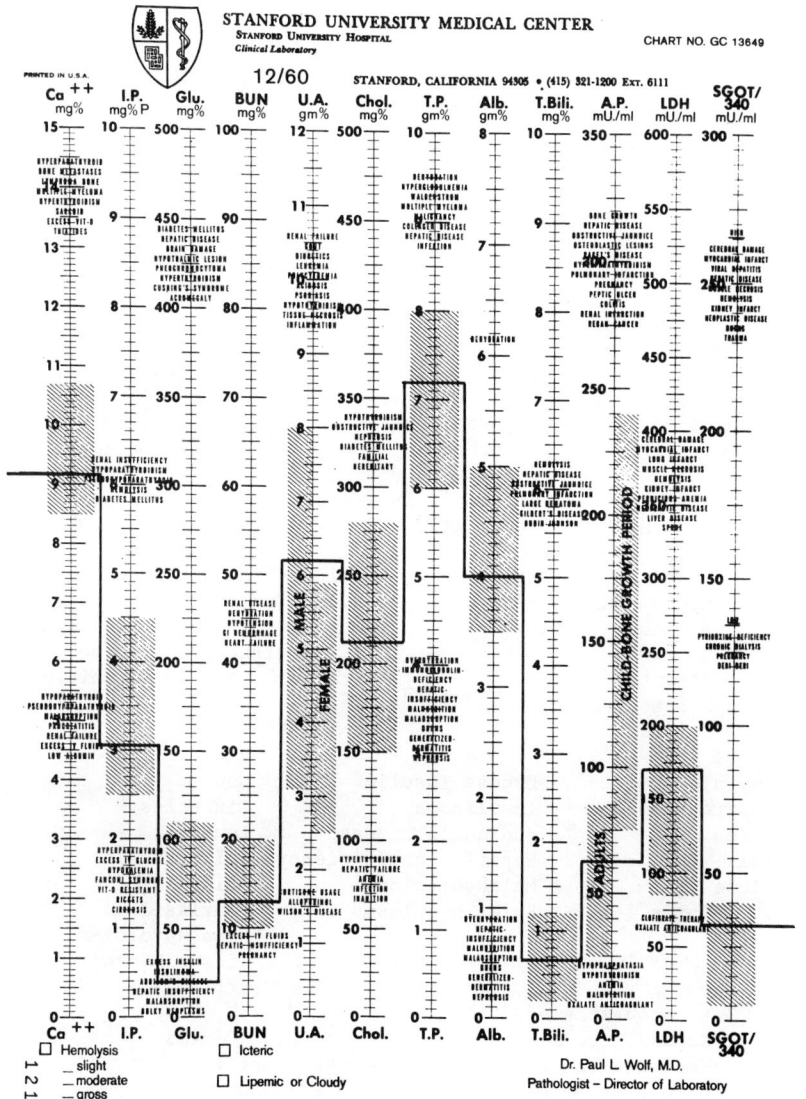

Fig. 22 - 80 BIOCHEMICAL PATTERN OF EXCESSIVE INSULIN UTILIZATION

One of the commonest causes for marked hypoglycemia is excessive utilization of insulin in diabetes mellitus. If this is not the cause for hypoglycemia, one should consider the possibility of insulinoma, bulky neoplasms such as sarcomas consuming glucose, or producing an insulin-like substance, severe liver disease, adrenal insufficiency, malabsorption, or metastatic neoplasms to the hypothalamus.

STANFORD UNIVERSITY MEDICAL CENTER
Stanford University Hospital
Clinical Laboratory

12/60 Chart

Ca++	I.P.	Glu.	BUN	U.A.	Chol.
		HIGH			HIGH
HIGH		Diabetes Mellitus			Hypothyroidism
		Hepatic Disease			Obstructive Jaundice
Hyperparathyroid		Brain Damage			Nephrosis
Bone Metastases		Hypothalmic Lesion			Diabetes Mellitus
Lymphoma Bone		Pheochromocytoma			Familial
Multiple Myeloma		Hyperthyroidism			Hereditary
Hyperthyroidism		Cushing's Syndrome			
Sarcoid	HIGH	Acromegaly	HIGH		LOW
Excess Vit-D					
Thiazides	Renal Insufficiency		Renal Disease	Hyperthyroidism
	Hypoparathyroidism		Dehydration	Hepatic Failure
	Pseudohypoparathyroid		Hypotension	Anemia
	Hemolysis		G.I. Hemorrhage	Infection
	Diabetes Mellitus		Heart Failure	Inanition
	Bone Growth			HIGH
LOW	Fracture	LOW		
				Renal Failure
Hypoparathyroid		Excess Insulin		Gout
Pseudohypoparathyroid		Insulinoma		Diuretics
Malabsorption		Addison's Disease		Leukemia
Pancreatitis		Hepatic Insufficiency		Polycythemia
Renal Failure		Malabsorption		Acidosis
Excess I.V. Fluids		Bulky Neoplasms		Psoriasis
				Hypothyroidism
		LOW		Tissue Necrosis
			LOW	Inflammation
	Hyperparathyroid			
	Excess I.V. Glucose		Excess I.V. Fluids	
	Hypokalemia		Hepatic Insufficiency	
	Fanconi Syndrome		Pregnancy	
	Vit-D Resistant-			
	Rickets			LOW
	Cirrhosis			
				Cortisone Usage
				Allopurinol
				Wilson's Disease

A magnified view of the causes for high and low values of each determination on the Stanford Hospital Clinical Laboratory SMA 12/60 Chart.

STANFORD UNIVERSITY MEDICAL CENTER
Stanford University Hospital
Clinical Laboratory

12/60 Chart (Continued):

T.P.	Alb.	T. Bili.	A.P.	LDH	SGOT
HIGH	HIGH			HIGH	

HIGH / HIGH / / / HIGH /

─hydration   Dehydration   Cerebral Damage
─perglobulinemia   Myocardial Infarct
─ldenstrom   Lung Infarct
─ltiple Myeloma   Muscle Necrosis
─lignancy   HIGH   Hemolysis
─llagen Disease   Kidney Infarct
─patic Disease   Hemolysis   Pernicious Anemia
─fection   Hepatic Disease   Neoplastic Disease   HIGH
                     Obstructive Jaundice   Liver Disease
                     Pulmonary Infarction   Sprue   Cerebral Damage
                     Large Hematoma   Myocardial Infarct
                     Gilbert's Disease   Viral Hepatitis
                     Dubin-Johnson   HIGH   Hepatic Disease
                                                             Muscle Necrosis
                     LOW   Bone Growth   Hemolysis
                                      Hepatic Disease   Kidney Infarct
                     Overhydration   Obstructive Jaundice   Neoplastic Disease
        LOW   Hepatic-   Osteoblastic Lesions   Burns
                     Insufficiency   Paget's Disease   Trauma
─erhydration   Malnutrition   Hyperparathyroidism
─munoglobulin-   Malabsorption   Pulmonary Infarction
   Deficiency   Burns   Pregnancy   LOW
   Hepatic-   Generalized-   Peptic Ulcer
─sufficiency   Dermatitis   Colitis   Pyridoxine Deficiency
─lnutrition   Nephrosis   Renal Infarction   Chronic Dialysis
─labsorption   Regan Cancer   Pregnancy
   Burns   Beri Beri
─eralized-
─rmatitis   LOW
─hrosis
                                      Hypophosphatasia
                                      Hypothyroidism
                                      Anemia
                                      Malnutrition
                                      Oxalate Anticoagulant

                                             LOW

                                      Clofibrate Therapy
                                      Oxalate Anticoagulant

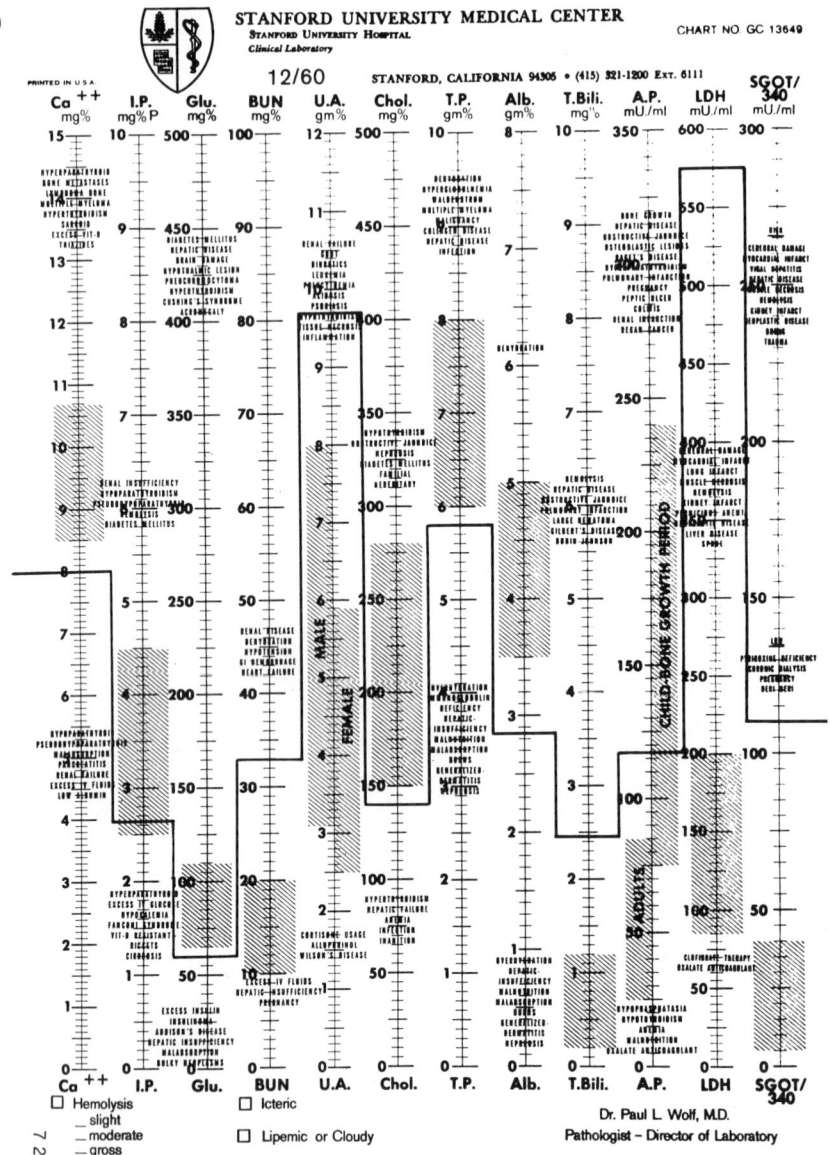

Fig. 22 - 81 BIOCHEMICAL PATTERN OF ACUTE LEUKEMIA

1). Acute leukemia is associated with extreme elevation of LDH and GOT derived from the large number of leukemic cells especially with necrosis of the malignant cells. 2). The elevated uric acid results from the large number of necrotic leukemic cells. 3). Glucose is low due to utilization by the malignant cells. 4). Cholesterol and protein are low due to utilization by the malignant cells and malnutrition. 5). Low cholesterol may result from

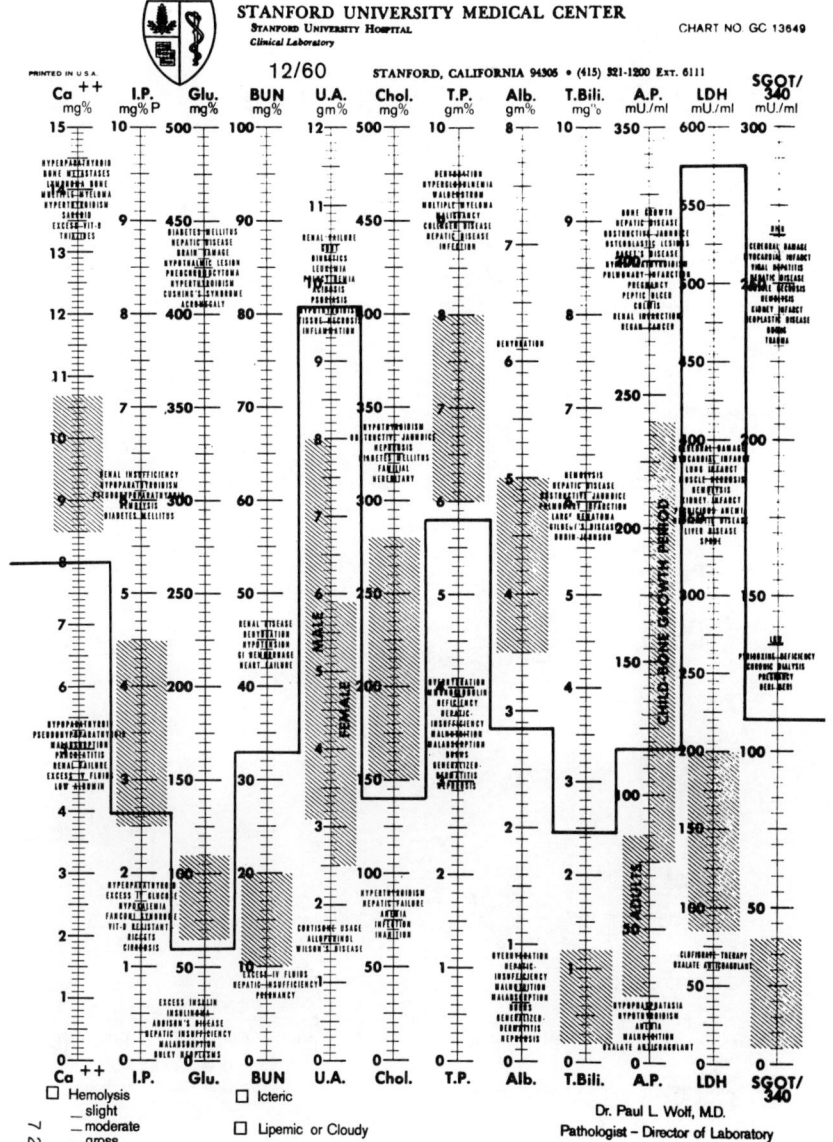

Fig. 22 - 81 (Continued):

anemia. 6). Hypocalcemia is related to the low serum protein. 7). Elevated BUN is caused by uric acid nephropathy, dehydration, leukemic infiltration, or renal tubular damage by muramidase. 8). Elevated bilirubin, alkaline phosphatase, LDH, and GOT are caused by fatty liver or leukemic liver infiltration. 9). Alkaline phosphatase elevation may be due to bone disease caused by bone marrow proliferation of leukemic cells.

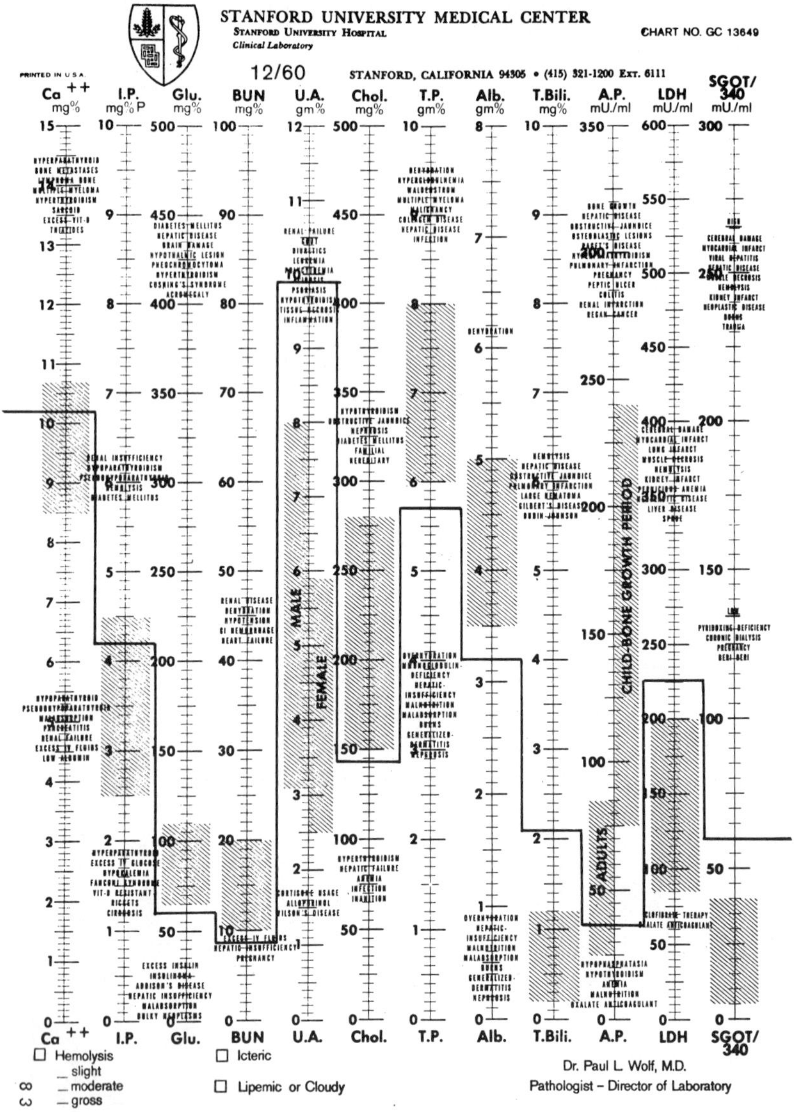

Fig. 22 - 82   BIOCHEMICAL PATTERN OF STRICT DIET WITH DEVELOPMENT
OF KETOACIDOSIS

1). A strict diet will cause hypoglycemia, low BUN, low cholesterol, and low serum proteins. 2). Utilization of protein and lipids for calories will cause tissue breakdown and hyperuricemia. Hyperuricemia also result from the ketoacidosis with inhibition of renal excreti 3). A fatty liver occurs with elevated bilirubin, LDH and GOT.

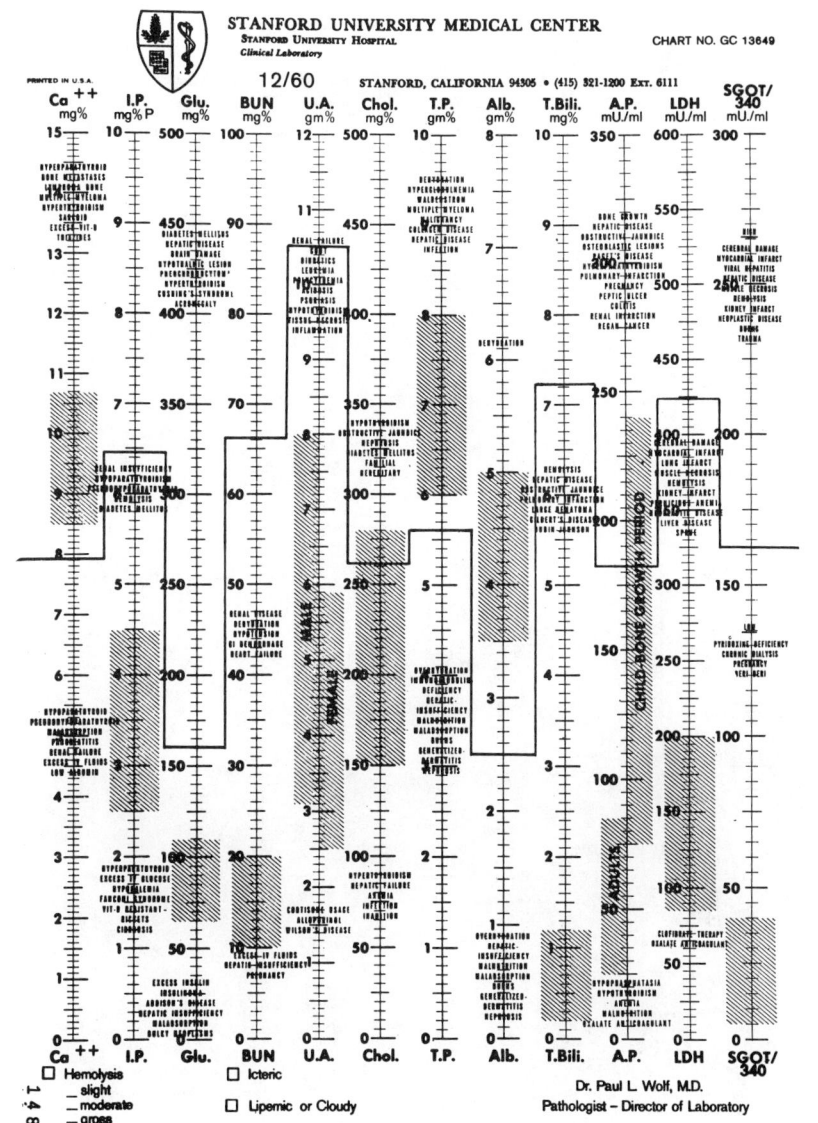

Fig. 22 - 83 BIOCHEMICAL PATTERN OF ECLAMPSIA

1). Eclampsia is characterized by hepatic necrosis and associated severe renal insufficiency. The hepatic lesion is necrosis of the peripheral portion of the lobule resulting in elevation of unconjugated and conjugated bilirubin, alkaline phosphatase, LDH, and GOT.
2). Albumin is decreased due to the liver disease and proteinuria and causes low calcium. 4). Acute renal failure causes elevated BUN, phosphate, uric acid, and glucose.

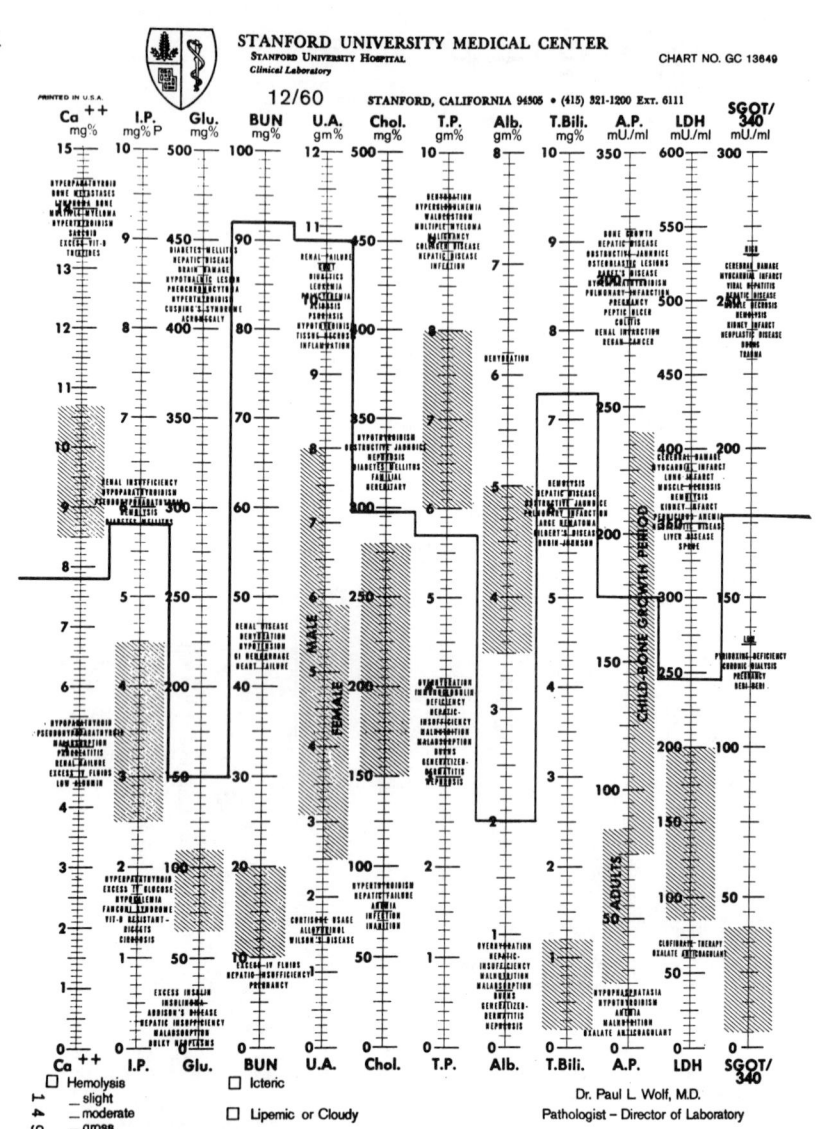

Fig. 22 - 84 BIOCHEMICAL PATTERN OF HEPATORENAL SYNDROME

1). The hepatorenal syndrome is usually found in the patient with obstructive jaundice who has surgery to relieve the extra-hepatic cause for the obstruction and subsequently develops acute renal failure. The renal failure may result from renal tubular disease caused by conjugated bile, amino acids, or hypoxia during surgery.
2). The obstructive jaundice causes elevation of conjugated bilirubin, alkaline phosphatase, cholesterol, LDH,

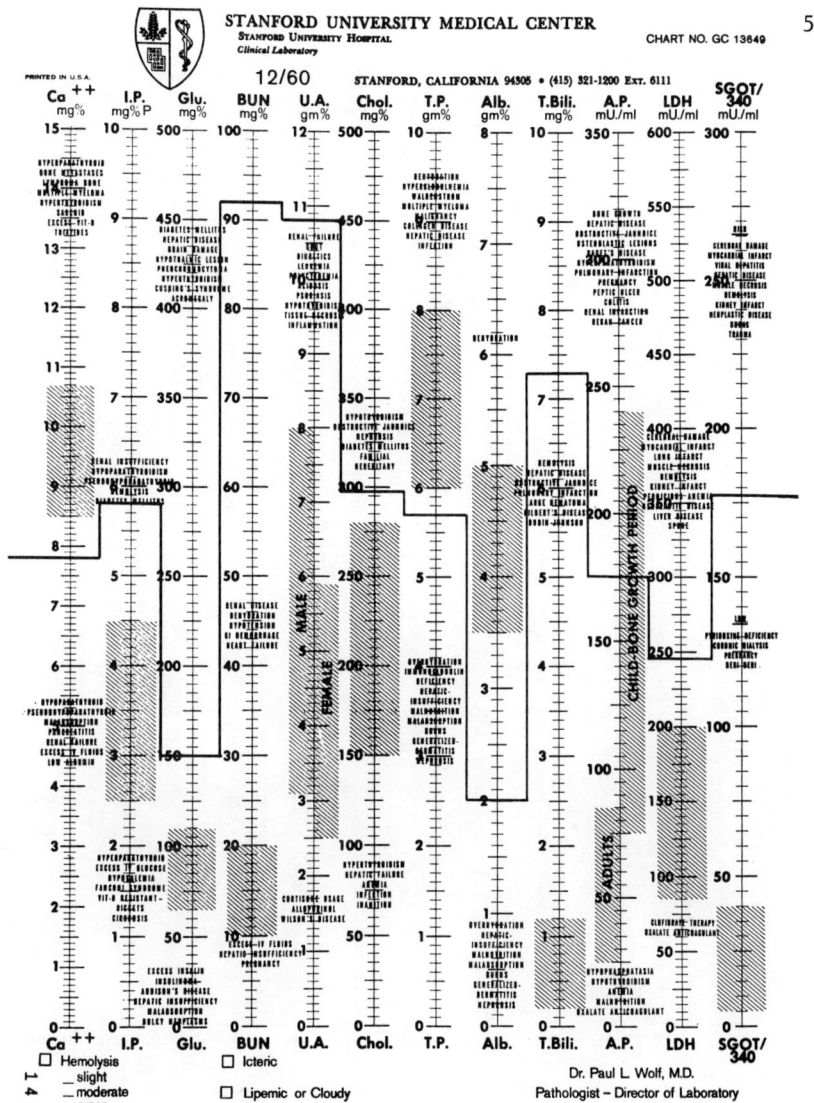

Fig. 22 - 84 (Continued):

and GOT. 3). Albumin is decreased due to the liver disease and proteinuria and causes low calcium.
4). Acute renal failure causes elevated BUN, phosphate, uric acid, and glucose.

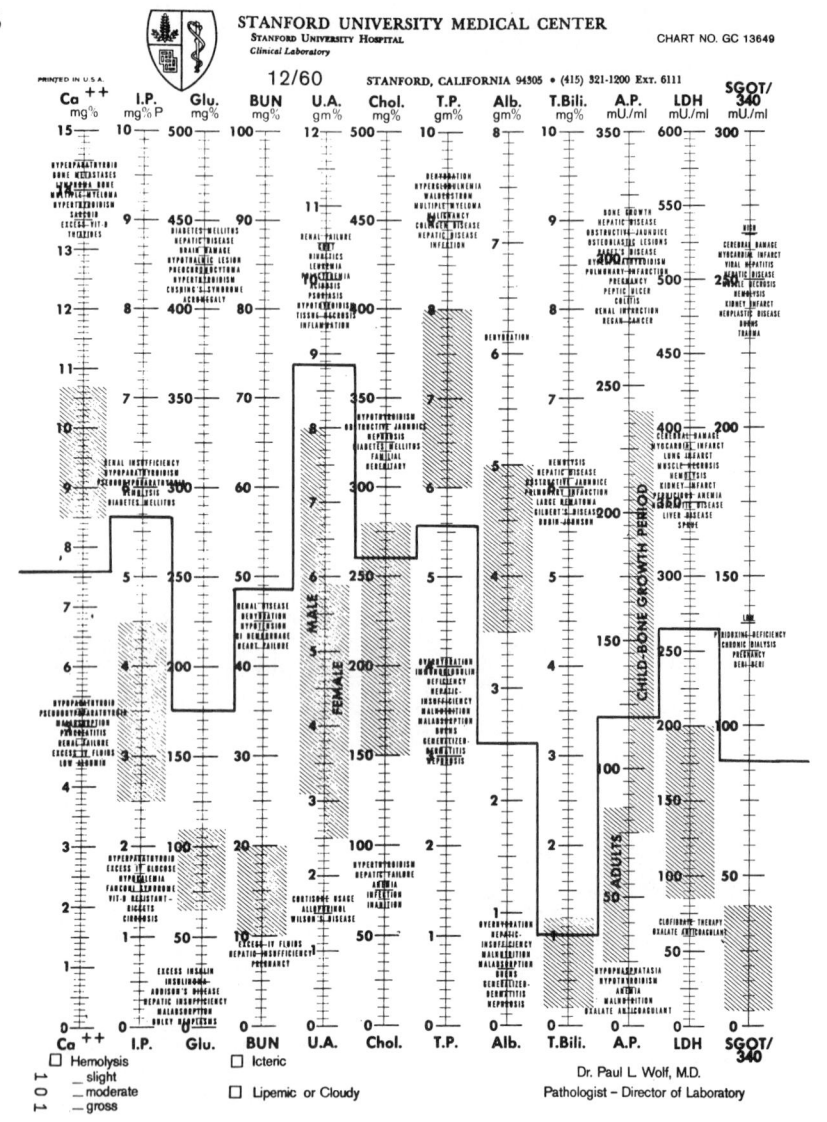

Fig. 22 - 85 BIOCHEMICAL PATTERN OF PRERENAL AZOTEMIA

1). Prerenal azotemia due to shock, congestive heart failure, or dehydration causes elevated BUN with a norma creatinine. The BUN:Creatinine ratio is 10:1. Prerenal azotemia causes a BUN:Creatinine ratio of greater than 10:1 ratio. Azotemia due to renal disease is characterized by a preservation of the ratio at 10:1. 2). Album is decreased due to marked proteinuria. 3). Elevated phosphate, uric acid, and LDH are caused by retention

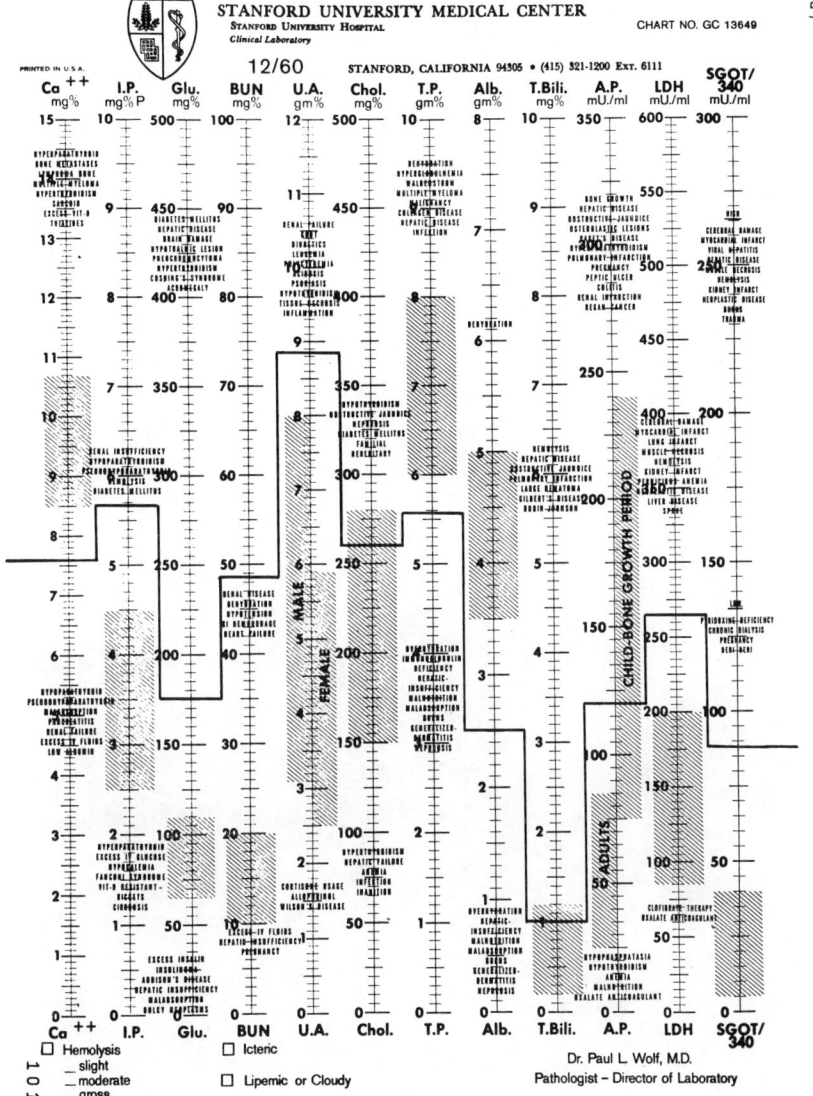

Fig. 22 - 85 (Continued):

in renal insufficiency. 4). Low calcium results from low albumin while elevated alkaline phosphatase is due to secondary hyperparathyroidism in renal failure. 5). Elevated glucose results from the anti-insulin action of the uremic toxins and also is due to non-specific reduction of the neocuproine chemical reagent in the 12/60 instrument by uric acid and creatinine. 6). GOT is elevated due to tissue damage from shock or heart failure.

508

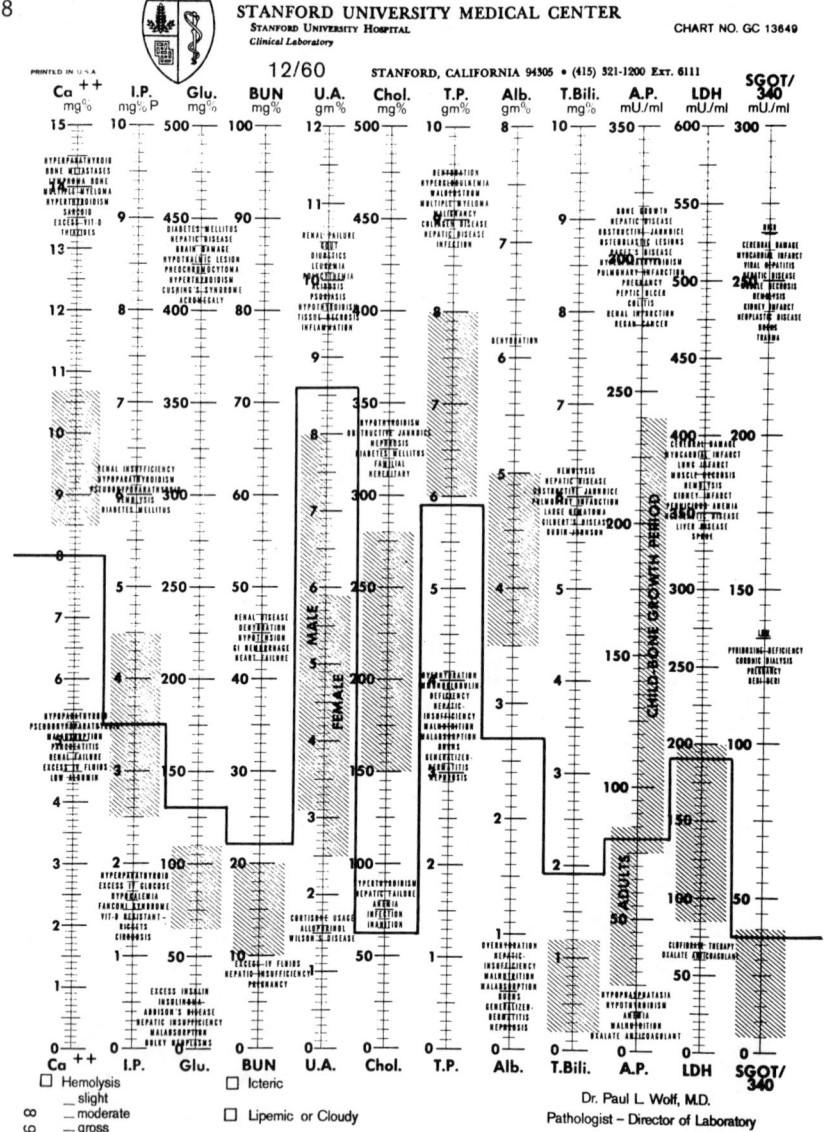

Fig. 22 - 86 BIOCHEMICAL PATTERN OF ACUTE BACTERIAL INFECTION

1). A fulminant bacterial infection causes a negative nitrogen balance which results from protein malnutrition, malabsorption, and increased utilization (metabolism and degradation). 2). Hypocalcium is present from the hypoalbuminemia. 3). Low cholesterol results from the same mechanism causing low protein, malnutrition, malabsorption, and increased utilization. 4). Elevated bilirubin is due to a fatty liver. 5). Dehydration causes

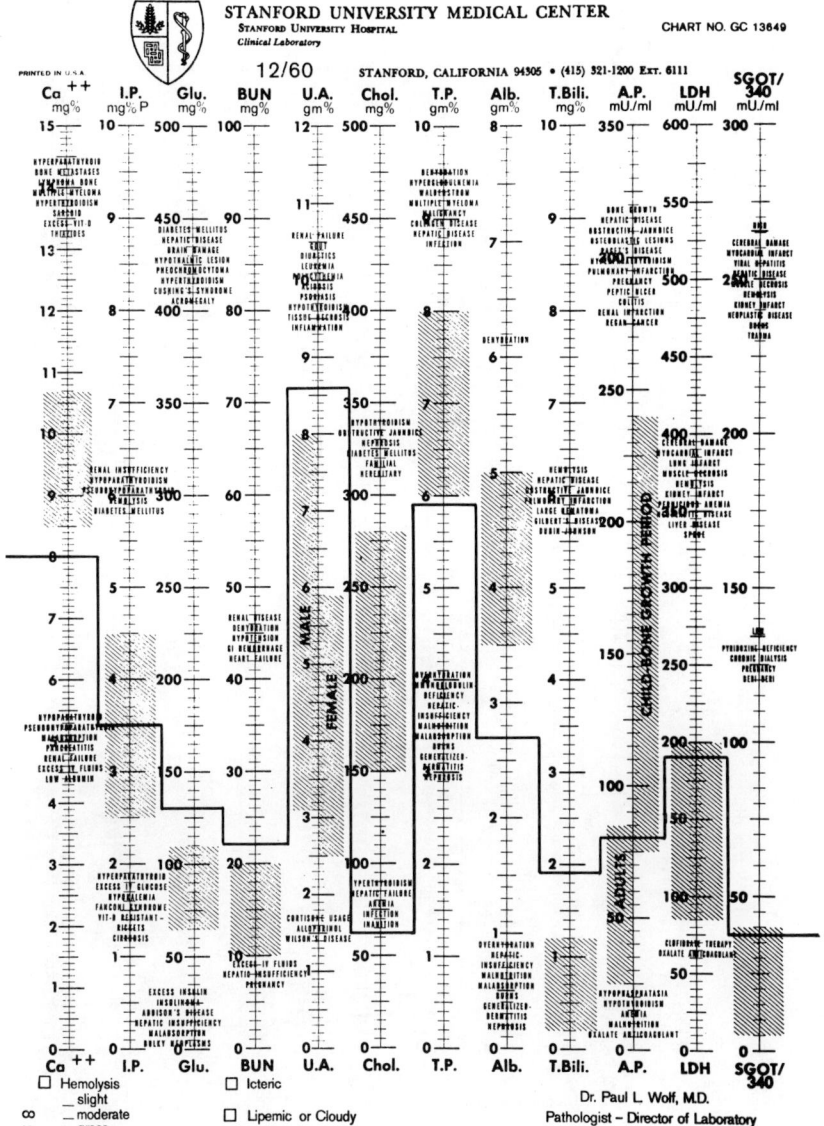

Fig. 22 - 86 (Continued):

hypovolemia and prerenal azotemia with elevated BUN, and uric acid also caused by the severe inflammation.

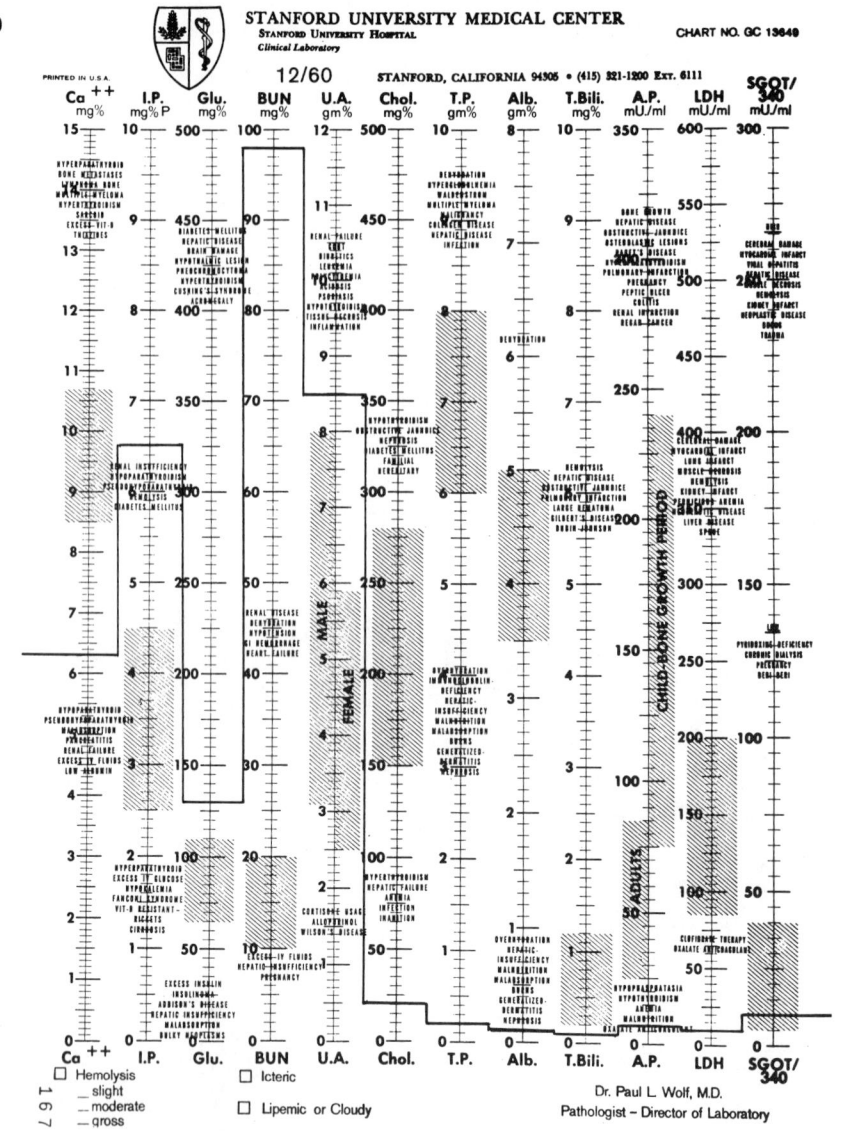

Fig. 22 - 87 BIOCHEMICAL PATTERN OF VITREOUS HUMOR - CHRONIC RENAL
               INSUFFICIENCY

The Medical Examiner or Hospital Pathologist may determ
that renal insufficiency may have caused death by perfo
ing a biochemical screening profile on the vitreous hum
of the eye. The profile indicates extreme elevation of
BUN with prominent elevation of phosphate and uric acid

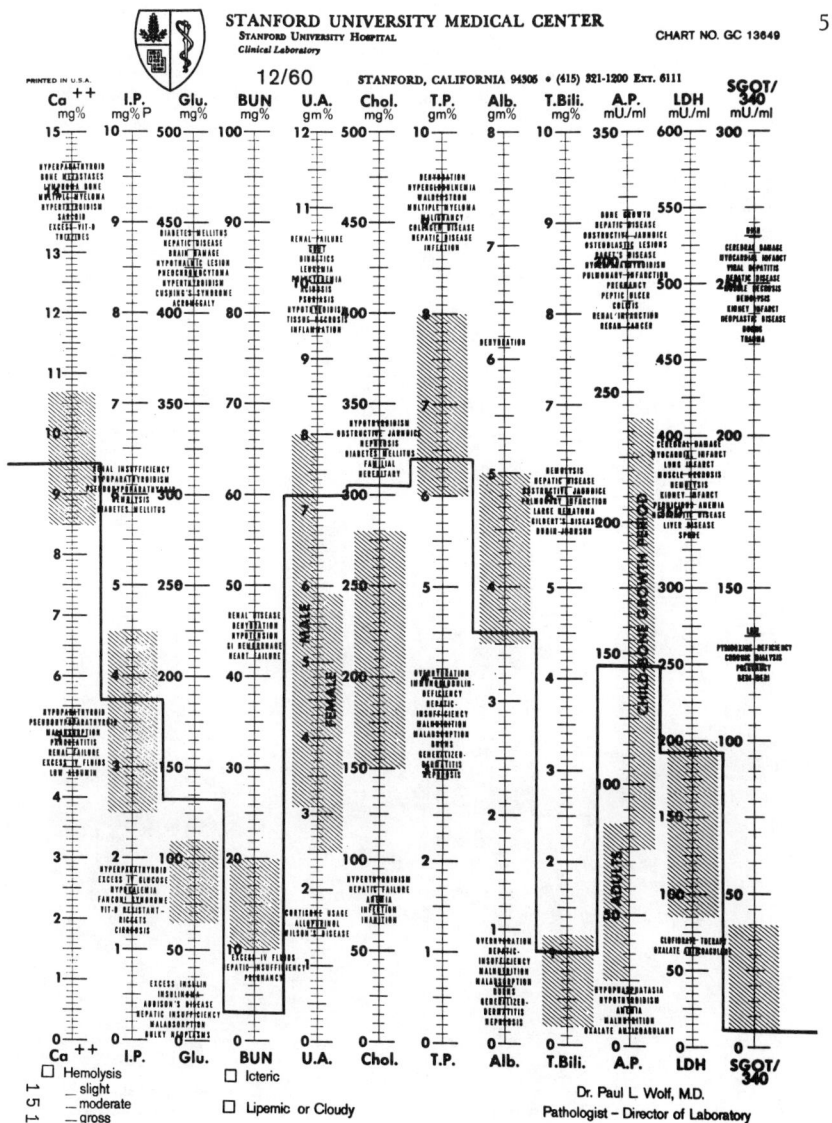

Fig. 22 - 88 BIOCHEMICAL PATTERN OF PREGNANCY - THIRD TRIMESTER

The biochemical pattern of a third trimester pregnancy includes hyperglycemia due to glucose intolerance induced by elevated estrogen - progesterone with accompanying elevation of cholesterol and triglycerides. The decreased BUN is due to the physiologic hydremia. The elevated alkaline phosphatase is derived from the placenta and is extremely heat stable similar to the Regan enzyme. It is inhibited by phenylalanine.

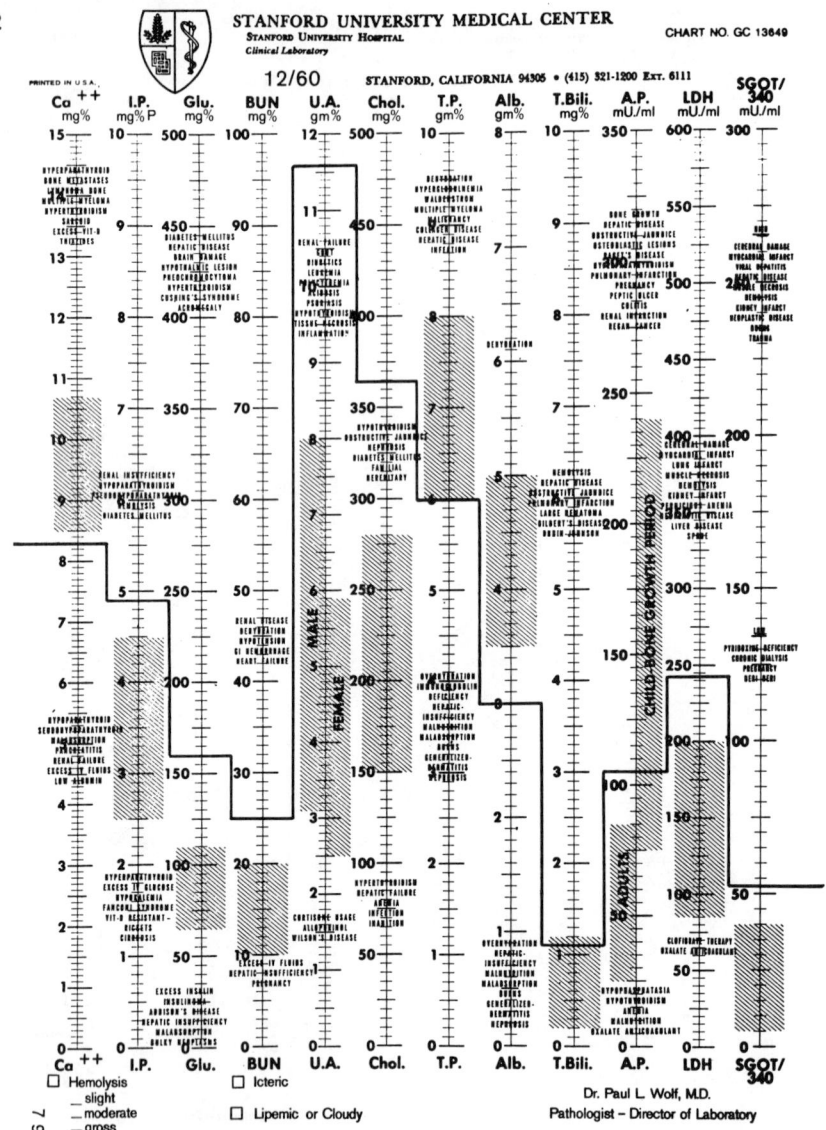

Fig. 22 - 89 BIOCHEMICAL PATTERN OF ACUTE GOUT

Uric acid elevation is a constant biochemical feature of gout unless an xanthine oxidase inhibitor such as allopurinol is used. Precipitation of uric acid in renal tubules will cause a uric acid obstructive nephropathy with elevated BUN, phosphate, glucose, alkaline phosphatase, LDH, and GOT. Proteinuria occurs with low serum proteins and calcium. An association exists between an elevated uric acid and increased cholesterol. See Fig. 22 - 90.

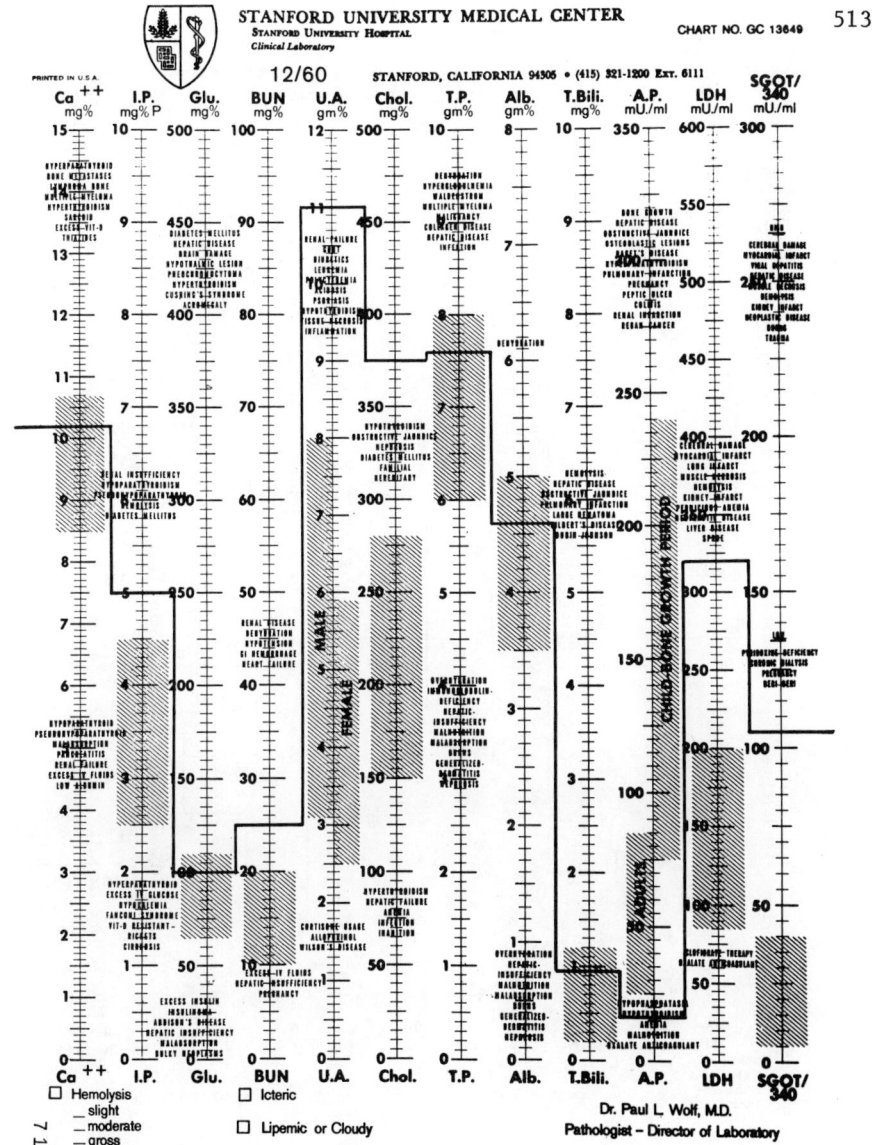

Fig. 22 - 90 BIOCHEMICAL PATTERN OF MYXEDEMA

1). Hypothyroidism is characterized by a high serum cholesterol because of the low metabolic rate. 2). Lower glomerular filtration rate results in high BUN, phosphate, and uric acid. 3). The elevated LDH and GOT result from the hypothyroid myopathy. 4). Low alkaline phosphatase is due to decrease in osteoblastic bone activity associated with myxedema.

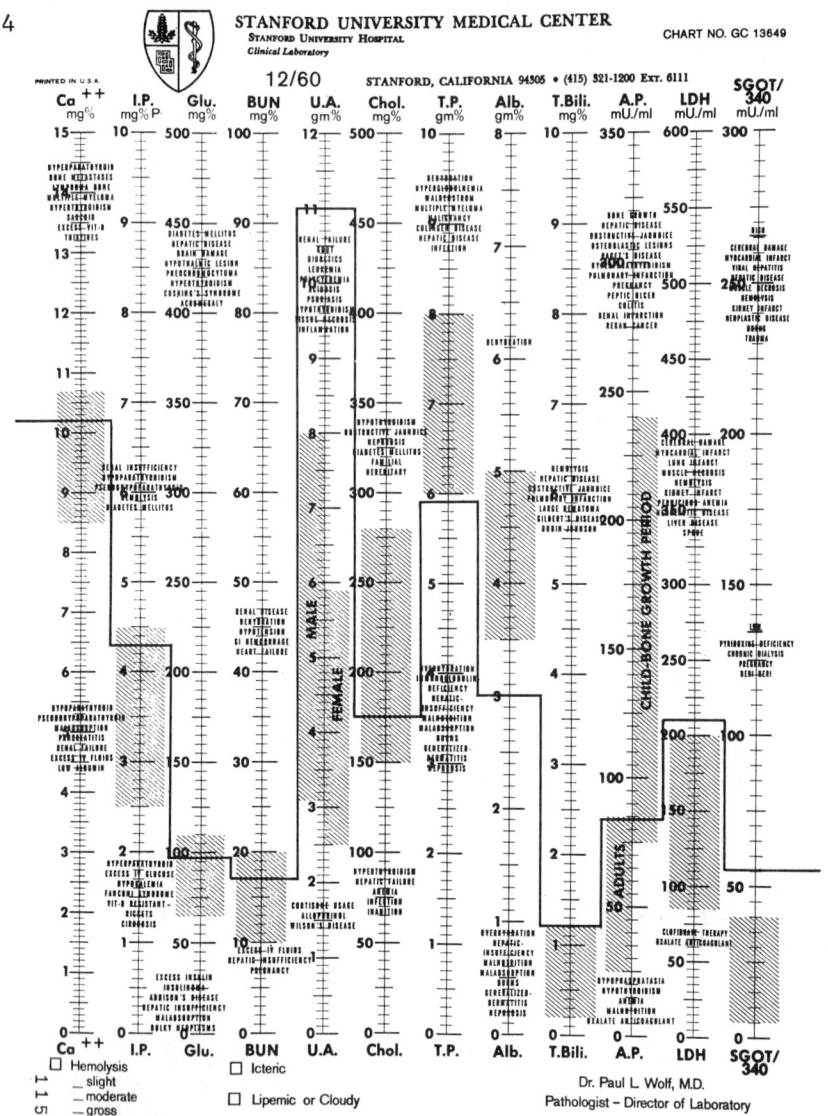

Fig. 22 - 91 BIOCHEMICAL PATTERN OF PSORIASIS

Psoriasis is characterized by excessive growth of epidermal cells with increased epidermal thickness. Hyperuricemia is a common abnormality associated with the epidermal cell hyperplasia. Severe generalized forms of the disease will result in loss of serum proteins from the skin. Slight elevation of LDH and GOT may be present in the generalized type due to release from hyperplastic epidermal cells. Methotrexate therapy of psoriasis may cause hepatic injury with elevation of bilirubin, LDH, GOT, and alkaline phosphatase.

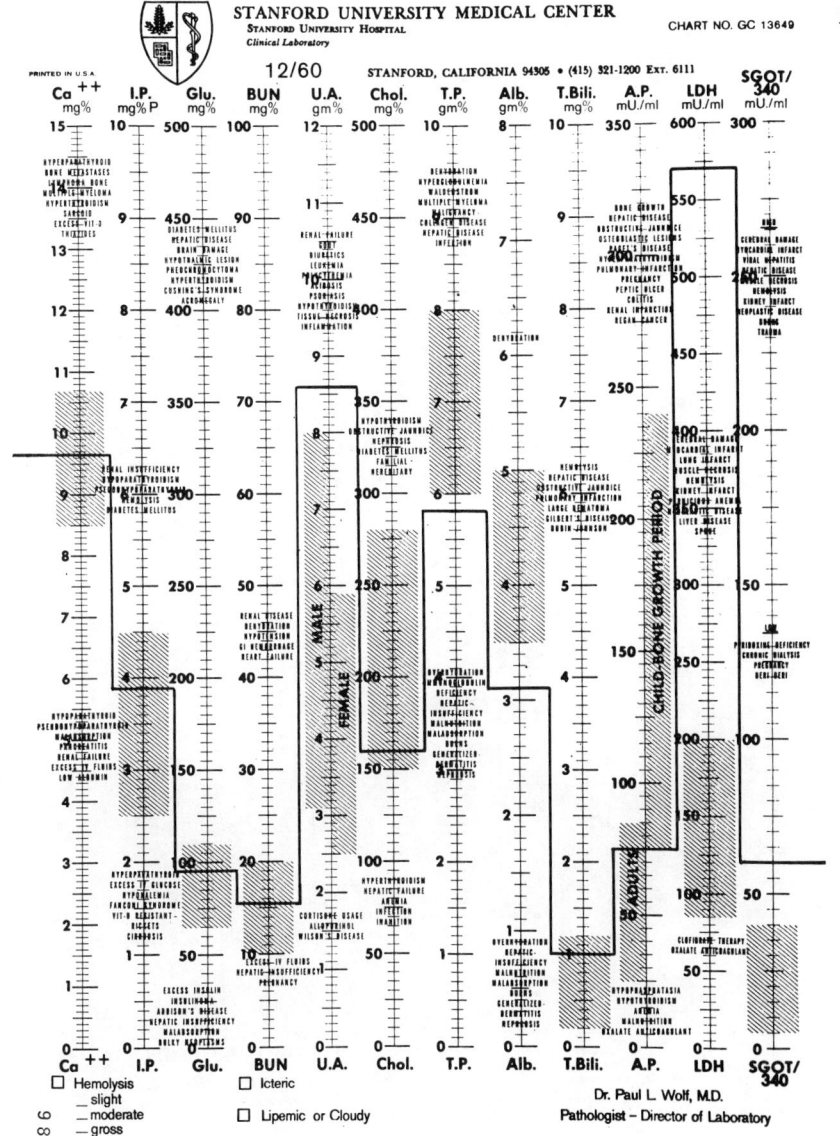

Fig. 22 - 92 BIOCHEMICAL PATTERN OF MALIGNANT NEOPLASM

1). A variety of malignant neoplasms will cause an elevation in serum LDH. Malignant cells contain much LDH and GOT and release these enzymes with proliferation and necrosis. There is no "malignant pattern". The common LDH isoenzymes which are increased are $LDH_{2,3,4}$ and at times $_5$. 2). The uric acid is elevated from necrosis of the neoplasm. 3). The serum protein is low due to inadequate production or consumption by the tumor.

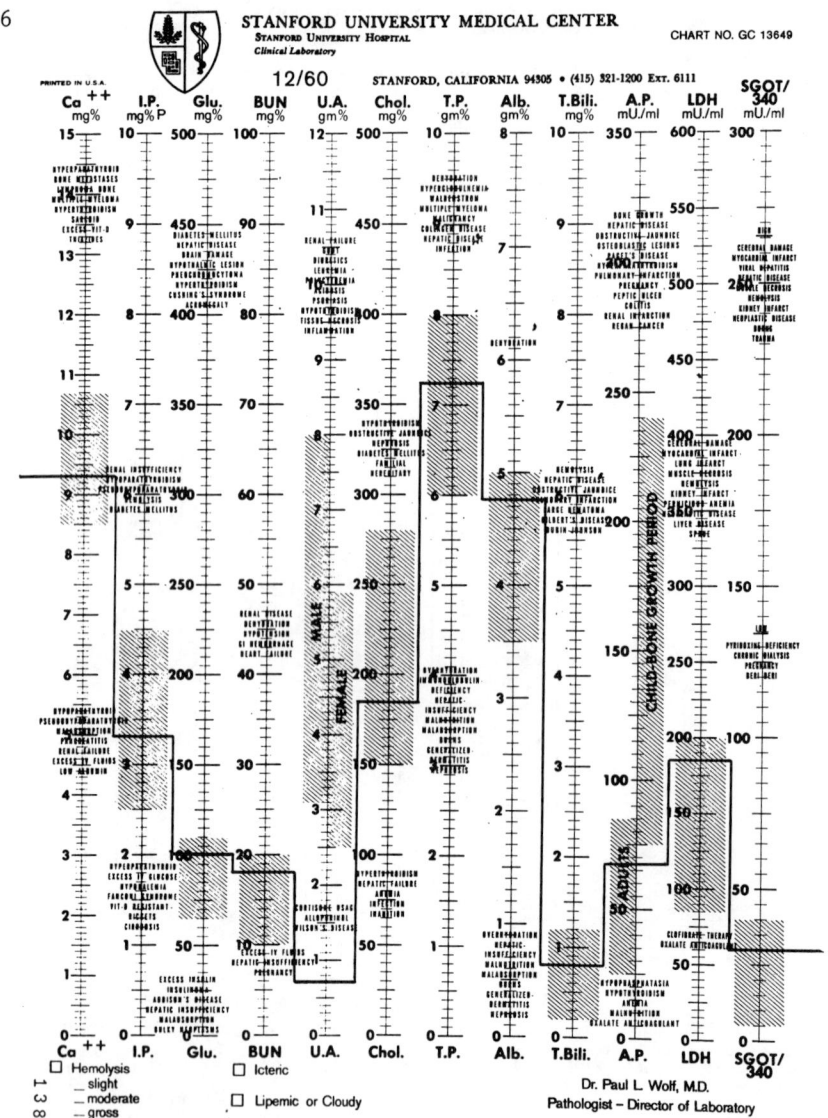

Fig. 22 - 93 BIOCHEMICAL PATTERN OF UTILIZATION OF ALLOPURINOL

Allopurinol is a xanthine oxidase inhibitor used clinically to lower serum uric acid. The oxidation of xanthine to uric acid is thus prevented. Consequently, serum uric acid is low. Other causes for low uric acid are Wilson's disease and excessive usage of cortisone and salicylates which promote excretion.

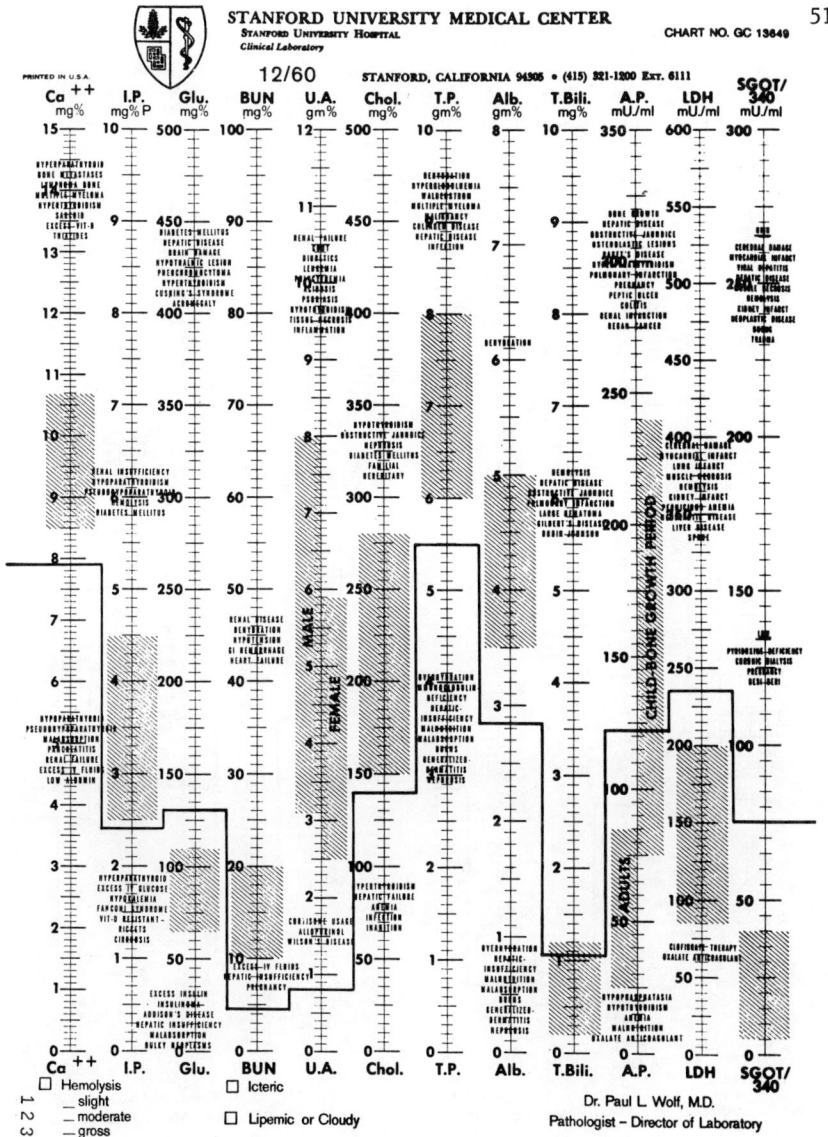

Fig. 22 - 94 BIOCHEMICAL PATTERN OF WILSON'S DISEASE

1). Wilson's disease is characterized by excessive copper deposition in liver, lentiform nuclei of brain and renal tubules. Ceruloplasmin is low while urinary copper is elevated. Cirrhosis results from excessive liver copper deposition with elevated alkaline phosphatase, LDH, GOT, and glucose. 2). Decreased production of albumin, BUN and cholesterol esters occur. 3). Hyperglycemia causes decreased phosphate. 4). Low calcium results

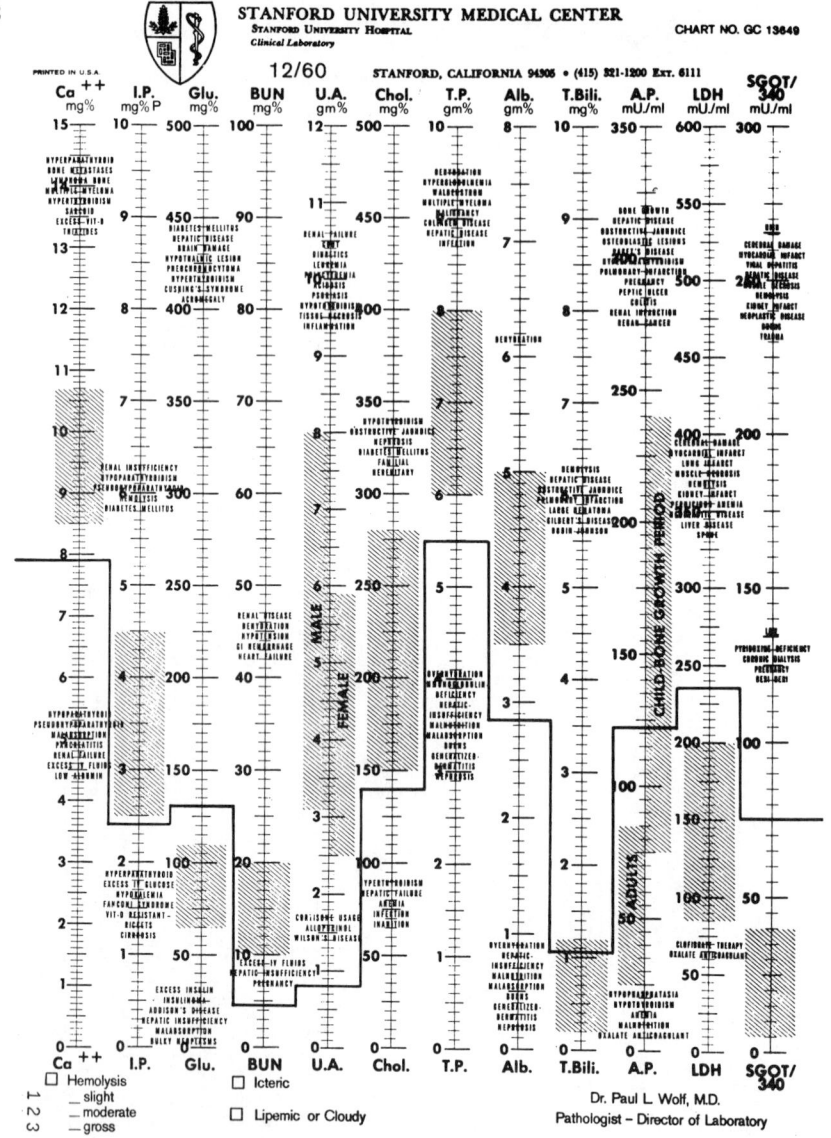

Fig. 22 - 94 (Continued):

from low serum protein. 5). Low uric acid results from renal tubular wastage from copper induced damage.

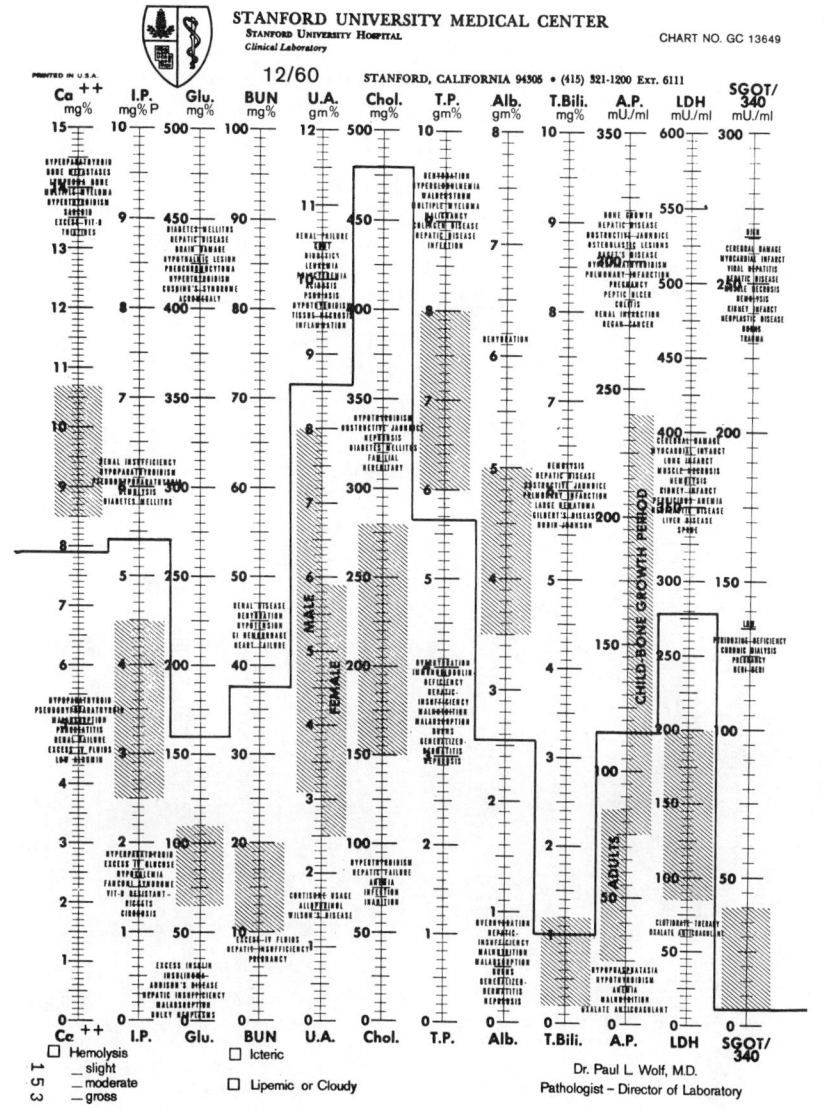

Fig. 22 - 95 BIOCHEMICAL PATTERN OF NEPHROTIC SYNDROME

1). Nephrosis will cause the BUN to be elevated due to decreased excretion. 2). Albumin is decreased due to marked proteinuria. 3). A decreased plasma lipoprotein lipase will result in elevated cholesterol which also is caused by excessive esterification of cholesterol by the liver cell in its attempt to compensate for the low Albumin. 4). Elevated phosphate, uric acid, and LDH are caused by retention in renal insufficiency. 5). Low

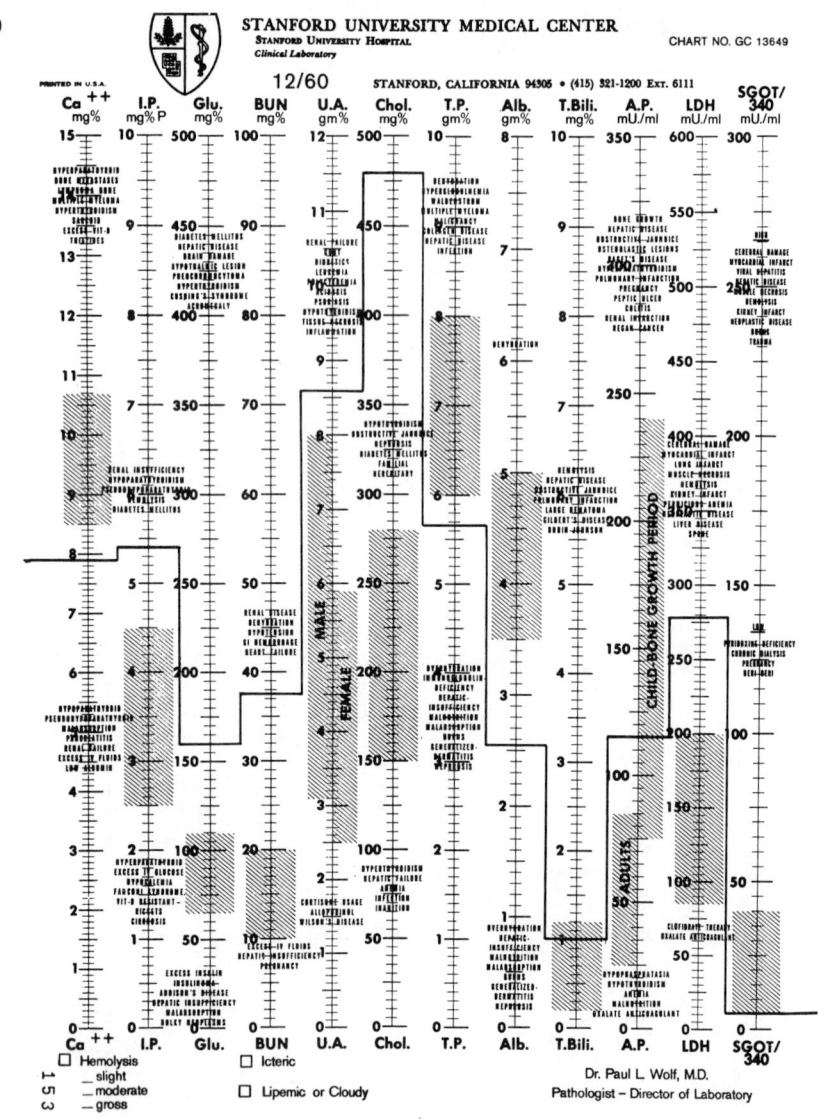

Fig. 22 - 95 (Continued):

calcium results from low albumin while elevated alkaline phosphatase is due to secondary hyperparathyroidism in renal failure. 6). Elevated glucose results from the anti-insulin action of the uremic toxins and also is due to non-specific reduction of the neocuproine chemical reagent in the 12/60 instrument by uric acid and creatinine.

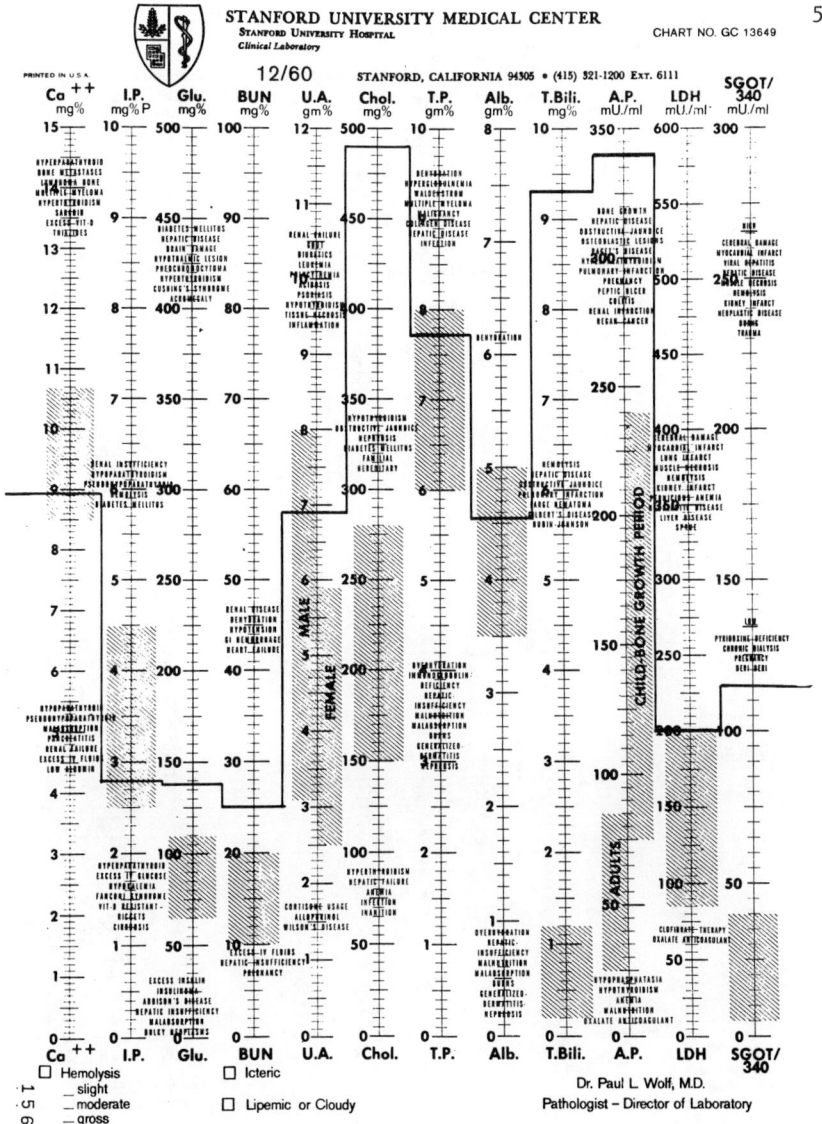

Fig. 22 - 96   BIOCHEMICAL PATTERN OF OBSTRUCTIVE JAUNDICE

Obstructive jaundice from a carcinoma of the head of the pancreas, bile ducts, or choledocholithiasis, or from an intrahepatic lesion will result in an elevated conjugated bilirubin, alkaline phosphatase, and cholesterol due to regurgitation. Furthermore, increased synthesis of conjugated bilirubin, alkaline phosphatase and cholesterol esters occurs with biliary tract obstruction. Increased GOT results from leakage from liver cells due to increase in

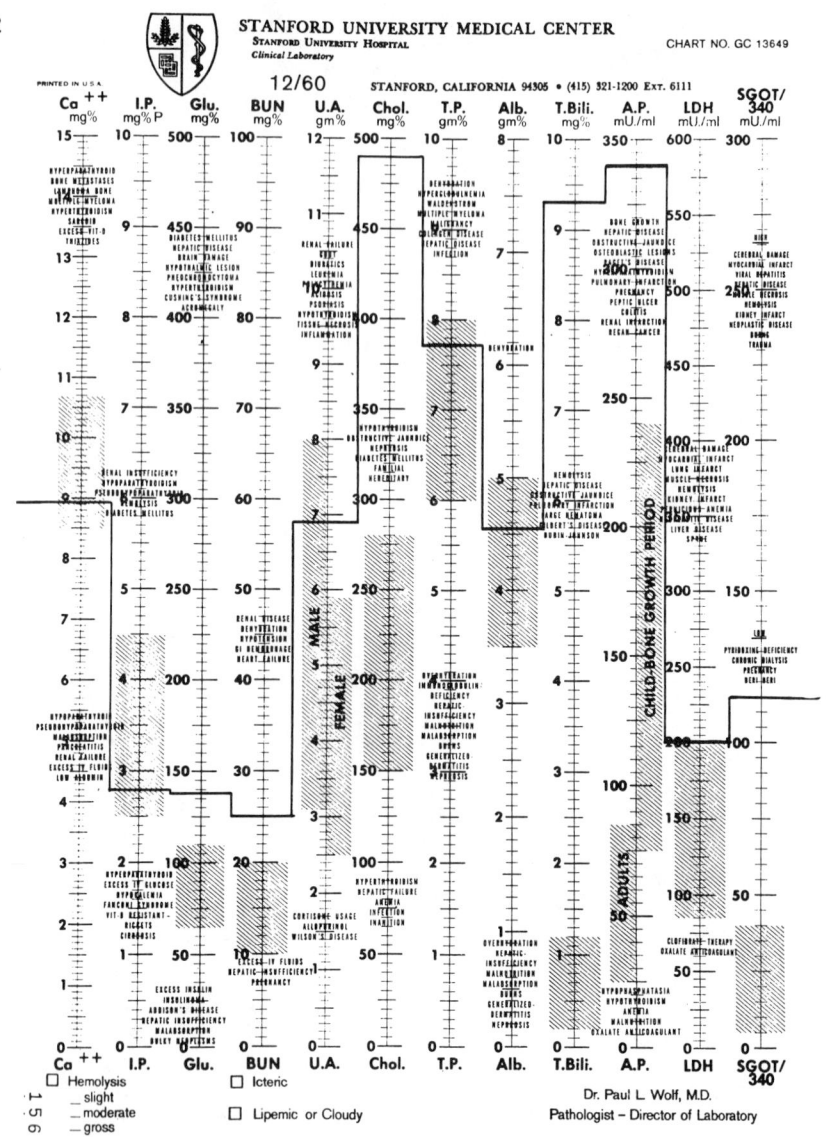

Fig. 22 - 96 (Continued):

biliary pressure or from cholangitis. The increased BUN is due to minimal renal disease associated with the obstructive biliary tract disease.

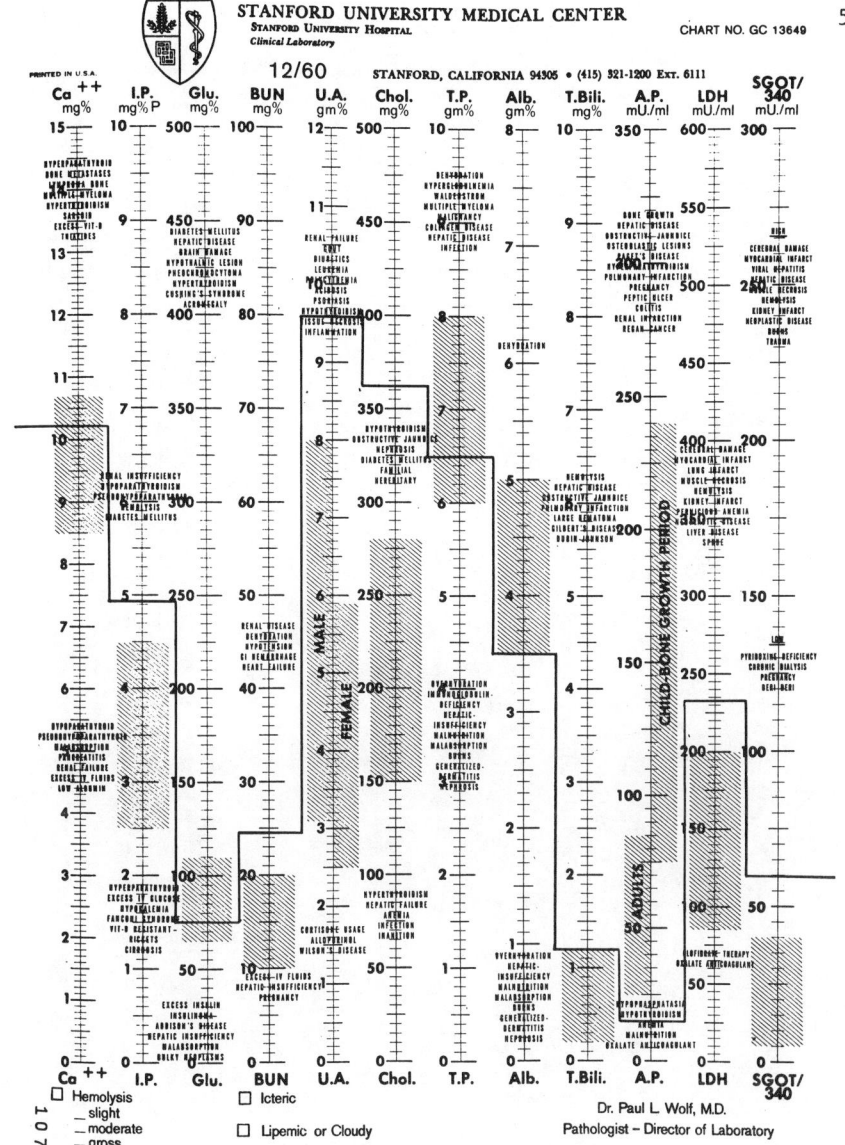

Fig. 22 - 97   BIOCHEMICAL PATTERN OF HYPOTHYROIDISM

1). Hypothyroidism is associated with an elevated cholesterol which is primarily caused by the markedly decreased metabolism. 2). A lowered glomerular filtration rate is present in hypothyroidism causing the elevated BUN, uric acid, and phosphate. 3). Hypothyroidism causes a low alkaline phosphatase in the child with inhibition of bone growth, since thyroid hormone is necessary for normal osteoblastic activity. 4). The elevated GOT and LDH result from the hypothyroid skeletal muscle myopathy.

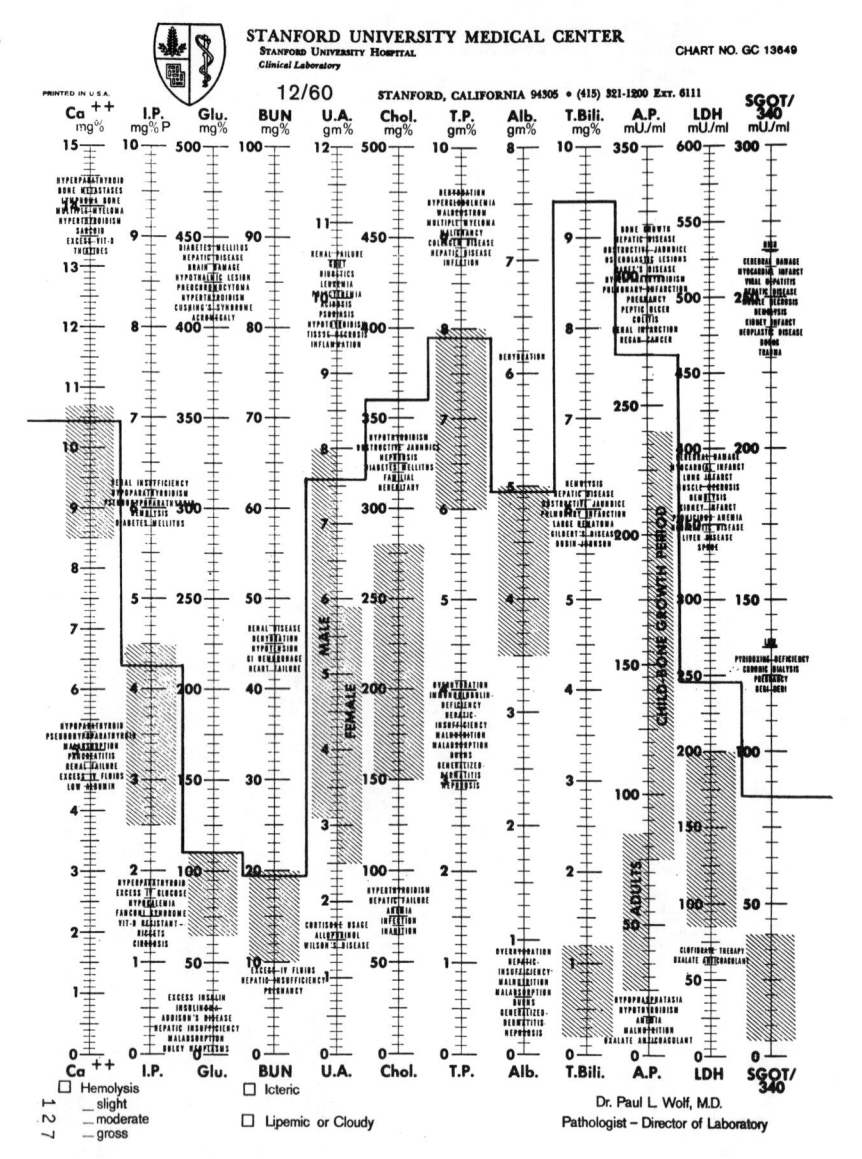

Fig. 22 - 98   BIOCHEMICAL PATTERN OF INFLUENCE OF BILIRUBIN ON CHOLESTEROL

See Fig. 22 - 96. One must subtract 5.0 mg. of cholesterol for each mg. of bilirubin above 1 because of artefactual elevation of cholesterol with an increase in serum bilirubin.

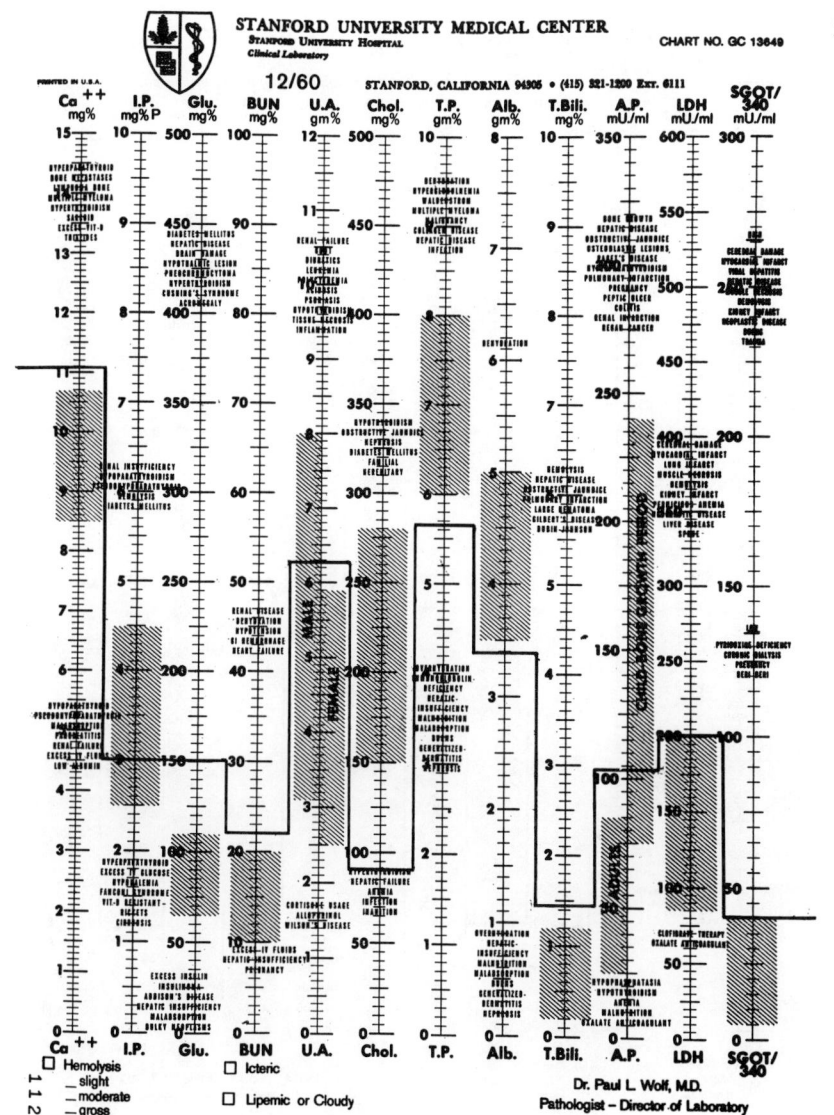

Fig. 22 - 99 BIOCHEMICAL PATTERN OF HYPERTHYROIDISM

Hyperthyroidism will cause a low serum cholesterol due to increased utilization resulting from the elevated metabolic rate. Increased glucose results from increased release of catecholamines with hepatic glycogenolysis. Increased utilization of protein causes decreased serum proteins and negative nitrogen balance. Hypercalcium is due to decreased bone osteoid with calcium release. Elevated alkaline phosphatase results. The BUN rises as does bilirubin due to fatty metamorphosis of kidney and liver.

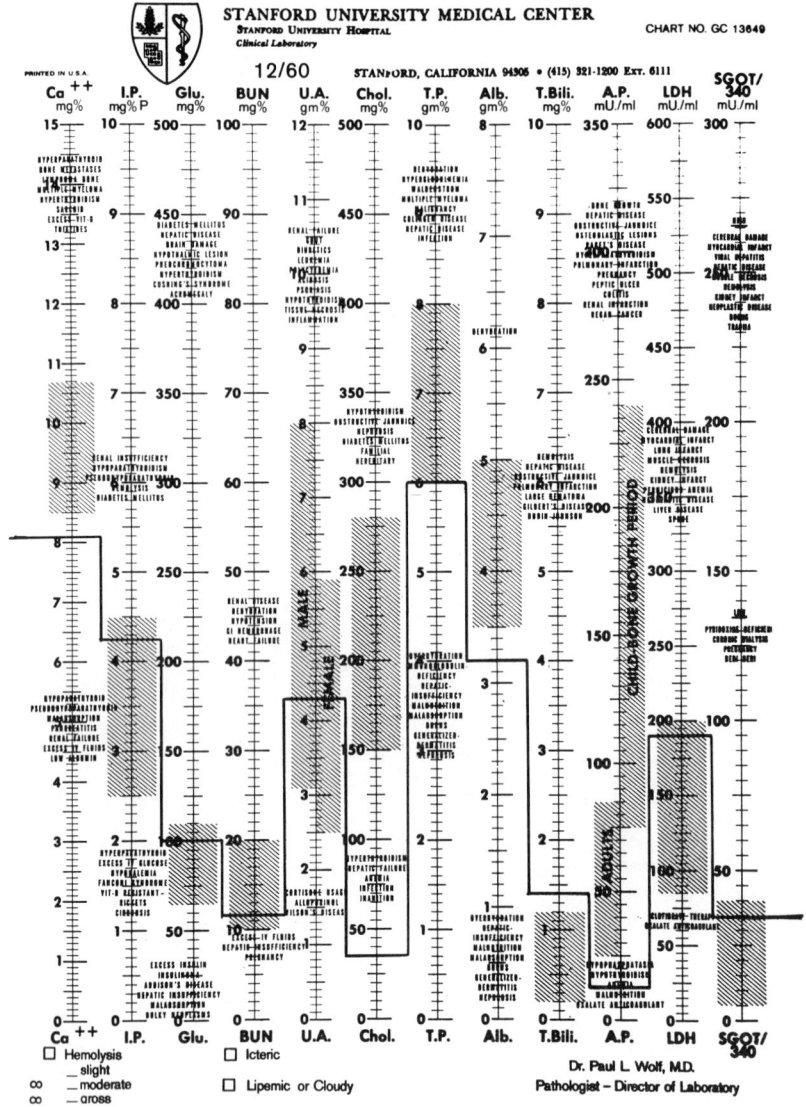

Fig. 22 - 100 BIOCHEMICAL PATTERN OF SEVERE ANEMIA

1). Anemia due to various causes may be associated with low serum protein which results from the same cause for the anemia, such as a neoplastic lesion. 2). The low serum calcium is due to the low serum protein. 3). The cholesterol is low for reasons which are not entirely understood. It may relate to the interchange of cholesterol between the erythrocyte and plasma. With decreased erythrocytes in anemia, the plasma cholesterol is lower.

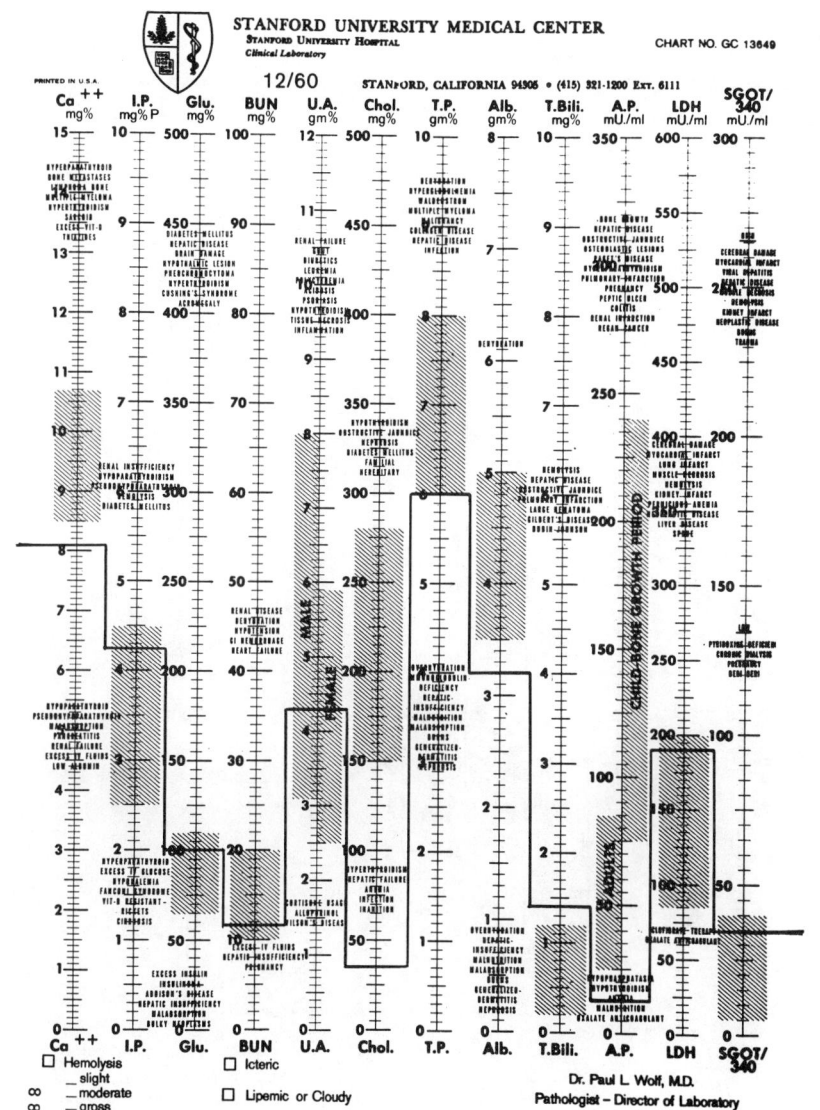

Fig. 22 - 100 (Continued):

4). Pernicious anemia causes a low alkaline phosphatase due to decreased osteoblastic activity. 5). Anemia causes a fatty liver with an increase in bilirubin.

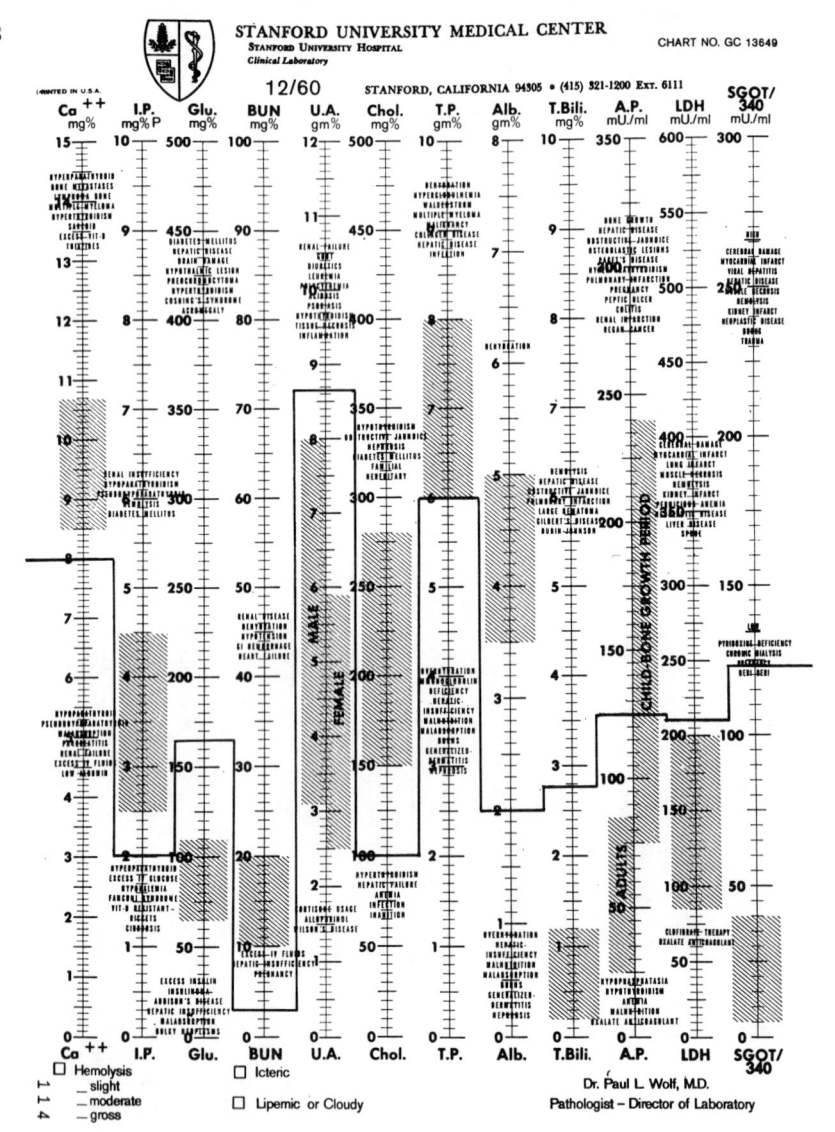

Fig. 22 -101 BIOCHEMICAL PATTERN OF CIRRHOSIS - LOW CHOLESTEROL

1). Cirrhosis is characterized by a low serum cholesterol due to decreased synthesis of cholesterol esters by the diseased hepatic cell. 2). Cirrhosis is characterized by minimal elevation of alkaline phosphatase, GOT, and LDH. 3). Bilirubin elevation is of the mixed type with elevation of unconjugated and conjugated fractions. 4). Serum albumin is low due to decreased synthesis resulting in decreased calcium due to decreased

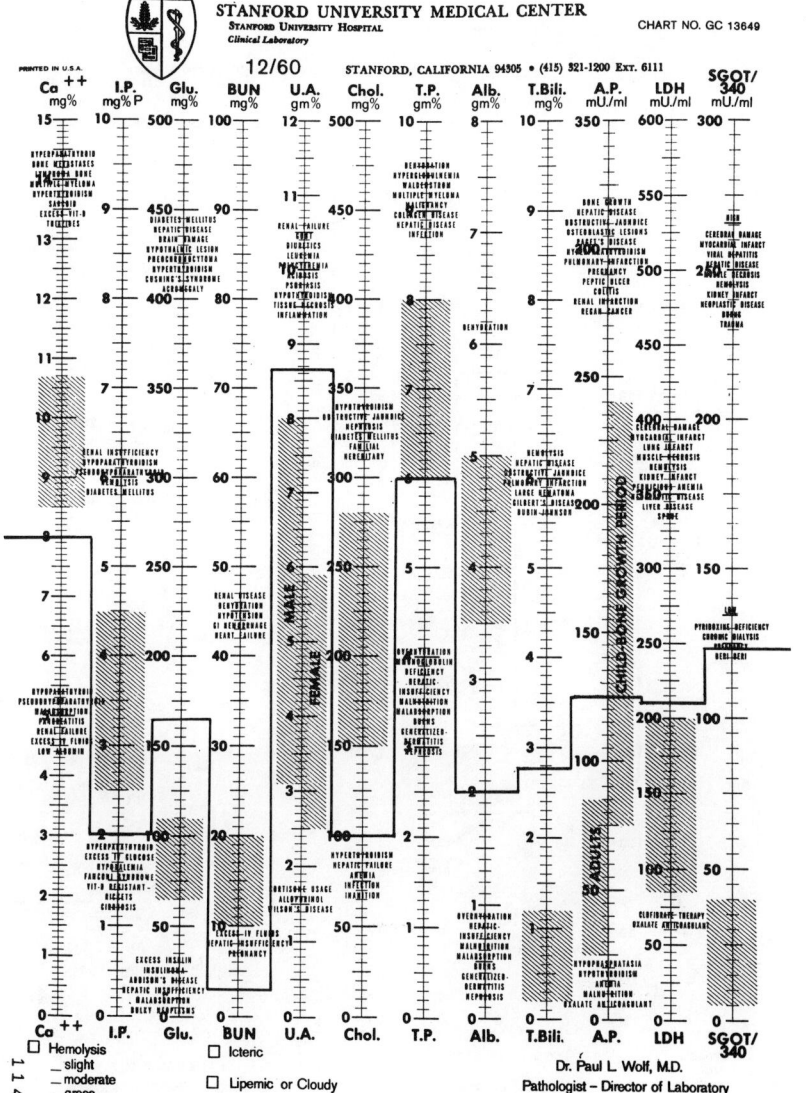

Fig. 22 - 101 (Continued):

binding capacity. 5). Decreased synthesis of BUN causes a low BUN. 6). Increased glycogenolysis causes hyperglycemia with low phosphate due to increased phosphorylation. Low phosphate also results from low potassium due to a renal tubule defect.

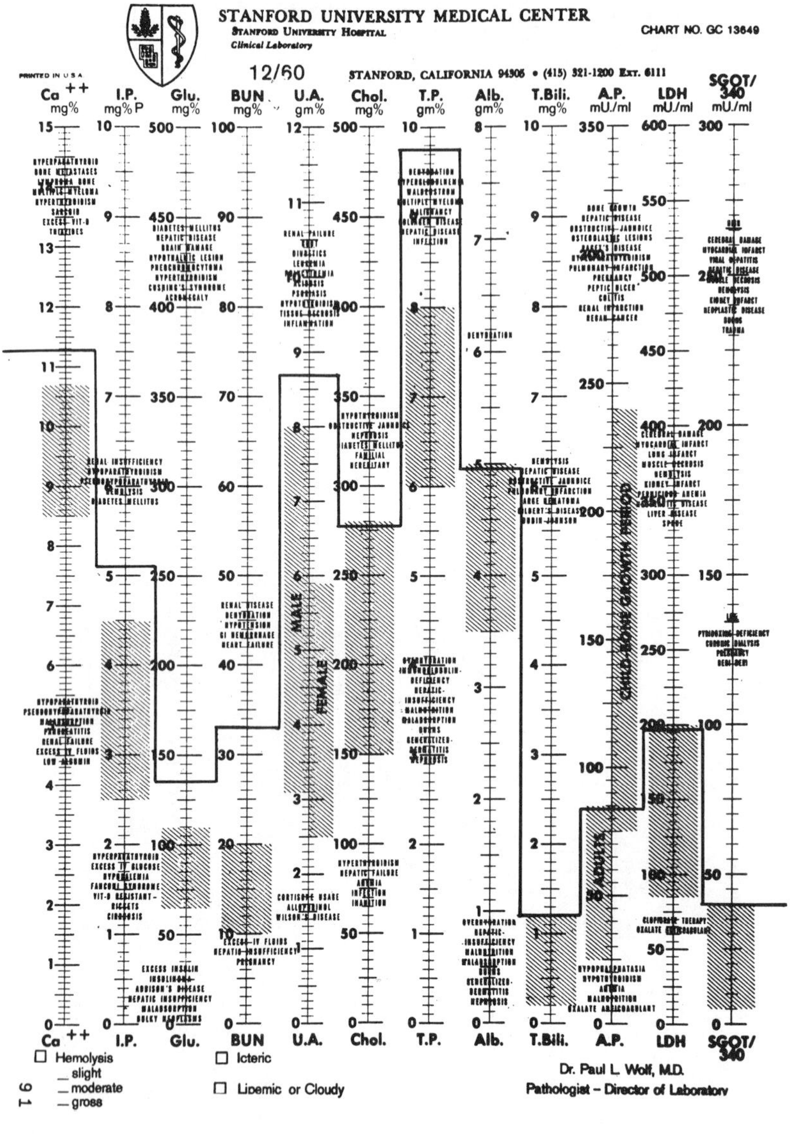

Fig. 22 - 102 BIOCHEMICAL PATTERN OF POLYCLONAL OR MONOCLONAL
GAMMAPATHY

1). Increase in gamma globulin may be due to monoclonal or polyclonal gammapathy. Common causes for polyclonal states are liver disease, chronic infections, neoplasms such as lymphoma or sarcoidosis. The monoclonal states are multiple myeloma, Waldenstrom's Macroglobulinemia, o Gaucher's disease. 2). Increase in calcium is due to increased binding of calcium by increase in protein.

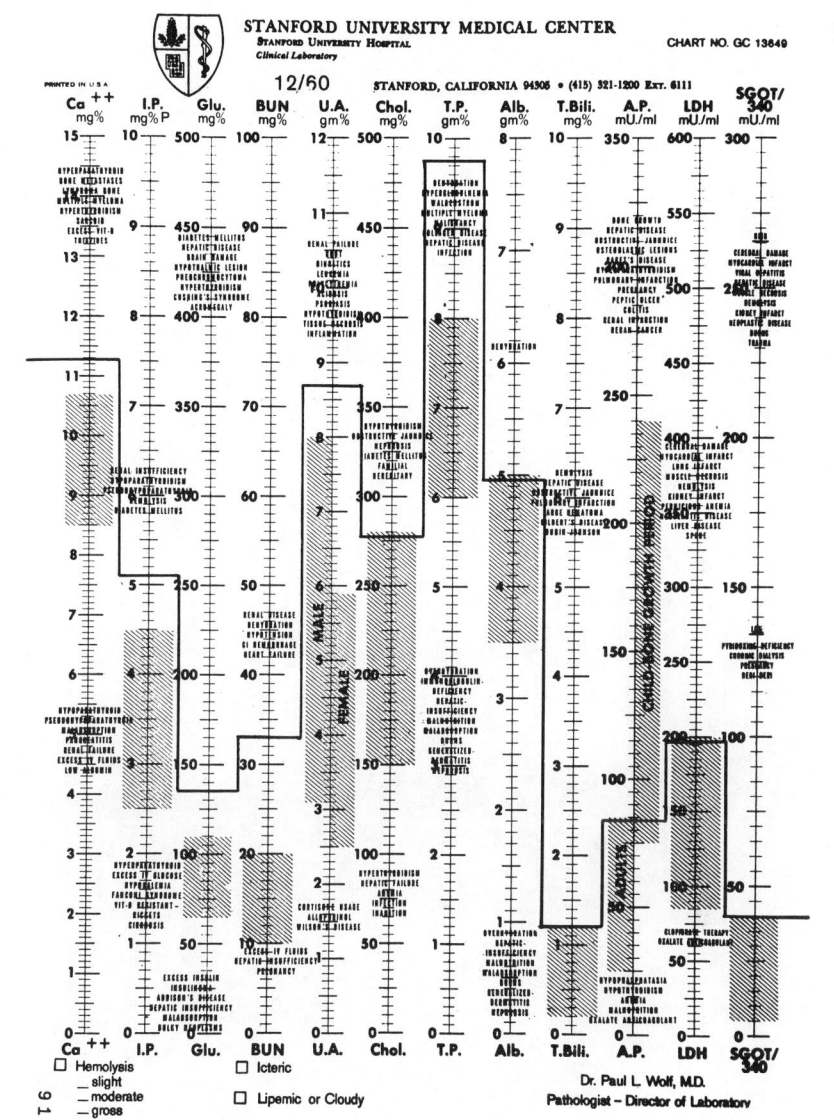

Fig. 22 -102 (Continued):

3). Renal disease results from deposition of gamma globulin in glomeruli and renal tubules depending on the etiology of the monoclonal or polyclonal gammapathy, with increase in BUN, phosphate, glucose, and uric acid.

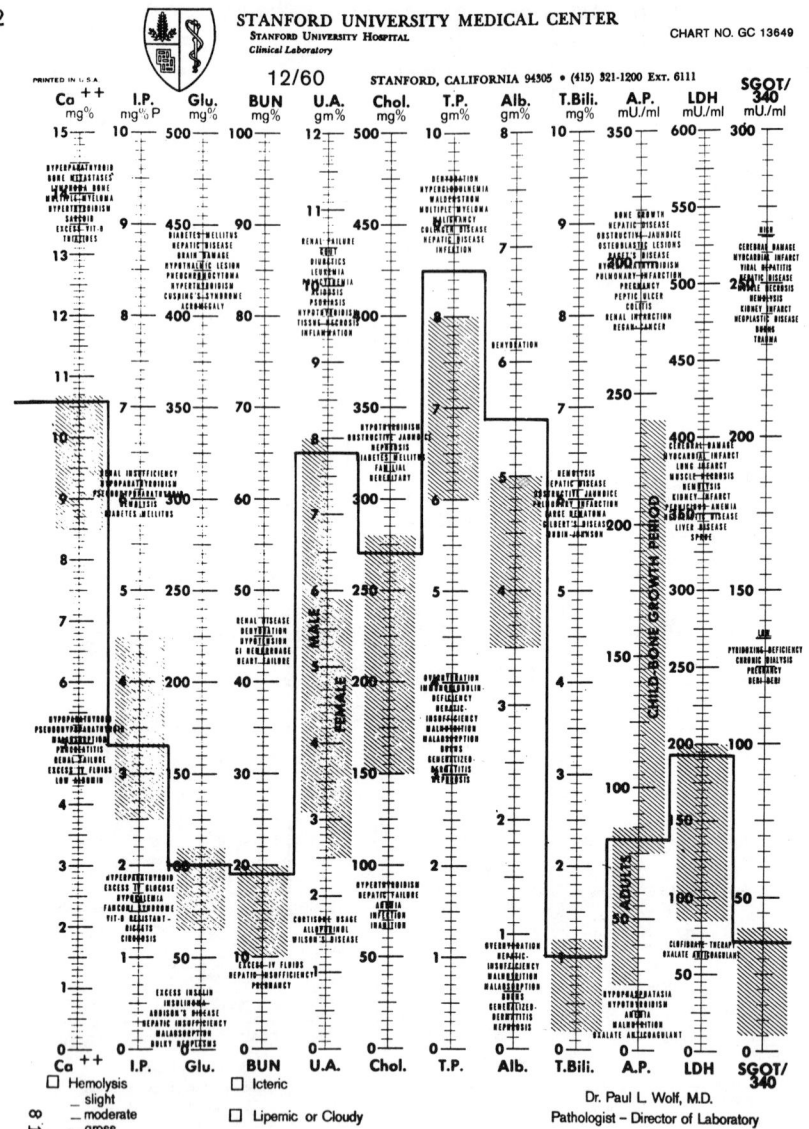

Fig. 22 - 103 BIOCHEMICAL PATTERN OF BSP DYE ARTIFACT ON SERUM

The presence of bromsulfophthalein dye in serum will cause a false elevation in the protein result in the colorimetric procedure.

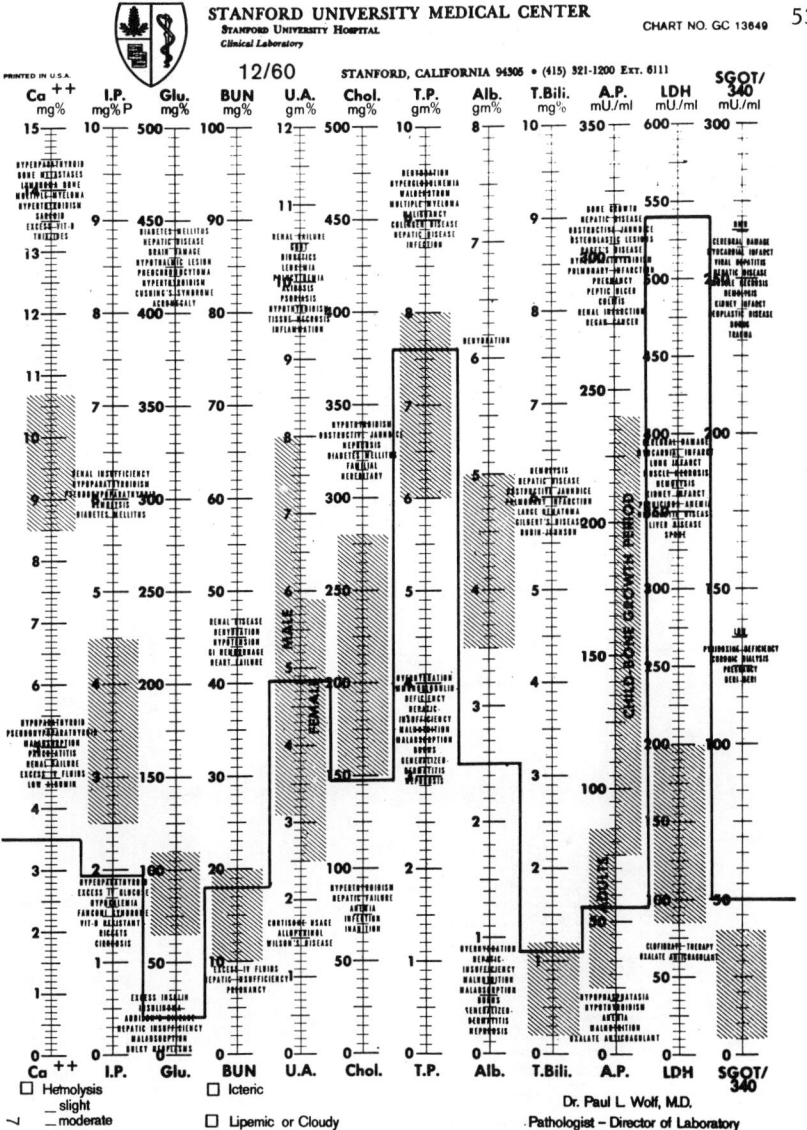

Fig. 22 -104 BIOCHEMICAL PATTERN OF EXUDATE

1). An exudate is a fluid which forms in the pericardial, pleural, or peritoneal cavity as a result of bacterial or viral or parasitic inflammation, necrosis of tissue, malignant neoplasms or collagen vascular disease. It has a high cell count from the inflammatory or neoplastic disease. It has the following biochemical pattern:
2). Protein greater than 2.5 gm.%. 3). Glucose lower than plasma due to consumption by the large number of cells.

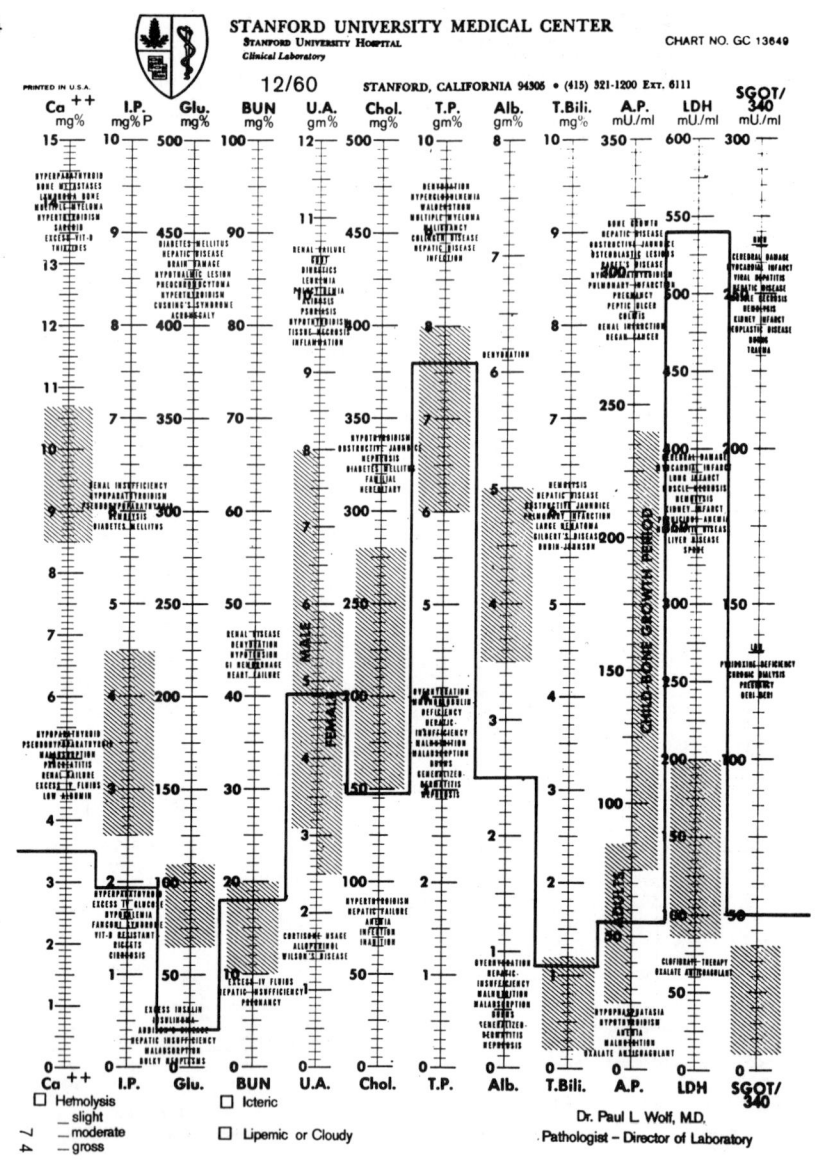

Fig. 22 -104 (Continued):

4). High cholesterol and triglycerides brought into the fluid by the cells. 5). LDH higher than plasma brought into the fluid by the large number of cells.

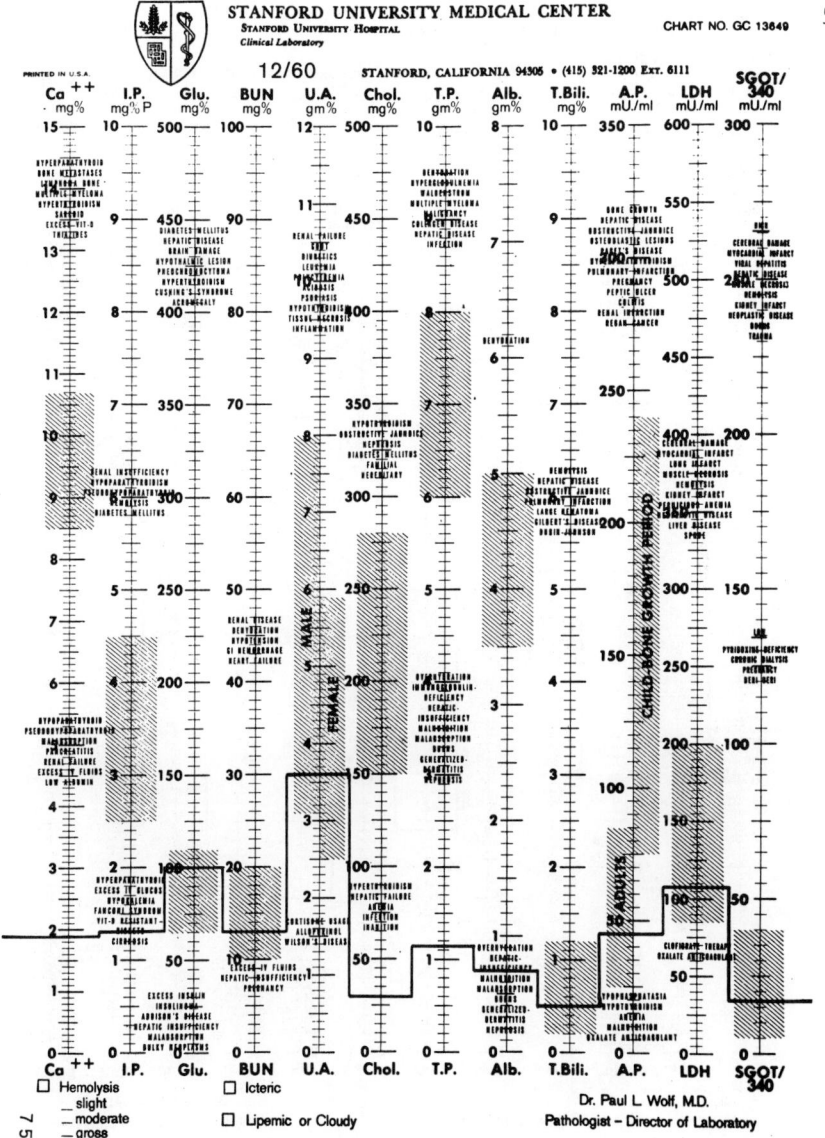

Fig. 22 - 105 BIOCHEMICAL PATTERN OF TRANSUDATE

1). A transudate is a fluid which forms in the pericardial, pleural, or peritoneal cavity as a result of low serum albumin, heart failure or extrinsic pressure on a large venous or lymphatic vessel. Biochemically it has the following pattern: 2). Protein of less than 2.5 gm.%. 3). Glucose similar to that of plasma. 4). Low cholesterol and triglycerides. 5). LDH similar to that of plasma. The cell count is minimal and consists of lymphocytes.

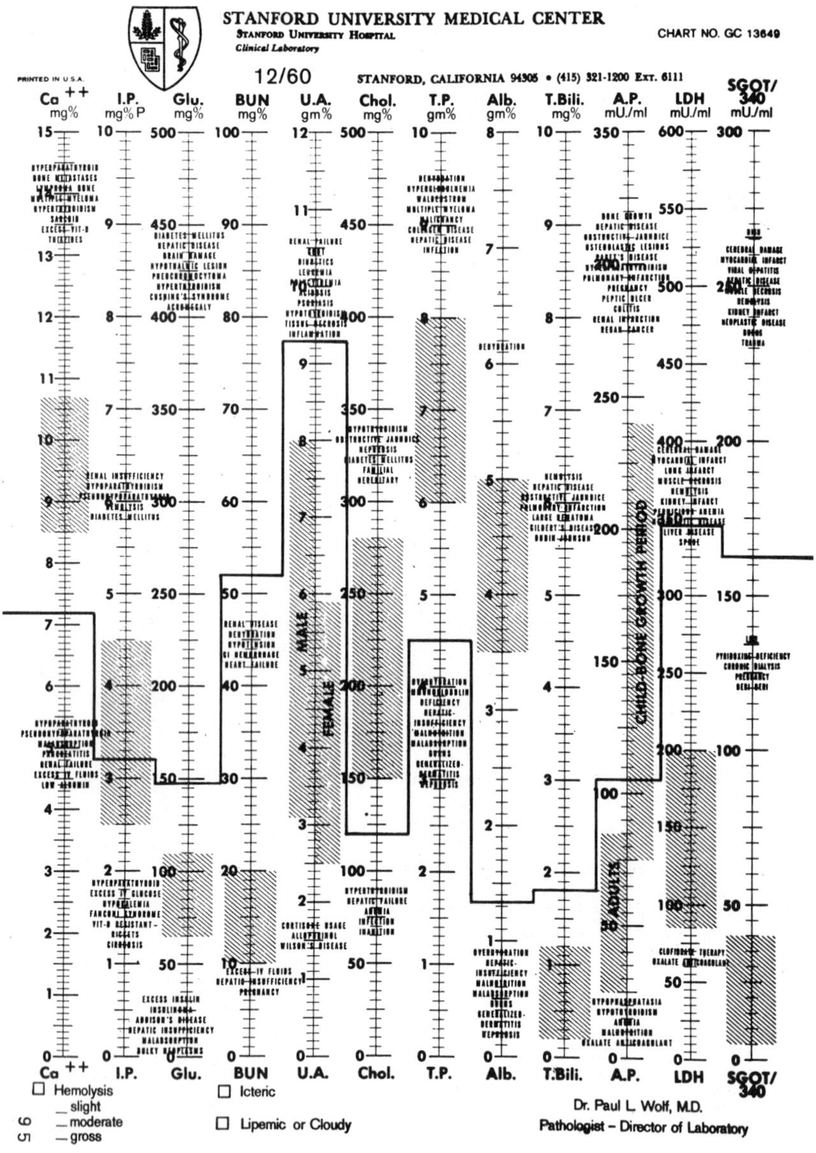

Fig. 22 - 106 BIOCHEMICAL PATTERN OF ACUTE, LARGE BURNS OR GENERALIZED BULLOUS DERMATITIS

1). An extensive burn or generalized bullous dermatitis will cause a large loss of serum proteins, including albumin from the skin. The serum calcium is lower due to low protein. 2). The skin contains GOT which is liberated from the burned skin. 3). Hemolysis and dehydration causes prerenal azotemia with elevated BUN, glucose, and uric acid. 4). The hyperglycemia is from

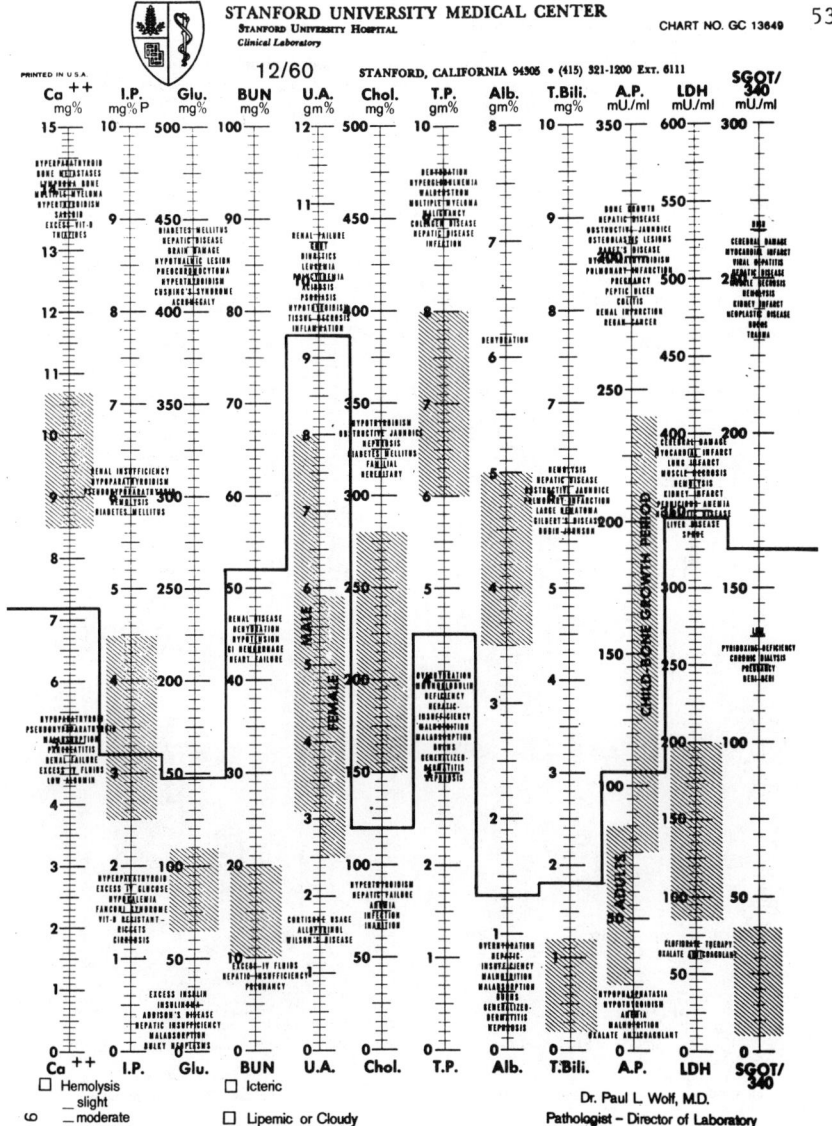

Fig. 22 - 106 (Continued):

the intravenous glucose and adrenal cortical stress.
5). Hemolysis from the heat injury causes elevated
bilirubin with the anemia causing the low cholesterol.
The low cholesterol also occurs as a result of lack of
lipid oral intake.

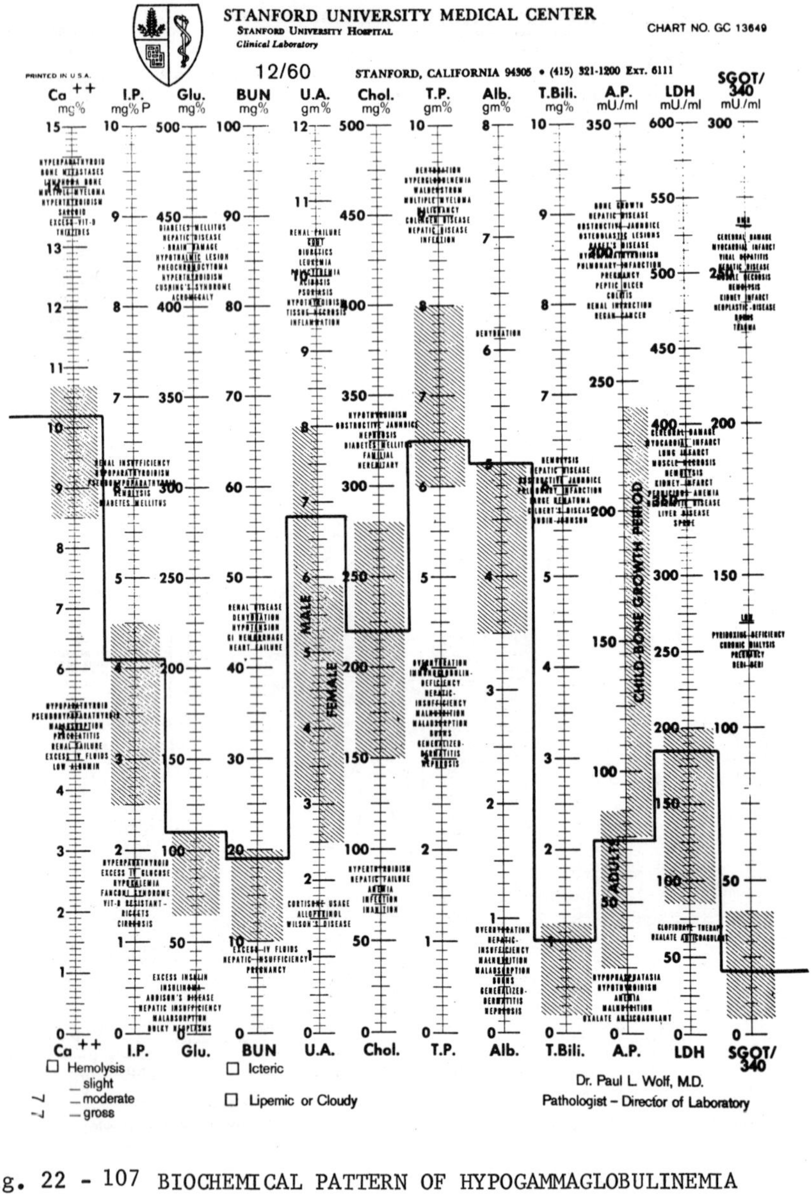

Fig. 22 - 107 BIOCHEMICAL PATTERN OF HYPOGAMMAGLOBULINEMIA

Hypogammaglobulinemia may be congenital, e.g. associated with thymic aplasia or acquired due to chemotherapy destruction of lymphopoietic tissue, neoplastic destruction of lymphopoietic tissue, or loss of gamma globulins in malabsorption, nephrosis or bullous dermatitis.

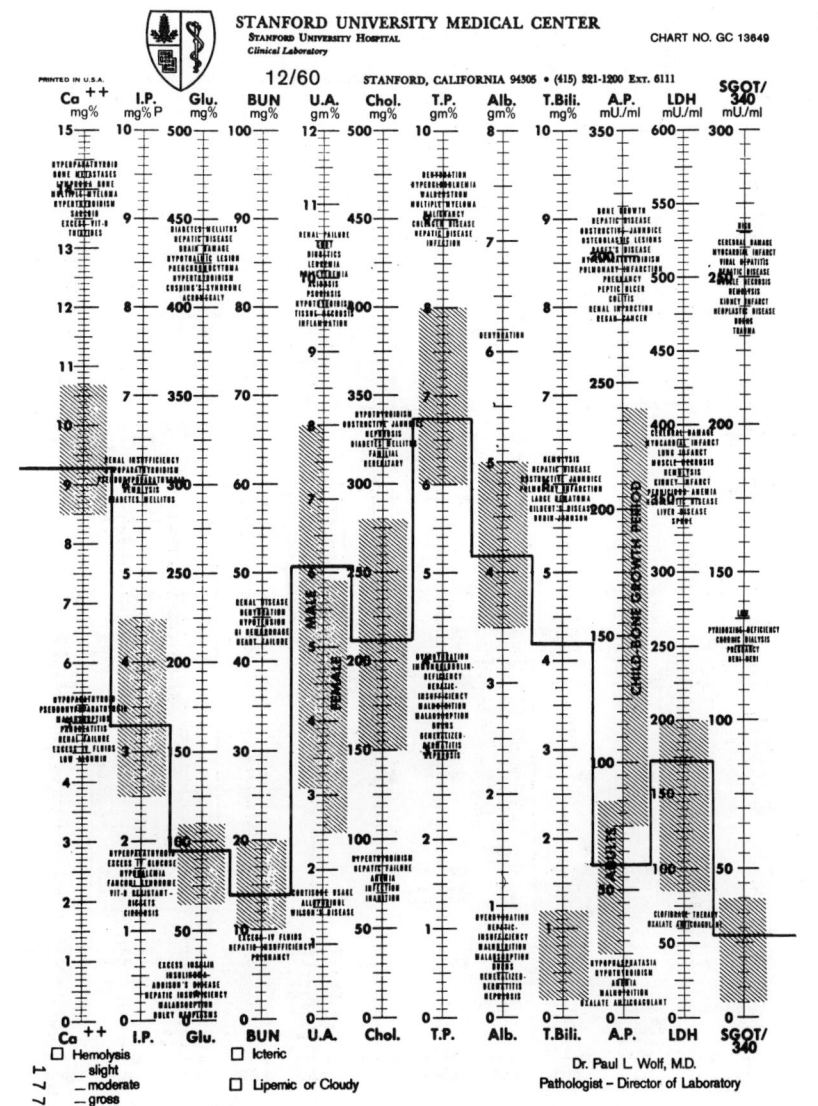

Fig. 22 - 108 BIOCHEMICAL PATTERN OF GILBERT'S SYNDROME

Gilbert's constitutional jaundice causes an elevated unconjugated bilirubin due to deficiency of hepatic cell mono and di-glucuronyl transferase. The Crigler-Najjar syndrome will result in the same biochemical pattern and the hepatic enzyme in this lesion is absent. Other causes for an elevated unconjugated bilirubin are hemolytic anemia, hemorrhagic pulmonary infarct, large hematoma, and utilization of numerous blood bank old units of blood.

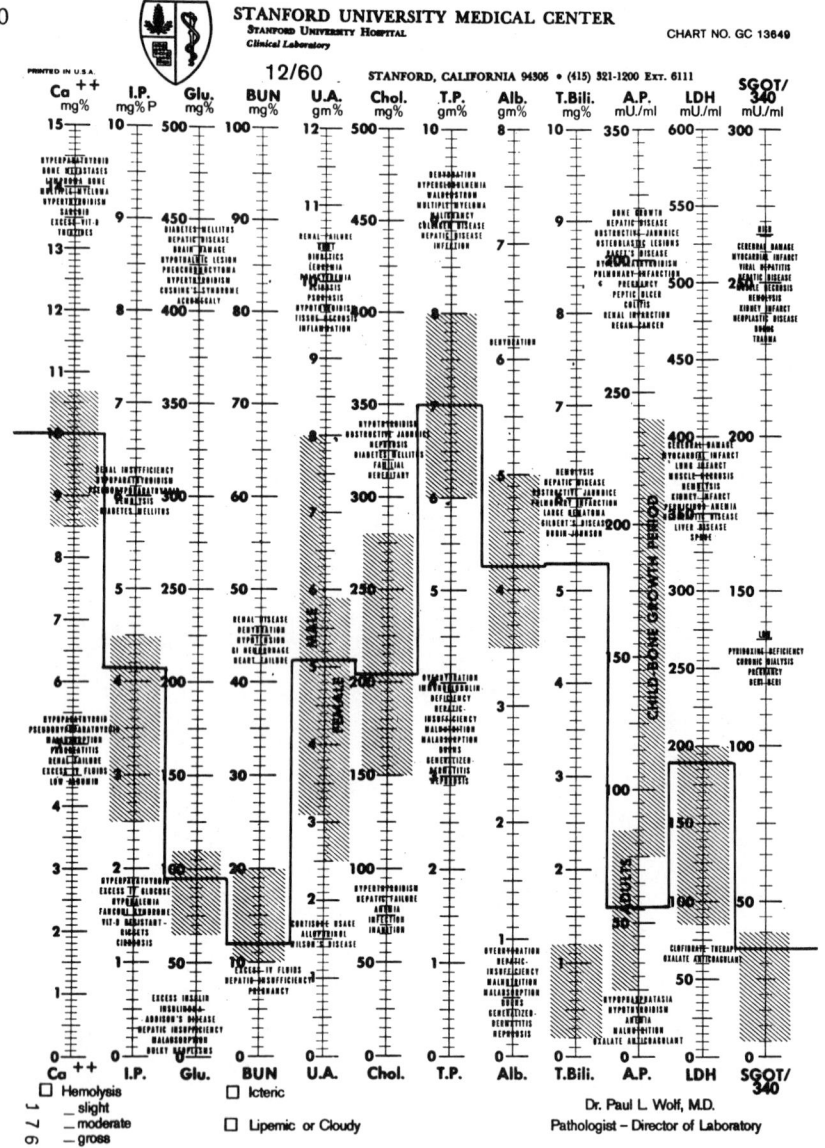

Fig. 22 - 109 BIOCHEMICAL PATTERN OF DUBIN - JOHNSON SYNDROME

The Dubin - Johnson syndrome is another cause for constitutional jaundice and in contrast to Gilbert's, the elevated bilirubin is the conjugated fraction.

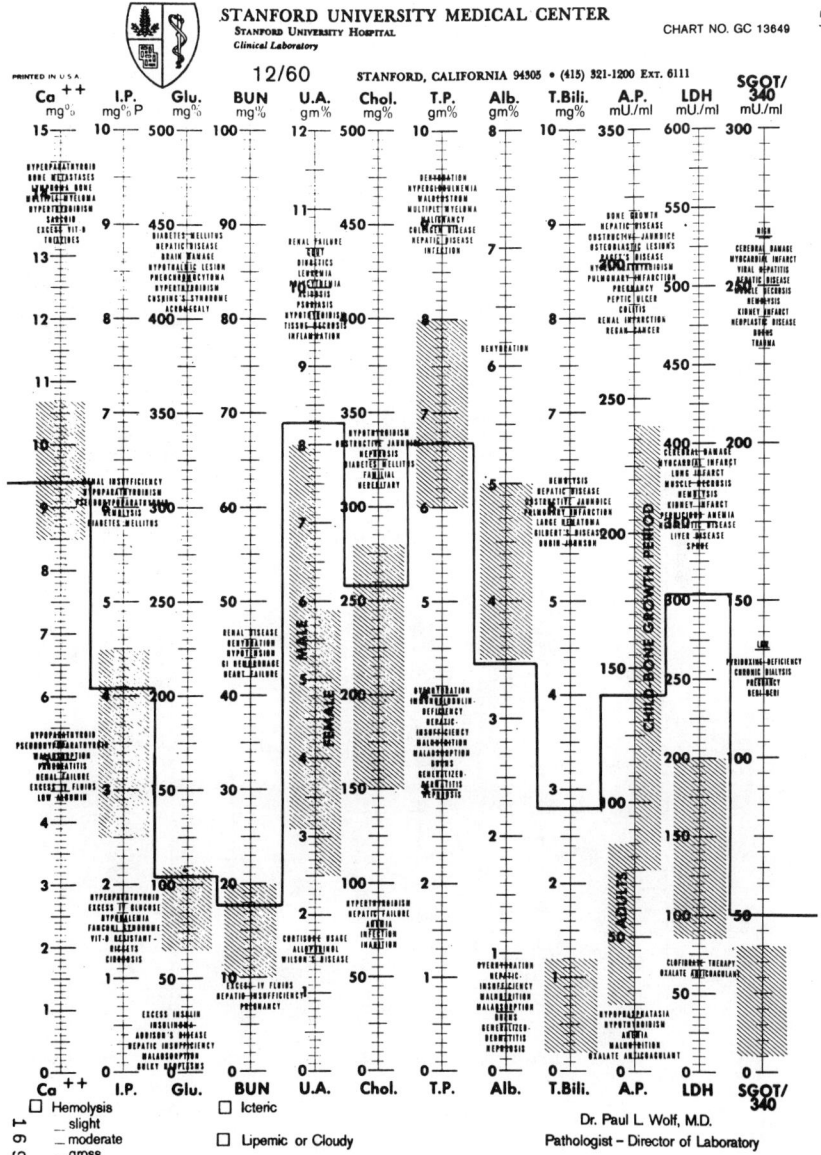

Fig. 22 - 110 BIOCHEMICAL PATTERN OF RECENT LUNG INFARCT

1). A recent infarct of the lung will result in an increase in unconjugated bilirubin from the hemolysis of the erythrocytes in the infarct. 2). Destruction of the lung alveolar cells releases uric acid and LDH$_3$ into the serum. 4). Elevated GOT is from the erythrocytes undergoing hemolysis. 5). The elevated alkaline phosphatase is derived from the necrosis of interalveolar vascular endothelium and from granulation tissue organizing the infarct.

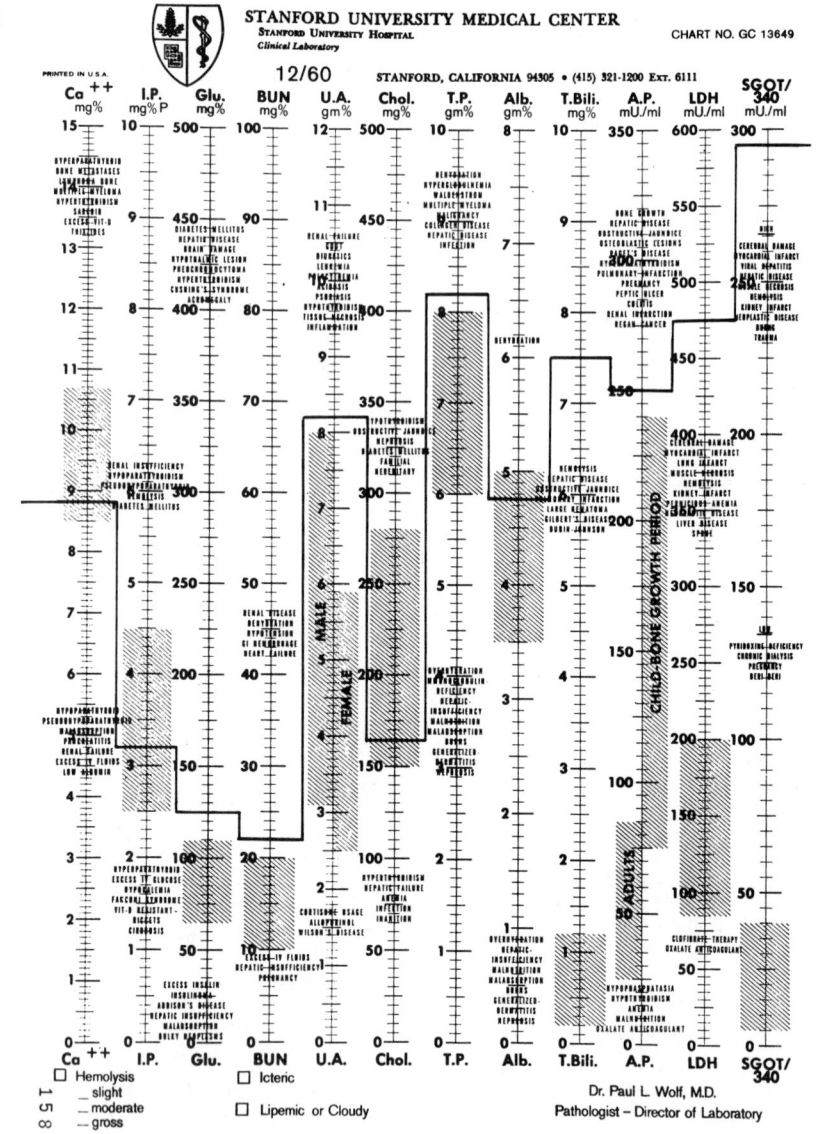

Fig. 22 - 111 BIOCHEMICAL PATTERN OF ACUTE VIRAL HEPATITIS

1). Acute viral hepatitis will result in hepatic cell necrosis with elevation of LDH_5 and GOT. 2). Hepatic cells are also swollen and filled with virus resulting in intrahepatic obstructive disease. Alkaline phosphatase is increased. 3). The increased bilirubin is of the mixed type with unconjugated and conjugated bilirubin elevated. 4). Hyperglobulinemia is of the polyclonal type with IgG and IgM elevation in response to

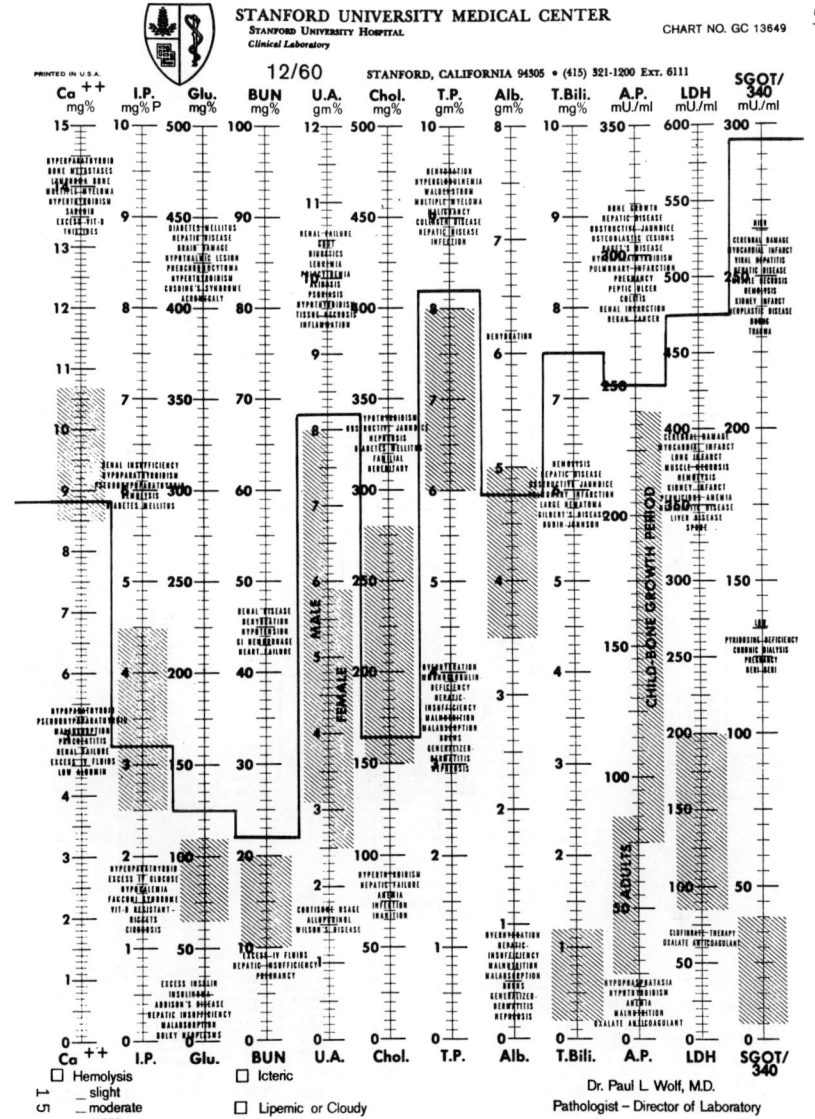

Fig. 22 - 111 (Continued):

the viral hepatitis. 5). Minimal renal insufficiency may be present associated with the hepatitis causing elevation of BUN, uric acid and glucose.

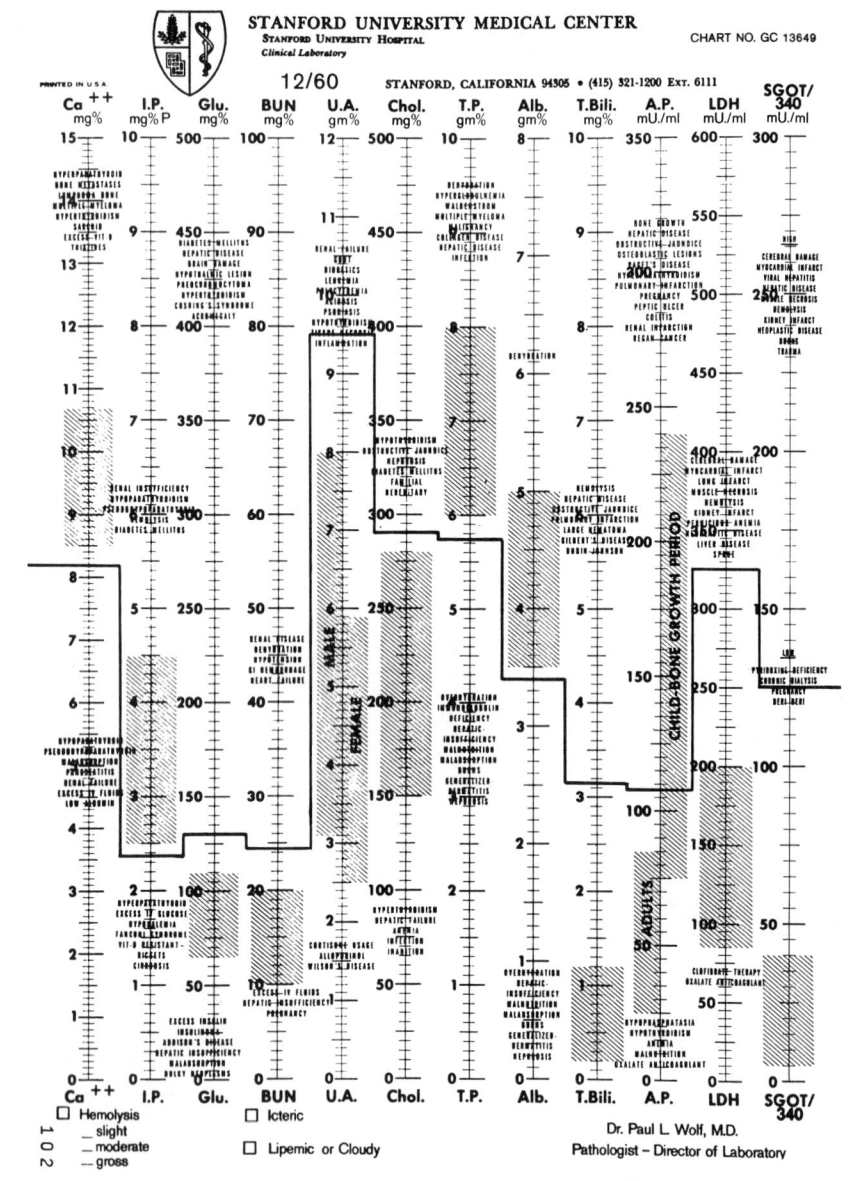

Fig. 22 - 112 BIOCHEMICAL PATTERN OF INFECTIOUS MONONUCLEOSIS WITH ACUTE HEPATITIS

1). Infectious mononucleosis is caused by the EB virus which affects not only the lymphopoietic tissue but also many organs such as the liver and kidney. 2). Elevation of LDH is due to the lymphoproliferative disease. 3). Involvement of the liver causes minimal hepatic necrosis with elevation of LDH and GOT. 4). Hepatic cells are also swollen and filled with virus resulting in intrahepati

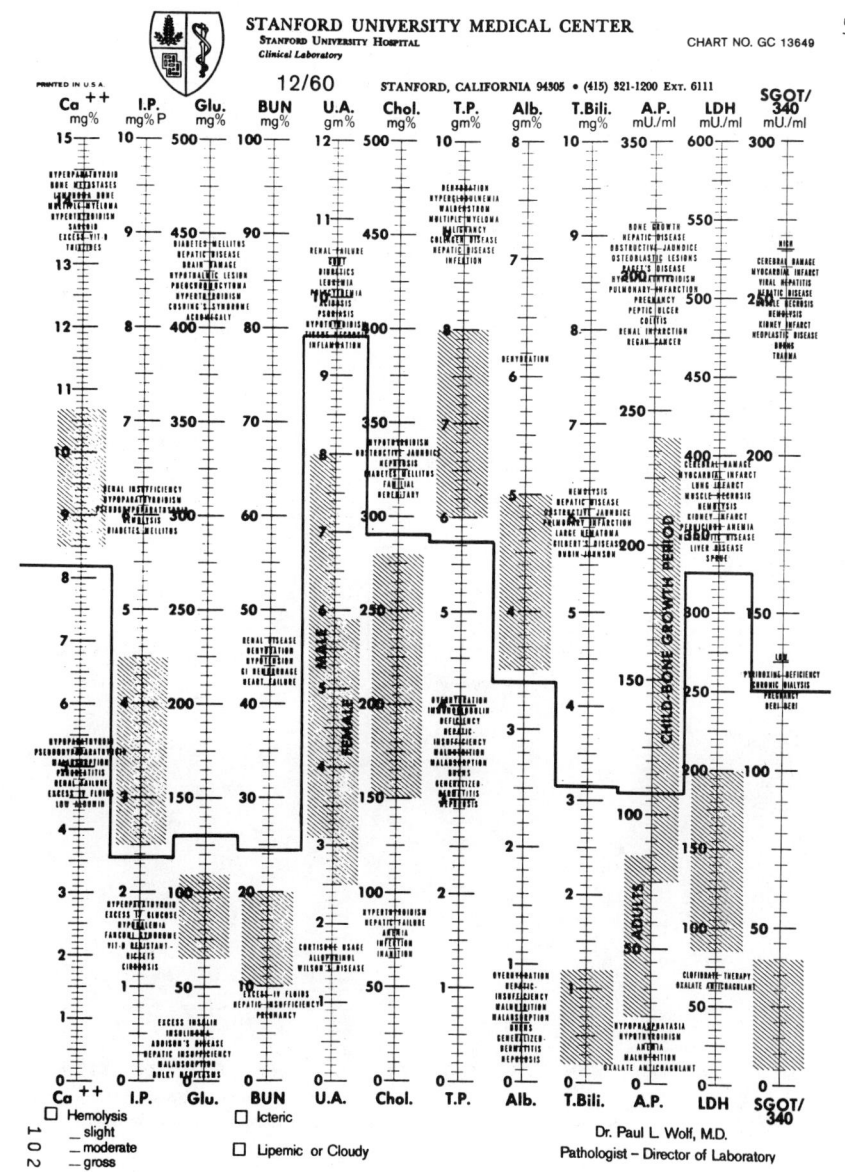

Fig. 22 - 112 (Continued):

obstructive disease. Alkaline phosphatase is increased. 5). The increased bilirubin is of the mixed type with unconjugated and conjugated bilirubin. 6). Serum proteins and consequently calcium are decreased due to the hepatitis. 7). Minimal renal insufficiency may be present associated with the hepatitis causing elevation of BUN, uric acid and glucose. 8). Phosphate is decreased due to the hyperglycemia resulting from liver and renal disease. 9). Increased cholesterol is due to intra-hepatic obstructive disease.

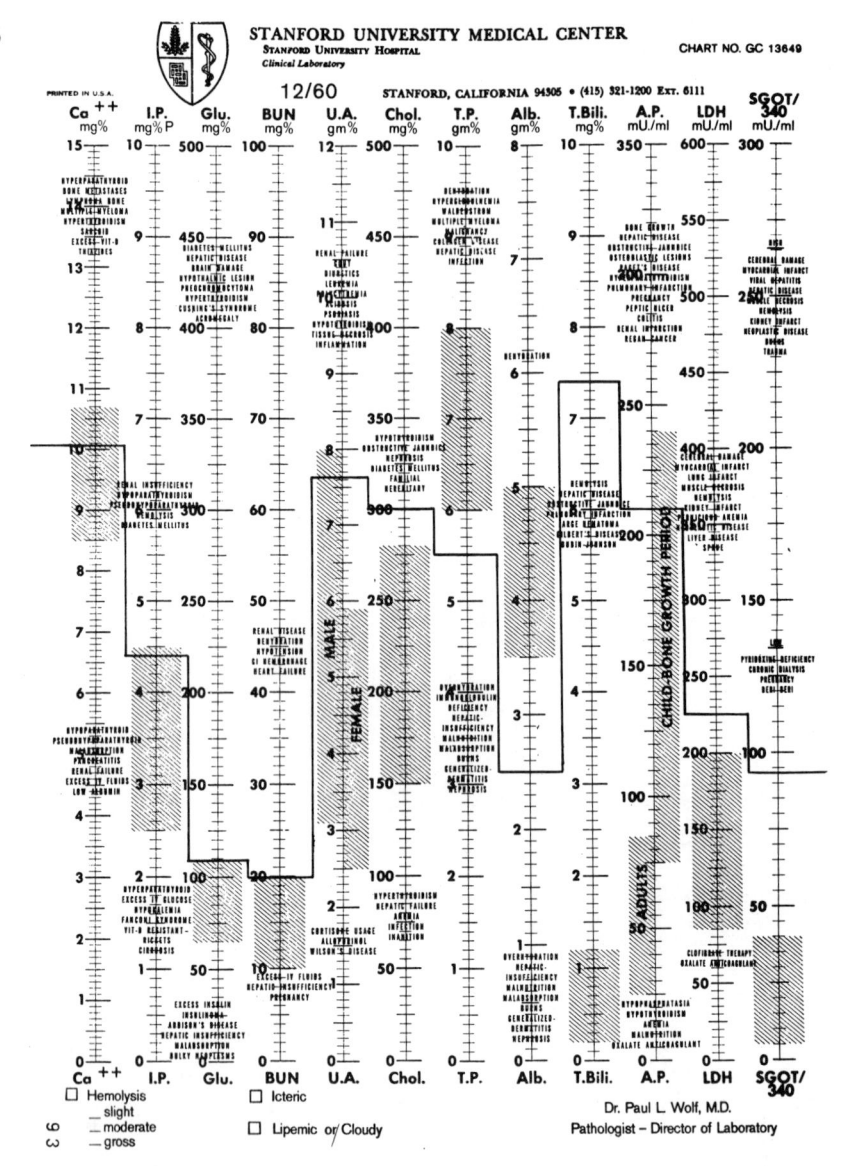

Fig. 22 - 113 BIOCHEMICAL PATTERN OF INFLUENCE OF BILIRUBIN ON SERUM ALBUMIN

See Fig. 22 - 83. An elevated bilirubin causes an artefactual lowering of albumin when albumin is determined by a dye binding procedure which is the Haba Dye Method on the SMA 12/60 Instrument. Bilirubin avidly binds to albumin in the serum. Elevated bilirubin may result in a spurious depression of one gram of albumin.

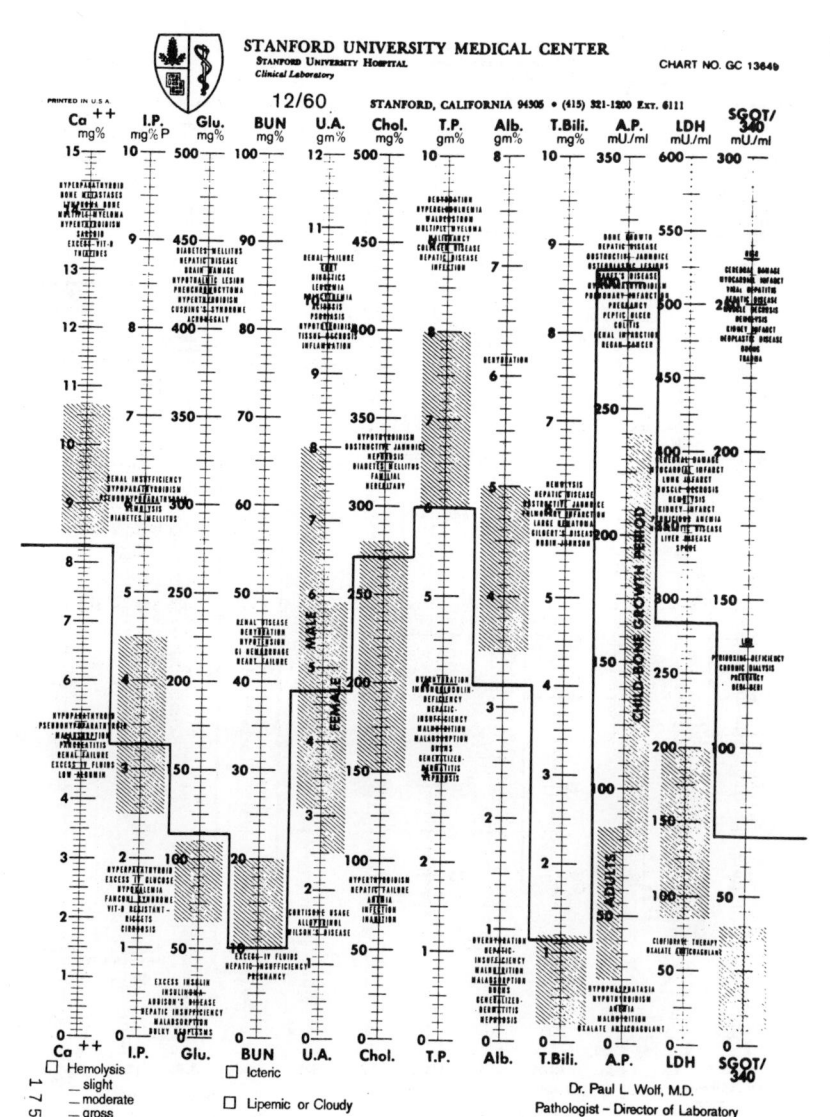

Fig. 22 - 114 BIOCHEMICAL PATTERN OF CANCER METASTATIC TO LIVER

An infiltrative lesion of the liver due to metastatic carcinoma, lymphoma, tuberculosis or sarcoidosis will result in elevated alkaline phosphatase and LDH with a normal bilirubin. The pattern may also be found in congestive heart failure.

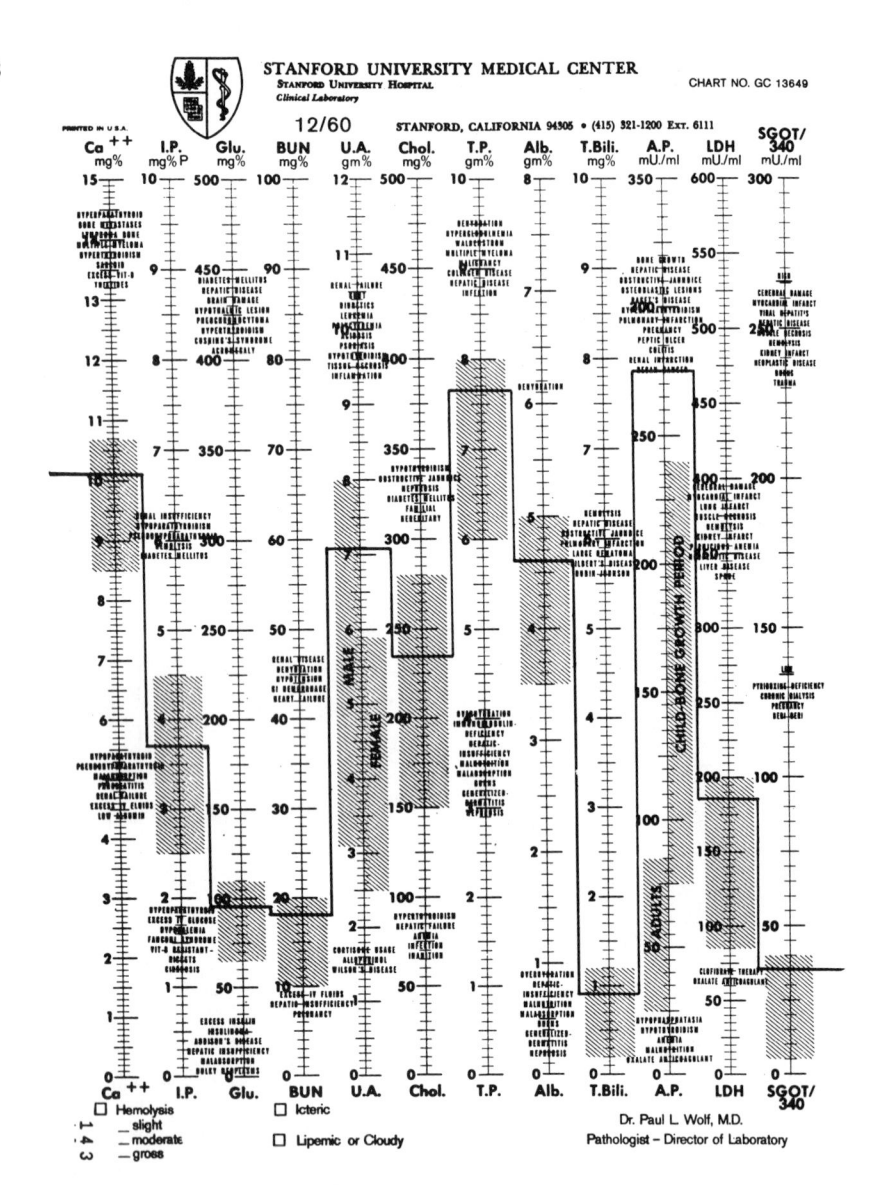

Fig. 22 - 115 BIOCHEMICAL PATTERN OF REGAN ENZYME - UNDIFFERENTIATED BRONCHOGENIC CANCER

The Regan isoenzyme is an alkaline phosphatase isoenzyme ectopically produced by a malignant neoplasm, such an undifferentiated bronchogenic carcinoma or ovarian malignant neoplasms. The isoenzyme is similar to the placental isoenzyme and is extremely heat stable and inhibited by phenylalanine.

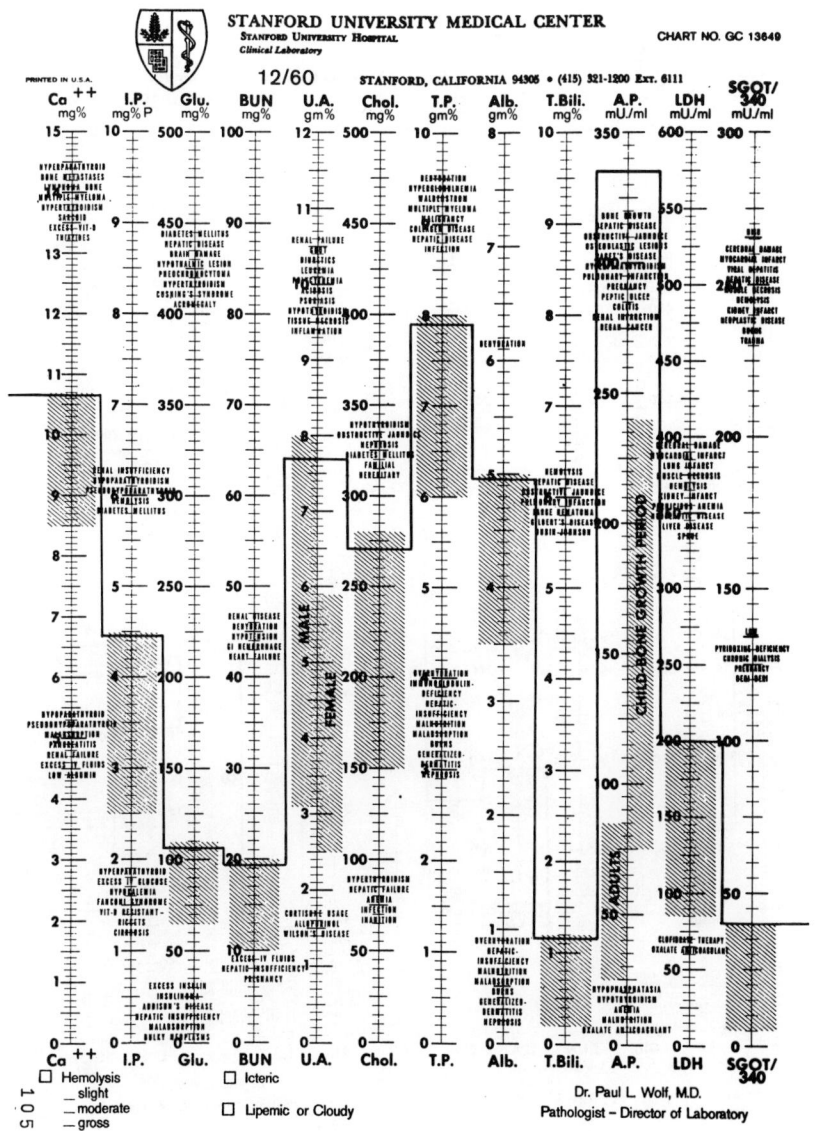

Fig. 22 - 116 BIOCHEMICAL PATTERN OF PAGET'S DISEASE OF BONE

Paget's disease of bone causes an osteoblastic bone lesion with resultant elevation of a heat labile alkaline phosphatase. The polyostotic lesion will cause a higher level than a monostotic lesion.

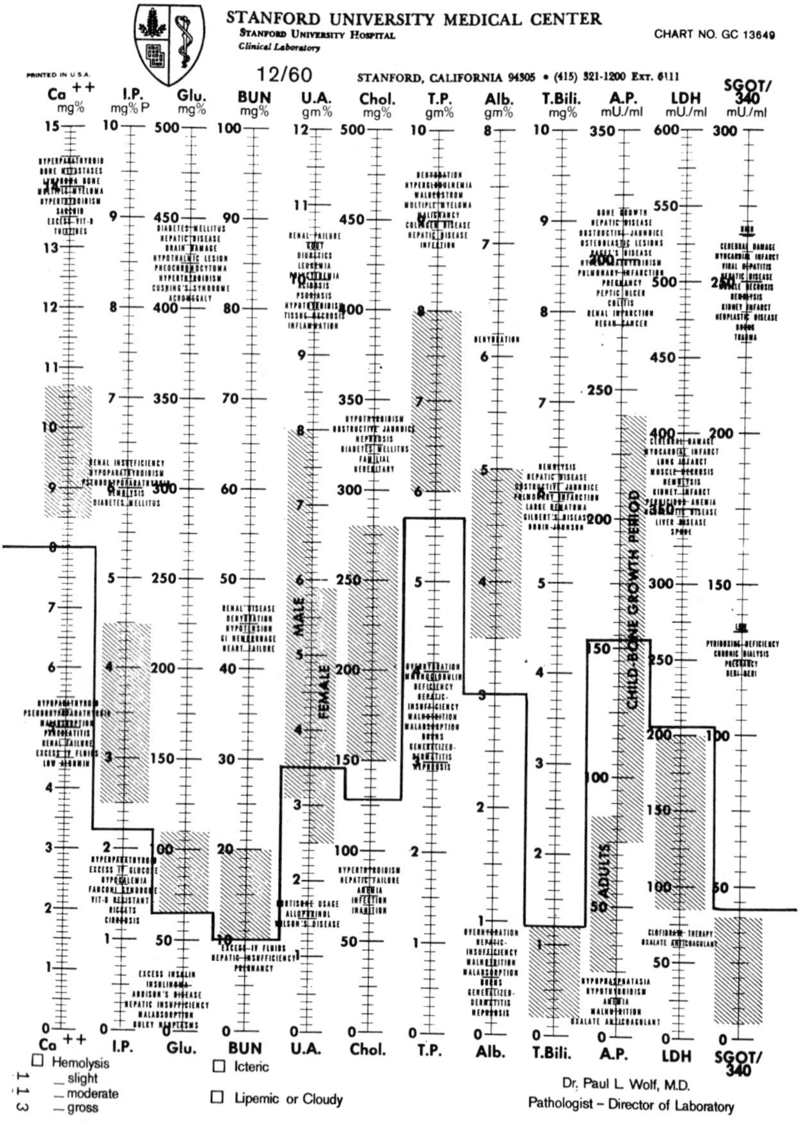

550

Fig. 22 - 117 BIOCHEMICAL PATTERN OF ULCERATIVE COLITIS

1). Ulcerative colitis causes a loss of serum proteins with resultant low calcium. 2). Malnutrition also contributes to low serum protein, calcium, phosphate, glucose, BUN, and cholesterol. 3). Anemia may contribute to low cholesterol. 4). Elevated alkaline phosphatase is caused by leakage of alkaline phosphatase from necrotic colonic mucosa, and is moderately heat stable and inhibited by phenylalanine. Another reason for elevation

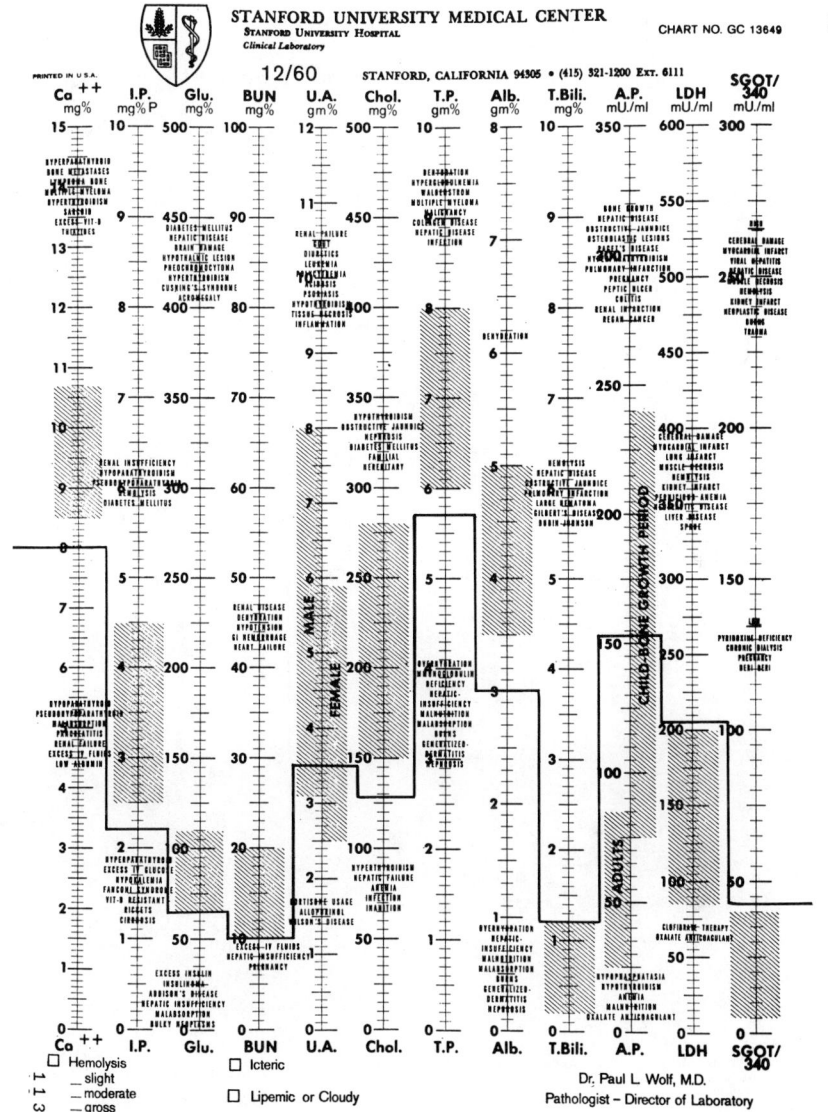

Fig. 22 - 117 (Continued):

of alkaline phosphatase in ulcerative colitis is sclerosing cholangitis or osteomalacia due to Vitamin D malabsorption. The bone phosphatase is heat labile.

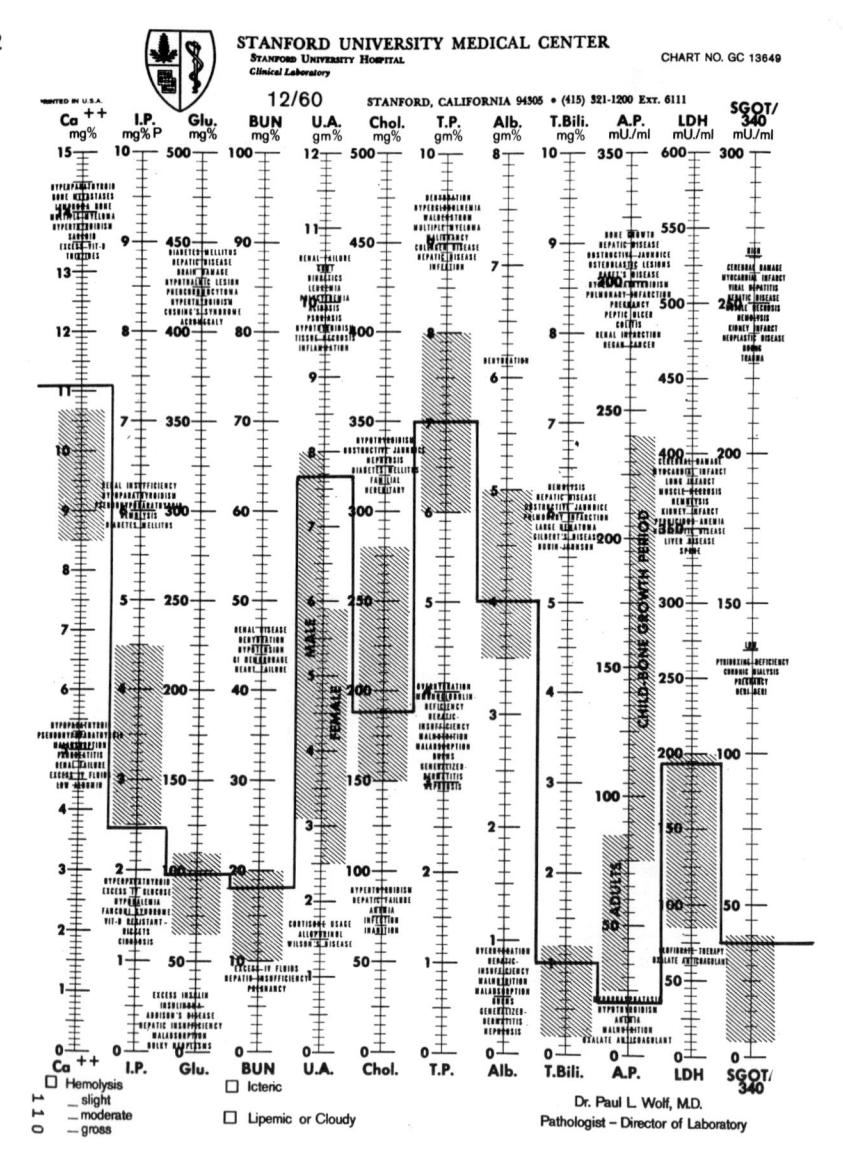

Fig. 22 - 118 BIOCHEMICAL PATTERN OF HYPOPHOSPHATASIA

Congenital hypophosphatasia will cause a bone disease which clinically may simulate the Ricket bone lesion. The prominent biochemical abnormality is low alkaline phosphatase in serum, tissues, and leukocytes. Calcium may rarely be elevated. Alkaline phosphatase in Rickets is usually prominently elevated.

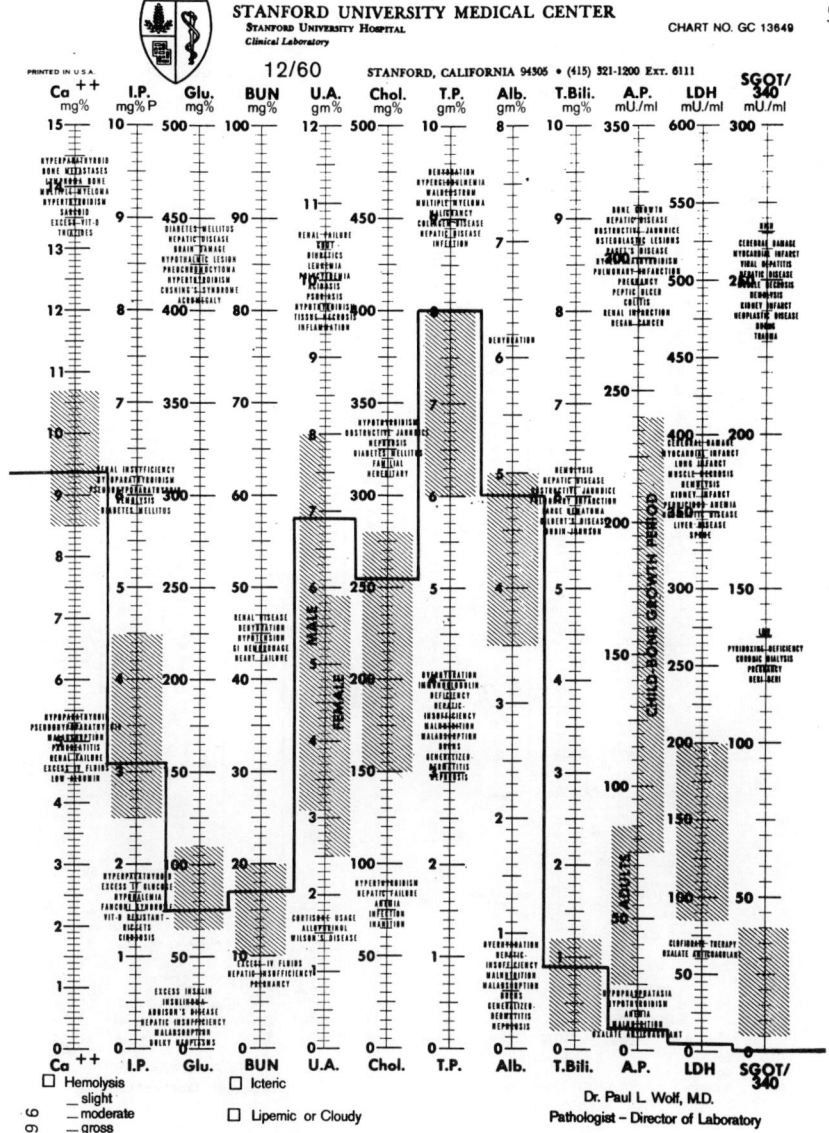

Fig. 22-119 BIOCHEMICAL PATTERN OF UTILIZATION OF OXALATE

When a plasma instead of serum is utilized for determination of enzymes, the anticoagulant, such as oxalate will inhibit enzyme function and cause extremely low values for the enzymes.

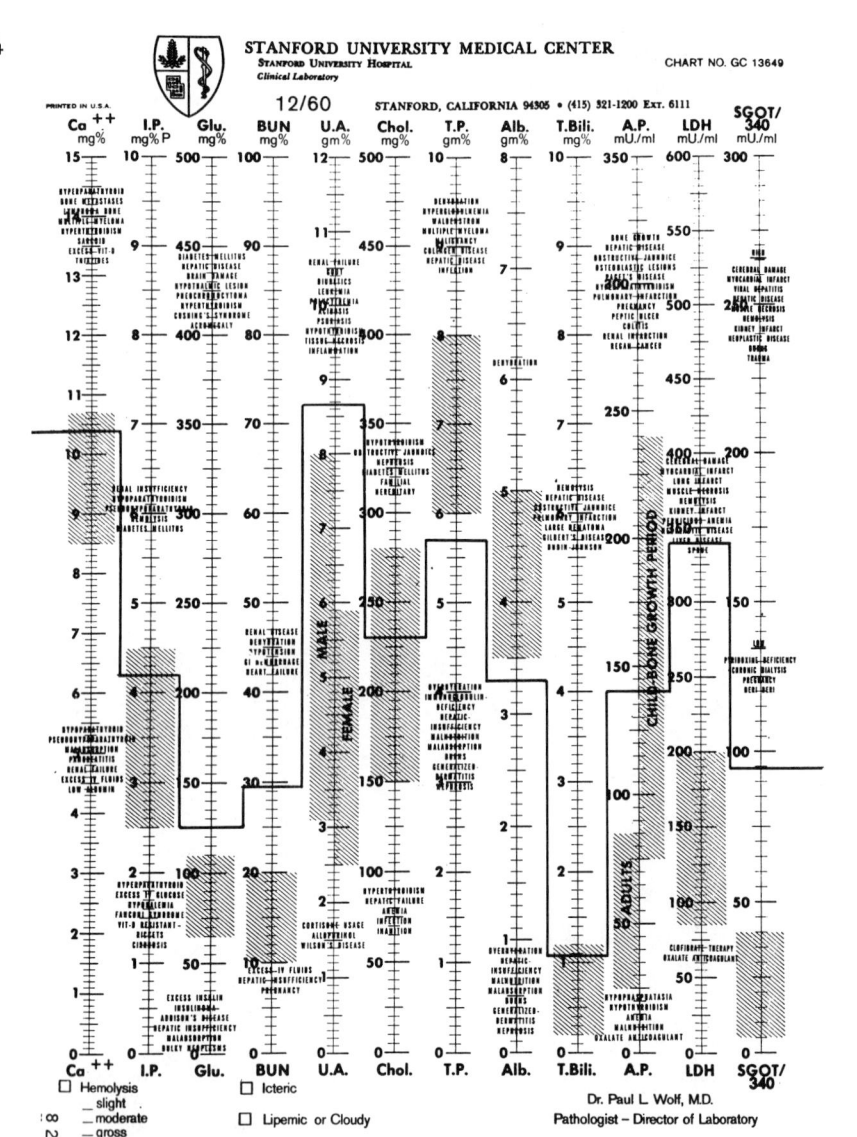

Fig. 22-120 BIOCHEMICAL PATTERN OF ACUTE RENAL INFARCTION

Acute infarction of the kidney causes a release of LDH and GOT from renal tubules. Approximately 5 to 7 days later alkaline phosphatase is also released from renal tubules. Mild renal insufficiency may occur with an increase in BUN, uric acid and glucose. Proteinuria may also be present.

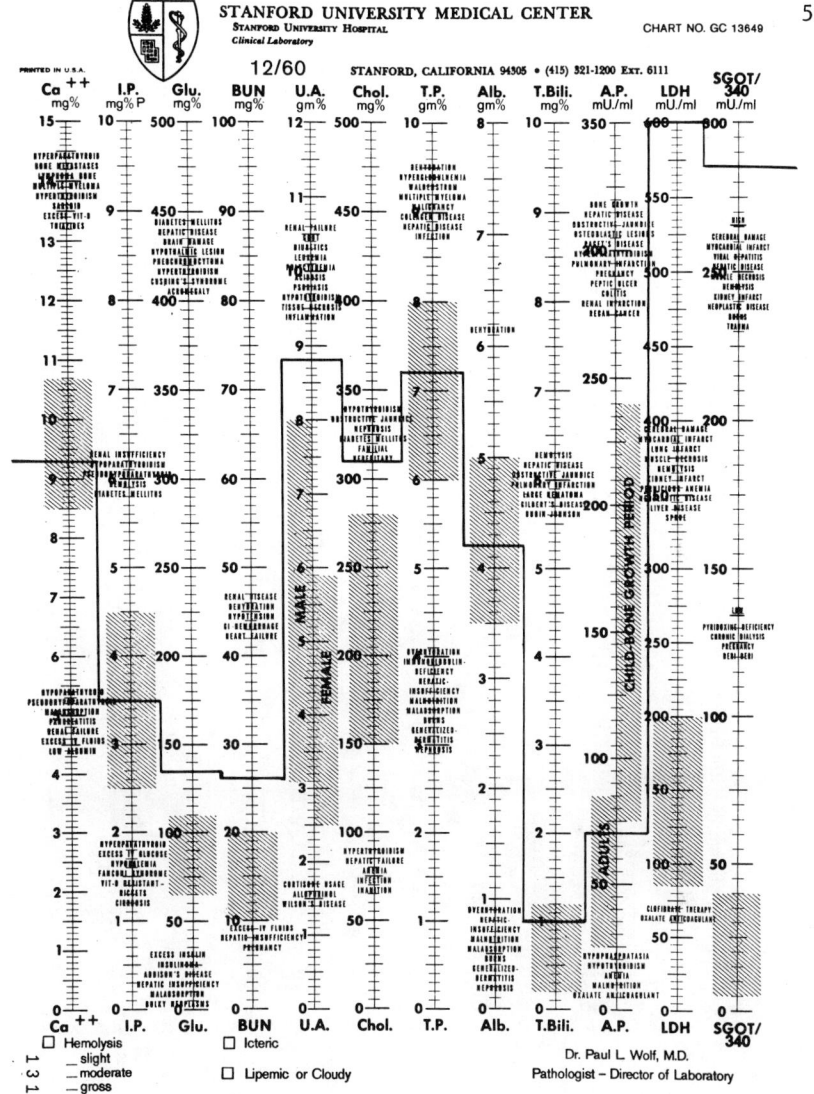

Fig. 22 - 121 BIOCHEMICAL PATTERN OF ACUTE MYOCARDIAL INFARCTION

1). Acute myocardial infarction is characterized by elevated LDH and GOT. The LDH and GOT rise within 18 - 24 hours. After the myocardial necrosis, GOT elevation persists for 5 days while elevated LDH which is usually LDH_1, may persist 10 - 14 days. 2). The elevated glucose is from adrenal cortical stress or anoxia of pancreas from heart failure. 3). Uric acid and BUN elevation is due to myocardial tissue necrosis

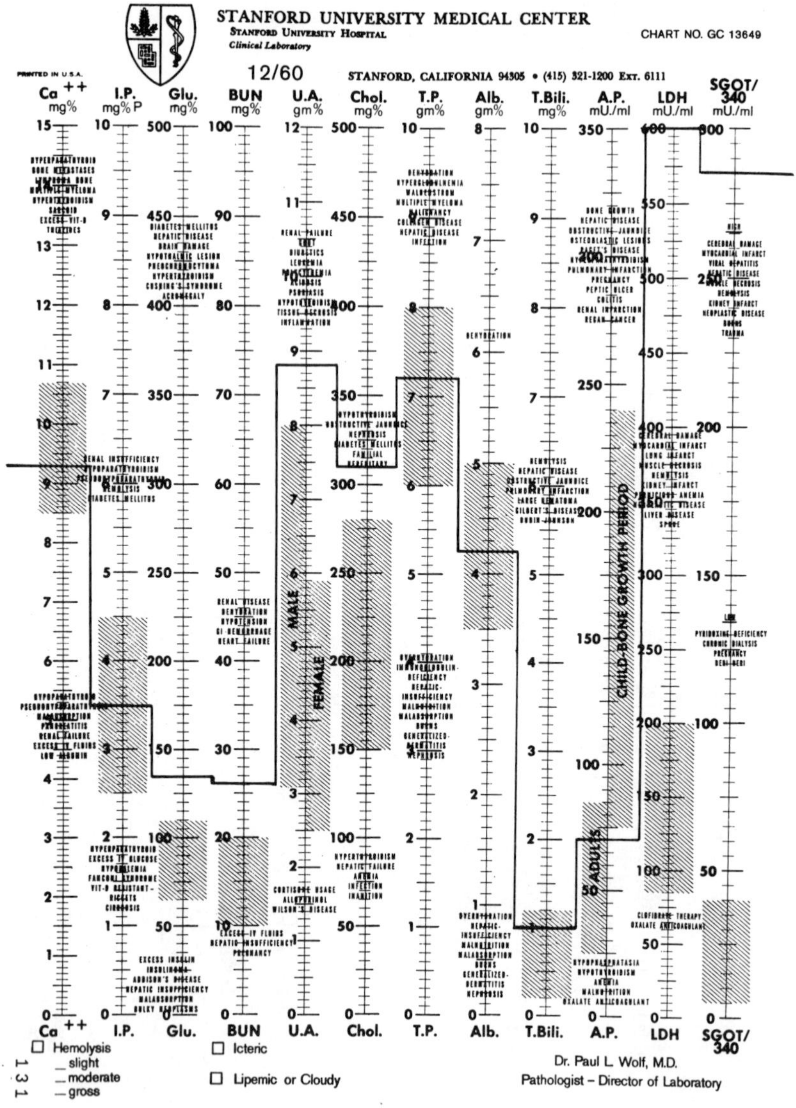

Fig. 22 -121 (Continued):

or from heart failure. 4). Elevated cholesterol is related to the arteriosclerotic heart disease, either hereditary or from various acquired causes. 5). The uric acid elevation may also be associated with elevated cholesterol. Uric acid levels are at times increased when cholesterol is high for poorly understood reasons.

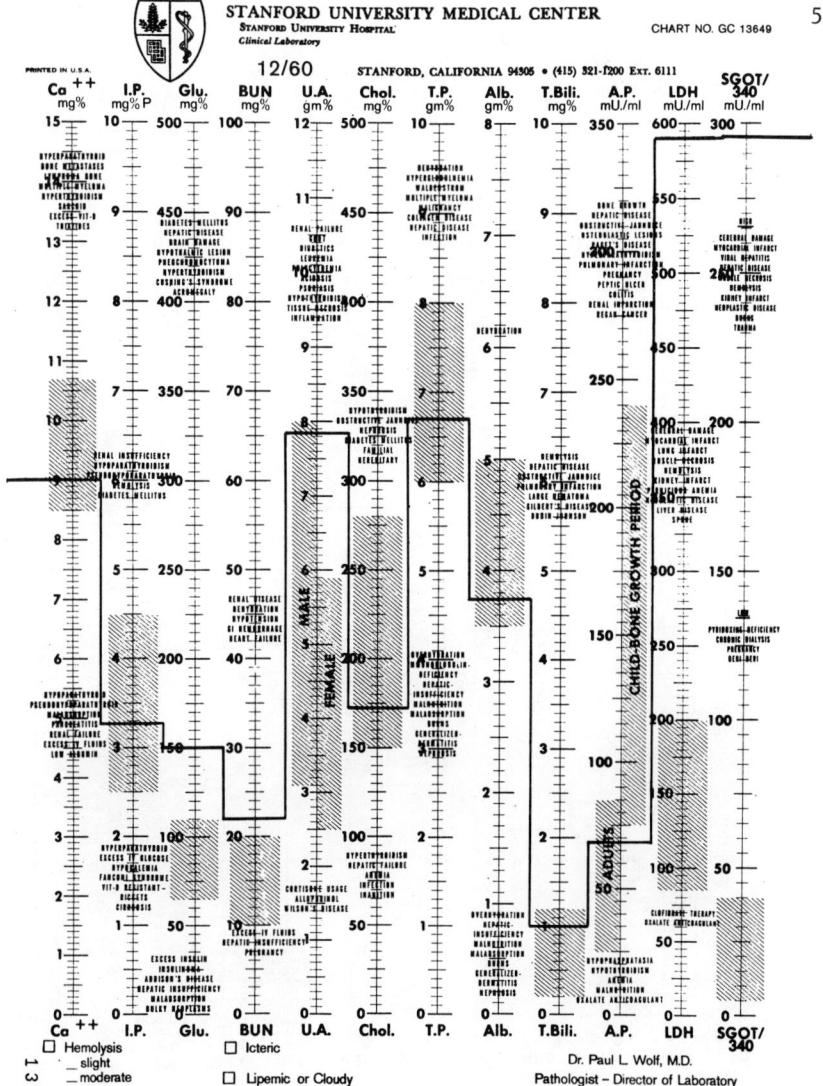

Fig. 22 -122 BIOCHEMICAL PATTERN OF RECENT CARDIAC SURGERY

LDH and GOT are elevated after surgery especially cardiac surgery. A release of these enzymes occurs from the surgical trauma to the heart and thoracic tissues, such as skeletal muscle. The elevated glucose is from the adrenal cortical stress of surgery or the intravenous glucose administered in the postoperative period.

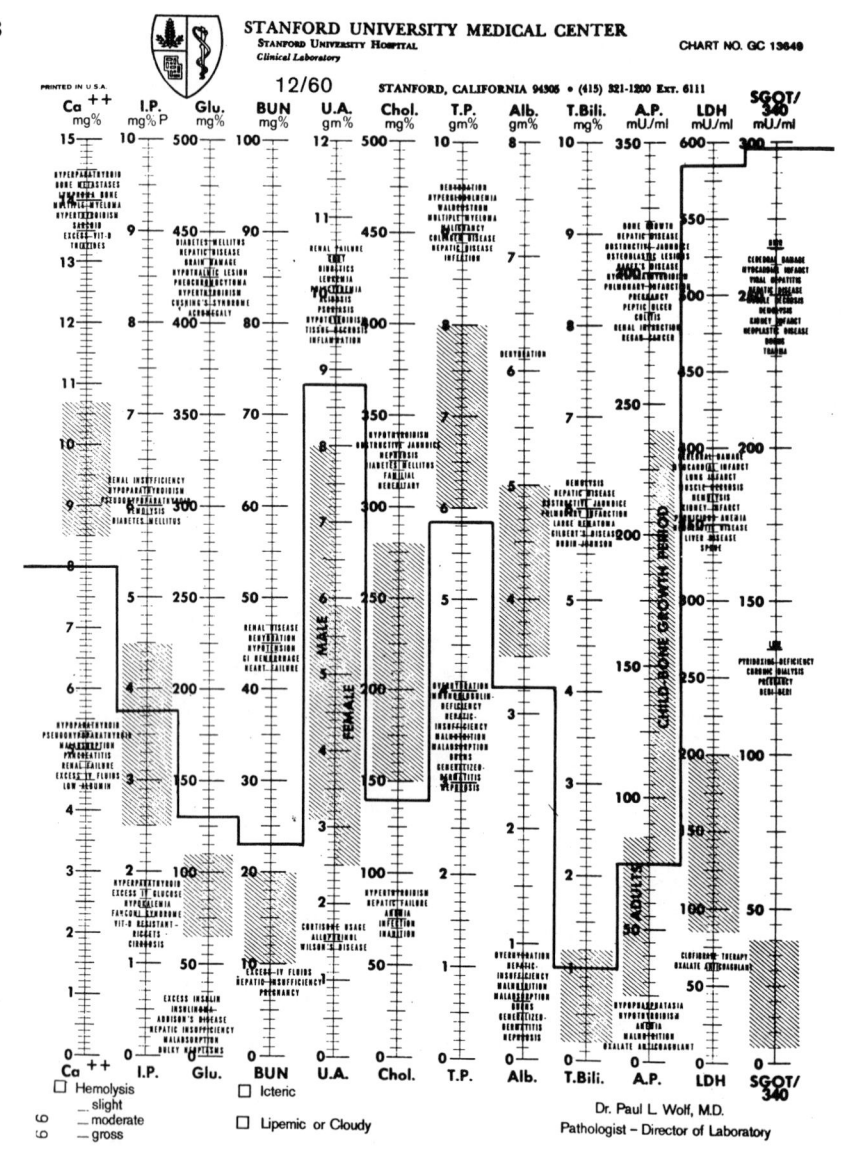

Fig. 22 - 123 BIOCHEMICAL PATTERN OF POSTOPERATIVE STATE OR TRAUMA

1). Trauma or a major surgical procedure will cause release of LDH and GOT from damaged tissues. 2). The elevated BUN and uric acid are associated with the tissue necrosis. 3). Hyperglycemia, lower serum cholesterol, and serum proteins are associated with postoperative adrenal cortical stress or intravenous glucose and lack of oral intake of protein and lipid.

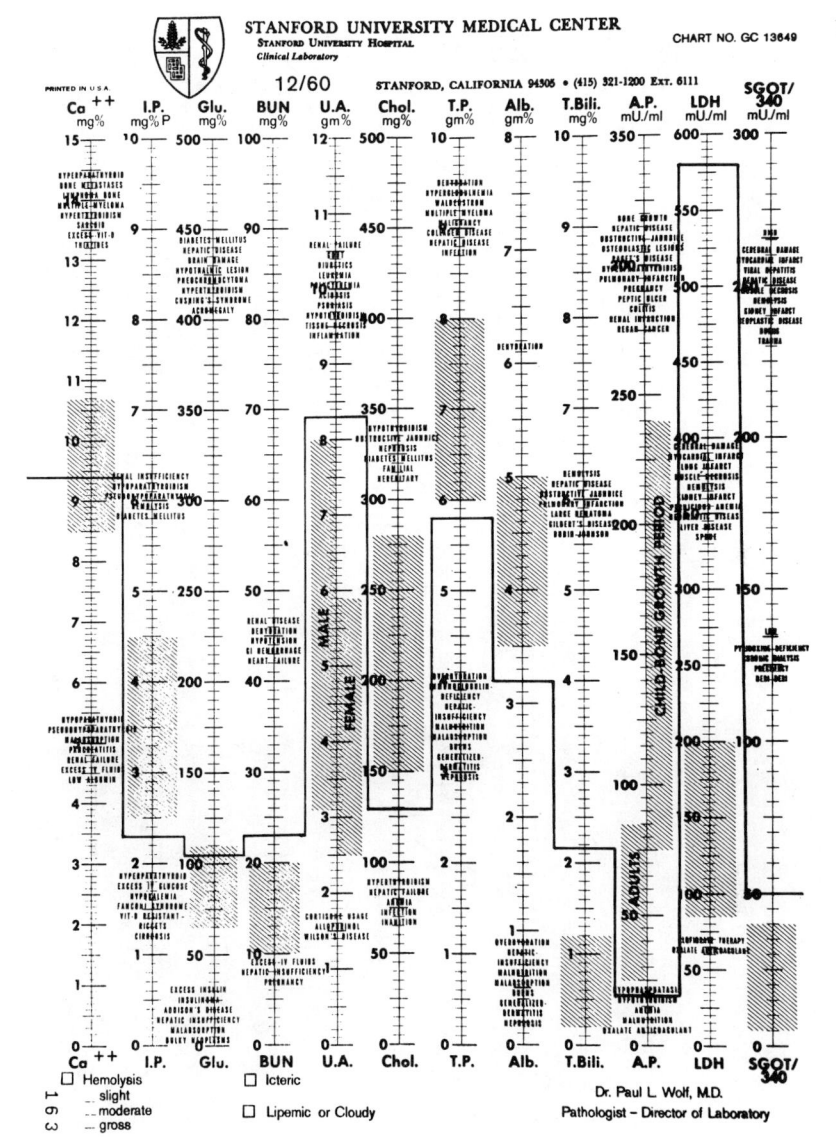

Fig. 22 -124   BIOCHEMICAL PATTERN OF PERNICIOUS ANEMIA

1). Pernicious anemia is associated with an elevated LDH, which is usually LDH_1, fast migrating type due to intra-bone marrow hemolysis of megaloblasts which contain much LDH. 2). The elevated bilirubin and GOT is from the hemolysis or fatty liver from the anemia. 3). The low cholesterol is due to the anemia as previously discussed. See Fig. 22 - 100. 6). The elevated BUN and uric acid are due to prerenal azotemia from the severe

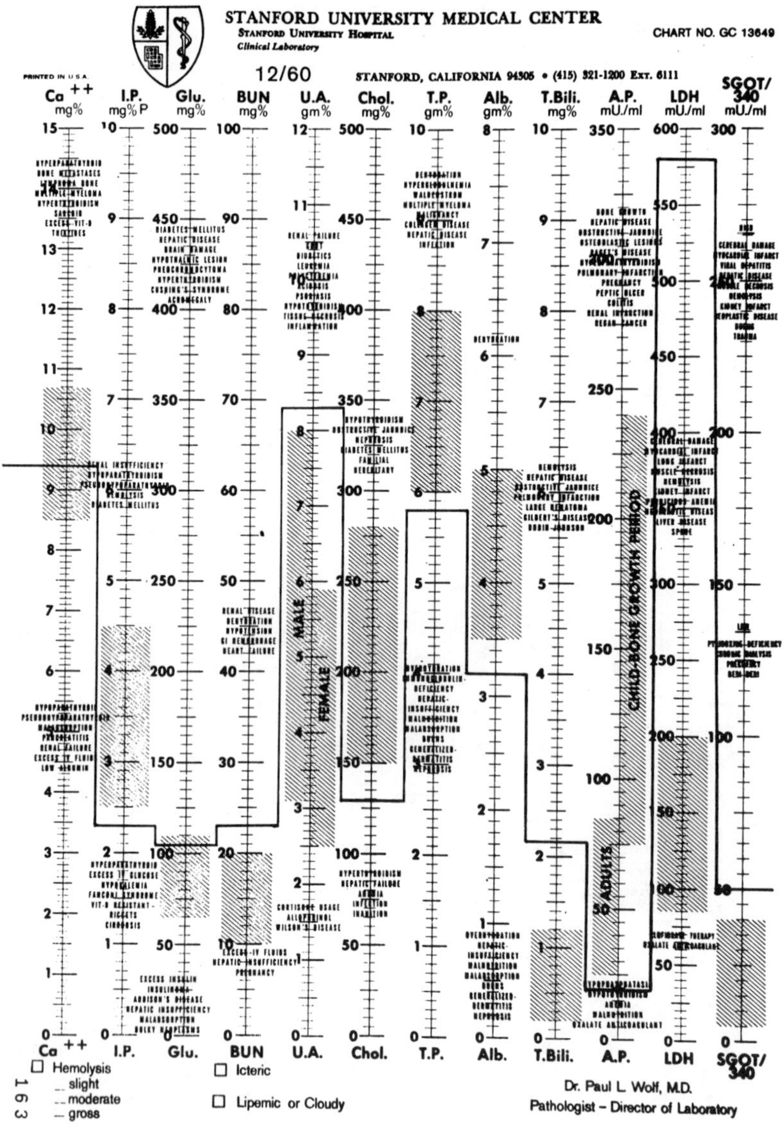

Fig. 22 - 124 (Continued):

anemia with heart failure or from fatty metamorphosis of renal tubules. 7). The low phosphate occurs from renal tubular wasting of phosphate as a result of hypokalemia. Hypokalemia results from a rapid shift of potassium into the normoblast with Vitamin B_{12} Therapy of the pernicious anemia. 8). The low alkaline phosphatase is due to decreased alkaline phosphatase with low serum Vitamin B_{12} levels.

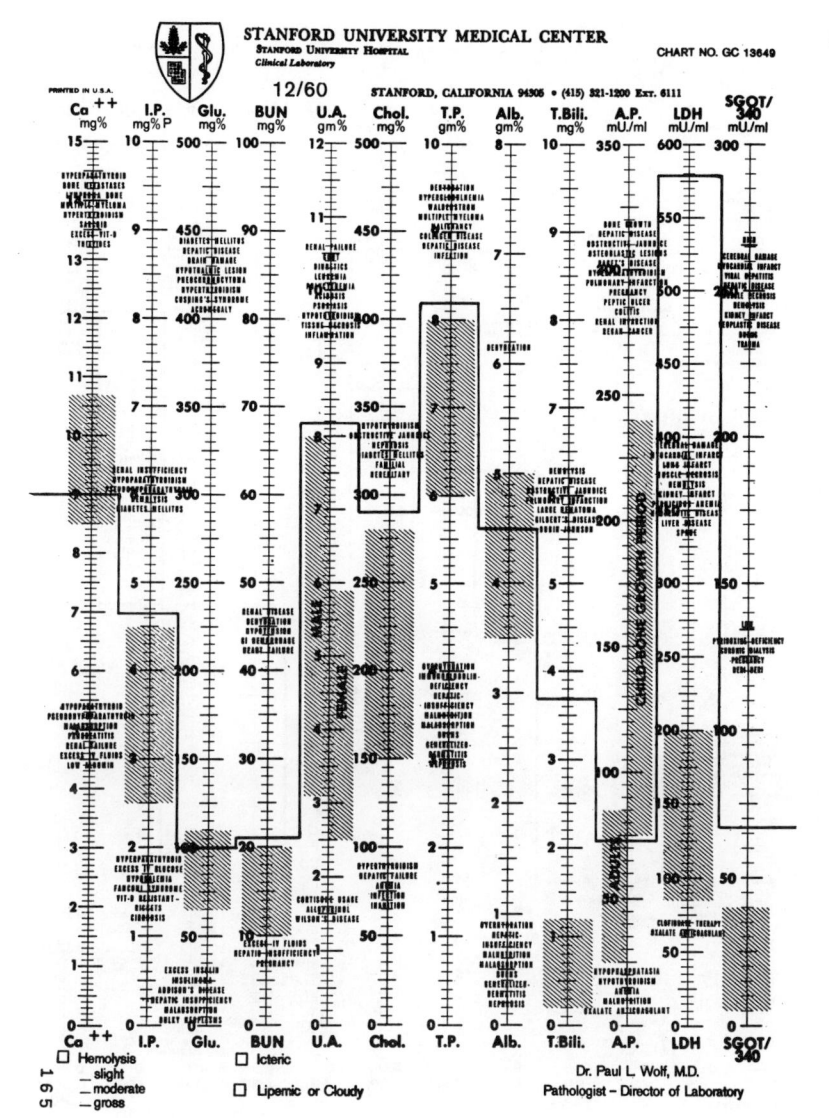

Fig. 22 - 125 BIOCHEMICAL PATTERN OF GERLACH'S RATIO LDH/GOT

Gerlach's ratio is applicable to hemolytic anemia. The normal ratio of LDH/GOT is 5:1. With hemolysis, the ratio becomes 10:1 or greater, since there is a greater amount of LDH than GOT in the erythrocyte. The elevation of phosphate, BUN, uric acid, cholesterol and protein is related to the hemolysis and release of these substances into the serum. The elevated bilirubin is the unconjugated type.

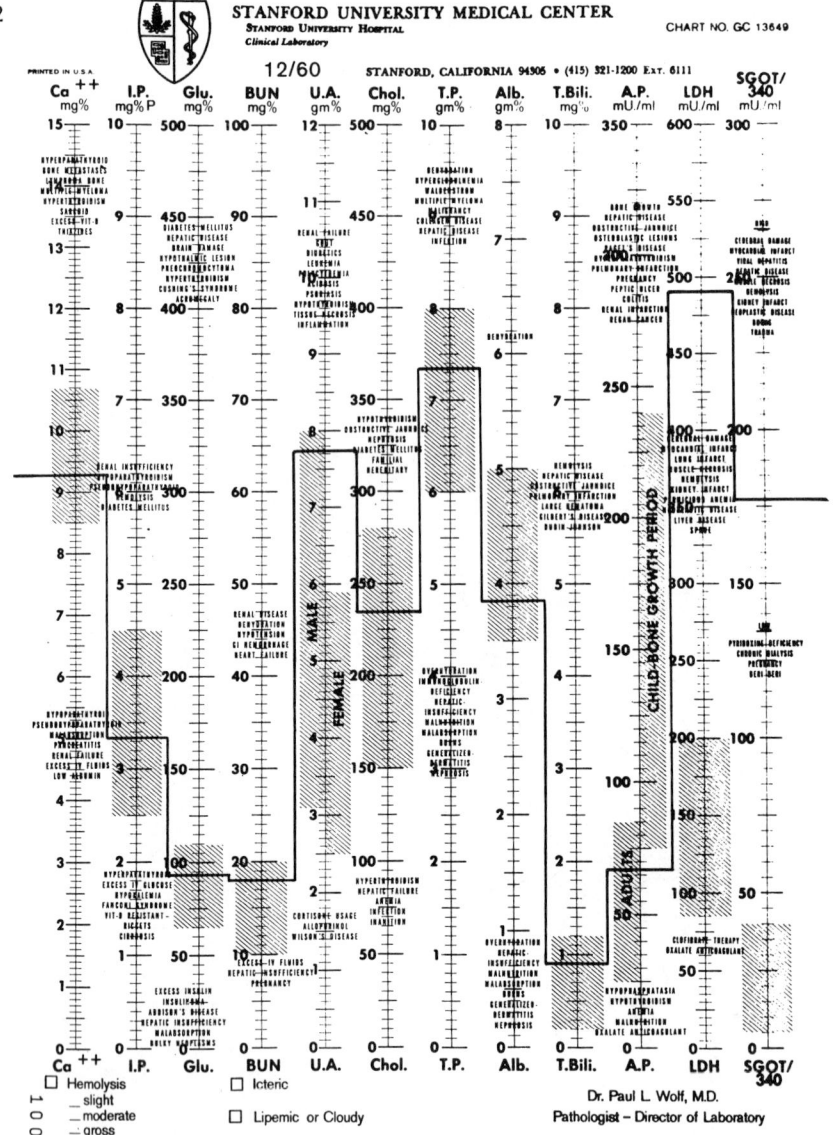

Fig. 22 - 126 BIOCHEMICAL PATTERN OF ACUTE MYOSITIS OR PRIMARY MYOPATH

In acute inflammatory myositis or primary myopathy due to rhabdomyolysis, dystrophy, or other lesions leading to skeletal muscle destruction, the LDH_5, GOT, aldolase, CPK and GPT are elevated due to release of these enzymes from the muscle fibers into the serum.

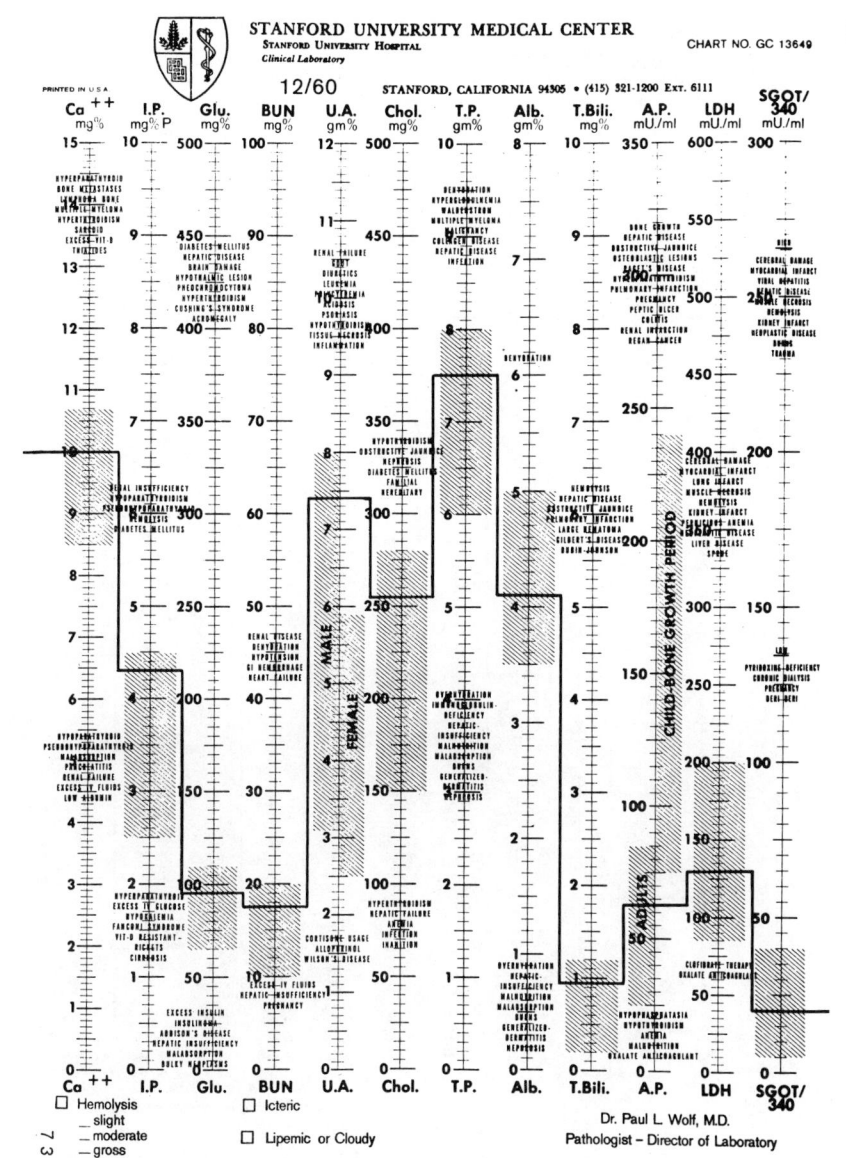

Fig. 22 - 127 BIOCHEMICAL PATTERN OF NEUROGENIC ATROPHY OR MYASTHENIA GRAVIS OF SKELETAL MUSCLE

In atrophy of skeletal muscle due to a neurogenic lesion or in myasthenia, all of the muscle enzymes LDH, GOT, GPT, CPK, and aldolase are normal.

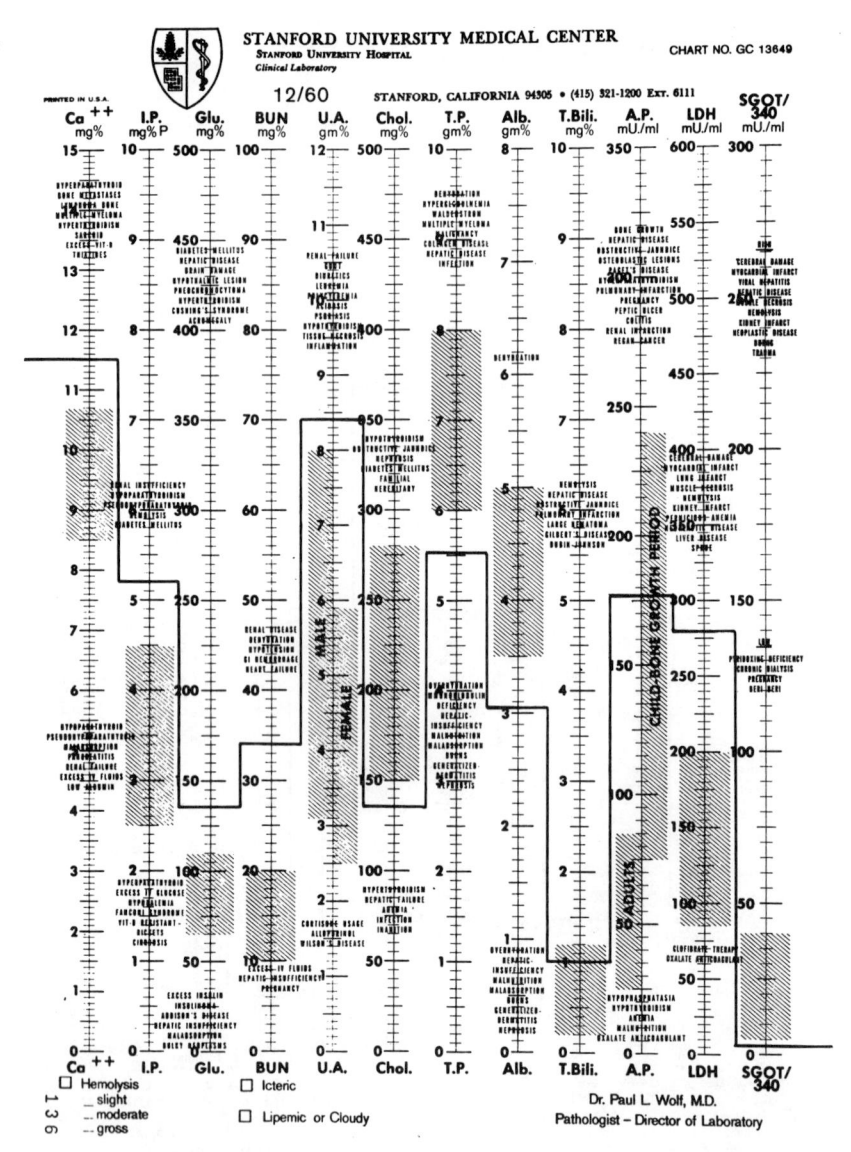

Fig. 22 - 128 BIOCHEMICAL PATTERN OF LONG TERM RENAL DIALYSIS

1). Long term renal dialysis causes a low GOT due to depletion of pyridoxine by the repeated dialysis. Pyridoxine is necessary for function of GOT. 2). Chronic renal failure usually causes an elevated phosphate due to retention with reciprocal depression of calcium. However, with repeated dialysis, the calcium may be elevated due to temporary correction of the abnormal biochemical values. 3). BUN is elevated due to retention

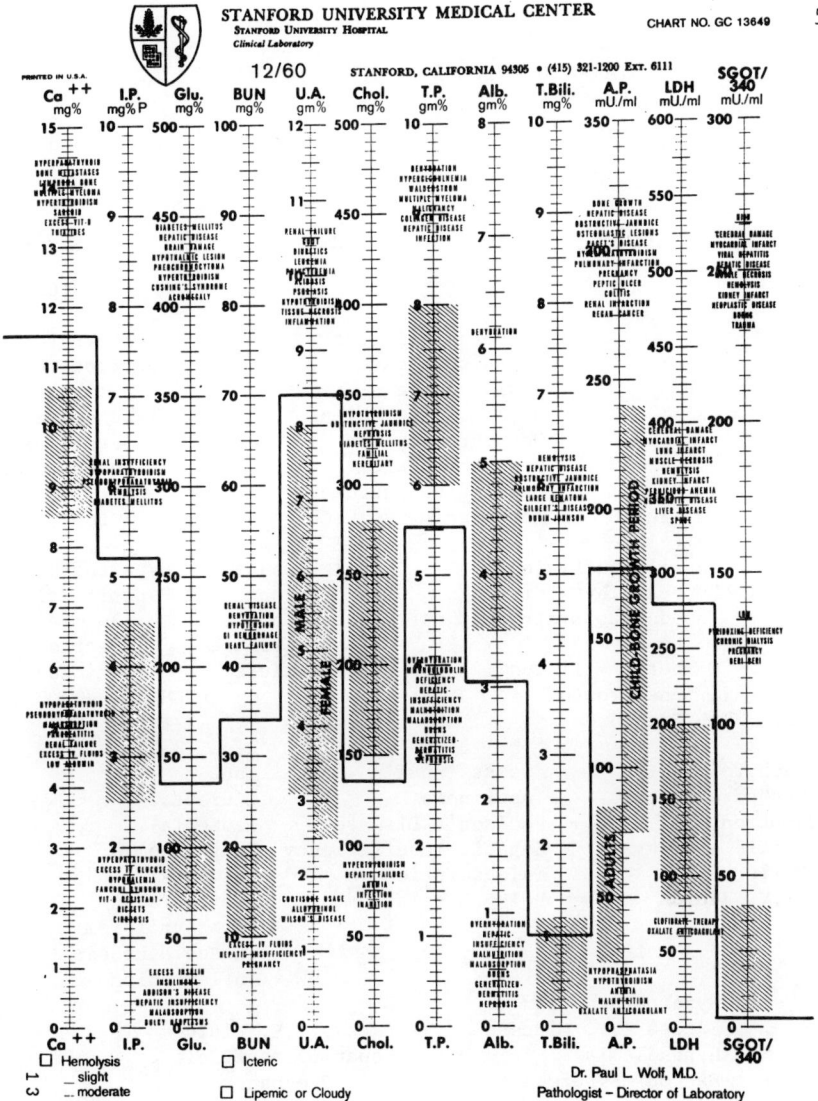

Fig. 22 -128 (Continued):

4). Albumin is decreased due to marked proteinuria.
5). Elevated phosphate, uric acid, and LDH are caused by retention in renal insufficiency. 6). Elevated glucose results from the anti-insulin action of the uremic toxins and also is due to non-specific reduction of the neocuproine chemical reagent in the 12/60 instrument by uric acid and creatinine. 7). Cholesterol is low due to anemia. 8). Alkaline phosphatase is elevated due to hyperparathyroidism.

STANFORD UNIVERSITY MEDICAL CENTER
Stanford University Hospital
Clinical Laboratory

12/60 Chart

Ca++	I.P.	Glu.	BUN	U.A.	Chol.
		HIGH			HIGH
		Diabetes Mellitus			Hypothyroidism
HIGH		Hepatic Disease			Obstructive Jaundice
Hyperparathyroid		Brain Damage			Nephrosis
Bone Metastases		Hypothalmic Lesion			Diabetes Melltius
Lymphoma Bone		Pheochromocytoma			Familial
Multiple Myeloma		Hyperthyroidism			Hereditary
Hyperthyroidism		Cushing's Syndrome			
Sarcoid	HIGH	Acromegaly	HIGH		LOW
Excess Vit-D					
Thiazides	Renal Insufficiency		Renal Disease		Hyperthyroidism
	Hypoparathyroidism		Dehydration		Hepatic Failure
	Pseudohypoparathyroid		Hypotension		Anemia
	Hemolysis		G.I. Hemorrhage		Infection
	Diabetes Mellitus		Heart Failure		Inanition
	Bone Growth			HIGH	
LOW	Fracture	LOW		Renal Failure	
Hypoparathyroid		Excess Insulin		Gout	
Pseudohypoparathyroid		Insulinoma		Diuretics	
Malabsorption		Addison's Disease		Leukemia	
Pancreatitis		Hepatic Insufficiency		Polycythemia	
Renal Failure		Malabsorption		Acidosis	
Excess I.V. Fluids		Bulky Neoplasms		Psoriasis	
				Hypothyroidism	
	LOW			Tissue Necrosis	
	Hyperparathyroid		LOW	Inflammation	
	Excess I.V. Glucose				
	Hypokalemia		Excess I.V. Fluids		
	Fanconi Syndrome		Hepatic Insufficiency		
	Vit-D Resistant-		Pregnancy		
	Rickets			LOW	
	Cirrhosis			Cortisone Usage	
				Allopurinol	
				Wilson's Disease	

A magnified view of the causes for high and low values of each determination on the Stanford Hospital Clinical Laboratory SMA 12/60 Chart.

STANFORD UNIVERSITY MEDICAL CENTER
Stanford University Hospital
Clinical Laboratory

12/60 Chart (Continued):

T.P.	Alb.	T. Bili.	A.P.	LDH	SGOT
HIGH	HIGH		HIGH		
Dehydration	Dehydration		Cerebral Damage		
Hyperglobulinemia			Myocardial Infarct		
Waldenstrom			Lung Infarct		
Multiple Myeloma		HIGH	Muscle Necrosis		
Malignancy			Hemolysis		
Collagen Disease		Hemolysis	Kidney Infarct		
Hepatic Disease		Hepatic Disease	Pernicious Anemia		
Infection		Obstructive Jaundice	Neoplastic Disease	HIGH	
		Pulmonary Infarction	Liver Disease		
		Large Hematoma	Sprue	Cerebral Damage	
		Gilbert's Disease		Myocardial Infarct	
		Dubin-Johnson	HIGH	Viral Hepatitis	
				Hepatic Disease	
				Muscle Necrosis	
	LOW	Bone Growth	Hemolysis		
		Hepatic Disease	Kidney Infarct		
LOW	Overhydration	Obstructive Jaundice	Neoplastic Disease		
	Hepatic-	Osteoblastic Lesions	Burns		
	Insufficiency	Paget's Disease	Trauma		
Overhydration	Malnutrition	Hyperparathyroidism			
Immunoglobulin-	Malabsorption	Pulmonary Infarction			
Deficiency	Burns	Pregnancy	LOW		
Hepatic-	Generalized-	Peptic Ulcer			
Insufficiency	Dermatitis	Colitis	Pyridoxine Deficiency		
Malnutrition	Nephrosis	Renal Infarction	Chronic Dialysis		
Malabsorption		Regan Cancer	Pregnancy		
Burns			Beri Beri		
Generalized-		LOW			
Dermatitis					
Nephrosis		Hypophosphatasia			
		Hypothyroidism			
		Anemia			
		Malnutrition			
		Oxalate Anticoagulant			

LOW

Clofibrate Therapy
Oxalate Anticoagulant

INDEX

Acetylcholinesterase, 313
Acholest test paper, 32, 231, 313
Acid phosphatase
  blood platelets, 265
  cancer of the prostate, 190, 323
  causes of decreased levels, 111
  cerebrospinal fluid, 110, 307
  childhood, 362
  cryptorchidism, 335
  estrogen therapy, 190
  ethanol inhibition, 116
  fever, 117
  formaldehyde inhibition, 114, 116
  Gaucher's disease, 111, 114, 193, 363, 391
  hemolytic anemia, 115
  hyperparathyroidism, 110
  Klinefelter's syndrome, 335
  methodology, 8
  Niemann-Pick disease, 111
  normal values, 11, 14
  orchiectomy, 112, 190
  osteogenic sarcoma, 193
  Paget's disease, 110, 193
  rectal examination, 112
  red blood cells, 266
  semen, 381
  sexual intercourse, 209
  synovial fluid, 382
  L-tartrate inhibition, 113, 191, 314
  thrombocytopenic purpura, 113, 193
  thromboembolism, 115
  trisomy, 116
  urine, 335
  vaginal fluid, 381
Acromegaly, panel, 476
Addison's disease, panel, 415, 416, 448

Adenosine triphosphatase, 276
Albumin from placenta, panel, 451
Alcohol dehydrogenase, 236
Aldolase
  adrenal gland, 176
  infants, 364
  method, 17
  muscular dystrophy, 176
  myasthenia gravis, 176
  neonatal period, 176
  neurogenic atrophy, 176
  newborns, 176, 297
  normal values, 18
  viral hepatitis, 176, 226
Alkaline phosphatase
  anticoagulants, 349
  Bessey-Lowry method, 22
  congenital hyperphosphatasia, 346
  causes of decreased levels, 109
  congestive heart disease, 218
  electrophoresis, 104
  endometrium, 101
  Ewing's sarcoma, 388
  heat stability, 105
  hypophosphatasia, 109
  hypothyroid, 110, 349
  increased synthesis, 218
  infants and children, 362
  infarction of the kidney, 109
  infectious mononucleosis, 218
  infiltrative liver disease, 217
  inhibition by urea, 104
  intestinal, 102
  intravenous serum protein from placenta, 103
  magnesium deficiency, 109
  malabsorption syndrome, 108
  neuroblastoma, 388
  obstructive jaundice, 215
  Osteitis deformans of children, 346

Index:

Alkaline phosphatase (Continued):
  osteomalacia, 102
  osteopetrosis, 105, 108, 346
  oxalate, 349
  Paget's disease, 106, 344
  pediatric patients, 100
  peptic ulcer, 102
  pernicious anemia, 110
  L-phenylalanine inhibition, 343
  placenta, 206
  plasma expanders, 207
  post-menopausal senile osteoporosis, 346
  pulmonary infarction, 109
  rickets, 106, 109, 345, 347
  secretors, 103
  storage, 5, 101
  subtotal gastric resection, 108
  ulcerative colitis, 102
  Vitamin B_{12} deficiency, 348
  Vitamin D deficiency, 106
Allantoin, 97
Allopurinol, panel, 516
Aminopeptidases
  urine, 332
Aminotripeptidase
  synovial fluid, 382
Amylase
  abdominal paracentesis, 131, 249
  acute choledocholithiasis, 315
  acute pancreatitis, 247
  alpha amylase, animal, 23, 247
  beta amylase, plants, 247
  causes of elevated levels, 129
  children, 369
  decreased serum levels, 134
  ectopic pregnancy, 131, 249, 316
  hemolytic anemia, 250
  hepatic necrosis, 132, 250
  intestinal obstruction, 317
  malabsorption, 250
  obstruction of the Sphincter of Oddi, 315
  pancreatitis, 23, 129

Amylase (Continued):
  parotid glands, 23, 251
  parotid sialoadenitis, 134
  peritoneal fluid, 380
  renal insufficiency, 132, 250
  thoracic duct, 380
  tubal ovarian abscess, 316
  urine, 23, 25, 131, 250, 328
Amyotonia, congenital, 354
Amyotrophic lateral sclerosis, 354
Andersen's disease Cori Type IV, 386
Anemia, panel, 526
Angina pectoris, 291
Antacid utilization, panel, 482
Antitrypsin
  alpha$_1$ globulin, 290
Antitryptic factors, 255
Arylsulfatase, 391
  urine, 330
Aspartate, 76, 86
ATPase
  histiocytic lymphoma, 388

Bacterial infection, panel, 508
Bessey-Lowry units, 8
Biliary atresia
  elevation of LAP_1 and LAP_2, 219
Biliary obstruction
  increased enzyme synthesis, 2
Bilirubin influence on serum albumin, panel, 546
Bilirubin influence on cholesterol, panel, 524
Blood-brain barrier, 302
Bodansky, 100
Bone fracture, recent, panel, 477
BSP dye artifact on serum, panel, 532
Bullous Dermatitis, panel, 536
BUN/Creatinine ratio, panel, 417, 506
Burnett's syndrome, panel, 443
Burns, panel, 536

Index:

Calcium, elevated, panel, 459
Calcium, elevation of protein,
   panel, 457
Cancer metastatic to bone, panel,
   438
Cancer metastatic to liver, panel,
   547
Carcinoma of the lung
   thrombocytosis, 288
Cardiac patients
   GOT/GPT ratio, 221
Catalase
   urine, 329
Central nervous system disease
   LDH_2 and LDH_3, spinal fluid, 162
Ceramide trihexoside
   Fabry's disease, 391
Cerebral damage, panel, 425, 492
Ceruloplasmin
   azide reagent, 26
   blue protein, 232
   estrogens, 234
   ferroxidase, 232
   green color of plasma, 234
   normal values, 25, 28
   oxidase activity, 28
   rheumatoid arthritis, 235
   Wilson's disease, 232, 364
Cervical cancer
   6-phosphogluconic acid dehydro-
      genase, 210
Charcot-Marie-Tooth, 354
Cholestasis, drug induced, 215
Cholestatic jaundice, panel, 493
Chronic pancreatitis
   provocative enzyme tests, 252
Choline, 32
Cholinesterase, 32, 33
   acetylthiocholine substrate, 29
   dithiobisnitrobenzoic acid, 29
   erythrocyte, 230
   nephrotic syndrome, 228
   nervous system, 230
   normal values, 31

Cholinesterase (Continued):
   thiocholine, 29
   viral hepatitis, 228
Choriocarcinoma, 161
Chylous effusion, 194, 377
Cirrhosis, panel, 430, 483
Cirrhosis - Low cholesterol,
   panel, 528
Congestive heart failure
   elevated LDH_5, 224
   panel, 491
Conn's syndrome, panel, 405, 406
Contamination of serum by detergent, panel, 460
Cortisone, excess, panel, 474
Creatine, 34
Creatine Phosphokinase
   amniotic fluid, 357
   arrhythmias, 171
   asymptomatic carriers, 172
   brain, 174
   carrier state in muscular dys-
      trophy, 357
   causes of decreased levels, 176
   central nervous system, 303
   cerebrospinal fluid, 175
   children, 366
   clofibrate, 296
   cysteine addition, 296
   diabetic acidosis, 296
   Duchenne's muscular dystrophy,
      357
   electrical defibrillation, 171
   hypoparathyroidism, 296
   hypothyroidism, 173
   intramuscular injections, 318
   low serum potassium, 296
   lung, 296
   malignant hyperthermia, 172
   muscular dystrophy, 171, 172
   myasthenia gravis, 172
   myocardial infarction, 295
   neurogenic muscle atrophy, 172
   normal values, 33, 36, 170

Index:

Creatine phosphokinase (Cont.):
  pregnant myometrium, 175, 358
  psychoses, 304
  schizophrenia, 174, 304
  storage, refrigerator, 5, 296
  umbilical cord blood, 175
Crigler-Najjar syndrome, panel, 539
Cushing's syndrome, panel, 405, 407, 447
Cysteine
  thiol-stimulated, 170
Cystic fibrosis, pancreas, 255
  exocrine pancreatic secretion, 371
Cystine aminopeptidase
  decline, 121
  urine, 332

Dehydration, panel, 424, 454
DeRitis
  GOT/GPT, 3, 221
  inflammation, less than one, 3
Diabetic ketoacidosis, panel, 402
Diabetes mellitus, panel, 487
Diet with ketoacidosis, panel, 502
Digitonin, 44
2,3-Diphosphoglycerate levels, 277,. 482
Disseminated intravascular coagulation, 264
Diuretic utilization, panel, 408
Drug induced liver disease, panel, 493
Dubin-Johnson syndrome, panel, 540
Duodenal secretions, 255
Dystrophia myotonica, 355

Eclampsia, panel, 503
Effusion-serum LDH ratio, 164
Emphysema
  alpha$_1$ anti-trypsin, 290

Enterokinase, 255
Enzyme
  cholelithiasis, 222
  chronic cholecystitis, 222
  clearance, 4
  content in tissues, 3
  excretion routes, 4
  half-life, 5
  increased synthesis in damaged cells, 4
  intravascular inactivation, 4
  small intestinal removal, 4
  ultraviolet light method, 5
  removal, vascular organs, 5
Enzymes, cerebrospinal fluid
  acid phosphatase, 307
  bacterial meningitis, 304
  demyelinating diseases, 306
  beta-glucuronidase, 307
  Korsakoff's psychosis, 306
  multiple sclerosis, 306
  neoplasms, 307
  ribonuclease, 306
  transketolase, 306
  tuberculous meningitis, 305
  viral meningocephalitis, 305
  Wernick's hemorrhagic encephalopathy, 306
Erythrocyte enzyme deficiencies, 272
Esterase
  leukemia, 388
Estrogen, increase, 429
Evaporation of specimen, panel, 455
Exudate, panel, 163, 194, 377, 533

Forbe's disease Cori Type III, 386

Galactosemia
  galactose-1-pyridyl transferase, 392
Alpha-Galactosidase
  Fabry's disease, 391

Index:

Beta-Galactosidase
  Hurler-Hunter syndrome, 391, 392
  synovial fluid, 382
Gangliosidosis
  GM_1 beta-galactosidase, 391
  GM_1 ganglioside, 391
Gastric Juice
  beta-glucuronidase, 317
  LDH, 162
Gastrointestinal hemorrhage,
  panel, 426
Gaucher's disease, panel, 458,
  530
Gerlach's ratio, 156
  Gerlach's LDH/GOT ratio, 264,
    314
  panel, 561
Gilbert's syndrome, panel, 539
Glucocerebrosidase
  Gaucher's disease, 391
Gluconic acid, 40
Gluconolactone, 40
Glucose excess, panel, 419, 427
Glucose oxidase, 40
  normal values, 43
Glucose-6-phosphate, 44
  normal serum level, 186
  viral hepatitis, 236
Glucose-6-phosphate dehydrogenase
  acetylphenylhydrazine, 185
  analgesics, 185
  antimalarials, 185
  assay methods, 275
  brilliant cresyl blue, 185
  cyanide and ascorbate test, 275
  diabetic acidosis, 185
  dye reduction test, 275
  Embden-Meyerhof Cycle, 183
  fluorescence screening test, 275
  glutathione stability test, 275
  Heinz body test, 275
  methemoglobin reduction test,
    275
  nitrofurantoin, 185

Glucose-6-phosphate dehydrogenase
  (Continued):
  normal values, 46
  Pentose-monophosphate shunt, 183
  reticulocytes, 183
  sulfa drugs, 185
  tetrazolium cytochemical method,
    275
  viral hepatitis, 185
Glucose phosphate isomerase, 278
Beta-Glucuronidase
  bilharziasis, 334
  carcinoma of the uterine cervix,
    199
  gastric juice, 380
  liver, 236
  malignancy, 387
  "premalignant" colonic polyps,
    387
  synovial fluid, 382
  urine, 332, 333, 335
  vaginal fluid, 381
Glucuronyl transferase, 539
Glutamic dehydrogenase
  half-life, 5
  liver, 225
  mitochondrial, 3, 47
  normal values, 49
Glutamic oxalacetic transaminase
  angina pectoris, 141
  angiography, 141, 292
  arrhythmia, 141
  burn, 146
  carbon tetrachloride necrosis,
    143
  cardiac catheterization, 292
  cardioversion, 292
  causes of decreased levels, 148
  causes of elevated levels, 140
  cerebrospinal fluid, 144
  choledocholithiasis, 222
  congestive heart failure, 292
  DeRitis ratio, 143
  diabetic acidosis, 148

Index:

Glutamic oxalacetic transaminase
  (Continued):
  direct-current electrical counter shock, 141
  drug effects, 148
  electrical defibrillation, 292
  erythromycin, 148, 370
  fatty metamorphosis, liver, 220
  half-life, 5
  heart disease, 293
  hepatic cell mitochondria, 221
  hydrazine, 148
  hydrazone, 148
  infants, 370
  infectious mononucleosis, 145, 222
  injections of drugs, 293
  intrabiliary pressure increase, 293
  leukemia, 146
  location in cell, 3
  myocarditis, 141
  muscular dystrophies, 147
  narcotics, 293
  normal values, 87
  pericarditis, 142
  phenothiazines, 148
  surgery on the myocardium, 294
  tachycardia, 141
  urine, 220
  viral hepatitis, 220
Glutamate pyruvate transaminase, 89
  $\frac{GOT + GPT}{GLDH}$, 221
  drugs, 148
  half-life, 5
  infants, 370
  normal values, 90
  viral hepatitis, 1
Gamma-Glutamyl-p-nitroanilide, 37
Gamma-Glutamyl transpeptidase, 37, 121
  liver disease, 226

Gamma-Glutamyl-transpeptidase
  (Continued):
  multiple myeloma, 388
  normal values, 39
  specific enzyme, 1
Glutathione deficiency, 276
Glutathione peroxidase, 276
Glutathione reductase, 276
Glutathione stability test, 184
Glyceraldehyde-3-phosphate dehydrogenase, 278
Glycerokinase, 91
Glycero-1-phosphate, 91
Gout, panel, 512

Heart failure, panel, 423
Heart surgery, recent, panel, 557
Hemodilution, panel, 466
Hemolysis, panel, 418, 485
Hemolytic anemia
  heat stable LDH_1 and LDH_2, 263
  hereditary, 272
  nonspherocytic, 272
  Primaquine, 44
Hepatic congestion
  elevation of GPT, 4
  elevation of sorbitol dehydrogenase, 4
Hepatitis
  LAP_1, 219
  LDH_5, 224
  neonatal, 219
Hepatitis associated antigen (HAA), 222
Hepatorenal syndrome, panel, 504
Hers' disease Cori Type VI, 386
Hexokinase, 277
Hodgkin's disease involving bone, panel, 440
Hurler's syndrome
  chondroiten sulfate B, 386
  beta-galactosidase deficiency, 386
  GM_3, GM_2 and GM_1, 386

Index:

Hurler's syndrome (Continued):
  heparin sulfate, 386
Hydatidiform mole, 161
Alpha-Hydroxybutyrate Dehydrogenase, 50, 156, 157, 295
  characteristics, heart muscle, 1
  normal values, 51
3-beta-Hydroxy steroid dehydrogenase
  adrenal, 388
Hyperalimentation, panel, 452
Hyperkalemia, panel, 418, 424
Hyperkalemic periodic paralysis, panel, 412
Hyperlipemia, panel, 421
Hypernatremia, panel, 425
Hyperparathyroidism, panel, 413, 437
Hyperthyroidism, panel, 525
Hyperventilation, panel, 429
Hypervitaminosis D, panel, 446
Hypogammaglobulinemia, panel, 538
Hypokalemia, panel, 405
Hypokalemic periodic paralysis, panel, 412
Hypoparathyroidism, panel, 463
Hypophosphatasia, 346, 347
  panel, 552
Hypothyroidism, panel, 523

Idiopathic thrombocytopenic purpura, 266
Inappropriate ADH, panel, 420
Indoxyl substrates, 390
Infectious mononucleosis with acute hepatitis, panel, 544
Iodometric method of Caraway, 23
Insulin, panel, 427
Insulin, excess, panel, 428, 497
Intravenous glucose, panel, 466
Isocitrate dehydrogenase, 53, 227
  heart, heat labile, 182
  hepatic disease, 182
  infants, 368

Isocitrate dehydrogenase (Cont.)
  megaloblastic anemia, 182
  mitochondria, 182
  normal range, 55, 183
Isomaltase, 367

Kallikrein
  urine, 330
Ketoacidosis, panel, 480, 487
Ketoglutarate, 76
King-Armstrong, 100

Lactase deficiency, 366, 367
Lactate, 56, 59
Lactic dehdyrogenase, 294
  A and B subunits, 154
  acute myocardial infarction, 295
  aerobic tissues, 154
  anaerobic tissues, 154
  causes of decreased levels, 164
  causes of elevated levels, 156
  diabetic glomerulosclerosis, 326
  differentiation of LDH_1 from LDH_5, 156
  effusion LDH:serum LDH ratio, 195, 378
  gastric juice, 317, 380
  H and M subunits, 154
  heart transplant
    Bernard, 157
    Cooley, 157
    Stinson, 157
  heat stability-lability, 155, 156
  hemolytic anemia, 161
  hemolytic anemia, cardiac valves, 317
  hepatic parenchymal cells, 224
  infants, 367
  inhibitors, 156
  liver isoenzyme, heat labile, 295
  half-life, 50 hours, 5
  megaloblastic anemia, 160, 263
  LDH_1 heat stable, 295

Index:

Lactic dehydrogenase (Cont.):
    LDH_1 inactivated by urea, 295
    LDH_5 inactivated by urea, 295
    LDH isoenzymes, normal values, 59, 66
    monomers, 154
    normal values, 58
    osteoarthritis, 381
    pregnancy, 161
    pulmonary infarction LDH_3, 157
    ratio of LDH_1/LDH_2, 295
    renal transplant patients, 325
    saliva, 255, 380
    synovial fluid, 381
    tetramer, 154
    Wacker method, 56
Leucine, 68
Leucine aminopeptidase, 118
    biliary atresia, 369
    electrophoresis, biliary atresia, 119
    Goldbarg and Rutenburg, 68
    infiltrative lesions of liver, 219
    jaundice of newborn, 118
    kinetic method of Nagel, 71
    neonatal giant cell hepatitis, 118, 369
    obstruction of bile ducts, 117
    Rose-Bengel test, 118
    ulcerative colitis, 120
Leukemia, panel, 500
Leukocyte alkaline phosphatase
    causes of increased levels, 268
    hypophosphatasia, 270
    myeloid cell, 267
    myelogenous leukemia, 270
    pregnancy, 101
    primary polycythemia, 268
    secondary polycythemia, 268
1-Leucyl-beta-naphthylamide, 68
2-Leucyl-beta-naphylamide, 118
Licorice use, panel, 410

Lipase, 195, 252
    acute pancreatitis, 135
    children, 369
    chronic pancreatitis, 136
    coconut oil substrate, 254
    decreased levels, 254
    elevated levels, 134, 253
    fat embolism of lungs, 135, 254, 289, 318
    lipoprotein, nephrotic syndrome, 236
    normal values, 72, 75
    olive oil substrate, 73, 254
    tributyrin substrate, 254
    triglyceride emulsion, 73
    turbidimetric method, 73
    urine, 135, 254
Liver disease
    BSP excretion, 217
    errors in enzymes, 237
    LDH_4 and LDH_5, 158
    no CPK, 4
Liver guanase, 225
Lung infarct, panel, 541
Lysozyme
    urine, 328

Macroamylase, 133, 251
Macroamylase macromolecule
    IgA or IgG immunoglobulin, 250
Malabsorption, panel, 409, 467
Malate dehydrogenase, 76
    normal values, 78
Malignant neoplasms
    LDH_2, LDH_3, LDH_4 and LDH_5, 159
    panel, 515
"Malignant Pattern", panel, 515
Maltose, 23
Maple syrup disease
    branched chain keto acid decarboxylase, 392
McArdle syndrome, 355, 357, 386
Metabolic acidosis, panel, 413
Metabolic acidosis - anion gap, panel, 401

Index:

Metabolic alkalosis, panel, 405, 414
Metastatic cancer to bone, panel, 414
Methemoglobin, 278
Milk alkali syndrome, panel, 443
Monamine oxidase
  hepatic fibrosis, 226
Monoclonal gammapathy, panel, 457, 530
Mucoviscidosis
  sweat chloride test, 255
Multiphasic testing, 396
Multiple myeloma, panel, 439
Muramidase
  damage to renal tubules, 272
  kidney transplant, 329
  lysosomal enzyme, 271
  monocytic leukemia, 265
  urine, 328
muscular atrophy
  serum enzymes, 355
Myasthenia gravis, 354
  panel, 563
Myocardial enzyme activity, 171
Myocardial infarction
  alpha-hydroxybutyrate dehydrogenase, 314
  extension of, 292
  LDH_1 elevated, 295
  panel, 555
Myocarditis, 292
Myokinase, 34
Myopathy
  serum enzymes, 355
Myositis, panel, 562
Myxedema, panel, 513

NADH-lactate substrate, 63
NADH-methemoglobin reductase, 278
Beta-Naphthylamine, 68
Narcotics
  contraction of the Sphincter of Oddi, 222

Necrosis of tissue, 3
Neocuproine, 422, 464, 507, 520
Nephrotic syndrome, panel, 422, 478, 519
Neurogenic atrophy, skeletal muscle, panel, 563
Niemann-Pick disease, 303, 391
Nitrate reductase
  urine, 329
P-Nitrophenol, 8, 20
P-Nitrophenyl phosphate, 8, 9, 12, 13, 20, 21
Non-fasting lipemic serum, panel, 456
Non-fasting specimen, panel, 431
5'-Nucleotidase, 117, 315
  obstructive jaundice, 219

Obstructive jaundice
  alkaline phosphatase, 102
  panel, 521
Old serum artifact, panel, 461
Oral contraceptive usage, panel, 494
Organic phosphate insecticides, 29, 231, 313
Ornithine carbamoyl transferase
  congenital deficiency, 228
  hyperammonemia, Type II, 392
Osteomalacia
  chronic liver disease, 215
  panel, 442
Oxalate usage, panel, 553
Oxalosuccinate, 53
Alpha-Oxobutyrate, 50
Alpha-Oxoglutarate, 47, 76, 86, 89
Oxytocinase, 121

Paget's disease, 107, 115
  panel, 549
Pancreatic fistula, 252
Pancreatic pseudocyst, 315
Pancreatitis
  hyperlipemia and hyperparathyroidism, 249

Index:

Pancreatitis (Continued):
  panel, 469
Paradoxical hyponatremia, panel, 423
Parathyroid adenoma, panel, 413
Parkinson's disease, 354
Paroxysmal nocturnal hemoglobinuria
  acetylcholinesterase, 278
Pentose phosphate, 44
Peptidases
  kidney, 332
Peritoneal amylase, 249
Pernicious anemia, panel, 559
Phenol oxidase
  malignant melanoma, 388
P-Phenylenediamine, 26
Phosphoenolpyruvate, 79, 277
Phosphoethanolamine, 109, 347
6-Phosphogluconate dehydrogenase, 276
6-Phosphogluconic acid, 44
6-Phosphogluconic acid dehydrogenase
  vaginal fluid, 199, 381
Phosphoglucose isomerase, 198
Phosphoglycerate kinase, 277
Phosphorylation, increased, panel, 484, 529
Physiological bone growth, panel, 475
Pickwickian syndrome, panel, 404
Placenta
  Leucine aminopeptidase, 208
Placental alkaline phosphatase
  phenylalanine, inhibited, 206
Plasminogen, kidney, 332
Polyclonal gammapathy, panel, 444, 457, 530
Polymyositis
  serum enzymes, 355
Pompe's disease
  Cori Type II, 386
  alpha-1, 4-glucosidase deficiency, 392

Pontacyl violet 6R, 27
Post-mortem vitreous humor - acute glomerulonephritis, panel, 433
Post-mortem vitreous humor - chronic renal insufficiency, panel, 510
Post-mortem vitreous humor - diabetes ketoacidosis, panel, 488
Post-mortem vitreous humor - Juvenile diabetes, panel, 432
Postoperative state, panel, 558
Pregnancy
  acid phosphatase, 110, 208
  creatine phosphokinase, 208
  cystine aminopeptidase third trimester, 208
  enzyme alterations, 209
  beta-glucuronidase, 208
  isocitric dehydrogenase, 208
  lactic dehydrogenase, 208
  oxytocinase, 208
  third trimester, panel, 429, 511
Prerenal azotemia, panel, 417, 506
Primary myopathy
  panel, 562
  serum enzymes, 355
Prostate gland
  hyperplasia, 190
  infarction, 190
Prostatic acid phosphatase, 14, 15
  inhibition by L-tartrate, 210
Pseudocholinesterase, 313
Pseudo-Conn's syndrome, panel, 410
Pseudochylous effusion, 194, 377
Pseudocysts of the pancreas, 249
Pseudohyponatremia, panel, 421, 422
Pseudohypoparathyroidism, panel, 463, 471
Pseudomyocardial infarction syndrome, 171
Pseudo-pseudohypoparathyroidism, panel, 473
Psoriasis, panel, 514
Pulmonary embolism
  alkaline phosphatase, 288

Index:

Pulmonary embolism (Continued):
  LDH_3, 287
  reflected ultrasound, 286
  unconjugated bilirubin, 286
Pulmonary scanning, 286
Pulmonary vascular tree
  angiography, 286
Pylonephritis
  beta-glucuronidase, 325
Pyridoxine
  decreased levels, panel, 460
Pyridoxine deficiency
  maternal blood GOT, 208
Pyridoxine, low, panel, 465, 479, 564
Pyruvate, 56, 59, 79
Pyruvate kinase, 276
  normal values, 82

Regan isoenzyme
  panel, 511, 548
  placental isoenzyme, 288, 342
Renal cortex infarction, 159
  LDH_1 and LDH_2, 159
Renal dialysis, panel, 459
Renal dialysis, long term, panel, 464, 564
Renal infarction, panel, 554
Renal insufficiency, panel, 416
Renal tubular acidosis, panel, 411, 481
Respiratory acidosis, panel, 404
Respiratory alkalosis, panel, 403, 429
Rheumatoid arthritis
  synovial fluid LDH_1, 163
Rickets, panel, 442

Sarcoidosis, panel, 444
Schistosomiasis
  urinary bladder, 334
Seabright-Bantam syndrome, panel, 472
Secretin, pancreozymin tests, 252

Serum protein, increased, panel, 457
Shock
  GOT increase, 220
Sodium tartrate, 13
D-Sorbitol, 83
Sorbitol dehydrogenase, 83, 227
  acute hepatitis, 181
  normal values, 85, 181
Sphingomyelinase, 391
Starch substrate, 23
Steroid-induced diabetes mellitus, panel, 474
Succinylcholine, 231, 313
Sucrase, 367
Sulfatase
  metachromatic leukodystrophy, 330, 391
  urine, 330
Sullivan test, 111, 190

L-Tartrate, 12, 114, 116
  inhibitor, 15
Tartrate acid buffer, 15
Tartrate-inhibited acid phosphatase, 15, 111
Tay-Sachs disease
  GM_2 ganglioside, 391
  hexosaminidase A, 391
Tetrazolium histochemical staining, 263
Thiazide diuretic excess utilization, panel, 450
Thiazide diuretic usage, panel, 496
Thoracic duct fluid, 379
Thorn syndrome, panel, 416
Thymolphthalein monophosphate, 116, 191
Thyrotoxicosis, panel, 449
Transaminase, normal values, 139
Transketolase
  Beriberi heart, 306
  inhibitor of, 306

Transketolase (Continued):
  uremic serum, 306
Transudate, 377
  panel, 535
Triglycerides, 91
  normal values, 96
Triphosphate isomerase, 277
Trypsin, 255
Tumor angiogenesis factor, 215
Tyrosinase
  malignant melanoma, 388

Ulcerative colitis, panel, 550
Uncontrolled diabetes mellitus,
  panel, 480
Uremia influence on glucose,
  panel, 489
Uric acid
  normal values, 99
Uricase, 97
Urinary enzyme inhibitors, 327
Urokinase
  urine, 330
Uropepsinogen
  urine, 328

Vaginal fluid
  enzymes, 209
  beta-glucuronidase, 211
Viral hepatitis, panel, 542
Von Gierke's Cori Type I, 386

Waldenstrom's macroglobulinemia,
  panel, 458, 530
Warburg technique, 5
Water intoxication, panel, 419
Werlhof's disease, 266
Wernig-Hoffman, 354
Wilson's disease, panel, 517
"Work" hypertrophy of liver,
  panel, 453
Wurster's red, 26

Xanthine oxidase inhibitor,
  panel, 512, 516